GEORGE WALTER ROCKLEIN
38 MASON DRIVE
MANHASSET, NEW YORK 11030

STANDARD SPECIFICATIONS

for

HIGHWAY BRIDGES

Adopted by

The American Association of State Highway Officials

ELEVENTH EDITION

1973

Published by the Association

General Offices

341 National Press Building

Washington, D.C. 20004

Copyright 1973

AMERICAN ASSOCIATION OF STATE HIGHWAY OFFICIALS
OFFICERS FOR 1973

President: Thomas F. Airis, Washintgon, D.C., 1973

First Vice President: Ward Goodman (Deceased March 15, 1973)

Regional Vice Presidents: Region I Raymond T. Schuler, New York, 1973
Region II A. B. Ratcliff, Jr. Louisiana, 1973
Region III Robert N. Hunter, Missouri, 1974
Region IV James A. Moe, California, 1974

Past Presidents: (Ex Officio) David H. Stevens, Maine
Charles E. Shumate, Colorado
Douglas B. Fugate, Virginia

Federal Highway Administrator: (Ex Officio)
R. R. Bartelsmeyer, Acting Administrator

Members: Wm. N. Price, Arizona, 1973
Jay W. Brown, Florida, 1973
George H. Andrews, Washington, (Interim) 1973
Wm. S. Ritchie, Jr. West Virginia, 1974
John C. Kohl, New Jersey, 1975
Ray Lappegaard, Minnesota, (Interim) 1973
R. H. Whitaker, New Hampshire, 1976
J. R. Coupal, Jr., Iowa, 1976

Treasurer: S. N. Pearman, South Carolina, 1976

Executive Director: Henrik E. Stafseth, Washington, D.C.

OPERATING SUB-COMMITTEE ON BRIDGES AND STRUCTURES 1973

Charles S. Matlock, Alaska, Chairman

W. Jack Wilkes, Federal Highway Administration, Secretary

ALABAMA, B. E. Higgins
ALASKA, Donald Halsted
ARIZONA, Martin Toney
ARKANSAS, Veral Pinkerton
CALIFORNIA, James E. McMahon, A. E. Elliott
COLORADO, Paul Chuvarsky
CONNECTICUT, Robert A. Norton, Edmund T. Koenig
DELAWARE, Robert C. McDowell
DISTRICT OF COLUMBIA, G. I. Sawyer, Bernard J. O'Donnell
FHWA, W. Jack Wilkes
FLORIDA, Tom Alberdi, Jr.
GEORGIA, Russell L. Chapman, Vernon W. Smith, Jr.
GUAM
HAWAII
IDAHO, Robert Jarvis
ILLINOIS, C. E. Thunman, Jr., Edward J. Kehl
INDIANA, Nelson W. Steinkamp, F. R. Richardson
IOWA, Charles Pestotnik
KANSAS, Charles Carpenter, E. E. Wilkinson
KENTUCKY, Charles G. Cook, J. T. Anderson
LOUISIANA, David Huval, Sidney L. Poleynard
MAINE, Theodore Karasopoulos
MARYLAND, Walter H. Biddle, H. H. Bowers
MASSACHUSETTS, John J. Aherne, Jr.
MICHIGAN, Nelson Jones
MINNESOTA, Leo A. Korth
MISSISSIPPI, Bennie D. Verrell, V. W. Guy
MISSOURI, W. D. Carney
MONTANA, Howard E. Stratton
NEBRASKA, Charles D. Smith
NEVADA, Allan Odell
NEW HAMPSHIRE, E. T. Swierz
NEW JERSEY, Warren J. Sunderland
NEW MEXICO, T. E. McCarty, J. A. Seibert
NEW YORK, Robert N. Kamp
NORTH CAROLINA, J. L. Norris, L. M. Temple
NORTH DAKOTA, Allen J. Anderson
OHIO, Robert Pfeifer
OKLAHOMA, Veldo Goins
OREGON, Walter S. Hart
PENNSYLVANIA, Bernard Kotalik
PUERTO RICO, Jose J. Muniz, Samuel Laboy
RHODE ISLAND, Samuel A. Engdahl, J. Norman Chopy

SOUTH CAROLINA, M. D. Moseley, W. E. Crum
SOUTH DAKOTA, K. C. Wilson
TENNESSEE, R. C. Odle, Henry Derthick
TEXAS, Wayne Henneberger
UTAH, Ray Behling
VERMONT, Wendell M. Smith
VIRGINIA, Fred G. Sutherland, E. H. Jones
WASHINGTON, C. S. Gloyd
WEST VIRGINIA, Harry H. Stephens, Robert C. Smith
WISCONSIN, William A. Kline
WYOMING, Charles Wilson

CONTENTS

		Page
Preface		xxxi
Introduction		xxxii
Division I	Design	1
Division II	Construction	247
Appendix A	Tables of Maximum Moments, Shears and Reactions, Standard Loadings	411
Appendix B	Truck Train and Equivalent Loadings	415
Appendix C	Formulas for Steel Columns	416
Appendix D	Position and Direction of Neutral Axis and Formulas for Stresses	433
Index		434

Division I

DESIGN

Article		Page
	Design Analysis	1
	SECTION 1—GENERAL FEATURES OF DESIGN	1
1.1.1	Bridge Locations	1
1.1.2	Bridge Waterways	1
	(A) Site Data	1
	(B) Hydrologic Analysis	2
	(C) Hydraulic Analysis	2
1.1.3	Pier Spacing, Orientation and Type	2
1.1.4	Culvert Waterway Openings	2
1.1.5	Culvert Location and Length	3
1.1.6	Width of Roadway and Sidewalk	3
1.1.7	Clearances	3
	(A) Navigational	3
	(B) Vehicular	4
	(C) Other	5
1.1.8	Curbs and Sidewalks	5
1.1.9	Railings	5
	(A) Traffic Railing	5
	(B) Pedestrian Railing	7
1.1.10	Roadway Drainage	8
1.1.11	Superelevation	8
1.1.12	Floor Surfaces	8
1.1.13	Blast Protection	8
1.1.14	Utilities	9
1.1.15	Roadway Width, Curbs and Clearances for Tunnels	9
	(A) Roadway Width	9
	(B) Clearance Between Walls	10
	(C) Curbs	10
	(D) Vertical Clearance	10
1.1.16	Roadway Width, Curbs and Clearances for Depressed Roadways	10
	(A) Roadway Width	10
	(B) Clearance Between Walls	10
	(C) Curbs	10

v

CONTENTS

Article		Page
1.1.17	Roadway Width, Curbs and Clearances For Underpasses	10
	(A) Widths	10
	(B) Vertical Clearance	12
	(C) Curbs	12

SECTION 2—LOADS 12

1.2.1	Loads	12
1.2.2	Dead Load	12
	(A) Unit Load on Culverts	13
	(B) Shear in Top Slabs	14
1.2.3	Live Load	14
1.2.4	Overload Provision	14
1.2.5	Highway Loadings	14
	(A) General	14
	(B) H Loadings	15
	(C) HS Loadings	16
	(D) Classes of Loadings	16
	(E) Designation of Loadings	18
	(F) Minimum Loading	18
1.2.6	Traffic Lanes	18
1.2.7	Standard Trucks and Lane Loads	18
1.2.8	Application of Loadings	19
	(A) Traffic Lane Units	19
	(B) Number and Position, Traffic Lane Units	19
	(C) Lane Loadings—Continuous Spans	19
	(D) Loading for Maximum Stress	19
1.2.9	Reduction in Load Intensity	20
1.2.10	Electric Railway Loading	20
1.2.11	Sidewalk, Curb and Railing Loading	20
	(A) Sidewalk Loading	20
	(B) Curb Loading	21
	(C) Railing Loading	21
1.2.12	Impact	22
	(A) Group A	22
	(B) Group B	23
	(C) Impact Formula	23
1.2.13	Longitudinal Forces	23
1.2.14	Wind Loads	24
	(A) Superstructure Design	24
	(B) Substructure Design	24
	(C) Overturning Forces	26
1.2.15	Thermal Forces	26
1.2.16	Uplift	26
1.2.17	Force of Stream Current, Floating Ice and Drift	27
1.2.18	Buoyancy	27
1.2.19	Earth Pressure	27
1.2.20	Earthquake Stresses	27
1.2.21	Centrifugal Forces	28
1.2.22	Loading Combinations	28

SECTION 3—DISTRIBUTION OF LOADS 29

1.3.1	Distribution of Wheel Loads to Stringers, Longitudinal Beams and Floor Beams	29
	(A) Position of Loads for Shear	29
	(B) Bending Moment in Stringers and Longitudinal Beams	30
	(C) Bending Moment in Floor Beams (Transverse)	32

CONTENTS

Article			Page
1.3.2	Distribution of Loads and Design of Concrete Slabs and Multi-Beam Precast Concrete Bridges		32
	(A)	Span Lengths	32
	(B)	Edge Distance of Wheel Load	33
	(C)	Bending Moment	34
	(D)	Edge Beams (Longitudinal)	35
	(E)	Distribution Reinforcement	35
	(F)	Shear and Bond Stress in Slabs	35
	(G)	Unsupported Edges, Transverse	35
	(H)	Cantilever Slabs	35
	(I)	Slabs Supported on Four Sides	36
	(J)	Median Slabs	37
1.3.3	Distribution of Wheel Loads Through Earth Fills		37
1.3.4	Distribution of Wheel Loads on Timber Flooring		37
	(A)	Flooring Transverse	37
	(B)	Flooring Longitudinal	37
	(C)	Continuous Flooring	38
1.3.5	Distribution of Loads and Design of Composite Wood-Concrete Members		38
	(A)	Distribution of Concentrated Loads for Bending Moment and Shear	38
	(B)	Distribution of Bending Moments in Continuous Spans	38
	(C)	Design	38
1.3.6	Distribution of Wheel Loads on Steel Grid Floors		39
	(A)	General	39
	(B)	Floors Filled with Concrete	39
	(C)	Open Floors	39
1.3.7	Moments, Shears and Reactions		40
	SECTION 4—SUBSTRUCTURES AND RETAINING WALLS		40
1.4.1	Allowable Stresses		40
1.4.2	Bearing Power of Foundation Soils, Determination of Bearing Power		40
1.4.3	Angles of Repose		41
1.4.4	Bearing Value of Piling		41
	(A)	General	41
	(B)	Case A. Capacity of Pile as a Structural Member	41
	(C)	Case B. Capacity of Pile to Transfer Load to the Ground	43
	(D)	Case C. Capacity of the Ground to Support the Load Delivered by the Pile	44
	(E)	Maximum Design Loads for Piles	45
	(F)	Uplift	45
	(G)	Group Pile Loading	45
1.4.5	Piles		46
	(A)	General	46
	(B)	Limitation of Use	46
	(C)	Design Loads	46
	(D)	Spacing, Clearances and Embedment	46
	(E)	Batter Piles	47
	(F)	Buoyancy	47
	(G)	Concrete Piles (Precast)	47
	(H)	Concrete Piles (Cast-in-Place)	48
	(I)	Steel H-piles	48
	(J)	Unfilled Tubular Steel Piles	49
	(K)	Steel Pile and Steel Pile Shell Protection	49

Article			Page
1.4.6	Footings		50
	(A)	Depth	50
	(B)	Anchorage	50
	(C)	Distribution of Pressure	50
	(D)	Spread Footings	50
	(E)	Internal Stresses in Spread Footings	50
	(F)	Reinforcement	52
	(G)	Transfer of Stress from Vertical Reinforcement	52
1.4.7	Abutments		52
	(A)	General	52
	(B)	Reinforcement for Temperature	53
	(C)	Wing walls	53
	(D)	Drainage	53
1.4.8	Retaining Walls		53
	(A)	General	53
	(B)	Base or Footing Slabs	53
	(C)	Vertical Walls	54
	(D)	Counterforts and Buttresses	54
	(F)	Reinforcement for Temperature	54
	(F)	Expansion and Contraction Joints	54
	(G)	Drainage	54
1.4.9	Piers		54
	(A)	General	54
	(B)	Pier Nose	55
1.4.10	Tubular Steel Piers		55
	(A)	Use	55
	(B)	Depth	55
	(C)	Piling	55
	(D)	Dimensions of Shell	55
	(E)	Splices and Joints	55
	(F)	Bracing	55
	SECTION 5—CONCRETE DESIGN		56
1.5.1	Allowable Stresses		56
	(A)	Standard Notations and Assumptions	56
	(B)	Strength of Concrete	56
	(C)	Allowable Stresses—Concrete	57
	(D)	Allowable Stresses—Reinforcement	57
1.5.2	General Assumptions		58
1.5.3	Span Lengths		59
1.5.4	Expansion		60
1.5.5	T-Beams		60
	(A)	Effective Flange Width	60
	(B)	Shear	60
	(C)	Isolated Beams	60
	(D)	Diaphragms	61
	(E)	Construction Joints	61
1.5.6	Reinforcement		61
	(A)	Spacing	61
	(B)	Covering	61
	(C)	Splicing	61
	(D)	End Anchorages and Hooks	62
	(E)	Extension of Reinforcement	63
	(F)	Structural Steel Shapes	63
	(G)	Interim Reinforcement for T-beams and Box Girders	63

Article				Page
1.5.6 cont.		(H)	Reinforcement for Temperature and Shrinkage	63
		(I)	Bundled Reinforcement	64
		(J)	Bond Stress in Flexural Members	64
1.5.7		Compression Reinforcement in Beams		64
1.5.8		Web Reinforcement		65
		(A)	General	65
		(B)	Calculation of Shear	65
		(C)	Bent-up Bars	65
		(D)	Vertical Stirrups	66
		(E)	Anchorage	66
1.5.9		Columns		67
		(A)	General	67
		(B)	Piers and Pedestals	68
		(C)	Spirally Reinforced Columns	68
		(D)	Tied Columns	69
		(E)	Bending Moments in Columns	70
		(F)	Combined Axial and Bending Stress	70
1.5.10		Concrete Arches		71
		(A)	Shape of Arch Rings	71
		(B)	Spandrel Walls	72
		(C)	Expansion Joints	72
		(D)	Reinforcement	72
		(E)	Waterproofing	72
		(F)	Drainage of Spandrel Fill	72
1.5.11		Viaduct Bents and Towers		72
1.5.12		Box Girders		73
		(A)	Effective Compression Flange Width	73
		(B)	Flange Thickness	73
		(C)	Flexure	73
		(D)	Shear	74
		(E)	Reinforcement	74
		(F)	Flange Reinforcement	74
		(G)	Diaphragms	74
		(H)	Flanges Supporting Pipes and Conduits	74
1.5.13		Bearings		75

LOAD FACTOR DESIGN

1.5.14		General		75
		(A)	Application	75
		(B)	Other Specifications	75
1.5.15		Notation		75
		(A)	Loads and Forces	75
		(B)	Dimensions and Constants	76
1.5.16		Materials Properties		77
		(A)	Concrete	77
		(B)	Reinforcement	78
1.5.17		Loads and Load Factor Equations		79
		(A)	Loads	79
		(B)	Load Factor Equations	79
1.5.18		Strength Provisions		79
		(A)	Assumptions	79
1.5.19		Capacity Modification Factors		80

CONTENTS

Article			Page
1.5.20	Flexure		80
	(A)	Rectangular sections with tension reinforcement only	80
	(B)	I- and T-sections	81
	(C)	Rectangular sections with compression reinforcement	81
	(D)	Other cross sections	82
1.5.21	Shear		82
	(A)	Shear stress	82
	(B)	Shear reinforcement	83
	(C)	Stress restrictions	83
	(D)	Shear reinforcement restrictions	83
	(E)	Shear stress in slabs and footings	84
1.5.22	Columns		84
	(A)	General	84
	(B)	Column Section Capacities	85
	(C)	Slenderness effects in columns	86
1.5.23	Bearing		88
1.5.24	Service Load Requirements		88
	(A)	Service Load Stresses	88
1.5.25	Fatigue		89
	(A)	Concrete	89
	(B)	Reinforcement	89
1.5.26	Flexural Stress Limitations		89
	(A)	General	89
	(B)	Bridges exposed to corrosive environments without a waterproof deck protection system	89
1.5.27	Deflections		89
	(A)	Superstructure depth recommendations	89
	(B)	Dead load deflections at falsework removal	90
	(C)	Long-time deflections caused by dead loads, creep and shrinkage	90
1.5.28	Overload		91
1.5.29	Development of Reinforcement		91
	(A)	General	91
	(B)	Positive moment reinforcement	92
	(C)	Negative moment reinforcement	92
	(D)	Special members	93
	(E)	Development length of deformed bars in tension	93
	(F)	Development length of deformed bars in compression	93
	(G)	Development length of bundled bars	94
	(H)	Standard hooks in tension	94
	(I)	Combination development length	94
	(J)	Mechanical anchorage	94
	(K)	Anchorage of shear reinforcement	94
	SECTION 6—PRESTRESSED CONCRETE		95
1.6.1	General		95
1.6.2	Notation		95
1.6.3	Design Theory		96
1.6.4	Basic Assumptions		97
1.6.5	Load Factors		97
1.6.6	Allowable Stresses		97
	(A)	Prestressing Steel	98
	(B)	Concrete	98
1.6.7	Loss of Prestress		99
	(A)	Friction Losses	99
	(B)	Prestress Losses	100

Article		Page
1.6.8	Flexure	101
1.6.9	Ultimate Flexural Strength	101
	(A) Rectangular Sections	101
	(B) Flanged Sections	101
	(C) Steel Stress	102
1.6.10	Maximum and Minimum Steel Percentage	102
	(A) Maximum Steel	102
	(B) Minimum Steel	102
1.6.11	Nonprestressed Reinforcement	102
1.6.12	Continuity	103
	(A) General	103
	(B) Cast-in-place Post-Tensioned Bridges	103
	(C) Bridges Composed of Simple-Span Precast Prestressed Girders Made Continuous	103
1.6.13	Shear	104
1.6.14	Composite Structures	105
	(A) General	105
	(B) Shear Transfer	105
	(C) Shear Capacity	105
	(D) Vertical Ties	105
	(E) Shrinkage Stresses	105
1.6.15	Anchorage Zones	106
1.6.16	Cover and Spacing of Steel	106
	(A) Minimum Cover	106
	(B) Minimum Spacing	107
	(C) Bundling	107
	(D) Size of Ducts	107
1.6.17	Post-Tensioning Anchorages and Couplers	107
1.6.18	Embedment of Prestressed Strand	107
1.6.19	Concrete Strength at Stress Transfer	108
1.6.20	Bearings	108
1.6.21	Span Lengths	108
1.6.22	Expansion and Contraction	108
1.6.23	T-Beams	108
	(A) Effective Flange Width	109
	(B) Construction Joints	109
	(C) Diaphragms	109
	(D) Isolated Beams	109
1.6.24	Box Girders	110
	(A) Lateral Distribution of Loads for Bending Moment	110
	(B) Effective Compression Flange Width	110
	(C) Flange Thickness	111
	(D) Minimum Bar Reinforcement for Cast-In-Place Post-Tensioned Box Girders	111
	(E) Shear	111
	(F) Diaphragms	111
	SECTION 7—STRUCTURAL STEEL DESIGN	111
1.7.1	Allowable Stresses	111
1.7.2	Allowable Stresses for Weld Metal	115
1.7.3	Fatigue Stresses	115
1.7.4	Pins, Rollers and Expansion Rockers	117
1.7.5	Fasteners (Rivets and Bolts)	121

Article		Page
1.7.6	Cast Steel, Ductile Iron Castings, Malleable Castings and Cast Iron	124
	(A) Cast Steel and Ductile Iron	124
	(B) Malleable Castings	124
	(C) Cast Iron	125
1.7.7	Bronze or Copper-alloy	125
1.7.8	Bearing on Masonry	125

DETAILS OF DESIGN 125

1.7.9	Effective Length of Span	125
1.7.10	Depth Ratios	125
1.7.11	Limiting Lengths of Members	126
1.7.12	Deflection	126
1.7.13	Minimum Thickness of Metal	127
1.7.14	Effective Area of Angles and Tee Sections in Tension	127
1.7.15	Outstanding Legs of Angles	128
1.7.16	Expansion and Contraction	128
1.7.17	Combined Stresses	128
1.7.18	Eccentric Connections	128
1.7.19	Splices and Connections	128
1.7.20	Strength of Connections	130
1.7.21	Diaphragms, Cross Frames and Lateral Bracing	131
1.7.22	Number of Main Members on Through Spans	131
1.7.23	Accessibility of Parts	132
1.7.24	Closed Sections and Pockets	132
1.7.25	Welding, General	132
1.7.26	Minimum Size of Fillet Welds	132
1.7.27	Maximum Effective Size of Fillet Welds	133
1.7.28	Effective Weld Areas	133
	(A) Butt Welds	133
	(B) Fillet Welds	133
1.7.29	Minimum Effective Length of Fillet Welds	133
1.7.30	Fillet Weld End Returns	133
1.7.31	Lap Joints	134
1.7.32	Seal Welds	134
1.7.33	Fillet Welds in Skewed Tee Joints	134
1.7.34	Fillet Welds in Holes and Slots	134
1.7.35	Size of Fasteners (Rivets or High Strength Bolts)	134
1.7.36	Spacing of Fasteners	134
1.7.37	Maximum Pitch of Sealing and Stitch Fasteners	135
	(A) Sealing Fasteners	135
	(B) Stitch Fasteners	135
1.7.38	Edge Distance of Fasteners	135
	(A) General	135
	(B) Special	136
1.7.39	Long Rivets	136
1.7.40	Links and Hangers	136
1.7.41	Location of Pins	137
1.7.42	Size of Pins	137
1.7.43	Pin Plates	137
1.7.44	Pins and Pin Nuts	137
1.7.45	Upset Ends	137
1.7.46	Eyebars	137
1.7.47	Packing of Eyebars	138
1.7.48	Forked Ends	138

Article		Page
	BEARINGS	138
1.7.49	Fixed Bearings	138
1.7.50	Expansion Bearings	138
1.7.51	Bronze or Copper-Alloy Sliding Expansion Bearings	138
1.7.52	Rollers	139
1.7.53	Sole Plates and Masonry Plates	139
1.7.54	Masonry Bearings	139
1.7.55	Anchor Bolts	139
1.7.56	Pedestals and Shoes	140
	FLOOR SYSTEM	140
1.7.57	Stringers	140
1.7.58	Floorbeams	140
1.7.59	Cross Frames	140
1.7.60	Expansion Joints	140
1.7.61	End Connections of Floorbeams and Stringers	141
1.7.62	End Floorbeams	141
1.7.63	End Panel of Skewed Bridges	141
1.7.64	Sidewalk Brackets	141
	ROLLED BEAMS	142
1.7.65	Rolled Beams, General	142
1.7.66	Bearing Stiffeners	142
1.7.67	Cover Plates	142
	PLATE GIRDERS	143
1.7.68	Plate Girders, General	143
1.7.69	Flanges	143
	(A) Welded Girders	143
	(B) Riveted or Bolted Girders	143
1.7.70	Thickness of Web Plates	144
	(B) Girders Not Stiffened Longitudinally	144
	(A) Girders Stiffened Longitudinally	146
1.7.71	Transverse Intermediate Stiffeners	146
1.7.72	Longitudinal Stiffeners	148
1.7.73	Bearing Stiffeners	149
	(A) Welded Girders	149
	(B) Riveted or Bolted Girders	150
1.7.74	Camber	150
	TRUSSES	150
1.7.75	Trusses, General	150
1.7.76	Truss Members	151
1.7.77	Secondary Stresses	151
1.7.78	Diaphragms	152
1.7.79	Camber	152
1.7.80	Working Lines and Gravity Axes	152
1.7.81	Portal and Sway Bracing	152
1.7.82	Fillers, Development, Maximum Numbers, Etc.	152
1.7.83	Perforated Cover Plates and Lacing Bars	153
	(A) Perforated Cover Plates	153
	(B) Lacing Bars	153
1.7.84	Gusset Plates	154
1.7.85	Half-Through Truss Spans	155
1.7.86	Fastener Pitch in Ends of Compression Members	155

CONTENTS

Article		Page
1.7.87	Net Section of Riveted or High Strength Bolted Tension Members	155
1.7.88	Compression Members—Thickness of Metal	156
1.7.89	Stay Plates	158
	RIBBED ARCHES	158
1.7.90	Thickness of Web Plates, Solid Rib Arches	158
	BENTS AND TOWERS	159
1.7.91	Bents and Towers, General	159
1.7.92	Single Bents	159
1.7.93	Batter	159
1.7.94	Bracing	159
1.7.95	Bottom Struts	160
	COMPOSITE GIRDERS	160
1.7.96	Composite I-Girders, General	160
1.7.97	Shear Connectors	161
1.7.98	Effective Flange Width	161
1.7.99	Stresses	161
1.7.100	Shear	162
	(A) Horizontal Shear	162
	(B) Vertical Shear	165
1.7.101	Deflection	165
1.7.102	Composite Box Girders, General	165
1.7.103	Lateral Distribution of Loads for Bending Moment	166
1.7.104	Design of Web Plates	166
	(A) Vertical Shear	166
	(B) Secondary Bending Stresses	166
1.7.105	Design of Bottom Flange Plates	167
	(A) Tension Flanges	167
	(B) Compression Flanges Unstiffened	167
	(C) Compression Flanges Stiffened Longitudinally	167
	(D) Compression Flanges Stiffened Longitudinally and Transversely	168
	(E) Compression Flange Stiffeners, General	169
1.7.106	Design of Flange to Web Welds	170
1.7.107	Diaphragms	170
1.7.108	Lateral Bracing	170
1.7.109	Access and Drainage	170
	HYBRID GIRDERS	170
1.7.110	Hybrid Girders, General	170
1.7.111	Allowable Stresses	171
	(A) Bending	171
	(B) Shear	171
	(C) Fatigue	171
1.7.112	Plate Thickness Requirements	173
1.7.113	Bearing Stiffener Requirements	173
	HEAT-CURVED ROLLED BEAMS AND WELDED PLATE GIRDERS	173
1.7.114	Scope	173
1.7.115	Minimum Radius of Curvature	173
1.7.116	Camber	173

Article			Page
	LOAD FACTOR DESIGN		174
1.7.117	Scope		174
1.7.118	Notation		174
1.7.119	Loads		175
1.7.120	Design Theory		175
1.7.121	Assumptions		176
1.7.122	Design Strength for Steel		176
1.7.123	Maximum Design Loads		176
1.7.124	Symmetrical Beams and Girders		177
	(A)	Compact Systems	177
	(B)	Braced Non-Compact Sections	178
	(C)	Transition	179
	(D)	Unbraced Sections	179
	(E)	Transversely Stiffened Girders	180
	(F)	Longitudinally Stiffened Girders	181
1.7.125	Unsymmetrical Beams and Girders		182
	(A)	General	182
	(B)	Unsymmetrical Sections with Transverse Stiffeners	183
	(C)	Longitudinally Stiffened Unsymmetrical Sections	183
1.7.126	Composite Beams and Girders		183
1.7.127	Positive Moment Sections of Composite Beams and Girders		183
	(A)	Compact Sections	183
	(B)	Non-compact Sections	185
	(C)	General	185
1.7.128	Negative Moment Sections of Composite Beams and Girders		185
1.7.129	Composite Box Girders		185
	(A)	Maximum Strength	185
	(B)	Lateral Distribution	186
	(C)	Web Plates	186
	(D)	Tension Flanges	186
	(E)	Compression Flanges	186
	(F)	Diaphragms	187
1.7.130	Shear Connectors		188
	(A)	General	188
	(B)	Design of Connectors	188
	(C)	Maximum Spacing	188
1.7.131	Hybrid Girders		188
1.7.132	Noncomposite Hybrid Girders		189
	(A)	Compact Sections	189
	(B)	Braced Non-compact Sections	189
	(C)	Unbraced Noncompact Sections	189
	(D)	Transversely Stiffened Girders	189
1.7.133	Composite Hybrid Girders		190
1.7.134	Compression Members		190
	(A)	Axial Loading	190
	(B)	Combined Axial Load and Bending	191
1.7.135	Splices, Connections and Details		192
	(A)	Connectors	192
	(B)	Connections	193
1.7.136	Overload		194
	(A)	Noncomposite Beams	194
	(B)	Composite Beams	195
	(C)	Friction Joints	195

Article			Page
1.7.137	Fatigue		195
	(A)	General	195
	(B)	Composite Construction	195
	(C)	Hybrid Beams and Girders	196
1.7.138	Deflection		196

ORTHOTROPIC-DECK BRIDGES 196

1.7.139	Orthotropic-Deck Bridges, General		196
1.7.140	Wheel-Load Contact Area		196
1.7.141	Effective Width of Deck Plate		197
	(A)	Ribs and Beams	197
	(B)	Girders	197
1.7.142	Allowable Stresses		197
	(A)	Local Bending Stresses in Deck Plate	197
	(B)	Bending Stresses in Longitudinal Ribs	197
	(C)	Bending Stresses in Transverse Beams	197
	(D)	Intersections of Ribs, Beams, and Girders	198
1.7.143	Thickness of Plate Elements		198
	(A)	Longitudinal Ribs and Deck Plate	198
	(B)	Girders and Transverse Beams	198
1.7.144	Maximum Slenderness of Longitudinal Ribs		198
1.7.145	Diaphragms		198
1.7.146	Stiffness Requirements		199
	(A)	Deflections	199
	(B)	Vibrations	199
1.7.147	Wearing Surface		199
1.7.148	Closed Ribs		199

SECTION 8—CORRUGATED METAL AND STRUCTURAL PLATE PIPES AND PIPE-ARCHES 200

1.8.1	General		200
1.8.2	Design		200
	(A)	Seam Strength	201
	(B)	Handling and Installation Strength	201
	(C)	Failure of the Conduit Wall	201
	(D)	Deflection or Flattening	203
1.8.3	Chemical and Mechanical Requirements		204
	(A)	Aluminum—Corrugated Metal Pipe and Pipe-Arch	204
	(B)	Aluminum—Structural Plate Pipe and Pipe-Arch	204
	(C)	Steel—Corrugated Metal Pipe and Pipe-Arch	204
	(D)	Steel—Structural Plate Pipe and Pipe-Arch	205
1.8.4	Abrasive or Corrosive Conditions		205
1.8.5	Rivets & Bolts		205
1.8.6	Multiple Structures		205
1.8.7	Sloped Ends—Skewed		206
1.8.8	Maximum Depths of Cover		206

SECTION 9—STRUCTURAL PLATE ARCHES 206

1.9.1	General	206
1.9.2	Ratio, Rise to Span	206
1.9.3	Minimum Height of Cover	206
1.9.4	Scour Conditions	207
1.9.5	Multiple Arches	207
1.9.6	Substructure Design	207

Article			Page
	SECTION 10—TIMBER STRUCTURES		207
1.10.1	Allowable Stresses		207
	(A)	Allowable Unit Stresses for Stress-Grade Lumber	207
	(B)	Allowable Unit Stresses for Glued Laminated Timber	207
	(C)	Allowable Unit Stresses for Normal Loading Conditions	208
	(D)	Allowable Unit Stresses for Permanent Loading	208
	(E)	Allowable Unit Stresses for Wind, Earthquake or Short Time Loading	209
	(F)	Combined Stresses	209
1.10.2	Formulas for the Computation of Stresses in Timber		209
	(A)	Horizontal Shear in Beams	209
	(B)	Secondary Stresses in Curved Glued Laminated Members	217
	(C)	Compression or Bearing Perpendicular to Grain	221
	(D)	Simple Solid Column Design	223
	(E)	Spaced Column Design	224
	(F)	Safe Load on Round Columns	225
	(G)	Notched Beams	226
	(H)	Bearing on Inclined Surfaces	226
	(I)	Timber Connectors	226
	(J)	Size Factor	227
	(K)	Lateral Stability	227
1.10.3	General		230
1.10.4	Bolts		230
1.10.5	Washers		230
1.10.6	Hardware for Seacoast Structures		230
1.10.7	Columns and Posts		230
1.10.8	Pile and Framed Bents		230
	(A)	Pile Bents	230
	(B)	Framed Bents	231
	(C)	Sills and Mud Sills	231
	(D)	Caps	231
	(E)	Bracing	231
	(F)	Pile Bent Abutments	231
1.10.9	Trusses		231
	(A)	Joints and Splices	231
	(B)	Floor Beams	232
	(C)	Hangers	232
	(D)	Eyebars and Counters	232
	(E)	Bracing	232
	(F)	Camber	233
1.10.10	Floors and Railings		233
	(A)	Stringers	233
	(B)	Bridging	233
	(C)	Nailing Strips	233
	(D)	Flooring	233
	(E)	Retaining Pieces	233
	(F)	Wheel Guards	234
	(G)	Drainage	234
	(H)	Railings	234
1.10.11	Fire Stops		234
	SECTION 11—LOAD CAPACITY RATING OF EXISTING BRIDGES		234
1.11.1	Overload Under Permit		234

Article		Page
1.11.2	Impact	235
1.11.3	Adjustable Loads	235
1.11.4	Stress Analysis	235
1.11.5	Allowable Stresses	235
	SECTION 12—ELASTOMERIC BEARINGS	238
1.12.1	General	238
1.12.2	Design	238
	SECTION 13—STEEL TUNNEL LINER PLATES	240
1.13.1	General	240
1.13.2	Loads	240
1.13.3	Design	242
1.13.4	Joint Strength	242
1.13.5	Handling and Installation Strength	242
1.13.6	Critical Buckling of Liner Plate Wall	243
1.13.7	Deflection or Flattening	243
1.13.8	Chemical and Mechanical Requirements	243
	(A) Chemical Composition	243
	(B) Minimum Mechanical Properties of Flat Plate Before Cold Forming	244
1.13.9	Sectional Properties	244
1.13.10	Coatings	244
1.13.11	Bolts	245
1.13.12	Safety Factors	245

Division II
CONSTRUCTION

	SECTION 1—EXCAVATION AND FILL	247
2.1.1	General	247
2.1.2	Preservation of Channel	247
2.1.3	Depth of Footings	248
2.1.4	Preparation of Foundations for Footings	248
2.1.5	Cofferdam and Cribs	248
	(A) General	248
	(B) Protection of Concrete	248
	(C) Drawings Required	249
	(D) Removal	249
2.1.6	Pumping	249
2.1.7	Inspection	249
2.1.8	Back-fill	249
2.1.9	Filled Spandrel Arches	250
2.1.10	Approach Embankment	250
2.1.11	Classification of Excavation	250
2.1.12	Measurement and Payment	250
	SECTION 2—SHEET PILES	251
2.2.1	General	251
2.2.2	Timber Sheet Piles	251
2.2.3	Concrete Sheet Piles	251
2.2.4	Steel Sheet Piles	251
2.2.5	Measurement and Payment	252
	SECTION 3—BEARING PILES	252
2.3.1	Materials	252

Article			Page
2.3.2	Design and Conditions of Use		252
2.3.3	Preparation for Driving		252
	(A)	Excavation	252
	(B)	Caps	252
	(C)	Collars	253
	(D)	Pointing	253
	(E)	Splicing Piles	253
	(F)	Painting Steel Piles	253
2.3.4	Methods of Driving		253
	(A)	General	253
	(B)	Hammers for Timber and Steel Piles	253
	(C)	Hammers for Concrete Piles	253
	(D)	Additional Equipment	254
	(E)	Leads	254
	(F)	Followers	254
	(G)	Water Jets	254
	(H)	Accuracy of Driving	254
2.3.5	Defective Piles		254
2.3.6	Determination of Bearing Values (See also Article 1.4.4)		255
	(A)	Loading Tests	255
	(B)	Timber Pile Formulas	255
	(C)	Concrete and Steel Piles	256
2.3.7	Test Piles		256
2.3.8	Order Lists for Piling		256
2.3.9	Storage and Handling of Timber Piles		256
2.3.10	Cutting off Timber Piles		257
2.3.11	Cutting off Steel or Steel Shell Piles		257
2.3.12	Capping Timber Piles		257
2.3.13	Manufacture of Precast Concrete Piles		257
	(A)	General	257
	(B)	Class of Concrete	257
	(C)	Form Work	257
	(D)	Reinforcement	257
	(E)	Casting	258
	(F)	Finish	258
	(G)	Curing	258
2.3.14	Storage and Handling of Precast Concrete Piles		258
2.3.15	Manufacture of Cast-in-Place Concrete Piles		258
	(A)	General	258
	(B)	Inspection of Metal Shells	258
	(C)	Class of Concrete	259
	(D)	Reinforcement	259
	(E)	Placing Concrete	259
2.3.16	Extensions or "Build-ups"		259
2.3.17	Painting Steel Piles and Steel Pile Shells		259
2.3.18	Measurement and Payment		259
	(A)	General	259
	(B)	Method A	259
	(C)	Method B	260
	(D)	Falsework and Defective Piles	261
	(E)	Additional Requirements	261
2.3.19	Payment for Test Piles		262
2.3.20	Payment for Loading Tests		262
	SECTION 4—CONCRETE MASONRY		262
2.4.1	General		262

Article				Page
2.4.2	Materials			262
	(A)	Cement		262
	(B)	Water and Admixtures		264
	(C)	Fine Aggregate		264
	(D)	Coarse Aggregates		265
2.4.3	Care and Storage of Concrete Aggregates			265
2.4.4	Storage of Cement			265
2.4.5	Classes of Concrete			266
2.4.6	Composition of Concrete			267
2.4.7	Sampling and Testing			268
2.4.8	Measurement of Materials			268
2.4.9	Mixing Concrete			269
	(A)	General		269
	(B)	Mixing at Site		269
	(C)	Truck Mixing		269
	(D)	Partial Mixing at the Central Plant		270
	(E)	Plant Mix		270
	(F)	Time of Hauling and Placing Mixed Concrete		270
	(G)	Hand Mixing		270
	(H)	Delivery		271
	(I)	Retempering		271
2.4.10	Handling and Placing Concrete			271
	(A)	General		271
	(B)	Culverts		273
	(C)	Girders, Slabs and Columns		274
	(D)	Arches		275
2.4.11	Pneumatic Placing			275
2.4.12	Pumping			275
2.4.13	Depositing Concrete Under Water			275
2.4.14	Construction Joints			276
	(A)	General		276
	(B)	Bonding		276
2.4.15	Rubble or Cyclopean Concrete			277
2.4.16	Concrete Exposed to Sea Water			277
2.4.17	Concrete Exposed to Alkali Soils or Alkali Water			278
2.4.18	Falsework and Centering			278
2.4.19	Forms			278
2.4.20	Removal of Falsework, Forms and Housing			279
2.4.21	Concreting in Cold Weather			280
2.4.22	Curing Concrete			280
2.4.23	Expansion and Fixed Joints and Bearings			281
	(A)	Open Joints		281
	(B)	Filled Joints		281
	(C)	Premolded Expansion Joint Fillers		281
	(D)	Steel Joints		281
	(E)	Water Stops		281
	(F)	Sheet Copper		282
	(G)	Bearing Devices		282

FINISHING CONCRETE SURFACES ... 282

2.4.24	General			282
2.4.25	Class 1, Ordinary Surface Finish			283
2.4.26	Class 2, Rubbed Finish			283
2.4.27	Class 3, Tooled Finish			283
2.4.28	Class 4, Sand Blasted Finish			284

Article			Page
2.4.29	Class 5, Wire Brushed or Srubbed Finish		284
2.4.30	Class 6, Floated Surface Finish		284
	(A)	Striking Off	284
	(B)	Floating	284
	(C)	Longitudinal Floating	284
	(D)	Transverse Floating	284
	(E)	Straightedging	285
	(F)	Final Finishing	285
2.4.31	Sidewalk Finish		285
2.4.32	Pneumatically Applied Mortar		285
	(A)	General	285
	(B)	Proportions	286
	(C)	Water Content	286
	(D)	Mixing	286
	(E)	Nozzle Velocity	286
	(F)	Nozzle Position	286
	(G)	Rebound Sand	286
	(H)	Forms	286
	(I)	Joints	286
	(J)	Bond	286
	(K)	Curing	287
	(L)	Reinforcement	287
2.4.33	Prestressed Concrete		287
	(A)	General	287
	(B)	Supervision	287
	(C)	Equipment	287
	(D)	Concrete	287
	(E)	Steam Curing	288
	(F)	Transportation and Storage	288
	(G)	Pretensioning Method	288
	(H)	Post-tensioning Method	289
	(I)	Grouting of Bonded Steel	289
	(J)	Prestressing Reinforcement	289
	(K)	Testing Prestressing Reinforcement and Anchorages	290
2.4.34	Measurement and Payment		290
	SECTION 5—REINFORCEMENT		291
2.5.1	Material		291
	(A)	Bar Reinforcement	291
	(B)	Wire and Wire Mesh	291
	(C)	Bar Mat Reinforcement	292
	(D)	Structural Shapes	292
2.5.2	Order Lists		292
2.5.3	Protection of Material		292
2.5.4	Fabrication		292
2.5.5	Placing and Fastening		293
2.5.6	Splicing		293
2.5.7	Lapping		293
2.5.8	Substitutions		294
2.5.9	Measurement		294
2.5.10	Payment		294
	SECTION 6—ASHLAR MASONRY		295
2.6.1	Description		295
2.6.2	Materials		295
	(A)	Ashlar Stone	295
	(B)	Mortar	295

Article		Page
2.6.3	Size of Stone	295
2.6.4	Surface Finishes of Stone	296
2.6.5	Dressing Stone	296
2.6.6	Stretchers	296
2.6.7	Headers	296
2.6.8	Cores and Backing	297
2.6.9	Mixing Mortar	297
2.6.10	Laying Stone	297
	(A) General	297
	(B) Face Stone	297
	(C) Stone Backing and Cores	298
	(D) Concrete Cores and Backing	298
2.6.11	Leveling Courses	298
2.6.12	Resetting	298
2.6.13	Dowels and Cramps	298
2.6.14	Copings	299
2.6.15	Arches	299
2.6.16	Pointing	299
2.6.17	Measurement and Payment	300
	SECTION 7—MORTAR RUBBLE MASONRY	300
2.7.1	Description	300
2.7.2	Materials	300
	(A) Rubble Stone	300
	(B) Mortar	300
2.7.3	Size	300
2.7.4	Headers	300
2.7.5	Shaping Stone	301
2.7.6	Laying Stone	301
2.7.7	Copings, Bridge Seats and Backwalls	301
2.7.8	Arches	302
2.7.9	Pointing	302
2.7.10	Measurement and Payment	302
	SECTION 8—DRY RUBBLE MASONRY	303
2.8.1	Description	303
2.8.2	Materials	303
2.8.3	Size of Stone	303
2.8.4	Headers	303
2.8.5	Shaping Stone	303
2.8.6	Laying Stone	303
2.8.7	Copings, Bridge Seats and Backwalls	303
2.8.8	Measurement and Payment	304
	SECTION 9—BRICK MASONRY	304
2.9.1	Description	304
2.9.2	Materials	304
	(A) Brick	304
	(B) Mortar	304
2.9.3	Construction	304
2.9.4	Copings, Bridge Seats and Backwalls	305
2.9.5	Measurement and Payment	305
	SECTION 10—STEEL STRUCTURES FABRICATION	305
2.10.1	Type of Fabrication	305

CONTENTS

Article			Page
2.10.2	Quality of Workmanship		305
2.10.3	Materials		306
	(A)	Structural Steel	306
	(B)	Steel Forgings and Steel Shafting	310
	(C)	Steel Castings	310
	(D)	Iron Castings	311
	(E)	Ductile Iron Castings	311
	(F)	Malleable Castings	312
	(G)	Bronze Castings and Copper-Alloy Plates	312
	(H)	Sheet Lead	312
	(I)	Sheet Zinc	312
	(J)	Galvanizing	312
	(K)	Canvas and Red Lead for Bedding Masonry Plates and Equivalent Bearing Areas	313
	(L)	Preformed Fabric Pads	313
2.10.4	Storage of Materials		313
2.10.5	Straightening Material and Curving Rolled Beams and Welded Girders		313
	(A)	Straightening Material	313
	(B)	Curving Rolled Beams and Welded Girders	313
2.10.6	Finish		315
2.10.7	Rivet and Bolt Holes		316
	(A)	Holes for Rivets, High-Strength Bolts and Unfinished Bolts	316
	(B)	Holes for Ribbed Bolts, Turned Bolts or Other Approved Bearing-Type Bolts	316
2.10.8	Punched Holes		316
2.10.9	Reamed or Drilled Holes		316
2.10.10	Subpunching and Reaming of Field Connections		317
2.10.11	Accuracy of Punched and Drilled Holes		317
2.10.12	Accuracy of Reamed and Drilled Holes		317
2.10.13	Fitting for Riveting and Bolting		317
2.10.14	Shop Assembling		318
	(A)	Full Truss or Girder Assembly	318
	(B)	Progressive Truss or Girder Assembly	318
	(C)	Full Chord Assembly	318
	(D)	Progressive Chord Assembly	319
	(E)	Special Complete Structure Assembly	319
2.10.15	Drifting of Holes		319
2.10.16	Match-Marking		319
2.10.17	Rivets		319
2.10.18	Field Rivets		319
2.10.19	Bolts and Bolted Connections		320
	(A)	General	320
	(B)	Unfinished Bolts	320
	(C)	Turned Bolts	320
	(D)	Ribbed Bolts	320
2.10.20	Connections Using High Strength Bolts		320
	(A)	General	320
	(B)	Bolts, Nuts and Washers	321
	(C)	Bolted Parts	321
	(D)	Installation	321
	(E)	Inspection	323
2.10.21	Riveting		324

CONTENTS

Article			Page
2.10.22	Plate Cut Edges		325
	(A)	Edge Planing	325
	(B)	Visual Inspection and Repair of Plate Cut Edges	325
2.10.23	Welds		328
	(A)	General	328
	(B)	Filler Metal	328
	(C)	Preheat and Interpass Temperature	330
	(D)	Qualification of Welders, Welding Operators and Tackers	331
	(E)	Procedure Qualification	332
	(F)	Inspection of Welds	333
	(G)	Stud Shear Connectors	335
2.10.24	Oxygen Cutting		341
2.10.25	Facing of Bearing Surfaces		342
2.10.26	Abutting Joints		342
2.10.27	End Connection Angles		342
2.10.28	Lacing Bars		343
2.10.29	Finished Members		343
2.10.30	Web Plates		343
2.10.31	Bent Plates		343
2.10.32	Fit of Stiffeners		344
2.10.33	Eyebars		344
2.10.34	Annealing and Stress Relieving		344
2.10.35	Pins and Rollers		345
2.10.36	Boring Pin Holes		345
2.10.37	Pin Clearances		345
2.10.38	Threads for Bolts and Pins		345
2.10.39	Pilot and Driving Nuts		346
2.10.40	Notice of Beginning of Work		346
2.10.41	Facilities for Inspection		346
2.10.42	Inspector's Authority		346
2.10.43	Working Drawings and Identification of Steel During Fabrication		346
	(A)	Working Drawings	346
	(B)	Identification of Steels During Fabrication	346
2.10.44	Weighing of Members		348
2.10.45	Full Size Tests		348
2.10.46	Marking and Shipping		348
2.10.47	Painting		349
	ERECTION		349
2.10.48	Orthotropic-Deck Bridges		349
	(A)	Protection of Deck Plate After Sand Blasting	349
	(B)	Dimensional Tolerance Limits	349
2.10.49	Erection of Structure		351
2.10.50	Plans		351
2.10.51	Plant		351
2.10.52	Delivery of Materials		351
2.10.53	Handling and Storing Materials		351
2.10.54	Falsework		351
2.10.55	Methods and Equipment		352
2.10.56	Bearings and Anchorages		352
2.10.57	Straightening Bent Material and Cambering		352
	(A)	Straightening Bent Material	352
	(B)	Cambering	353

Article		Page
2.10.58	Assembling Steel	353
2.10.59	Riveting	353
2.10.60	Pin Connections	354
2.10.61	Misfits	354
2.10.62	Removal of Old Structure and Falsework	354
2.10.63	Method of Measurement	355
2.10.64	Basis of Payment	357

SECTION 11—BRONZE OR COPPER-ALLOY BEARING AND EXPANSION PLATES 358

2.11.1	General	358
2.11.2	Materials	358
	(A) Bronze Bearing and Expansion Plates	358
	(B) Rolled Copper-Alloy Bearings and Expansion Plates	358
	(C) Metal Powder Sintered Bearings and Expansion Joints (Oil Impregnated)	358
2.11.3	Bronze Plates	358
2.11.4	Copper-Alloy Plates	358
2.11.5	Placing	358
2.11.6	Measurement and Payment	358

SECTION 12—STEEL GRID FLOORING 359

2.12.1	General	359
2.12.2	Materials	359
	(A) Steel	359
	(B) Protective Treatment (Shop Coat)	359
	(C) Concrete	359
	(D) Skid Resistance	359
2.12.3	Arrangement of Sections	359
2.12.4	Provision for Camber	360
2.12.5	Field Assembly	360
2.12.6	Connection to Supports	360
2.12.7	Welding	360
2.12.8	Repairing Damaged Galvanized Coatings	360
2.12.9	Concrete Filler	361
2.12.10	Painting	361
2.12.11	Measurement and Payment	361

SECTION 13—RAILINGS 361

2.13.1	General	361
2.13.2	Materials	362
2.13.3	Line and Grade	362

METAL RAILING 362

2.13.4	Construction	362
2.13.5	Painting	362

CONCRETE RAILING 362

2.13.6	General	362
2.13.7	Materials	362
2.13.8	Railings Cast-in-Place	363
2.13.9	Precast Rails	363

Article		Page
2.13.10	Surface Finish	363
2.13.11	Expansion Joints	363
	STONE AND BRICK RAILING	364
2.13.12	General	364
	WOOD RAILING	364
2.13.13	General	364
2.13.14	Measurement and Payment	364
	SECTION 14—PAINTING METAL STRUCTURES	364
2.14.1	General	364
2.14.2	Material	364
	(A) Shop Coat (Prime Coat)	364
	(B) First Field Coat	365
	(C) Second Field Coat (Finish Coat)	365
2.14.3	Number of Coats and Color	365
2.14.4	Mixing of Paint	365
2.14.5	Weather Conditions	365
2.14.6	Application	366
	(A) General	366
	(B) Brushing	366
	(C) Spraying	366
	(D) Inaccessible Surfaces	366
2.14.7	Removal of Paint	366
2.14.8	Thinning Paint	366
2.14.9	Painting Galvanized Surfaces	367
2.14.10	Cleaning of Surfaces	367
	(A) General	367
	(B) Method A—Hand Cleaning	367
	(C) Method B—Blast Cleaning	367
	(D) Method C—Flame Cleaning	367
	(E) Surfaces Inaccessible After Assembly	368
2.14.11	Shop Painting	369
2.14.12	Field Painting	369
	SECTION 15—PROTECTION OF EMBANKMENTS AND SLOPES	370
2.15.1	General	370
	MATERIAL	370
2.15.2	Materials	370
	CONSTRUCTION	371
2.15.3	Loose Riprap for Slopes	371
2.15.4	Mortar Riprap for Slopes	372
2.15.5	Stone Riprap for Foundation Protection	373
2.15.6	Concrete Riprap in Bags	373
2.15.7	Concrete Slab Riprap	373
	FILTER MATERIAL	374
2.15.8	Filter or Bedding Material	374

Article		Page
	MEASUREMENT	374
2.15.9	Measurement	374
	PAYMENT	375
2.15.10	Payment	375
	SECTION 16—CONCRETE CRIBBING	376
2.16.1	General	376
2.16.2	Construction	377
2.16.3	Measurement and Payment	377
	SECTION 17—WATERPROOFING	377
2.17.1	General	377
2.17.2	Materials	377
	(A) Mortar	377
	(B) Asphalt	377
	(C) Pitch	378
	(D) Fabric	378
	(E) Tar for Absorptive Treatment	378
	(F) Tar Seal Coat	378
	(G) Joint Fillers	378
	(H) Inspection and Delivery	379
2.17.3	Storage of Fabric	379
2.17.4	Preparation of Surface	379
2.17.5	Application—General	379
2.17.6	Application—Details	380
2.17.7	Damage Patching	381
2.17.8	Protection Course	381
2.17.9	Measurement and Payment	381
	SECTION 18—DAMPPROOFING	381
2.18.1	General	381
2.18.2	Materials	382
2.18.3	Preparation of Surface	382
2.18.4	Application	382
2.18.5	Measurement and Payment	382
	SECTION 19—NAME PLATES	382
2.19.1	General Requirements	382
	SECTION 20—TIMBER STRUCTURES	383
2.20.1	Materials	383
	(A) Lumber and Timber (Solid sawn or glued laminated)	383
	(B) Structural Shapes	383
	(C) Castings	383
	(D) Hardware	383
	(E) Paint for Timber Structures	384
	(F) Timber Connectors	385
2.20.2	Timber Connectors	386
2.20.3	Storage of Material	389
2.20.4	Workmanship	389

Article		Page
2.20.5	Treated Timber	389
	(A) Handling	389
	(B) Framing and Boring	389
	(C) Cuts and Abrasions	390
	(D) Bolt Holes	390
	(E) Temporary Attachment	390
2.20.6	Untreated Timber	390
2.20.7	Treatment of Pile Heads	390
	(A) General	390
	(B) Method A—Zinc Covering	390
	(C) Method B—Fabric Covering	391
2.20.8	Holes for Bolts, Dowels, Rods and Lag Screws	391
2.20.9	Bolts and Washers	391
2.20.10	Countersinking	391
2.20.11	Framing	392
2.20.12	Pile Bents	392
2.20.13	Framed Bents	392
	(A) Mud Sills	392
	(B) Concrete Pedestals	392
	(C) Sills	392
	(D) Posts	392
	(E) Design and Construction	393
2.20.14	Caps	393
2.20.15	Bracing	393
2.20.16	Stringers	393
2.20.17	Plank Floors	393
2.20.18	Laminated or Strip Floors	394
2.20.19	Composite Wood-Concrete Decks	394
	(A) Slab Spans	394
	(B) "T" Beams	394
2.20.20	Wheel Guards and Railing	395
2.20.21	Trusses	395
2.20.22	Truss Housings	395
2.20.23	Erection of Housing and Railings	395
2.20.24	Painting	395
2.20.25	Measurement and Payment	395
	SECTION 21—PRESERVATIVE TREATMENTS FOR TIMBER	396
2.21.1	General	396
2.21.2	Materials	396
2.21.3	Identification and Inspection	396
	SECTION 22—TIMBER CRIBBING	397
2.22.1	Material	397
	(A) Timber	397
	(B) Logs	397
2.22.2	Preparation	397
2.22.3	Dimensions	397
	(A) Timber	397
	(B) Logs	397
2.22.4	Construction	398
	(A) Foundation	398
	(B) Mud Sills	398
	(C) Face Logs or Timbers	398

Article			Page
2.22.4 cont.	(D)	Ties	398
	(E)	Fastening	398
2.22.5	Filling		399
2.22.6	Measurement and Payment		399

SECTION 23—CONSTRUCTION AND INSTALLATION OF CORRUGATED METAL AND STRUCTURAL PLATE PIPES, PIPE-ARCHES, AND ARCHES 399

2.23.1	General		399
2.23.2	Forming and Punching of Corrugated Structural Plates and Sheets for Pipe		400
	(A)	Structural Plate Pipe	400
	(B)	Corrugated Metal Pipe	400
	(C)	Elongation	400
2.23.3	Assembly		400
2.23.4	Bedding		401
2.23.5	Pipe Foundation		401
2.23.6	Sidefill		401
2.23.7	Bracing		402
2.23.8	Camber		402
2.23.9	Arch Substructures and Headwalls		402
2.23.10	Cover Over Pipe During Construction		403
2.23.11	Workmanship and Inspection		404
2.23.12	Method of Measurement		404
2.23.13	Basis of Payment		404

SECTION 24—WEARING SURFACES 404

2.24.1	Description		404
2.24.2	Orthotropic Deck Bridges		405
	(A)	Material	405
	(B)	Placement	405
	(C)	Inspection	405

SECTION 25—ELASTOMERIC BEARINGS 405

2.25.1	Description	405
2.25.2	Materials	405
2.25.3	Manufacturing Requirements	406
2.25.4	Tolerances	408
2.25.5	Quality Assurance	408

SECTION 26—CONSTRUCTION OF TUNNEL USING STEEL TUNNEL LINER PLATES 408

2.26.1	Scope	408
2.26.2	Description	409
2.26.3	Forming and Punching of Liner Plates	409
2.26.4	Installation	409
2.26.5	Measurement	410
2.26.6	Payment	410

INDEX TO FIGURES

Figures		Page
1.1.7	Clearance Diagram for Bridges	4
1.1.9	Railings	6
1.1.15	Clearance Diagram for Tunnels—Two Lane Highway Traffic	9
1.1.17	Clearance Diagram for Underpasses	11
1.2.5A	Standard H Trucks	15
1.2.5B	H Lane and HS Lane Loadings	16
1.2.5C	Standard HS Trucks	17
1.7.3A	Fatigue Stresses	117
1.7.3B	Fatigue Stresses	120
1.7.19	Splice Details	130
1.7.70	Web Thickness and Girder Depth	145
1.7.71A	Transverse Stiffener Spacing	148
1.7.71B	Web Plate without Stiffeners	149
1.7.111A	Flange Stress Reduction Factors, $\alpha = .72$	172
1.7.111B	Flange Stress Reduction Factor, $\alpha = .50$	172
1.10.2	Spaced Column, Connector Joined	229
1.13.1	Diagram for Coefficient C_d for Tunnels in Soil	241
2.10.22B	Discontinuities	327
2.10.23A	Procedure Qualification Fillet Weld Test	334
2.10.23B	Stud Shear Connector	338
2.10.23C	Tensile Test Fixture	339
2.10.23D	Bend Test Fixture and Failures	339
2.23	Pipe Bedding, Foundation & Sidefill	403

INDEX TO TABLES

1.3.1B	Distribution of Loads	31
1.3.1C	Bending Moment in Floor Beams	33
1.5.27	Recommended Minimum Thickness for Constant Depth Members	90
1.5.27A	Long-time Deflections	91
1.5.29	Standard Hooks in Tension	94
1.7.1	Allowable Stresses	112
1.7.3A	Stress Cycles	115
1.7.3B	Fatigue Stresses	118
1.8.2	Minimum Longitudinal Seam Strengths	202
1.10.1	Allowable Unit Stresses for Structural Lumber Visually Graded	210
1.10.1A	Allowable Unit Stresses for Structural Glued Laminated Timber	218
1.10.1B	Allowable Unit Stresses for Structural Glued Laminated Timber	222
1.11.5	Allowable Stresses	236
2.10.3A	Bolt and Nut Dimensions	309
2.10.3B	Washer Dimensions	309
2.10.20A	Bolt Tension	322
2.10.20B	Nut Rotation from Snug Tight Condition	323
2.10.22B	Discontinuities	326
2.10.43	Identification Color Codes	348
2.20.1	Typical Dimensions of Timber Connectors	387

PREFACE

to

Eleventh Edition

Major changes and revisions to this edition are as follows:

1. The Interim Specifications of 1970, 1971, and 1972, have been adopted and included, together with twelve items which were balloted and adopted in November 1972.

2. The section on Railings (Article 1.1.9) has been expanded and modified to provide for improved safety performance.

3. Load Factor design concepts are now permitted for both structural steel and reinforced concrete structures as an alternate to existing design criteria. This load factor concept has also been included in the prestressed concrete design provisions for ultimate strength.

4. Section 7 of Design contains provisions for heat curving and for design of orthotropic deck bridges.

5. Section 10 has been extensively modified to provide new design requirements for Timber structures.

6. Updated provisions for Steel Tunnel Liner Plates and Slope Protection have also been included.

INTRODUCTION

The compilation of these specifications began with the organization, in 1921, of the Committee on Bridges and Structures of the American Association of State Highway Officials. During the period from 1921, until printed in 1931, the specifications were gradually developed, and as the several divisions were approved from time to time, they were made available in mimeographed form for use of the State Highway Departments and other organizations. A complete specification was available in 1926 and it was revised in 1928. Though not in printed form, the specifications were valuable to the bridge engineering profession during the period of development.

The first edition of the Standard Specifications was published in 1931, and it was followed by the 1935, 1941, 1944, 1949, 1953, 1957, 1961, 1965, and 1969 revised editions. The present and eleventh edition constitutes a revision of the 1969 Specifications, including those changes adopted since the publication of the tenth edition and those through 1972. The constant research and development in steel, concrete, and timber structures practically dictates the necessity of revising the specifications every four years, and the 1973 edition continues this trend.

Among the important revisions in the 1973 edition is the adoption of load factor criteria as an alternate design method. Other modifications are outlined in the Preface.

Interim Specifications are usually published late in the calendar year, and a revised edition of this book is published every four years. The Interim Specifications have the same status as standards of the American Association of State Highway Officials, but are tentative revisions approved by at least two-thirds of the Committee on Bridges and Structures. These revisions are voted on by the Association Member Departments prior to the publication of each new edition of this book, and if approved by at least two-thirds of the members, they are included in the new edition as standards of the Association. Members of the Association are the fifty State Highway Departments, the District of Columbia, Puerto Rico and the Federal Highway Administration. Each member has one vote.

Annual Interim Specifications are generally used by the State Highway Department after their adoption by the Bridge Committee. Orders for these annual Interim Specifications should be sent to the Executive Director of the Association at 341 National Press Building, Washington, D.C. 20004.

The Specifications for Highway Bridges are intended to serve as a standard or guide for the preparation of State specifications and for reference by bridge engineers.

Primarily, the specifications set forth minimum requirements which are consistent with current practice, and certain modifications may be necessary to suit local conditions. They apply to ordinary highway bridges and supplemental specifications may be required for unusual types and for bridges with spans longer than 500 feet.

Specifications of the American Society for Testing and Materials, the American Welding Society, the American Wood Preservers Association and the National Lumber Manufacturers' Association are referred to or are recognized. Numerous research bulletins are noted for references.

The American Association of State Highway Officials wishes to express its sincere appreciation to the above organizations, as well as to those Universities and representatives of industry whose research efforts and consultations have been most helpful in continued improvement of these specifications.

Extensive references have been made to the Standard Specifications for Highway Materials published by the American Association of State Highway Officials including equivalent ASTM specifications which have been reproduced in the Association's Standard Specifications by permission of the American Society for Testing and Materials.

Attention is also directed to the following publications prepared by the Bridge Committee and published subsequent to the 1969 edition of these specifications:

    Construction Manual for Highway Bridges
    and Incidental Structures—1973 Edition
    Standard Specifications for Moveable
    Highway Bridges—1970 Edition
    Specifications for the Design and
    Construction of Structural Supports for
    Highway Luminaires—1971 Edition
    Manual for Maintenance Inspection
    of Bridges—1970 Edition

The following have served as chairmen of the Committee since its inception in 1921: Messrs. E. F. Kelley, who pioneered the work of the Committee, Albin L. Gemeny, R. B. McMinn, Raymond Archiband, G. S. Paxson, and Mr. E. M. Johnson. The late Mr. Ward Goodman would have completed eight years as Chairman of the Bridge Committee this year, but his untimely death in March 1973, resulted in the appointment of Mr. Charles Matlock. Mr. Goodman was also serving as Vice President of AASHO. The Committee expresses its sincere appreciation of the work of these men and of those active members of the past, whose names, because of retirement, are no longer on the roll.

Suggestions for the improvement of the specifications are welcomed. They should be sent to the Chairman, Committee on Bridges and Structures, AASHO, 341 National Press Bldg., Washington, D.C. 20004. Inquiries as to the intent or application of the specifications should be sent to the same address.

ABBREVIATIONS

AASHO—American Association of State Highway Officials
  ACI—American Concrete Institute
  AITC—American Institute of Timber Construction
  ASCE—American Society of Civil Engineers
  ASTM—American Society for Testing and Materials
  ANSI—American National Standards Institute
  AWS—American Welding Society
AWPA—American Wood Preservers Association
    CS—Commercial Standards
  NDS—National Design Specifications for Stress Grade Lumber and Its Fastenings
 NLMA—National Lumber Manufacturers' Association
  SAE—Society of Automotive Engineers
  WPA—Western Pine Association
WWPA—Western Wood Products Association

Division I

DESIGN

DESIGN ANALYSIS

In any case where the specifications provide for an empirical formula as a design convenience, a rational analysis based on a theory accepted by the Committee on Bridges and Structures of the American Association of State Highway Officials, with stresses in accordance with the specifications, will be considered as compliance with the specifications.

Section 1—GENERAL FEATURES OF DESIGN

1.1.1 — BRIDGE LOCATIONS

Selecting favorable stream crossings should be considered in the preliminary route determination to minimize construction, maintenance and replacement costs. Natural stream meanders should be studied and, if necessary, channel changes, river training works and other construction which would reduce erosion problems and prevent possible loss of the structure should be considered. Foundations of bridges placed across channel changes should be designed for possible deepening and widening of the relocated channel. On wide flood plains, the lowering of approach fills to provide overflow sections designed to pass unusual floods over the highway is a means of preventing loss of structures. Where relief bridges are needed to maintain the natural flow distribution and reduce backwater, caution must be exercised in proportioning the size and in locating such structures to avoid undue scour or changes in the course of the main river channel.

1.1.2 — BRIDGE WATERWAYS

The determination of adequate waterway openings for stream crossings is essential to the design of safe and economical bridges. Hydraulic studies of bridge sites are a necessary part of the preliminary design of a bridge and reports of such studies should include applicable parts of the following outline:

(A) Site Data

    1. Maps, stream cross sections, aerial photographs.
    2. Complete data on existing bridges, including dates of construction and performance during past floods.
    3. Available highwater marks with dates of occurrence.
    4. Information on ice, debris and channel stability.
    5. Factors affecting water stages such as high water from other streams, reservoirs, flood control projects and tides.

(B) Hydrologic Analysis

1. Compile flood data applicable to estimating floods at site, including both historical floods and maximum floods of record.
2. Plot flood-frequency curve for site.
3. Determine distribution of flow and velocities at site for flood discharges to be considered in design of structure.
4. Plot stage-discharge curve for site.

(C) Hydraulic Analysis

1. Compute backwater and mean velocities at bridge opening for various trial bridge lengths and selected discharges.
2. Estimate scour depth at piers and abutments of proposed structures.

Usually, bridge waterways are sized to pass a design flood of a magnitude and frequency consistent with the type or class of highway. In the selection of the waterway opening, consideration should be given to the amount of upstream ponding, the passage of ice and debris and possible scour of the bridge foundations. Where floods exceeding the design flood have occurred, or where superfloods would cause extensive damage to adjoining property or the loss of a costly structure, a larger waterway opening may be warranted. Due consideration should be given to any Federal, State and local requirements.

Relief openings, spur-dikes, debris deflectors and channel training works should be used where needed to minimize the effect of adverse flood flow conditions. Where scour is likely to occur, protection against damage from scour should be provided for in the design of bridge piers and abutments. Embankment slopes adjacent to structures subject to erosion should be adequately protected by rip-rap, flexible mattresses, retards, spur dikes or other appropriate construction. Clearing of brush and trees along embankments in the vicinity of bridge openings should be avoided to prevent high flow velocities and possible scour. Borrow pits should not be located in areas which would increase velocities and the possibility of scour at bridges.

1.1.3 — PIER SPACING, ORIENTATION AND TYPE

Piers shall be located to meet navigational clearance requirements and to give a minimum interference to flood flow. In general, piers should be placed parallel with the direction of the stream current at flood stage. Adequate provision should be made for drift and ice by increasing span lengths and vertical clearances, by selecting proper pier types and by using debris deflectors. Special precautions against scour are required when large cofferdams are placed in unstable stream beds.

1.1.4 — CULVERT WATERWAY OPENINGS

Culverts, as distinguished from bridges, are usually smaller in waterway opening, covered with embankment material and composed of structure around the entire perimeter of the culvert barrel. Some culverts are supported on spread footings when the natural stream bed can serve as the bottom of the waterway without undermining the foundation.

Criteria for design discharges, allowable headwater depth and outlet velocities will vary, depending upon the class of highway, hazards to traffic, risks of flooding adjacent property, risks of damaging embankments, stream bed material and other factors. Based on conditions at and in the vicinity of the culvert site, it may be determined that the culvert can be designed to operate satisfactorily under submergence. Generally, designs for culverts to operate under submergence are limited to floods of infrequent occurrence. Culverts should be designed to resist the hydraulic forces to be encountered and should be protected from undermining by means of adequate aprons, wingwalls, cutoff walls or other appropriate devices. Adjacent embankments should be protected against erosion as necessary by rip-rap or other suitable means.

1.1.5 — CULVERT LOCATION AND LENGTH

In general, culverts are located in natural stream channels. If foundations are poor, bearing piles, selected backfill material, a cambered profile or slip-collars can be provided to assure good alignment and a culvert that is structurally sound.

The length of culverts should be sufficient to prevent the embankment material from encroaching on a culvert end. Headwalls and endwalls with cutoff walls and aprons are used to protect the fill slopes and stream beds from erosion and to secure the culvert end against hydraulic forces. If headwalls and endwalls are required, they should be designed not to protrude above the ground line. Culvert openings shall be placed a minimum of 30 feet from the edge of the traffic lanes or protection provided by guardrail or other means. Where feasible, culverts shall be continuous across medians to avoid the traffic hazard presented by additional openings. Where needed, debris control devices should be constructed to prevent clogging. If backfill and embankment materials are subject to piping, consideration should be given to the use of cutoff walls or impervious material placed at the entrance.

1.1.6 — WIDTH OF ROADWAY AND SIDEWALK

The width of roadway shall be the clear width measured at right angles to the longitudinal center line of the bridge between the bottoms of curbs or if curbs are not used, the clear width shall be the minimum measured between the nearest faces of the bridge railing.

The width of the sidewalk shall be the clear width, measured at right angles to the longitudinal center line of the bridge, from the extreme inside portion of the handrail to the bottom of the curb or guard-timber, except that if there is a truss, girder, or parapet wall adjacent to the roadway curb, the width shall be measured to its extreme walk side portion.

1.1.7 — CLEARANCES

(A) Navigational

Permits for the construction of crossings over navigable streams, except those streams that have been placed in the 'advance approval' category by the Commandant, U. S. Coast Guard, must be obtained from the U. S. Coast Guard and other appropriate agencies. Requests for such permits from the U. S. Coast Guard should be addressed to the appropriate District Commander.

(B) Vehicular

The horizontal clearance shall be the clear width and the vertical clearance the clear height for the passage of vehicular traffic as shown in Figure 1.1.7.

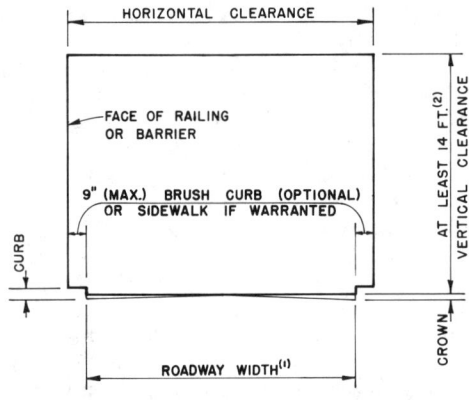

CLEARANCE DIAGRAM FOR BRIDGES

FIGURE 1.1.7

(1) The roadway width shall generally equal the full shoulder width of the approach roadway section. Where curbed roadway sections approach a structure, the same section shall be carried across the structure. The minimum horizontal clearance for low traffic speed, low traffic volume bridges shall be 8'-0" greater than the approach travelled way.

For recommendations as to roadway widths for various volumes of traffic see AASHO "A Policy on Design Standards—Interstate System", "Geometric Design Standards for Highways Other Than Freeways", "A Policy on Geometric Design of Rural Highways", and/or "A Policy on Arterial Highways in Urban Areas".

(2) Vertical clearance on State trunk highways and interstate systems in rural areas shall be at least 16 feet over the entire roadway width, to which an allowance should be added for resurfacing. On State trunk highways and interstate routes through urban areas a 16-foot clearance shall be provided except in highly developed areas. A 16-foot clearance should be provided in both rural and urban areas where such clearance is not unreasonably costly and where needed for defense requirements. Vertical clearance on all other highways shall be at least 14 feet over the entire roadway width to which an allowance should be added for resurfacing.

(C) Other

The channel openings and clearance shall be cleared with other agencies having jurisdiction over such matters. Channel openings and clearances in general shall conform in width, height, and location to all Federal, State and local requirements.

1.1.8 — CURBS AND SIDEWALKS

The face of the curb is defined as the vertical or sloping surface on the roadway side of the curb. Horizontal measurements of roadway and curb width are given from the botttom of the face, or, in the case of stepped back curbs, from the bottom of the lower face for roadway width. Maximum width of brush curbs, if used, shall be 9 inches.

Where curb and gutter sections are used on the roadway approach, at either or both ends of the bridge, the curb height on the bridge may match the curb height on the roadway approach, or if preferred, it may be made higher than the approach curb. Where no curbs are used on the roadway approaches, the height of the bridge curb above the roadway shall be not less than 8 inches, and preferably not more than 10 inches.

Where sidewalks are warranted for pedestrian traffic on urban expressways, they shall be separated from the bridge roadway by the use of a traffic or combination railing as shown in Figure 1.1.9.

1.1.9 — RAILINGS

Railing shall be provided at the edge of structures for the protection of traffic and for the protection of pedestrians if pedestrian walkways are provided.

Where pedestrian walkways are provided adjacent to roadways on other than urban expressways, a traffic railing or barrier may be provided between the two with a pedestrian railing outside.

(A) Traffic Railing

While the primary purpose of traffic railing is to contain the average vehicle using the structure, consideration should also be given to protection of the occupants of a vehicle in collision with the railing, to protection of other vehicles near the collision, to vehicles or pedestrians on roadways being overcrossed, and to appearance and freedom of view from passing vehicles.

Materials for traffic railing shall be concrete, metal, timber or a combination. Metal materials with less than 10 percent tested elongation shall not be used.

Traffic railings should provide a smooth, continuous face of rail on the traffic side with the posts set back from the face of rail. Structural continuity in the rail members, including anchorage of ends, is

HIGHWAY BRIDGES

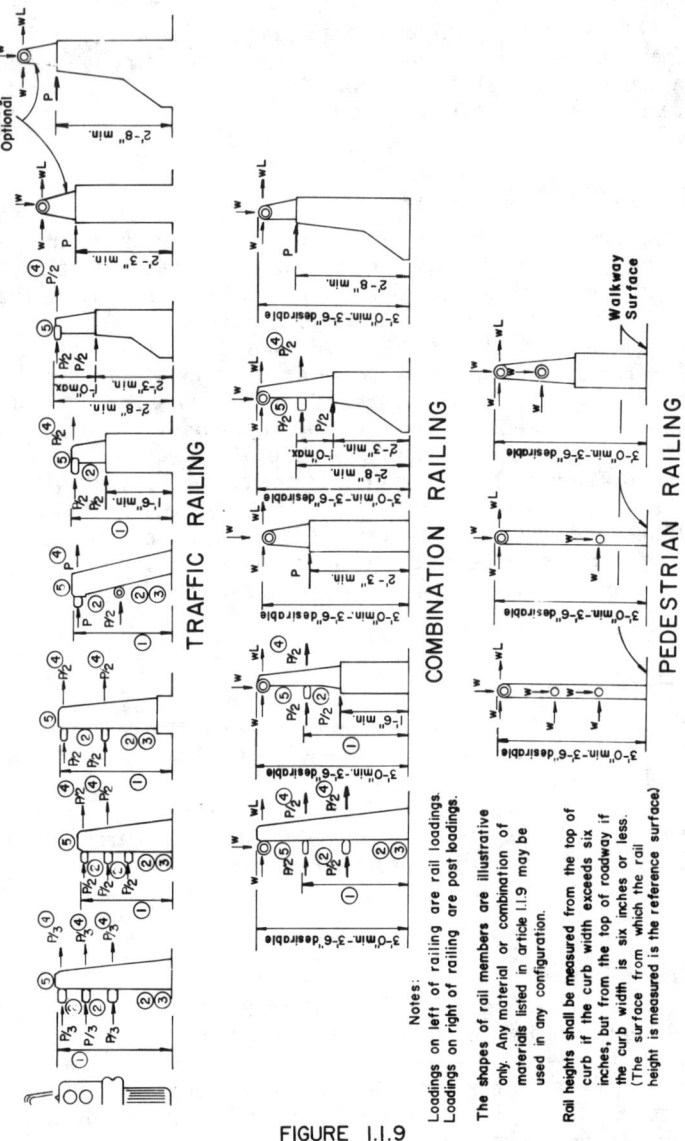

FIGURE 1.1.9

Notes for Fig. 1.1.9:

1. Traffic railings and the traffic railing portions of combination railings shall have a minimum height of 2'-3". When the height of the top of the top traffic rail exceeds 2'-9", the total transverse load distributed to the traffic rails shall be equal CP except that the maximum load applied to a rail need not exceed P.
2. The lower rail should be centered between 15 and 20 inches above the reference surface. The maximum clear vertical opening below the lower rail or between succeeding rails shall not exceed 15 inches.

essential. The railing system shall be able to resist the applied loads at all locations.

Protrusions or depressions at rail joints shall be acceptable provided their thickness or depth is no greater than the wall thickness of the rail member or ⅜", whichever is less.

The height of traffic railing shall be no less than 2'-3" measured from the top of the roadway or curb to the top of the upper rail member, except that parapets designed with sloping traffic faces intended to allow vehicles to ride up them under low angle contacts shall be at least 2'-8" in height. This sloping face parapet height may be reduced to 2'-3" provided a traffic railing is mounted on top of the parapet at a height not exceeding 3'-3". The lower element of a traffic or combination railing should consist of either a parapet projecting at least 18 inches above the reference surface or a rail centered between 15 and 20 inches above the reference surface. The roadway surface is the reference surface unless there is a curb or sidewalk projecting more than 6 inches from the traffic face of the railing, in which case the surface of the curb or sidewalk is the reference surface. The maximum clear vertical opening below the lower rail or between succeeding rails shall not exceed 15 inches. (See Figure 1.1.9) Railings other than those shown in Figure 1.1.9 are permissible provided the total applied loading is not less than 10 kips.

Careful attention shall be given to the treatment of railing at the bridge ends. Exposed rail ends, posts, and sharp changes in the geometry of the railing shall be avoided. A smooth transition by means of a continuation of the bridge barrier, guard rail anchored to the bridge end, or other effective means shall be provided to protect the traffic from direct collision with the bridge rail ends.

(B) Pedestrian Railing

Railing components shall be proportioned commensurate with the type and volume of anticipated pedestrian traffic, taking account of appearance, safety and freedom of view from passing vehicles.

---

3 The traffic faces of all traffic rails must be within one inch of a vertical plane through the traffic face of the rail closest the traffic. Rails a greater distance behind this plane or centered lower than 15 inches above the reference surface shall not be considered traffic rails for the purpose of distributing the transverse loading P or CP. Rails not considered traffic rails may be considered in determining maximum vertical clear opening (see Note 2) provided they are designed for a transverse loading equal to that applied to an adjacent traffic rail or P/2, whichever is the lesser.

4 A load equal to ½ the transverse load on a post shall simultaneously be applied longitudinally, divided among not more than four posts in a continuous rail length. Each traffic post shall also be designed to resist an independently applied inward load equal to ¼ the outward transverse load.

5 The attachment of each rail required in a traffic railing shall be designed to resist a vertical load equal to ¼ the transverse design load of the rail. The vertical load is to be applied alternately upward and downward. The attachment shall also be designed to resist an inward transverse load equal to ¼ the transverse rail design load.

Nomenclature:

$P = 10,000$ lbs.

$L =$ post spacing (ft.)

$w = 50$ lbs. per linear foot

$C = 1 + \dfrac{h - 33}{18}$; but shall not be less than 1

$h =$ height of top of top traffic rail (in.)

Materials for pedestrian railing may be concrete, metal, timber or a combination.

The minimum height of pedestrian railing shall be 3'-0" (a preferred height is 3'-6") measured from the top of the walkway to the top of the upper rail member.

1.1.10 — ROADWAY DRAINAGE

The transverse drainage of the roadway should be accomplished by providing a suitable crown in the roadway surface and longitudinal drainage should be accomplished by camber or gradient. Water flowing downgrade in a gutter section should be intercepted and not permitted to run onto the bridge. Short, continuous span bridges, particularly overpasses, may be built without inlets and the water from the bridge roadway carried downslope by open or closed chutes near the end of the bridge structure. Longitudinal drainage on long bridges is accomplished by means of scuppers or inlets which should be of sufficient size and number to drain the gutters adequately. Downspouts, where required, should be of rigid corrosion-resistant material not less than 4 inches in least dimension and should be provided with cleanouts. The details of deck drains should be such as to prevent the discharge of drainage water against any portion of the structure and to prevent erosion at the outlet of the downspout. Overhanging portions of concrete deck should be provided with a drip bead or notch.

1.1.11 — SUPERELEVATION

The superelevation of the floor surface of a bridge on a horizontal curve shall be provided in accordance with the standard practice of the commission for the highway construction, except that the superelevation shall not exceed 0.10 foot per foot width of roadway.

1.1.12 — FLOOR SURFACES

All bridge floors shall have skid-resistant characteristics.

1.1.13 — BLAST PROTECTION

On bridges over steam railroad tracks, metal likely to be injured by locomotive gases, and concrete surfaces less than 20 feet above the tracks, shall be protected by blast plates. The blast plate shall be centered on a line normal to the plane of the two rails at the center line of the track, thus taking into account the direction of blast due to superelevation. The plates shall be not less than 4 feet wide and shall consist of wrought-iron, cast-iron, a corrosion and blast resisting alloy or asbestos-board shields, so supported that they may be readily replaced. The thickness of plates and other parts in direct contact with locomotive blast shall be not less than $\frac{3}{4}$ inch for cast-iron, $\frac{3}{8}$ inch for wrought-iron or alloy, $\frac{1}{2}$ inch for plain asbestos-board and $\frac{7}{16}$ inch for corrugated asbestos-board. Bolts shall be not less than $\frac{5}{8}$ inch in diameter. Pockets which may hold locomotive gases shall be avoided as far as practicable. All fastenings shall be galvanized or of corrosive resistant material.

1.1.14 — UTILITIES

Where required, provisions shall be made for trolley wire supports and poles, pillars for lights, electric conduits, telephone conduits, water pipes, gas pipes and sanitary sewers.

1.1.15 — ROADWAY WIDTH, CURBS AND CLEARANCES FOR TUNNELS
(See Figure 1.1.15)

(A) Roadway Width

The horizontal clearance shall be the clear width and the vertical clearance the clear height for the passage of vehicular traffic as shown in Figure 1.1.15.

Unless otherwise provided, the several parts of the structure shall be constructed to secure the following limiting dimensions or clearances for traffic:

The clearances and width of roadway for 2-lane traffic shall be not less than those shown in Figure 1.1.15. The roadway width shall be

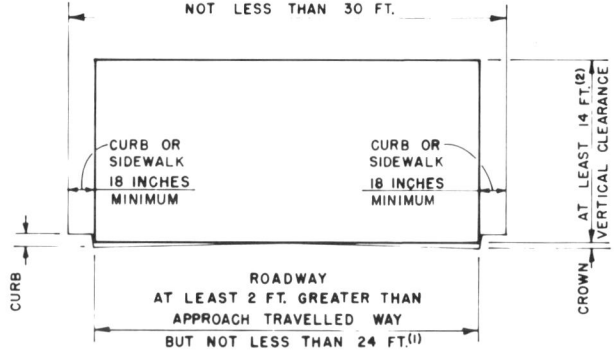

CLEARANCE DIAGRAM FOR TUNNELS
TWO LANE HIGHWAY TRAFFIC

FIGURE 1.1.15

(1) For heavy traffic roads, roadway widths greater than the above minima are recommended.
  If traffic lane widths exceed 12 feet, the roadway width may be reduced 2'-0" from that calculated from Fig. 1.1.15.
  For recommendations as to roadway widths for various volumes of traffic see AASHO "A Policy on Design Standards—Interstate System", "Geometric Design Standards for Highways Other Than Freeways", "A Policy on Geometric Design of Rural Highways", and/or "A Policy on Arterial Highways in Urban Areas".

(2) Vertical clearance on State trunk highways and interstate systems in rural areas shall be at least 16 feet over the entire roadway width, to which an allowance should be added for resurfacing. On State trunk highways and interstate routes through urban areas a 16-foot clearance shall be provided except in highly developed areas. A 16-foot clearance should be provided in both rural and urban areas where such clearance is not unreasonably costly and where needed for defense requirements. Vertical clearance on all other highways shall be at least 14 feet over the entire roadway width to which an allowance should be added for resurfacing.

increased at least ten feet and preferably twelve feet for each additional traffic lane.

(B) Clearance Between Walls

The minimum width between walls of two-lane tunnels shall be 30 feet.

(C) Curbs

The width of curbs shall be not less than 18 inches. The height of curbs shall be as specified for bridges.

(D) Vertical Clearance

The vertical clearance, between curbs, shall be not less than 14 feet.

1.1.16 — ROADWAY WIDTH, CURBS AND CLEARANCES FOR DEPRESSED ROADWAYS

(A) Roadway Width

The clear width between curbs shall be not less than that specified for tunnels.

(B) Clearance Between Walls

The minimum width between walls for depressed roadways carrying two lanes of traffic shall be 30 feet.

(C) Curbs

The width of curbs shall be not less than 18 inches. The height of curbs shall be as specified for bridges.

1.1.17 — ROADWAY WIDTH, CURBS AND CLEARANCES FOR UNDERPASSES
(See Figure 1.1.17)

(A) Widths

The pier columns or walls for grade separation structures shall generally be located a minimum of 30 feet from the edges of the through traffic lanes. Where the practical limits of structure costs, type of a structure, volume and design speed of through traffic, span arrangement, skew and terrain make the 30 foot offset impractical, the pier or

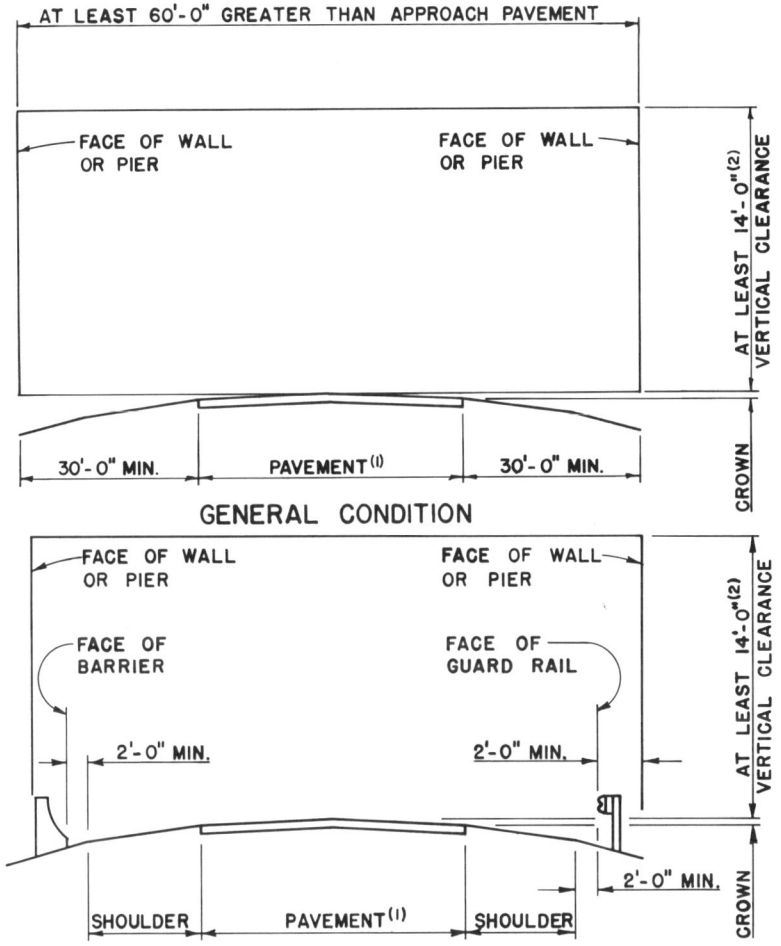

CLEARANCE DIAGRAMS FOR UNDERPASSES

FIGURE 1.1.17

SEE ARTICLE 1.1.17(A) FOR GENERAL REQUIREMENTS.

(1) For recommendations as to roadway widths for various volumes of traffic see AASHO "A Policy on Design Standards—Interstate System", "Geometric Design Standards for Highways Other Than Freeways", "A Policy on Geometric Design of Rural Highways", and/or "A Policy on Arterial Highways in Urban Areas".

(2) Vertical clearance on State trunk highways and interstate systems in rural areas shall be at least 16 feet over the entire roadway width, to which an allowance should be added for resurfacing. On State trunk highways and interstate routes through urban areas a 16-foot clearance shall be provided except in highly developed areas. A 16-foot clearance should be provided in both rural and urban areas where such clearance is not unreasonably costly and where needed for defense requirements. Vertical clearance on all other highways shall be at least 14 feet over the entire roadway width to which an allowance should be added for resurfacing.

wall may be placed closer than 30 feet and protected by the use of guard rail or other barrier devices. The guard rail shall be independently supported with the roadway face at least 2'-0" from the face of pier or abutment. The face of the guard rail or other device shall be at least 2'-0" outside the normal shoulder line.

(B) Vertical Clearance

A vertical clearance of not less than 14 feet shall be provided between curbs, or if curbs are not used, over the entire width that is available for traffic.

(C) Curbs

Curbs, if used, shall match those of the approach roadway section.

Section 2—LOADS

1.2.1 — LOADS

Structures shall be proportioned for the following loads and forces when they exist:
Dead load.
Live load.
Impact or dynamic effect of the live load.
Wind loads.
Other forces, when they exist, as follows:
  Longitudinal forces, centrifugal force, thermal forces, earth pressure, buoyancy, shrinkage stresses, rib shortening, erection stresses, ice and current pressure, and earthquake stresses.

Members shall be proportioned using allowable stresses and design limitations for the appropriate material.

Upon the stress sheets a diagram or notation of the assumed loads shall be shown and the stresses due to the various loads shall be shown separately.

Where required by design conditions, concrete placing sequence shall be indicated on the plans or in the special provisions.

The loading combinations shall be in accordance with Article 1.2.22.

1.2.2 — DEAD LOAD

The dead load shall consist of the weight of the structure complete, including the roadway, sidewalks, car tracks, pipes, conduits, cables and other public utility services.

The snow and ice load is considered to be offset by an accompanying decrease in live load and impact and shall not be included except under special conditions.

If a separate wearing surface is to be placed when the bridge is constructed, or if placement of a separate wearing surface is anticipated in the future by the department, adequate allowance shall be made for its weight in the design dead load. Otherwise provision for a future wearing surface is not required.

Special consideration shall be given to the necessity for a separate wearing surface for those regions where the use of chains on tires or studded snow tires, is anticipated.

Where the abrasion of concrete is not anticipated, the traffic may bear directly on the concrete slab. If considered desirable, ¼ inch or more may be added to the slab for a wearing surface.

The following weights are to be used in computing the dead load:

	Weight per cubic foot, pounds
Steel or cast steel	490
Cast iron	450
Aluminum alloys	175
Timber (treated or untreated)	50
Concrete, plain or reinforced	150
Compacted sand, earth, gravel or ballast	120
Loose sand, earth and gravel	100
Macadam or gravel, rolled	140
Cinder filling	60
Pavement, other than wood block	150
Railway rails, guard rails, and fastenings (per linear foot of track)	200
Stone masonry	170
Asphalt plank, 1 inch thick	9 lbs. per square foot

(A) Unit Load on Culverts

Earth pressures or loads on culverts may be computed ordinarily as the weight of earth directly above the structure. For box culverts, and culverts with cast-in-place inverts or footings, the weight of the earth may be taken at 70 percent of its actual weight. This will have the effect of increasing the allowable design dead load stresses 40% more than allowed for live load. For flexible and rigid pipes, not cast-in-place, the weight of the earth may be taken at 83% of its actual weight. This will have the effect of increasing the allowable design dead load stresses 20% more than allowed for live load.

For definite conditions of bedding and backfill, the principles of soil mechanics may be applied. The following are recommended formulas for these conditions:

(1) Culvert in trench on unyielding subgrade, or culvert untrenched on yielding foundation.

$$P = WH$$

(2) *Culvert untrenched on unyielding foundation (such as rock or piles).

$$P = W(1.92H - 0.87B) \text{ for } H > 1.7B \tag{1A}$$

$$P = 2.59 \, BW \, (e^k - 1) \text{ for } H < 1.7B \text{ where } k = \frac{0.385H}{B} \tag{1B}$$

where P = the unit pressure in pounds per square foot due to earth backfill.

B = width in feet of trench, or in case there is no trench, the overall width of the culvert.

H = depth in feet of fill over culvert.

W = effective weight per cubic foot of fill material, which may be taken as 70 percent, or 83 percent, of actual weight in accordance with above stated provisions.

e = 2.7182818 = base of natural logarithms, abstract number.

(B) Shear in Top Slabs

The maximum shear in the top slabs of culverts under embankments shall be assumed to occur at a distance, "d", out from the wall or abutment; "d" being equal to the depth from the top of the slab to the centroid of the tension reinforcement.

The shear in bottom slabs shall be computed as specified for footings in Article 1.4.6.

1.2.3. — LIVE LOAD

The live load shall consist of the weight of the applied moving load of vehicles, cars and pedestrians.

1.2.4 — OVERLOAD PROVISION

The following provision for overload shall apply to all loadings except the H 20 and HS 20 loadings.

Provision for infrequent heavy loads shall be made by applying in any single lane an H or HS truck as specified, increased 100 per cent, and without concurrent loading of any other lanes. Combined dead, live and impact stresses resulting from such loading shall not be greater than 150 per cent of the allowable stresses prescribed herein. The overload shall apply to all parts of the structure affected, except the roadway deck.**

1.2.5 — HIGHWAY LOADINGS

(A) General

The highway live loadings on the roadways of bridges or incidental structures shall consist of standard trucks or of lane loads which are equivalent to truck trains. Two systems of loading are provided,

---

* Note: Formulas 1A and 1B have been derived from Iowa Engineering Experiment Station Bulletin 96, "The Theory of External Loads on Closed Conduits in the Light of the Latest Experiments", by Anson Marston, Director, February 19, 1930.

** For orthotropic bridges, the roadway deck consists of the deck plate and stiffening ribs.

the H loadings and the HS loadings, the corresponding HS loadings being heavier than the H loadings.

(B) H Loadings

The H loadings are illustrated in Figures 1.2.5A and 1.2.5B. They consist of a two-axle truck or the corresponding lane loading. The H

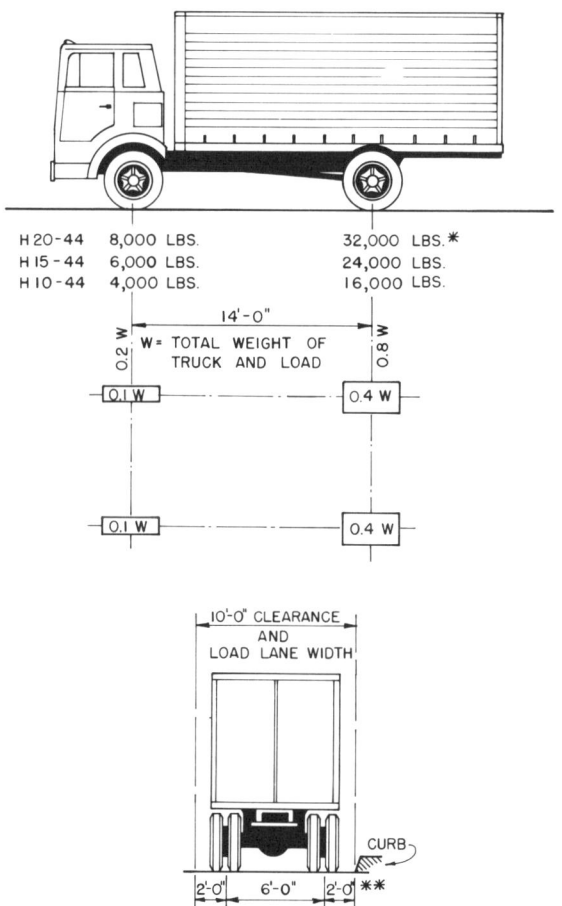

STANDARD H TRUCKS

FIGURE 1.2.5A

* In the design of timber floors and orthotropic steel decks (excluding transverse beams) for H20 loading, one axle load of 24,000 pounds or two axle loads of 16,000 pounds each, spaced 4 feet apart may be used, whichever produces the greater stress, instead of the 32,000 pound axle shown.

** For slab design, the center line of wheels shall be assumed to be 1 foot from face of curb. (See Art. 1.3.2(B))

loadings are designated H followed by a number indicating the gross weight in tons of the standard truck.

(C) HS Loadings

The HS loadings are illustrated in Figures 1.2.5B and 1.2.5C. They consist of a tractor truck with semi-trailer or of the corresponding lane loading. The HS loadings are designated by the letters HS followed by a number indicating the gross weight in tons of the tractor truck. The variable axle spacing has been introduced in order that the spacing of axles may approximate more closely the tractor trailers now in use. The variable spacing also provides a more satisfactory loading for continuous spans, in that heavy axle loads may be so placed on adjoining spans as to produce maximum negative moment.

(D) Classes of Loadings

Highway loadings shall be of five classes: H 20, H 15, H 10, HS 20 and HS 15. Loadings H 15 and H 10 are 75 per cent and 50 per

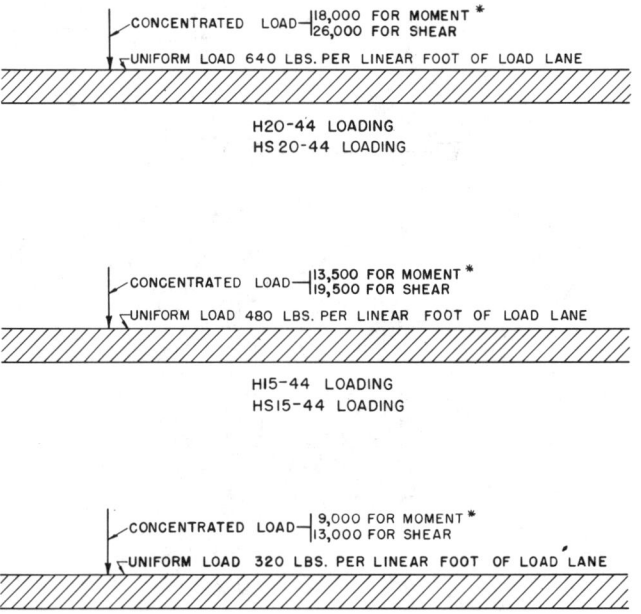

H LANE AND HS LANE LOADINGS
FIGURE 1.2.5 B

* For the loading of continuous spans involving lane loading refer to Article 1.2.8(C) which provides for an additional concentrated load.

cent, respectively, of loading H 20. Loading HS 15 is 75 per cent of loading HS 20. If loadings of weights other than those designated are desired, they shall be obtained by proportionately changing the weights shown for both the standard truck and the corresponding lane loads.

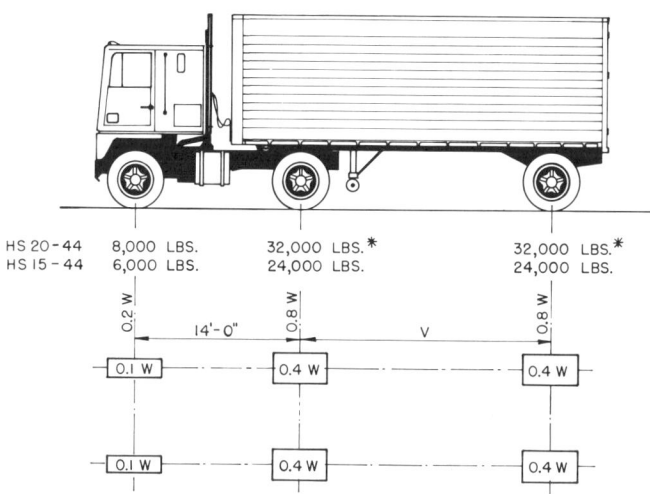

W = COMBINED WEIGHT ON THE FIRST TWO AXLES WHICH IS THE SAME AS FOR THE CORRESPONDING H TRUCK.
V = VARIABLE SPACING – 14 FEET TO 30 FEET INCLUSIVE. SPACING TO BE USED IS THAT WHICH PRODUCES MAXIMUM STRESSES.

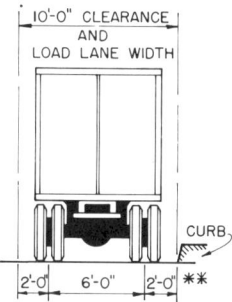

STANDARD HS TRUCKS

FIGURE 1.2.5C

* In the design of timber floors and orthotropic steel decks (excluding transverse beams) for HS20 loading, one axle load of 24,000 pounds or two axle loads of 16,000 pounds each, spaced 4 feet apart may be used, whichever produces the greater stress, instead of the 32,000 pound axle shown.
** For slab design the center line of wheels shall be assumed to be 1 foot from face of curb. (See Art. 1.3.2(B))

(E) Designation of Loadings

The policy of affixing the year to loadings to identify them was instituted with the publication of the 1944 edition in the following manner:

H10 Loading, 1944 Edition shall be designated H10-44
H15 Loading, 1944 Edition shall be designated H15-44
H20 Loading, 1944 Edition shall be designated H20-44
H15-S12 Loading, 1944 Edition shall be designated HS15-44
H20-S16 Loading, 1944 Edition shall be designated HS20-44

The affix remains unchanged until such time as the loading specification is revised. The same policy for identification shall be applied, for future reference, to loadings previously adopted by the American Association of State Highway Officials.

(F) Minimum Loading

For trunk highways, or for other highways which carry, or which may carry, heavy truck traffic, the minimum live load shall be the HS 15 designated herein.

1.2.6 — TRAFFIC LANES

The lane loading or standard trucks shall be assumed to occupy a width of 10 feet. These loads shall be placed in design traffic lanes having a width of $W = \frac{W_c}{N}$.

Where W_c = Roadway width between curbs without median on structure. Roadway width between median curb and outside curb, or between median barrier and outside curb with median on structure. If curbs are not used, roadway width is between faces of bridge railing.

N = Number of design traffic lanes as shown in the following table.
W = Width of design traffic lane.

W_c (in feet)	N
20 to 30 inc.	2
over 30 to 42 inc.	3
over 42 to 54 inc.	4
over 54 to 66 inc.	5
over 66 to 78 inc.	6
over 78 to 90 inc.	7
over 90 to 102 inc.	8
over 102 to 114 inc.	9
over 114 to 126 inc.	10

The lane loadings or standard trucks shall be assumed to occupy any position within their individual design traffic lane which will produce the maximum stress.

1.2.7 — STANDARD TRUCKS AND LANE LOADS

The wheel spacing, weight distribution, and clearance of the standard H and HS trucks shall be as shown in Figures 1.2.5A and 1.2.5C and corresponding lane loads shall be as shown in Figure 1.2.5B.

Each lane loading shall consist of a uniform load per linear foot of traffic lane combined with a single concentrated load (or two concentrated loads in the case of continuous spans—see Article 1.2.8(C)), so placed on the span as to produce maximum stress. The concentrated load and uniform load shall be considered as uniformly distributed over a 10-foot width on a line normal to the center line of the lane.

For the computation of moments and shears, different concentrated loads shall be used as indicated in Figure 1.2.5B. The lighter concentrated loads shall be used when the stresses are primarily bending stresses and the heavier concentrated loads shall be used when the stresses are primarily shearing stresses.

Note: The system of lane loads here defined (and illustrated in Fig. 1.2.5B) was developed in order to give a simpler method of calculating moments and shears than that based on wheel loads of the trucks.

Appendix B shows the truck train loadings of the 1935 Specifications of AASHO and the corresponding lane loadings.

In 1944 the HS series of trucks were developed. These approximate the effect of the corresponding 1935 truck preceded and followed by a train of trucks weighing ¾ as much as the basic truck.

1.2.8 — APPLICATION OF LOADINGS

(A) Traffic Lane Units

In computing stresses, each 10-foot lane loading or single standard truck shall be considered as a unit, and fractional load lane widths or fractional trucks shall not be used.

(B) Number and Position, Traffic Lane Units

The number and position of the lane loadings or truck loadings shall be as specified in Article 1.2.6 and, whether lane loading or truck loading, shall be such as to produce maximum stress, subject to the reduction specified in Article 1.2.9.

(C) Lane Loadings—Continuous Spans

The lane loadings shown in Figure 1.2.5B shall be modified as follows for the design of continuous spans. The lane loadings shall consist of the loads shown in Figure 1.2.5B and in addition thereto another concentrated load of equal weight shall be placed in one other span in the series in such position as to produce maximum negative moment. For maximum positive moment, only one concentrated load shall be used per lane, combined with as many spans loaded uniformly as required to produce maximum moment.

(D) Loading for Maximum Stress

The type of loading, whether lane loading or truck loading, to be used, and whether the spans be simple or continuous, shall be the loading which produces the maximum stress. The moment and shear tables given in Appendix A show which loading controls for simple spans. The axle spacing for HS trucks shall be varied between the specified limits to produce maximum stresses.

For continuous spans, the lane loading shall be continuous or discontinuous, as may be necessary to produce maximum stresses, and the concentrated load or loads as specified in paragraph (C) shall be placed in such position as to produce maximum stresses.

For continuous spans, only one Standard H or HS truck per lane shall be considered on the structure and placed so as to produce maximum positive and negative moments.

1.2.9 — REDUCTION IN LOAD INTENSITY

Where maximum stresses are produced in any member by loading any number of traffic lanes simultaneously, the following percentages of the resultant live load stresses shall be used in view of improbable coincident maximum loading:

	Per cent
One or two lanes	100
Three lanes	90
Four lanes or more	75

The reduction in intensity of floor beam loads shall be determined as in the case of main trusses or girders, using the width of roadway which must be loaded to produce maximum stresses in the floor beam.

1.2.10 — ELECTRIC RAILWAY LOADING

If highway bridges carry electric railway traffic, the railway loading shall be determined on the basis of the class of traffic which the bridge may be expected to carry. The possibility that the bridge may be required to carry railroad freight cars shall be given consideration.

1.2.11 — SIDEWALK, CURB, AND RAILING LOADING

(A) Sidewalk Loading

Sidewalk floors, stringers and their immediate supports, shall be designed for a live load of 85 pounds per square foot of sidewalk area. Girders, trusses, arches and other members shall be designed for the following sidewalk live loads per square foot of sidewalk area:

Spans 0 to 25 ft. in length .... 85 lbs.
Spans 26 to 100 ft. in length .... 60 lbs.
Spans over 100 ft. in length according to the formula

$$P = \left(30 + \frac{3000}{L}\right)\left(\frac{55 - W}{50}\right) \text{ in which}$$

P = live load per square foot (maximum, 60 lbs. per sq. ft.).
L = loaded length of sidewalk in feet.
W = width of sidewalk in feet.

In calculating stresses in structures which support cantilevered sidewalks, the sidewalk shall be considered as fully loaded on only one side of the structure if this condition produces maximum stress.

Pedestrian bridges shall be designed for a live load of 85 pounds per square foot of area walkway.

(B) Curb Loading

Curbs shall be designed to resist a lateral force of not less than 500 pounds per linear foot of curb, applied at the top of the curb, or at an elevation 10 inches above the floor if the curb is higher than 10 inches.

Where sidewalk, curb and traffic rail form an integral system, the traffic railing loading shall apply and stresses in curbs computed accordingly.

(C) Railing Loading*

(1) Traffic railing

Rail members and parapets shall be designed for a transverse load (P) of 10,000 lbs. or C times P divided between the various members that are centered 15 inches or more above the bridge floor (or top of curb wider than 6 inches) as shown in Figure 1.1.9.

All members that have this transverse load distributed to them shall have their roadside faces within one inch of a common vertical plane. Rail members offset more than one inch back of this plane or centered less than 15 inches above the bridge floor (or top of curb wider than 6 inches), and required because of the spacing requirements of Article 1.1.9(A), shall be designed for a transverse load equal to that applied to adjacent traffic rails, except that this loading need not exceed 5,000 lbs.

Rail members shall be designed for a moment, due to concentrated loads, at the center of the panel and at the posts of $P'L/6$ where P' is equal to P, $P/2$ or $P/3$ shown in Figure 1.1.9. The handrail members of combination railing shall be designed for a moment at center of panel and at posts of $0.1wL^2$. L is the post spacing.

Each attachment of a rail to a post shall be designed for vertical loads applied upward and downward, but not simultaneously, equal to $P'/4$ applied at the center line of the rail. Each rail attachment shall also be designed to resist an inward transverse load equal to ¼ the transverse rail design load.

Posts shall be designed for the transverse loading indicated in Figure 1.1.9, plus simultaneous longitudinal loading of ½ this amount.** When the tensile strength of the rail members is maintained through a series of post spaces, the longitudinal loading may be divided among as many as four posts in this continuous length. Each traffic post shall also be designed to resist an inde-

---

* Railing configurations which have been successfully tested by full scale impact tests are exempt from the provisions of this article.

** The designer is alerted to the possibility of heavy loads being applied at higher lever arms than normally encountered. Posts which have rails above the minimum traffic rail height of 2'-3" may have to be investigated for possible impact by vehicles with high centers of gravity such as tractor-trailers.

pendently applied roadward load equal to ¼ the outward transverse load.

The transverse force on concrete parapet and barrier walls shall be spread over a longitudinal length of 5 feet.

Railing loads shall be applied to the supporting slab in accordance with Article 1.3.2(H)(2). Railing and wheel loads are not to be applied simultaneously.

(2) Pedestrian railing

The minimum design loading for pedestrian railing shall be w=50 lbs. per lin. ft. acting simultaneously transversely and vertically on each longitudinal member. Rail members located more than 5'-0" above the walkway are excluded from these requirements.

Posts shall be designed for a transverse load of wl acting at the center of gravity of the upper rail, or for high rails, at 5'-0" maximum above the walkway.

(3) Design

Railings shall be designed by the elastic method to the allowable stresses for the appropriate material. For aluminium alloys 5154-H38, 6061-T6, 6063-T6, 6005-T5, and 6351-T5, the design stresses given in Tables 3.3.7, 8, 9, and 10 of the April 1969, "Specifications for Aluminum Bridge and other Highway Structures" published by the Aluminum Association shall be used. For Alloy A344-T4, 35% of the values listed in Table 3.3.8, and for Alloys A356-T61 and 356-T6, T7, 45% of the values listed in Table 3.3.8 shall be used for design. Aluminum railings shall be fabricated and built in accordance with the provisions of Section 6 of the above publication for riveted and bolted fabrication, and in accordance with Section 10 of the 1968 "Specifications for the Design and Construction of Structural Supports for Highway Signs" for welded fabrication.

The allowable unit stress for steel shall be as given by the AASHO "Standard Specifications for Highway Bridges" except as modified by "Section 6, Unit Stresses" of the AASHO "Specifications for the Design and Construction of Structural Supports for Highway Signs."

1.2.12 — IMPACT

Live load stresses produced by H or HS loadings shall be increased for items in Group A by allowance as stated herein for dynamic, vibratory and impact effects. Impact shall not be applied to items in Group B.

(A) Group A

(1) Superstructure, including steel or concrete supporting columns, steel towers, legs of rigid frames and generally those portions of the structure which extend down to the main foundation.

(2) The portion above the ground line of concrete or steel piles

which are rigidly connected to the superstructure as in rigid frame or continuous designs.

(B) Group B
   (1) Abutments, retaining walls, piers, piles, except Group A(2).
   (2) Foundation pressures and footings.
   (3) Timber structures.
   (4) Sidewalk loads.
   (5) Culverts and structures having cover of 3 feet or more.

(C) Impact Formula

The amount of this allowance or increment is expressed as a fraction of live load stress, and shall be determined by the formula:

$$I = \frac{50}{L + 125} \text{ in which}$$

I = impact fraction (maximum 30 per cent)
L = length in feet of the portion of the span which is loaded to produce the maximum stress in the member.

For uniformity of application the loaded length "L" shall be especially considered as follows:

For roadway floors, use the design span length.

For transverse members, such as floor beams, use the span length of member center to center of supports.

For computing truck load moments use the span length, except for cantilever arms use the length from moment center to the farthermost axle.

For shear due to truck loads use the length of the loaded portion of span from the point under consideration to the far reaction, except for cantilever arms use 30 per cent.

For continuous spans use the length of span under consideration for positive moment, and use the average of two adjacent loaded spans for negative moment.

   For culverts with cover 0' to 1'- 0" inc. I=30%
   "   "   "   " 1'-1" to 2'- 0" inc. I=20%
   "   "   "   " 2'-1" to 2'-11" inc. I=10%

1.2.13 — LONGITUDINAL FORCES

Provision shall be made for the effect of a longitudinal force of five per cent of the live load in all lanes carrying traffic headed in the same direction. All lanes shall be considered as loaded for bridges likely to become one directional in the future. The load used, without impact, shall be the lane load plus the concentrated load for moment specified in Article 1.2.8, with reduction for multiple-loaded lanes as specified in Article 1.2.9. The center of gravity of the longitudinal force shall be assumed to be located 6 feet above the floor slab and transmitted to the substructure through the superstructure.

The longitudinal force due to friction at expansion bearings or shear resistance at elastomeric bearings shall also be provided for in the design.

1.2.14 — WIND LOADS

The following wind load forces per square foot of exposed area shall be applied to all structures (see Article 1.2.22 for percentage of basic unit stress to be used under various combinations of loads and forces). The exposed area considered shall be the sum of the areas of all members, including floor system and railing, as seen in elevation at 90 degrees to the longitudinal axis of the structure. The forces and loads given herein are for a wind velocity of 100 miles per hour. For Group II loading, but not for Group III loading, they may be reduced or increased in the ratio of the square of the design wind velocity to the square of 100, provided the maximum probable wind velocity can be ascertained with reasonable accuracy, or there are permanent features of the terrain which make such changes safe and advisable. If change in the design wind velocity is made, the design wind velocity shall be shown on the plans.

(A) Superstructure Design

A moving uniformly distributed wind load of the following intensity shall be applied horizontally at right angles to the longitudinal axis of the structure in the design of the superstructure:

> For trusses and arches .... 75 pounds per square foot
> For girders and beams .... 50 pounds per square foot

The total force shall not be less than 300 pounds per linear foot in the plane of the loaded chord and 150 pounds per linear foot in the plane of the unloaded chord on truss spans, and not less than 300 pounds per linear foot on girder spans.

The above forces shall be used for Group II loading. For Group III loading there shall be added thereto a load of 100 pounds per linear foot applied at right angles to the longitudinal axis of the structure and 6 feet above the deck as a wind load on a moving live load. When a reinforced concrete floor slab or a steel grid deck is keyed to or attached to its supporting members, it may be assumed that the deck resists, within its plane, the shear resulting from the wind load on the moving live load.

(B) Substructure Design

Forces transmitted to the substructure by the superstructure and forces applied directly to the substructure by wind loads shall be assumed to be as follows:

(1) Forces from superstructure

The transverse and longitudinal forces transmitted by the superstructure to the substructure for varying angles of wind direction shall be as set forth in the following table. The skew angle is measured from the perpendicular to the longitudinal axis. The assumed wind direction shall be that which produces the maximum stress in the substructure being designed. The transverse and longitudinal forces shall be applied simultaneously at the elevation of the center of gravity of the exposed area of the superstructure.

Skew Angle of Wind	Trusses		Girders	
	Lateral Load Per Sq. Ft. of Area	Longitudinal Load Per Sq. Ft. of Area	Lateral Load Per Sq. Ft. of Area	Longitudinal Load Per Sq. Ft. of Area
(Degrees)	(Pounds)	(Pounds)	(Pounds)	(Pounds)
0	75	0	50	0
15	70	12	44	6
30	65	28	41	12
45	47	41	33	16
60	25	50	17	19

The loads listed above shall be used in Group II loading as given in Article 1.2.22.

For Group III loading, these loads may be reduced 70 per cent and there shall be added thereto, as a wind load on a moving live load, a load per linear foot as given in the following table:

Skew Angle of Wind (Degrees)	Lateral Load Per Lin. Ft. (Pounds)	Longitudinal Load Per Lin. Ft. (Pounds)
0	100	0
15	88	12
30	82	24
45	66	32
60	34	38

This load shall be applied at a point 6 feet above the deck.

For the usual girder and slab bridges having maximum span lengths of 125 feet, the following wind loading may be used in lieu of the more precise loading specified above:

> W (wind load on structure)
>> 50 pounds per square foot, transverse;
>> 12 pounds per square foot, longitudinal.
>> Both forces shall be applied simultaneously.
>
> WL (wind load on live load)
>> 100 pounds per linear foot, transverse;
>> 40 pounds per linear foot, longitudinal.
>> Both forces shall be applied simultaneously.

(2) Forces applied directly to the substructure

The transverse and longitudinal forces to be applied directly to the substructure for a 100 mile per hour wind shall be calculated from an assumed wind force of 40 pounds per square foot. For wind directions assumed skewed to the substructure this force shall be resolved into components perpendicular to the end and front elevations of the substructure according to the functions of the skew angle. The component perpendicular to the end elevation shall act on the exposed substructure area as seen in end elevation and the component perpendicular to the front elevation shall act on the exposed substructure area as seen in front elevation. These loads shall be assumed to act on horizontal lines at the centers of gravity of the exposed areas and shall be applied simultaneously with the wind loads from the superstructure. The above loads are for Group II loading and may be reduced 70 per cent for Group III loading, as indicated in Article 1.2.22.

(C) Overturning Forces

The effect of forces tending to overturn structures shall be calculated under Group II and Group III of Article 1.2.22, and there shall be added an upward force applied at the windward quarter point of the transverse superstructure width. This force shall be 20 pounds per square foot of deck and sidewalk plan area for Group II combination and 6 pounds per square foot for Group III combination. The wind direction shall be assumed to be at right angles to the longitudinal axis of the structure.

1.2.15 — THERMAL FORCES

Provision shall be made for stresses or movements resulting from variations in temperature. The rise and fall in temperature shall be fixed for the locality in which the structure is to be constructed and shall be figured from an assumed temperature at the time of erection. Due consideration shall be given to the lag between air temperature and the interior temperature of massive concrete members or structures.

The range of temperature shall generally be as follows:

Metal Structures
    Moderate climate, from 0 to 120 F.
    Cold climate, from −30 to 120 F.

Concrete Structures.	Temperature rise	Temperature fall
Moderate climate	30 F.	40 F.
Cold climate	35 F.	45 F.

1.2.16 — UPLIFT

Provision shall be made for adequate attachment of the superstructure to the substructure by engaging a mass of masonry equal to the largest force obtained under one of the following conditions:

    (a) 100% of the calculated uplift caused by any loading or combination of loadings in which the live plus impact loading is increased by 100%.
    (b) 150% of the calculated uplift at working load level.

Anchor bolts subject to tension or other elements of the structure stress under the above conditions shall be designed at 150% of the allowable basic stress.

1.2.17 — FORCE OF STREAM CURRENT, FLOATING ICE AND DRIFT

All piers and other portions of structures which are subject to the force of flowing water, floating ice, or drift shall be designed to resist the maximum stresses induced thereby.

The pressure of ice on piers shall be calculated at 400 pounds per square inch. The thickness of ice and height at which it applies shall be determined by investigation at the site of the structure.

The effect of flowing water on piers shall be calculated by the formula:

$P = KV^2$, where
$P =$ pressure in pounds per square foot,
$V =$ velocity of water in feet per second,
$K =$ a constant, being 1⅜ for square ends, ½ for angle ends where the angle is 30 degrees or less, and ⅔ for circular piers.

1.2.18 — BUOYANCY

Buoyancy shall be considered as it affects the design of either substructure, including piling, or the superstructure.

1.2.19 — EARTH PRESSURE

Structures which retain fills shall be proportioned to withstand pressure as given by Rankine's formula; provided, however, that no structure shall be designed for less than an equivalent fluid pressure of 30 pounds per cubic foot.

For rigid frames a maximum of one-half of the moment caused by earth pressure (lateral) may be used to reduce the positive moment in the beams, in the top slab, or in the top and bottom slab, as the case may be.

When highway traffic can come within a horizontal distance from the top of the structure equal to one-half its height, the pressure shall have added to it a live load surcharge pressure equal to not less than 2 feet of earth.

Where an adequately designed reinforced concrete approach slab supported at one end by the bridge is provided, no live load surcharge need be considered.

All designs shall provide for the thorough drainage of the backfilling material by means of weep holes and crushed rock, pipe drains or gravel drains, or by perforated drains.

1.2.20 — EARTHQUAKE STRESSES

In regions where earthquakes may be anticipated, provision shall be made to accommodate lateral forces from earthquakes as follows:

$$EQ = CD$$

where

$EQ =$ Lateral force applied horizontally in any direction at center of gravity of the weight of the structure.

D = Dead load of structure.
C = 0.02 for structures founded on spread footings on material rated as 4 tons or more per square foot.
 = 0.04 for structures founded on spread footings on material rated as less than 4 tons per square foot.
 = 0.06 for structures founded on piles.

Live load may be neglected.

1.2.21 — CENTRIFUGAL FORCES

Structures on curves shall be designed for a horizontal radial force equal to the following percentage of the live load, without impact, in all traffic lanes:

$$C = 0.00117\ S^2\ D = \frac{6.68\ S^2}{R}$$

where

C = the centrifugal force in percent of the live load, without impact.
S = the design speed, in miles per hour.
D = the degree of curve.
R = the radius of the curve, in feet.

The effects of superelevation shall be taken into account.

The centrifugal force shall be applied 6 feet above the roadway surface, measured along the center line of the roadway. The design speed shall be determined with regard to the amount of superelevation provided in the roadway. The traffic lanes shall be loaded in accordance with the provisions of Article 1.2.8.

Each design traffic lane shall be loaded with one standard truck (lane loading shall not be used in any case) placed in position for maximum loading.

When a reinforced concrete floor slab or a steel grid deck is keyed to or attached to its supporting members, it may be assumed that the deck resists, within its plane, the shear resulting from the centrifugal forces acting on the live load.

1.2.22 — LOADING COMBINATIONS

The following Groups represent various combinations of loads and forces to which a structure may be subjected. Each part of such structure, or the foundation on which it rests, shall be proportioned for all combinations of such of these forces as are applicable to the particular site or type, and at the percentage of the basic unit stress indicated for the various groups except that no increase in allowable unit stresses shall be permitted for members or connections carrying wind loads only. See Articles 1. 2. 1 to 1. 2. 21 for loads and forces.

The maximum section required shall be used.

		Percentage of Unit Stress
Group I	= D+L+I+E+B+SF	100%
Group II	= D+E+B+SF+W	125%
Group III	= Group I+LF+F+30% W+WL+CF	125%
Group IV	= Group I+R+S+T	125%
Group V	= Group II+R+S+T	140%
Group VI	= Group III+R+S+T	140%
Group VII	= D+E+B+SF+EQ	133⅓%
Group VIII	= Group I+ICE	140%
Group IX	= Group II+ICE	150%
D	= Dead Load	
L	= Live Load	
I	= Live Load Impact	
E	= Earth Pressure	
B	= Buoyancy	
W	= Wind Load on Structure	
WL	= Wind Load on Live Load—100 pounds per linear foot	
LF	= Longitudinal Force from Live Load	
CF	= Centrifugal Force	
F	= Longitudinal force due to friction or shear resistance (elastomeric bearings).	
R	= Rib Shortening	
S	= Shrinkage	
T	= Temperature	
EQ	= Earthquake	
SF	= Stream Flow Pressure	
ICE	= Ice Pressure	

Section 3—DISTRIBUTION OF LOADS

1.3.1 — DISTRIBUTION OF WHEEL LOADS TO STRINGERS, LONGITUDINAL BEAMS AND FLOOR BEAMS *

(A) Position of Loads for Shear

In calculating end shears and end reactions in transverse floor beams and longitudinal beams and stringers, no longitudinal distribution of the wheel load shall be assumed for the wheel or axle load adjacent to the end at which the stress is being determined.

---

* Provisions in this Article shall not apply to orthotropic deck bridges.

Lateral distribution of the wheel load shall be that produced by assuming the flooring to act as a simple span between stringers or beams. For loads in other positions on the span, the distribution for shear shall be determined by the method prescribed for moment, except that the calculation of horizontal shear in rectangular timber beams shall be in accordance with Article 1.10.2.

(B) Bending Moment in Stringers and Longitudinal Beams *

In calculating bending moments in longitudinal beams or stringers, no longitudinal distribution of the wheel loads shall be assumed. The lateral distribution shall be determined as follows:

(1) Interior Stringers and Beams

The live load bending moment for each interior stringer shall be determined by applying to the stringer the fraction of a wheel load (both front and rear) determined in Table 1.3.1(B).

(2) Outside Roadway Stringers and Beams

(a) Steel — Timber — Concrete T-beams

The dead load considered as supported by the outside roadway stringer or beam shall be that portion of the floor slab carried by the stringer or beam. Curbs, railings and wearing surface, if placed after the slab has cured, may be considered equally distributed to all roadway stringers or beams.

The live load bending moment for outside roadway stringers or beams shall be determined by applying to the stringer or beam the reaction of the wheel load obtained by assuming the flooring to act as a simple span between stringers or beams.

When the outside roadway beam or stringer supports the sidewalk live load as well as traffic live load and impact, the allowable stress in the beam or stringer may be increased 25% for the combination of dead load, sidewalk live load, traffic live load, and impact, providing the beam is of no less carrying capacity than would be required if there were no sidewalks.

In no case shall an exterior stringer have less carrying capacity than an interior stringer.

In the case of a span with concrete floor supported by 4 or

---

* In view of the complexity of the theoretical analysis involved in the distribution of wheel loads to stringers, the empirical method herein described is authorized for the design of normal highway bridges.

TABLE 1.3.1 (B)

Kind of Floor	Bridge designed for one traffic lane	Bridge designed for two or more traffic lanes
Timber:[1]		
Plank	S/4.0	S/3.75
Strip 4" thick or multiple layer floors over 5" thick....	S/4.5	S/4.0
Strip 6" or more thick......	S/5.0 If S exceeds 5' use footnote [2].	S/4.25 If S exceeds 6.5' use footnote [2].
Concrete:		
On Steel I-Beam Stringers [3] and Prestressed Concrete Girders	S/7.0 If S exceeds 10' use footnote [2].	S/5.5 If S exceeds 14' use footnote [2].
On Concrete T-Beams.......	S/6.5 If S exceeds 6' use footnote [2].	S/6.0 If S exceeds 10' use footnote [2].
On Timber Stringers........	S/6.0 If S exceeds 6' use footnote [2].	S/5.0 If S exceeds 10' use footnote [2].
Concrete box girders [4].......	S/8.0 If S exceeds 12' use footnote [2].	S/7.0 If S exceeds 16' use footnote [2].
On Steel Box Girders........	(See Art. 1.7.103)	
On Prestressed Concrete Spread Box Beams	[See Art. 1.6.24(A)]	
Steel grid:		
(less than 4" thick)	S/4.5	S/4.0
(4" or more)	S/6.0 If S exceeds 6.0' use footnote [2].	S/5.0 If S exceeds 10.5' use footnote [2].

S = average stringer spacing in feet.

[1] Splined and dowelled timber flooring shall have the same distribution as strip floors of equivalent thickness.

[2] In this case the load on each stringer shall be the reaction of the wheel loads, assuming the flooring between the stringers to act as a simple beam.

[3] "Design of I-Beam Bridges" by N. M. Newmark—Proceedings, ASCE, March 1948.

[4] The sidewalk live load (see Article 1.2.11) shall be omitted for interior and exterior box girders designed in accordance with the wheel load distribution indicated herein.

more steel stringers, the fraction of the wheel load shall not be less than:

$\dfrac{S}{5.5}$ where S = 6' or less.

$\dfrac{S}{4.0 + 0.25S}$ where S is more than 6' and less than 14'.

When S is 14' or more, use footnote 2, Article 1.3.1(B)(1)

S = distance in feet outside and adjacent interior stringers.

(b) Concrete Box Girders

The dead load considered as supported by the exterior girder shall be determined in the same manner as for steel, timber, concrete T-beams, as given in (2)(a) above.

The wheel load distribution to the exterior girder shall be $W_e/7$.

W_e = width of exterior girder. The width to be used in determining the wheel line distribution to the exterior girder shall be the top slab width as measured from the midpoint between girders to the outside edge of the slab. The cantilever dimension of any slab extending beyond the exterior girder shall preferably not exceed S/2.

(3) Total Capacity of Stringers and Beams

The combined design load capacity of all the beams and stringers in a span shall not be less than required to support the total live and dead load in the span.

(C) Bending Moment in Floor Beams (Transverse)

In calculating bending moments in floor beams no transverse distribution of the wheel loads shall be assumed.

If longitudinal stringers are omitted and the floor is supported directly on floor beams, the beams shall be designed for loads determined in accordance with Table 1.3.1(C).

1.3.2 — DISTRIBUTION OF LOADS AND DESIGN OF CONCRETE SLABS * AND MULTI-BEAM PRECAST CONCRETE BRIDGES **

(A) Span Lengths   (See also Article 1.5.3)

For simple spans the span length shall be the distance center to center of supports but not to exceed clear span plus thickness of slab.

---

* The slab distribution set forth herein is based, substantially, upon the "Westergaard" theory. The following references are furnished concerning the subject of slab design:

Public Roads, March, 1930, "Computation of Stresses in Bridge Slabs Due to Wheel Loads," by H. M. Westergaard.

University of Illinois Bulletin No. 303, "Solutions for Certain Rectangular Slabs Continuous Over Flexible Supports," by Vernon P. Jensen; Bulletin 304, "A Distribution Procedure

TABLE 1.3.1 (C)

Kind of floor	Fraction of wheel load to each floor beam
Plank	$\dfrac{S}{4}$
Strip 4 inches in thickness, wood block on 4-inch plank subfloor or multi-thickness plank more than 5 inches thick............	$\dfrac{S}{4.5}$
Strip 6 inches or more in thickness...........	$\dfrac{S}{5}$***
Concrete	$\dfrac{S}{6}$***
Steel grid (less than 4 inches thick).........	$\dfrac{S}{4.5}$
Steel grid (4 inches or more)...............	$\dfrac{S}{6}$***

S = spacing of beams in feet.
Spline and doweled flooring shall have the same distribution as strip floors of equivalent thickness.

The following effective span lengths shall be used in calculating distribution of loads and bending moments for slabs continuous over more than two supports:

Slabs monolithic with beams or walls (without haunches), S = clear span.

Slabs supported on steel stringers, S = distance between edges of flanges plus ½ of the stringer flange width.

Slabs supported on timber stringers, S = clear span plus ½ thickness of stringer.

(B) Edge Distance of Wheel Load

In designing slabs the center line of wheel load (axle load/2) shall be assumed to be one foot from the face of the curb. If curbs or sidewalks are not used, the wheel load shall be assumed to be one foot from the face of the rail. Combined dead, live and impact stresses shall be not greater than the allowable stresses.

In designing sidewalks, slabs and supporting members, a wheel load located on the sidewalk shall be assumed to be one foot from the face of the rail. Combined dead, live and impact stresses for this loading shall be not greater than 150 percent of the allowable stresses. Wheel loads shall not be applied on sidewalks protected by a traffic barrier.

---

for the Analysis of Slabs Continuous Over Flexible Beams," by Nathan M. Newmark; Bulletin 315, "Moments in Simple Span Bridge Slabs with Stiffened Edges," by Vernon P. Jensen; and Bulletin 346, "Highway Slab Bridges with Curbs: Laboratory Tests and Proposed Design Method."

** A multi-beam bridge is constructed with precast reinforced or prestressed concrete beams which are placed side by side on the supports. The interaction between the beams is developed by continuous longitudinal shear keys and lateral bolts which may, or may not, be prestressed.

*** If S exceeds denominator, the load on the beam shall be the reaction of the wheel loads assuming the flooring between beams to act as a simple beam.

(C) Bending Moment

Bending moment per foot width of slab shall be calculated according to methods given under Cases A and B, unless the more exact methods referred to in the footnote to Article 1.3.2 are used.

In Cases A and B:

S = Effective span length, in feet, as defined under "Span Lengths" Articles 1.3.2(A) and 1.5.3
E = Width of slab in feet over which a wheel is distributed
P = Load on one rear wheel of truck (P_{15} or P_{20})
P_{15} = 12,000 pounds for H15 loading
P_{20} = 16,000 pounds for H20 loading

Case A—Main Reinforcement Perpendicular to Traffic
(Spans 2 to 24 feet, inclusive)

The live load moment for simple spans shall be determined by the following formulas (impact not included):

HS20 Loading:

$$\frac{(S+2)}{32} P_{20} = \text{Moment in foot-pounds per foot width of slab}$$

HS15 Loading:

$$\frac{(S+2)}{32} P_{15} = \text{Moment in foot-pounds per foot width of slab}$$

In slabs continuous over three or more supports, a continuity factor of 0.8 shall be applied to the above formulas for both positive and negative moment.

Case B—Main Reinforcement Parallel to Traffic

Distribution of wheel loads, E = 4 + 0.06S, maximum 7.0 feet. Lane loads are distributed over a width of 2E. Longitudinally reinforced slabs shall be designed for the appropriate HS loading.

For simple spans, the maximum live load moment per foot width of slab, without impact, is closely approximated by the following formulas:

HS20 Loading:

Spans up to and including 50 feet: LLM = 900S foot-pounds
Spans 50 feet to 100 feet:
LLM = 1000 (1.30S − 20.0) foot-pounds

HS15 Loading:

Use ¾ of the values obtained from the formulas for HS20 loading.

Moments in continuous spans shall be determined by suitable analysis using the truck or appropriate lane loading.

The lateral distribution of wheel loads for multi-beam precast concrete bridges, conventional or prestressed, shall not exceed that specified for slabs under Case B—Main Reinforcement Parallel to Traffic.

(D) Edge Beams (Longitudinal)

Edge beams shall be provided for all slabs having main reinforcement parallel to traffic. The beam may consist of a slab section additionally reinforced, a beam integral with and deeper than the slab, or an integral reinforced section of slab and curb.

It shall be designed to resist a live load moment of 0.10 PS, where

P = Wheel load, in pounds (P_{15} or P_{20})
S = Span length, in feet

This formula gives the simple span moment. Values for continuous spans may be reduced 20 per cent unless a greater reduction results from a more exact analysis.

(E) Distribution Reinforcement

Reinforcement shall be placed in the bottoms of all slabs transverse to the main steel reinforcement, to provide for the lateral distribution of the concentrated live loads, except that this specification will not apply on culverts or bridge slabs when the depth of fill over the slab exceeds two feet. The amount shall be the percentage of the main reinforcement steel required for positive moment as given by the following formulas:

For main reinforcement parallel to traffic:

$$\text{Percentage} = \sqrt{\frac{100}{S}} \quad \text{Maximum } 50\%$$

For main reinforcement perpendicular to traffic:

$$\text{Percentage} = \sqrt{\frac{220}{S}} \quad \text{Maximum } 67\%$$

where S = the effective span length, in feet.

For main reinforcement perpendicular to traffic the specified amount of distribution reinforcement shall be used in the middle half of the slab span, and in the outer quarters of the slab span not less than 50 percent of the above amount shall be used.

(F) Shear and Bond Stress in Slabs

Slabs designed for bending moment in accordance with the foregoing shall be considered satisfactory in bond and shear.

(G) Unsupported Edges, Transverse

The design assumptions of this article do not provide for the effect of loads near unsupported edges. Therefore, at the ends of the bridge and at intermediate points where the continuity of the slab is broken, the edges shall be supported by diaphragms or other suitable means. The diaphragms shall be designed to resist the full moment and shear produced by the wheel loads which can come on them.

(H) Cantilever Slabs

(1) Truck Loads

Under the following formulas for distribution of loads on cantilever slabs, the slab is designed to support the load independent of edge support along the end of the cantilever. The distribution given includes the effect of wheels on parallel elements.

Case A—Reinforcement Perpendicular to Traffic

Each wheel on the element perpendicular to traffic shall be distributed according to the following formula:

$$E = 0.8X + 3.75$$

Moment per foot of slab $= \dfrac{P}{E} X$ foot-pounds in which X=distance in feet from load to point of support.

Case B—Reinforcement Parallel to Traffic

The distribution for each wheel load on the element parallel to traffic shall be as follows:

$$E = 0.35X + 3.2 \text{ but shall not exceed 7.0 feet.}$$

Moment per foot of slab $= \dfrac{P}{E} X$ foot-pounds.

(2) Railing loads

It shall be assumed that the effective length of slab resisting post loadings is equal to $E = 0.8X + 3.75$ feet, where no parapet is used and is equal to $E = 0.8X + 5.0$ feet where a parapet is used, where X is the distance in feet from the center of the post to the point under investigation. Railing loading shall be applied in accordance with Article 1.2.11(C).

(I) Slabs Supported on Four Sides

In the case of slabs supported along four edges and reinforced in both directions, the proportion of the load carried by the short span of the slab shall be assumed as given by the following equations:

For load uniformly distributed, $p = \dfrac{b^4}{a^4 + b^4}$

For load concentrated at center, $p = \dfrac{b^3}{a^3 + b^3}$

where
    p = proportion of load carried by short span.
    a = length of short span of slab.
    b = length of long span of slab.

Where the length of the slab exceeds 1½ times its width, the entire load shall be assumed to be carried by the transverse reinforcement.

The distribution width, E, for the load taken by either span shall be determined as provided for other slabs. Moments obtained shall be used in designing the center half of the short and long slabs. The reinforcement steel in the outer quarters of both short and long spans may be reduced 50 per cent. In the design of the supporting beams, consideration shall be given to the fact that the loads delivered to the supporting beams are not uniformly distributed along the beams.

(J) Median Slabs

Raised median slabs shall be designed in accordance with the provisions of this article with truck loadings so placed as to produce maximum stresses. Combined dead, live and impact stresses may be not greater than 150 percent of the allowable stresses. Flush median slabs shall be designed without any overstress.

1.3.3 — DISTRIBUTION OF WHEEL LOADS THROUGH EARTH FILLS

When the depth of fill is 2 feet or more, concentrated loads shall be considered as uniformly distributed over a square, the sides of which are equal to 1¾ times the depth of fill.

The shear produced by such loads shall be calculated as provided for dead loads.

When such areas from several concentrations overlap, the total load shall be considered as uniformly distributed over the area defined by the outside limits of the individual areas, but the total width of distribution shall not exceed the total width of the supporting slab. For single spans, the effect of live load may be neglected when the depth of fill is more than 8 feet and exceeds the span length; for multiple spans it may be neglected when the depth of fill exceeds the distance between faces of end supports or abutments. When the depth of fill is less than 2 feet the wheel load shall be distributed as in slabs with concentrated loads. When the calculated live load and impact moment in concrete slabs, based on distribution of the wheel load through fills as herein outlined, exceeds the live load and impact moment calculated according to Article 1.3.2, then the latter moment shall be used.

1.3.4 — DISTRIBUTION OF WHEEL LOADS ON TIMBER FLOORING

For the calculation of bending moments in timber flooring each wheel load shall be distributed as follows:

(A) Flooring Transverse

In direction of span:
> Over width of tire (10 inches for H10; 15 inches for H15; and 20 inches for H20 loading.)

Normal to direction of span:
> Plank floor: width of plank
> Laminated floor: 15 inches
> Splined or doweled floor, not less than 5½ inches thick: 4 times thickness.

For transverse flooring the span shall be taken as the clear distance between stringers plus one-half the width of one stringer, but shall not exceed the clear span plus the floor thickness.

(B) Flooring Longitudinal

In direction of span:
> Point loading

Normal to direction of span:
> Plank floor: width of plank

Laminated floor: width of wheel plus thickness of floor
Splined or doweled floor, not less than 5½ inches thick: width of wheel plus twice thickness of floor

For longitudinal flooring the span shall be taken as the clear distance between floor beams plus one-half the width of one beam but shall not exceed the clear span plus the floor thickness.

(C) Continuous Flooring

If the flooring is continuous over more than two spans the maximum bending moment shall be assumed as being 80 per cent of that obtained for a simple span.

1.3.5 — DISTRIBUTION OF LOADS AND DESIGN OF COMPOSITE WOOD-CONCRETE MEMBERS

(A) Distribution of Concentrated Loads for Bending Moment and Shear

For freely supported or continuous slab spans of composite wood-concrete construction, as described in Article 2.20.19(A), the wheel loads shall be distributed over a transverse width of 5 feet for bending moment and a width of 4 feet for shear.

For composite T-beams of wood and concrete, as described in Article 2.20.19(B), the effective flange width shall not exceed that given in Article 1.7.98. Shear connectors shall be capable of resisting both vertical and horizontal movement.

(B) Distribution of Bending Moments in Continuous Spans

Both positive and negative moments shall be distributed in accordance with the following table:

Maximum Bending Moments—Per Cent of Simple Span Moment

SPAN	Maximum Uniform Dead Load Moments				Maximum Live Load Moments			
	Wood Subdeck		Composite Slab		Concentrated Load		Uniform Load	
	Pos.	Neg.	Pos.	Neg.	Pos.	Neg.	Pos.	Neg.
Interior	50	50	55	45	75	25	75	55
End	70	60	70	60	85	30	85	65
2-Span *	65	70	60	75	85	30	80	75

* Continuous beam of 2 equal spans.

Impact should be considered in computing stresses for concrete and steel, but neglected for wood.

(C) Design

The combination in a structural member of two elements having different mechanical properties requires the formulation of a design premise. Such a formulation as follows is based on the elastic properties of the materials:

$\dfrac{E_c}{E_w} = 1$ for slab in which the net concrete thickness is less than half the over-all depth of the composite section

$\dfrac{E_c}{E_w} = 2$ for slab in which the net concrete thickness is at least half the over-all depth of the composite section

$\dfrac{E_s}{E_w} = 18.75$ (for Douglas fir and Southern pine)

in which
E_c = modulus of elasticity of concrete
E_w = modulus of elasticity of wood
E_s = modulus of elasticity of steel

1.3.6 — DISTRIBUTION OF WHEEL LOADS ON STEEL GRID FLOORS *

(A) General

The grid floor shall be designed as continuous. Simple span moments may be used and reduced as provided in Article 1.3.2.

The formulas for distribution of loads provided herein are based upon there being adequate transfer of the load normal to the main elements. Reinforcement for this purpose shall consist of transverse bars or shapes welded to the main steel. The strength and details of the transverse reinforcement shall meet with the approval of the Engineer.

(B) Floors Filled with Concrete

The distribution and bending moment shall be as specified for concrete slabs, Article 1.3.2. The following items specified in that article shall also apply to concrete filled steel grid floors:

Edge beams (longitudinal).

Unsupported edges (transverse).

Span lengths.

The strength of the composite steel and concrete slab shall be determined by means of the "transformed area" method. The allowable stresses shall be as set forth in Articles 1.5.1 and 1.7.1.

(C) Open Floors

A wheel load shall be distributed, normal to the main bars, over a width equal to 1¼ inches per ton of axle load plus twice the distance center to center of main bars. The portion of the load assigned to each main bar shall be applied to the bar uniformly over a length equal to the rear tire width (20 inches for H20, 15 inches for H15).

The strength of the section shall be determined by the moment of inertia method. The allowable stresses shall be as set forth in Article 1.7.1.

Edges of open grid steel floors shall be supported by suitable means as required. These supports may be longitudinal or transverse, or both, as may be required to properly support all edges.

When investigating for fatigue, use minimum cycles of maximum stress.

---

* Provisions in this Article shall not apply to orthotropic deck bridges.

1.3.7 — MOMENTS, SHEARS AND REACTIONS

Maximum moments, shears, and reactions are given in tables, Appendix A, for H15, H20, HS15 and HS20 loadings. They are calculated for the standard truck or the lane loading applied to a single lane on the basis of freely supported spans. It is indicated in the table whether the standard truck or the lane loading produces the maximum stress.

Section 4—SUBSTRUCTURES AND RETAINING WALLS

1.4.1 — ALLOWABLE STRESSES

Concrete, steel or timber substructures and retaining walls shall be designed for the unit stresses indicated in Section 5, Section 7 or Section 10.

1.4.2 — BEARING POWER OF FOUNDATION SOILS
DETERMINATION OF BEARING POWER *

When required by the Engineer, the bearing power of the soil in excavated foundation pits shall be determined by loading tests.

The following tabulation of the bearing power of broad basic groups of materials may be used as an aid to the judgment in the absence of more definite information:

Material	Safe bearing power, Tons per square foot	
	Min.	Max.
Alluvial soils	½	1
Clays	1	4
Sand, confined	1	4
Gravel	2	4
Cemented sand and gravel	5	10
Rock	5	...

Loading tests have a limited depth influence and may not disclose long-time consolidation.

When the consolidation of foundation soils causes the settlement of the backfill against an abutment or the settlement of the soil under an abutment which is placed on piles driven through a fill, the load transmitted may result in overloading the piles.

When the hydraulic gradient is increased as in excavating material from below the water table, foundation soils may be loosened by the upward flow of water. Such a condition should be guarded against.

---

* For methods of estimating bearing capacities of foundation soils and computing settlements of piers and abutments, reference should be made to "Soil Mechanics in Engineering Practice," by Terzaghi & Peck, Edition of 1948, published by John Wiley & Son, New York, N. Y.

1.4.2   DESIGN

Intrusion failures should be prevented by requiring a base course between rip-rap and fine soils and by requiring proper gradation of drainage backfill behind abutments.

1.4.3 — ANGLES OF REPOSE

Earth, Loam	30° to 45°	Gravel	30° to 40°
Dry Sand	25 to 35	Cinders	25 to 40
Moist Sand	30 to 45	Coke	30 to 45
Wet Sand	15 to 30	Coal	25 to 35
Compact earth	35 to 40		

In the absence of exact data, which has been determined by field investigation and soil analysis, the angle of repose of the material shall be assumed to be the minimum given in the table.

1.4.4 — BEARING VALUE OF PILING

(A) General

The design loads for piles shall not be greater than the minimum value which shall be determined for Case A, Case B and Case C; where Case A is the capacity of the pile as a structural member, Case B is the capacity of the pile to transfer its load to the ground and Case C is the capacity of the ground to support the load delivered to it by the pile or piles. The values assignable to each of the three cases shall be determined by making subsurface investigations or tests of sufficient extent to justify the assumed design values used for the particular condition of support under consideration.

In determining the bearing value of piles for use in designing, consideration shall be given to all information available relative to the subsurface conditions. Consideration shall also be given to:

(1) The difference between the supporting capacity of a single pile and a group of piles.
(2) The capacity of the underlying strata to support the load of the pile group.
(3) The effect of driving additional piles and the effect of their loads on adjacent structures.
(4) Possibility of scour and its effect.

(B) Case A. Capacity of Pile as a Structural Member

(1) Structural Columns

Piles shall be designed as structural columns. Timber piles shall be designed in accordance with Article 1.10.2, using the allowable unit stresses given in Article 1.10.1 for lumber and in the following table for round piles.

Round Timber Piles

Species	Allowable unit working stress pounds per sq. in. Compression parallel to grain for normal duration of loading
Ash, white	1200
Beech	1300
Birch	1300
Chestnut	900
Cypress, Southern	1200
Cypress, Tidewater red	1200
Douglas fir, coast type	1200
Douglas fir, inland	1100
Elm, rock	1300
Elm, soft	850
Gum, black and red	850
Hemlock, Eastern	800
Hemlock, West Coast	1000
Hickory	1650
Larch	1200
Maple, hard	1300
Oak, red and white	1100
Pecan	1650
Pine, Lodgepole	800
Pine, Norway	850
Pine, Southern	1200
Pine, Southern, dense	1400
Poplar, yellow	800
Redwood	1100
Spruce, Eastern	850
Tupelo	850

Concrete piles shall be designed in accordance with Article 1.5.1, steel piles in accordance with Article 1.7.1, and concrete-filled pipe piles in accordance with Article 1.5.1, except that the allowable unit stresses may be increased 20% provided the shell thickness is not less than ¼ inch. The area of the shell shall be included in determining the value of p, (percentage of reinforcement). Where corrosion may be expected $\frac{1}{16}$ inch shall be deducted from the shell thickness to allow for reduction in section by corrosion. The allowable stresses of Articles 1.5.1, 1.7.1 and 1.10.1 may be used in all cases where all of the stresses to which the piles may be subjected have been included. These stresses may be increased in accordance with Article 1.2.22. For trestle piles or other piles without lateral support designed for dead load and live load only and where temperature, traction, water pressure and other forces are not considered, the allowable unit stresses specified in Articles 1.5.1, 1.7.1 and 1.10.1 shall be decreased 20%.

(2) Required Subsurface Investigations

Subsurface investigations shall be made which will determine the probable depth of scour or flotation of material and the condition of lateral support of the pile.

(C) Case B. Capacity of Pile to Transfer Load to the Ground
(1) Point-bearing Piles

A pile shall be considered to be a point-bearing pile when placed or driven on or into a material which is capable of developing the pile load by direct bearing at the point with reasonable factor of safety.

The allowable load at the tip of the pile shall not exceed the following:

- (a) For round timber piles, use values tabulated in Article 1.4.4(B) for allowable compression parallel to grain.

   For sawn timber piles, use those values applicable to "wet condition" for allowable compression parallel to grain, in accordance with Article 1.10.1.
- (b) For concrete piles, $0.33f'_c$ in accordance with Article 1.5.1(B).
- *(c) For concrete-filled piles, $0.40f'_c$ in accordance with Article 1.5.1(B) applied to the total actual area of the concrete and steel.
- *(d) For steel H-piles and unfilled tubular steel piles, 9,000 psi over the cross sectional area of the pile tip, not including the area of any pile tip reinforcement.

(2) Friction Piles

A pile shall be considered to be a friction pile if its point does not rest on or in a material which is capable of developing the pile load by direct bearing at the point.

The load-carrying capacity of friction piles shall be determined by one or more of the following methods:

- (a) Driving and loading test piles. The safe allowable load shall be as defined by Article 2.3.6(A).
- (b) Pile-driving experience in the vicinity. When piles are designed on the basis of experience in the vicinity, due consideration will be given to the variation in pile types and lengths, and in the variation of the soil strata. Where possible, the complete driving records of all piles in the vicinity shall be examined and compared to the driving records of the project piles.
- (c) Adequate tests of the soil strata through which the pile is to be driven. These tests should be projected and compared, if possible, to tests of similar material through which piles of known capacity have been driven.

(3) Required Subsurface Investigations

- (a) Point-bearing piles. Sufficient borings shall be made to determine the presence, position, and thickness of

---

* NOTE: The limitation in (c) and (d) govern except where the point bearing capacity of the piles is determined by loading test piles.

the material which is capable of developing point-bearing, and the log of borings shall show the nature of the overlying strata in order that the extent of lateral support may be determined. If the point-bearing stratum is of doubtful thickness and quality, the borings shall be made to such sufficient depth below this stratum that the capacity of a friction pile may be determined.

(b) Friction piles. Borings shall be made to an elevation well below the expected elevation of the pile tips and accurate logs of these borings shall be made. In those cases where the piles are to be designed on the basis of soil tests, undisturbed samples shall be taken on all strata which will have appreciable influence on the capacity of the pile.

(c) Combination point-bearing and friction piles. Piles shall be classified as either (1) point-bearing or (2) friction. Those cases where adequate strength is developed by both point bearing and friction may be designed under either of these classifications.

(D) Case C. Capacity of the Ground to Support the Load Delivered by the Pile

Preference shall be given to the determination of maximum loads on piles by test loading or by satisfactory subgrade investigation.

The capacity of the ground to support the load delivered by the pile shall be determined from the results of the applicable subsurface investigations:

(1) Point-bearing Piles

Sufficient borings shall be made to determine the thickness and quality of the stratum in which the point bearing is developed. If that stratum is of sufficient thickness and is underlain by a firm material, no reduction will be made for group action of piles. In general, piles should not rest on a thin stratum of hard material which is underlain by a thick stratum of soft or yielding material, but where this condition cannot be avoided, group action should be considered and the design loads reduced accordingly.

(2) Friction Piles

Borings shall be carried well below the tips of the piles in order to determine the characteristics of the underlying material. In most cases a study of those borings will suffice to determine whether or not the underlying soil will support the loads delivered to it, but in doubtful or special cases, especially large foundation areas and important footings, the material should be investigated more thoroughly by soil mechanics methods.

A single row of piles shall not be considered as a group provided that they are not spaced closer center to center than 2½ times the nominal diameter or dimension. In those cases where piles are driven in groups into plastic material, the design load shall be determined by the loading of a group of piles or definite allowance shall be made for the difference between the supporting capacity of a single pile and a group of piles. (Refer to (G))

(E) Maximum Design Loads for Piles

In those cases where it is not feasible to make the required subsurface investigations or test loads, the maximum assumed design load for piles shall be as given in the table below. These values may be increased for certain combinations of loads as specified in Article 1.2.22.

The assumed pile loads shall be substantiated by determining the allowable load by formula, when the piles are driven, as provided in Article 2.3.6(B).

Types of Piles

Size or Diameter at Butt,* Inches	Timber Tons	Concrete Tons	Steel (Friction) Tons	Steel Point-Bearing
8	—	—	16	9000 pounds per sq. in. of point area, not including the area of any pile tip reinforcement
10	20	20	20	
12	24	24	24	
14	28	28	28	
16	32	32	—	
20	—	40	—	
24	—	50	—	

* Timber piles, diameter to be measured 3 feet from butt.

(F) Uplift

Friction piles may be considered to resist an intermittent but not sustained uplift equivalent to 40 per cent of the above loads providing proper provision is made for the anchorage at the top and sufficient skin friction is developed and in no case shall it exceed the weight of material (buoyancy considered) surrounding the embedded portion of the pile.

(G) Group Pile Loading

Where the capacity of a group of friction piles driven into plastic material is not determined by test loading, the following Converse-Labarre formula is suggested to determine the reduction of a single pile load for a group pile load:

$$E = 1 - \Phi \frac{(n-1)m + (m-1)n}{90mn}$$

Where

E = the efficiency or the decimal fraction of the single pile value to be used for each pile in the group.
n = the number of piles in each row.
m = the number of rows in each group.
d = the average diameter of the pile.
s = center to center spacing of piles.
Tan $\Phi = d/s$
Φ is numerically equal to the angle expressed in degrees.

1.4.5 — PILES

(A) General

In general, piling shall be used when footings cannot, at a reasonable expense, be founded on rock or other solid foundation material. At locations where unusual erosion may occur and the soil conditions permit the driving of piles, they, preferably, shall be used as a protection against scour, even though the safe bearing resistance of the natural soil is sufficient to support the structure without piling.

In general, the penetration for any pile shall be not less than 10 feet in hard material and not less than ⅓ the length of the pile nor less than 20 feet in soft material.

For foundation work, no piling shall be used to penetrate a very soft upper stratum overlying a hard stratum unless the piles penetrate the hard material a sufficient distance to rigidly fix the ends.

(B) Limitation of Use

Untreated timber piles may be used for temporary construction, revetments, fenders and similar work, and in permanent construction under the following conditions:

(1) For foundation piling when the cutoff is below permanent ground water level.
(2) For trestle construction when it is economical to do so, though treated piles are preferable.
(3) They shall not be used where they will, or may be, exposed to marine borers.

The limitations of use of treated timber piles are given in Division II, Section 21.

(C) Design Loads

The design loads for piles shall be according to Article 1.4.4.

Piles shall be designed to carry the entire superimposed load, no allowance being made for the supporting value of the material between the piles.

The supporting power of piles shall be determined by the application of test loads or by the use of formulas as specified in Article 2.3.6(B).

(D) Spacing, Clearances and Embedment

Footing areas shall be so proportioned that pile spacing shall be not less than 2 feet 6 inches center to center. When the tops of foundation piles are incorporated in a concrete footing, the distance from the side of any pile to the nearest edge of the footing shall be not less than 9 inches.

The tops of piling in general shall project not less than 12 inches into the concrete after all damaged pile material has been removed, but in special cases it may be reduced to 6 inches.

Where a reinforced concrete beam is cast-in-place and used as a bent cap supported by piling, concrete cover at the sides of piles shall be a minimum of six inches. The piles shall project at least six inches, and preferably nine inches, into the cap; provided, however, concrete piles may project a lesser distance into the cap if the projection of the pile reinforcement is sufficient to provide for adequate bond.

(E) Batter Piles

When the lateral resistance to the soil surrounding the piles is inadequate to counteract the horizontal forces transmitted to the foundation or when increased rigidity of the entire structure is required, batter piles shall be used in the foundation.

(F) Buoyancy

The effect of hydrostatic pressure shall be considered in the design as provided in Article 1.2.18.

(G) Concrete Piles (Precast)

Precast concrete piles shall be of approved size and shape. If a square section is employed, the corners shall be chamfered at least one inch. Piles, preferably, shall be cast with a driving point and for hard driving, preferably shall be shod with a metal shoe of approved pattern. Piling may be either of uniform section or tapered. In general, tapered piling shall not be used for trestle construction except for that portion of the pile which lies below the ground line; nor shall tapered piles be used in any location where the piles are to act as columns. In general, concrete piles shall have a cross sectional area, measured above the taper, of not less than 140 square inches and when they are to be used in salt water they shall have a cross sectional area of not less than 220 square inches.

The diameter of tapered piles measured 2 feet from the point shall be not less than 8 inches. In all cases the diameter shall be considered as the least dimension through the center. The point in all cases, where steel points are not used, shall be not less than 6 inches in diameter and the pile shall be beveled, tapered or sloped uniformly from the point to 2 feet from the point.

Vertical reinforcement shall be provided consisting of not less than four bars spaced uniformly around the perimeter of the pile. It shall be at least 1½ per cent of the total cross section measured above the taper, except that if more than four bars are used, the number may be reduced to four in the bottom 4 feet of the pile.

The full length of vertical steel shall be enclosed with spiral reinforcement or equivalent hoops.

The spiral reinforcement at the ends of the pile shall have a pitch of 3 inches, and gage of not less than No. 5 (U.S. Steel Wire Gage). In addition the top 6 inches of pile shall have five turns of spiral winding at one-inch pitch.

For the remainder of the pile the vertical steel shall be enclosed with spiral reinforcement No. 5 gage (U.S. Steel Wire Gage) with not more than 6-inch pitch, or with ¼-inch round hoops spaced not more than 6-inch centers.

The reinforcement shall be placed at a clear distance from the face of the pile of not less than 2 inches and when the piles are for use in salt water or alkali soils this clear distance shall be not less than 3 inches.

In computing stresses due to handling, the computed static loads shall be increased by 50 per cent as an allowance for impact and shock.

(H) Concrete Piles (Cast-in-Place)

Cast-in-place concrete piles shall be, in general, cast in metal shells which shall remain permanently in place. However, other types of cast-in-place concrete piles, plain or reinforced, cased or uncased, may be used if, in the opinion of the Engineer, the soil conditions permit their use and if their design and the method of placing are satisfactory to him.

Cast-in-place concrete piles may be of either uniform section or tapered, or a combination thereof. The minimum size, measured at the butt, or above the taper, and embedment of reinforcement shall be as specified for precast piles, except that foundation piles may have a minimum butt cross-section area of 100 square inches. The minimum diameter at tip of pile shall be 8 inches.

Cast-in-place piling shall be reinforced when specified or shown on the plans. Cast-in-place foundation piling, carrying axial loads only and where the possibility of lateral forces being applied to the piles is insignificant, need not be reinforced when the soil provides adequate lateral support. Those portions of cast-in-place piling which are not supported laterally shall be designed as reinforced concrete columns in accordance with Article 1.5.9 and the reinforcing steel shall extend ten feet below the plane where the soil provides adequate lateral restraint. Where the shell is more than 0.12 inch in thickness, it may be considered as reinforcement.

Sufficient reinforcement shall be provided at the junction of the pile with the superstructure to make a suitable connection.

The metal shall be of sufficient thickness and strength so that the shell will hold its original form and show no harmful distortion after it and adjacent shells have been driven and the driving core, if any, has been withdrawn. The design of the shell shall be approved by the **Engineer** before any driving is done.

(I) Steel H-piles

(1) Thickness of Metal

Steel piles shall have a minimum thickness of web of .400 inch. Splice plates shall be not less than ⅜ inch thick.

(2) Splices

Piles shall be spliced to develop the net section of pile. The flanges and web shall be either spliced by butt welding or with plates, welded, riveted or bolted. The bolted splices shall only be used on projects where a small number of piling are required and where facilities for riveting or welding are not available.

Splices shall be detailed on the contract plans.

(3) Caps

In general, caps are not required for steel piles embedded in concrete. Reference is made to Research Report No. 1, "Investigation of the Strength of the Connection between a Concrete Cap and the Embedded end of the Steel H-Pile"—Department of Highways, State of Ohio, for a discussion of this subject and for the results of tests pertinent to it.

(4) Scour

If heavy scour is anticipated, consideration shall be given to design of the portion of the pile which would be exposed, as a column.

(5) Lugs, Scabs, and Core-stoppers

These devices may be used to increase the bearing power of the pile where necessary. They may consist of structural shapes, welded, riveted or bolted, of plates welded between the flanges, or of timber or concrete blocks securely fastened.

(J) Unfilled Tubular Steel Piles

(1) Thickness of Metal

Piles shall have minimum wall thickness not less than indicated in the following table:

Outside Diameter	Less than 14 inches	14 inches and over
	.25 inch	.375 inch

(2) Splices

Piles shall be spliced to develop the full section of the pile. The piles shall be spliced either by butt welding or by the use of welded sleeves. Splices shall be detailed on the contract plans.

(3) Driving

Tubular steel piles may be driven either closed or open ended. Closure plates should not extend beyond the perimeter of the pile.

(4) Column Action

Where the piles are to be used as part of a bent structure or where heavy scour is anticipated that would expose a portion of the pile, the pile shall be investigated for column action.

The provisions of Article 1.4.5(K) shall apply to unfilled tubular steel piles.

(K) Steel Pile and Steel Pile Shell Protection

Where conditions of exposure warrant, concrete encasement shall be used on steel piles and steel shells or $\frac{1}{16}$ inch depth of thickness shall be deducted from all exposed surfaces in computing the area of steel in the piles or shells.

1.4.6 — FOOTINGS

(A) Depth

The depths of footings shall be determined with respect to the character of the foundation materials and the possibility of undermining. Except where solid rock is encountered or in other special cases, the footings of all structures, other than culverts, which are exposed to the erosive action of stream currents, preferably, shall be founded at a depth of not less than 4 feet below the permanent bed of the stream. Stream piers and arch abutments, preferably, shall be founded at a depth of not less than 6 feet below stream bed. The above preferred minimum depths shall be increased as conditions may require.

Footings not exposed to the action of stream currents shall be founded on a firm foundation and below frost.

Footings for culverts shall be carried to an elevation sufficient to secure a firm foundation, or a heavy reinforced floor shall be used to distribute the pressure over the entire horizontal area of the structure. In any location liable to erosion, aprons or cut-off walls shall be used at both ends of the culvert and, where necessary, the entire floor area between the wing walls shall be paved. Baffle walls or struts across the unpaved bottom of a culvert barrel shall not be used where the stream bed is subject to erosion. When conditions require, culvert footings shall be reinforced longitudinally.

(B) Anchorage

Footings on inclined smooth solid rock surfaces which are not restrained by an overburden of resistant material, shall be effectively anchored by means of anchor bolts, dowels, keys or other suitable means.

(C) Distribution of Pressure

All footings shall be designed to keep the maximum soil pressures within safe bearing values. In order to prevent unequal settlement, footings shall be designed to keep the pressure as nearly uniform as practicable. In footings having unequal pressures and requiring piling, the spacing of the piles shall be such as to secure as nearly equal loads on each pile as may be practicable.

(D) Spread Footings

Spread footings which act as cantilevers may be decreased in thickness from the junction of the footing slab with column or wall toward the edge of the footing, provided sufficient section is maintained at all points to provide the necessary resistance to diagonal tension and bending stresses. This decrease in section may be accomplished by sloping the upper surface of the footing or by means of vertical steps. Stepped footings shall be cast monolithically.

(E) Internal Stresses in Spread Footings

Spread footings shall be considered as under the action of downward forces, due to the superimposed loads, resisted by an upward pressure exerted by the foundation materials and distributed over the area of the footings as determined by the eccentricity of the resultant of the downward forces. Where piles are used under footings, the upward reaction of the foundation shall be considered as a series of concentrated loads applied at the pile centers, each pile being assumed to carry its computed proportion of the total footing load.

When a single spread footing supports a column, pier or wall, this footing shall be assumed to act as a cantilever. When two or more piers or columns are placed upon a common footing, the footing slab shall be designed for the actual conditions of continuity and restraint.

Footings shall be designed for the bending stress, diagonal tension stress and bond at the critical section designated herein.

The critical section for bending shall be taken at the face of the column, pedestal or wall. In the case of columns other than square or rectangular, the critical section shall be taken at the side of the concentric square of equivalent area. For footings under masonry walls, where bond between the wall and footing is reduced to friction value, the critical section shall be taken as midway between the middle and the face of the wall. For footings under metallic column bases, the critical section shall be taken as midway between the face of the column and the edge of the metallic base. The load shall be considered as uniformly distributed over the column, pedestal or wall, or metallic column base.

The critical section for bond shall be taken at the same plane as for bending, and the shear used for computing bond shall be based on the same loading and section as for bending. Bond should also be investigated at planes where changes of section or of reinforcement occur.

The critical section for diagonal tension in footings on soil or rock shall be considered as the concentric vertical section through the footing at a distance "d" from each face of the column, pedestal, or wall; "d" being equal to the depth from the top of the section to the centroid of the longitudinal tension reinforcement.

The critical section for diagonal tension in footings supported on piles shall be considered as the concentric vertical section through the footing at a distance, d/2, from each face of the column, pedestal or wall, and any piles whose centers are at or outside this section shall be considered in computing the diagonal tension.

In sloped or stepped footings, stresses should be investigated at sections where the depth changes outside the critical section as defined above.

Bending need not be considered unless the projection of the footing is more than two-thirds of the depth.

In plain concrete footings, the stresses shall be computed on the basis of a monolithic section having a depth measured from the top of the footing to a plane 2 inches above the bottom of the footing. The maximum fibre stress due to bending shall not exceed that specified in Article 1.5.1 and the average shearing stress on a concentric vertical section through the footing at a distance (d minus 2 inches) from each face of the column, pedestal or wall, shall not exceed the shearing stresses specified in Article 1.5.1 for beams without web reinforcement and with longitudinal bars not anchored.

(F) Reinforcement

Footing slabs shall be reinforced for bending stresses and, where necessary, for diagonal tension. The computed stress in the bar shall be developed in bond.

The reinforcement for square footings shall consist of two or more bands of bars. The reinforcement necessary to resist the bending moment in each direction in the footing shall be determined as for a reinforced concrete beam; the effective depth of the footing shall be the depth from the top to the plane of the reinforcement. The required reinforcement shall be spaced uniformly across the footing, unless the footing width is greater than the side of the column or pedestal plus twice the effective depth of the footing, in which case the width over which the reinforcement is spread may equal the width of the column or pedestal plus twice the effective depth of the footing plus one-half the remaining width of the footing. In order that no considerable area of the footing shall remain unreinforced, additional bars shall be placed outside of the width specified, but such bars shall not be considered as effective in resisting the calculated bending moment. For the extra bars a spacing double that used for the reinforcement within the effective belt may be used.

(G) Transfer of Stress from Vertical Reinforcement

The stresses in the vertical reinforcement of columns or walls shall be transferred to the footings by extending the reinforcement into them a sufficient distance to develop the strength of the bars in bond, or by means of dowels anchored in the footings and overlapping or fastened to the vertical bars in such manner as to develop their strength. If the dimensions of the footings are not sufficient to permit the use of straight bars, the bars may be hooked or otherwise mechanically anchored in the footings.

1.4.7 — ABUTMENTS

(A) General

Abutments shall be designed to withstand earth pressure as specified in Article 1.2.19, the weight of abutment and superstructure, live load over any portion of the superstructure or approach fill, wind forces, longitudinal force when the bearings are fixed, and longitudinal forces due to friction or shear resistance of bearings. The design shall be investigated for any combination of these forces which may produce the most severe condition of loading.

Abutments shall be designed to be safe against overturning about the toe of the footing, against sliding on the footing base and against crushing of foundation material or overloading of piles at the point of maximum pressure.

In computing stresses in abutments, the weight of filling material directly over an inclined or stepped rear face, or over a reinforced concrete spread footing extending back from the face wall, may be considered as part of the effective weight of the abutment. In the case of a spread footing, the rear projection shall be designed as a cantilever supported at the abutment stem and loaded with the full weight of the superimposed material, unless a more exact method is used.

The cross section of stone masonry or plain concrete abutments shall be proportioned to avoid the introduction of tensile stress in the material.

(B) Reinforcement for Temperature

Except in gravity abutments, not less than ⅛ square inch of horizontal reinforcement per foot of height shall be provided near exposed surfaces not otherwise reinforced, to resist the formation of temperature and shrinkage cracks.

(C) Wing Walls

Wing walls shall be of sufficient length to retain the roadway embankment to the required extent and to furnish protection against erosion. The wing lengths shall be computed on the basis of the required roadway slopes.

Where deflection joints are not used, reinforcement rods or other suitable rolled sections preferably shall be spaced across the junction between all wing walls and abutments to thoroughly tie them together. Such bars shall extend into the masonry on each side of the joint far enough to develop the strength of the bar as specified for bar reinforcement, and shall vary in length so as to avoid planes of weakness in the concrete at their ends. If bars are not used, an expansion joint shall be provided at this point in which the wings shall be mortised into the body of the abutment.

(D) Drainage

The filling material behind abutments shall be effectively drained by weep holes with French drains, placed at suitable intervals.

1.4.8 — RETAINING WALLS

(A) General

Retaining walls shall be designed to withstand earth pressure, including any live load surcharge, and the weight of the wall, in accordance with the general principles specified above for abutments.

Stone masonry and plain concrete walls shall be of the gravity type. Reinforced concrete walls may be of either the cantilever, counterforted, buttressed, or cellular types.

(B) Base or Footing Slabs

The rear projection or heel of base slabs shall be designed to support the entire weight of the superimposed materials, unless a more exact method is used.

The base slabs of cantilever walls shall be designed as cantilevers supported by the wall.

The base slabs of counterforted and buttressed walls shall be designed as fixed or continuous beams of spans equal to the distance between counterforts or buttresses.

(C) Vertical Walls

The vertical stems of cantilever walls shall be designed as cantilevers supported at the base.

The vertical or face walls of counterforted and buttressed walls shall be designed as fixed or continuous beams. The face walls shall be securely anchored to the supporting counterforts or buttresses by means of adequate reinforcement.

(D) Counterforts and Buttresses

Counterforts shall be designed as T-beams. Buttresses shall be designed as rectangular beams. In connection with the main tension reinforcement of counterforts there shall be a system of horizontal and vertical bars or stirrups to effectively anchor the face walls and base slab. These stirrups shall be anchored as near the outside faces of the face walls, and as near the bottom of the base slab as practicable.

(E) Reinforcement for Temperature

Except in gravity walls, not less than ⅛ square inch of horizontal reinforcement per foot of height shall be provided near exposed surfaces not otherwise reinforced, to resist the formation of temperature and shrinkage cracks.

(F) Expansion and Contraction Joints

Contraction joints shall be provided at intervals not exceeding 30 feet and expansion joints at intervals not exceeding 90 feet, for gravity or reinforced concrete walls.

(G) Drainage

The filling material behind all retaining walls shall be effectively drained by weep holes with French drains, placed at suitable intervals. In counterforted walls there shall be at least one drain for each pocket formed by the counterforts.

1.4.9 — PIERS

(A) General

Piers shall be designed to withstand the dead and live loads superimposed thereon; wind pressures acting on the pier and superstructure; the forces due to stream current, floating ice and drift; and longitudinal forces at the fixed ends of spans.

Where necessary, piers shall be protected against abrasion by facing them with granite, vitrified brick, timber or other suitable material within the limits of damage by floating ice or debris.

(B) Pier Nose

In streams carrying ice or drift, the pier nose shall be designed as an ice breaker. When a steel angle or other metal nosing is used it shall be effectively secured to the masonry by means of suitable anchors.

1.4.10 — TUBULAR STEEL PIERS

(A) Use

Preferably, tubular steel piers shall not be used and they shall never be used in locations where they will be subjected to lateral earth pressure. In special cases their use may be permitted, in which cases the following requirements shall apply.

(B) Depth

The general requirements governing the depths of foundations as above set forth shall govern in the case of tubular steel piers except that steel tubes resting upon gravel foundation without piling shall in no case be carried to a depth less than 8 feet below the permanent bed of the stream and to such additional depth as may be necessary to eliminate all danger of undermining.

(C) Piling

Piles used in connection with tubular piers shall extend into the concrete filling a sufficient distance to thoroughly brace the tubes. In general, these piles shall extend not less than 6 to 8 feet above the bottom of the concrete.

(D) Dimensions of Shell

The minimum thickness of the metal in the shells of tubular piers shall be $5/16$ inch. This thickness shall be increased where necessary to secure strength and rigidity for placing the shell. In all cases the pier shall be designed for safe pile or soil bearing values as specified herein, but when the diameter required by these values is greater than that required for the superstructure bearing, the diameter may be reduced at any splice point. The minimum diameter of steel cylinders used for piers shall be 42 inches.

(E) Splices and Joints

All horizontal joints shall be butt joints. Vertical joints may be lapped if the corners of the plates are properly scarfed. When field splicing is necessary the lower section of the tube shall extend at least 2 feet above the water line when in position.

(F) Bracing

Adequate bracing connecting the tubes of cylinder piers shall be provided. In general, this bracing shall consist of a steel or concrete girder diaphragm effectively secured to the tubes. The depth of this diaphragm shall be as great as conditions will permit.

Section 5—CONCRETE DESIGN*

1.5.1 — ALLOWABLE STRESSES

(A) Standard Notations and Assumptions

A_v = total area of web reinforcement in tension within a distance "s" (measured in a direction parallel to that of the main reinforcement), or the total of all bars bent up in any one plane.
b = width of beam.
d = effective depth, or depth from compression surface of beam to centroid of tension reinforcement.
D = nominal diameter of bar, inches.
f_c = permissible extreme fiber stress in compression.
f'_c = unit ultimate compressive strength of concrete as determined by cylinder tests at the age of 28 days.
f_v = tensile unit stress in web reinforcement.
j = ratio of lever arm of resisting couple to depth "d."
n = ratio of modulus of elasticity of steel to that of concrete. See Article 1.5.2(4).
s = spacing of web reinforcing bars, measured at the neutral axis in a direction parallel to that of the main reinforcement.
u = bond stress per unit area of the surface of the bar.
v = shearing unit stress.
V = total shear.
V' = external shear on any section after deducting shear carried by concrete.
Σo = sum of perimeters of bars
α = angle between web bars and axis of beam.
  Coefficients:
    Thermal, .000006.   Shrinkage, .0002

(B) Strength of Concrete

The proportions for concrete mixes specified in Article 2.4.6 were selected on the basis of meeting the following minimum requirements for strength for 28 day cylinders for the various classes of concrete here recognized.

Class of Concrete	Minimum Compressive Strength at 28 days
A or A(AE)	3,000
B or B(AE)	2,200
C or C(AE)	1,500
X or X(AE)	3,000
Y or. Y(AE)	3,000

---

* Note: The ratios and values in this section apply to concrete made with conventional hard rock aggregate. Values applicable to lightweight aggregate concrete should be established by adequate investigation.

The basic value used in design under these specifications is 3,000 pounds per sq. in. at 28 days and if another value is used it shall be substantiated by test data and stipulated on the plans, along with the resultant allowable stresses. In no case shall the ultimate strength upon which allowable stresses are based exceed 4,500 pounds per sq. in., except for prestressed concrete.

(C) Allowable Stresses — Concrete

(1) Flexure

Extreme fiber in compression	$f_c = 0.4\ f'_c$
Extreme fiber in tension, plain concrete, primarily in footings	$f_c = 0.03\ f'_c$
Extreme fiber in tension, reinforced concrete	None

(2) Shear

Beams without web reinforcement:

Longitudinal bars not anchored or plain concrete footings	$0.02\ f'_c$ (max. 75 psi.)
Longitudinal bars anchored	$0.03\ f'_c$ (max. 90 psi.)
Beams with web reinforcement	$V = 0.075\ f'_c bjd$
Horizontal shear in shear keys between slab and stem of T-beams and box girders	$0.15\ f'_c$

(3) Bond on Piles (in Seals)

Timber, steel or concrete piles, 10 lbs. per square inch. (Providing the pile has the resistance to the pull thereby induced.)

(4) Bearing on Bridge Seats

Refer to Article 1.7.8

(5) Columns — See Article 1.5.9.

(D) Allowable Stresses — Reinforcement

Specified Yield Strength		
AASHO M 31 (ASTM A 615)	Grade 40	40,000 psi
AASHO M 31 (ASTM A 615)	Grade 60	60,000 psi
Steel Reinforcement:	Grade 40	Grade 60
Tension in flexural members	20,000 psi	24,000 psi

Tension in web reinforcement	20,000 psi	24,000 psi
Compression in columns See Article 1.5.9(F)	16,000 psi	20,000 psi
Compression in Beams See Article 1.5.7	20,000 psi	24,000 psi

Bond (for tension bars conforming to AASHO M 31, ASTM A 615)

(1) Sizes #3 through #11
    Top bars* $\dfrac{3.4\sqrt{f_c'}}{D}$ 350 psi max

    Bars other than top bars $\dfrac{4.8\sqrt{f_c'}}{D}$ 500 psi max

(2) Size #14 and #18
    Top bars* $2.1\sqrt{f_c'}$

    Bars other than top bars $3\sqrt{f_c'}$

(3) For all deformed compression bars:

                                                     $6.5\sqrt{f_c'}$ 400 psi max

1.5.2 — GENERAL ASSUMPTIONS

The design of reinforced concrete members under these specifications shall be based on the following assumptions:

(1) Calculations are made with reference to unit working stresses and safe loads, as elsewhere specified herein, rather than with reference to ultimate strength and ultimate loads.

(2) A plane section before bending remains plane after bending.

(3) The modulus of elasticity of concrete in compression is constant within the limits of working stresses; the distribution of compressive stress in flexure is, therefore, linear.

(4) The ratio "n" shall be assumed as follows:

	Values of $n = \dfrac{E_s}{E_c}$	
Ultimate strength of concrete, f'_c, Lbs. per sq. in.	For computations of strength	For computations of deflection
2,000 to 2,400.............	15	
2,500 to 2,900.............	12	
3,000 to 3,900.............	10	8
4,000 to 4,900.............	8	
5,000 or more.............	6	

---

* Top bars, in reference to bond, are horizontal bars so placed that more than 12 inches of concrete is cast in the member below the bar.

In computing the ultimate deflection of slabs and beams, the value of the modulus of elasticity of concrete should be assumed as one-thirtieth that of steel in order to allow for the effect of plastic flow and shrinkage.

(5) Concrete shall be assumed as offering no tensile resistance.

(6) The bond between concrete and metal reinforcement is assumed to remain unbroken throughout the range of working stresses. Under compression the two materials are therefore stressed in proportion to their moduli of elasticity.

(7) Initial stress in the reinforcement, due to contraction or expansion of the concrete, is neglected, except in the design of reinforced concrete columns.

(8) For the determination of external reactions, moments, shears, and deflections, moments of inertia of rigid frame and continuous structures shall be computed for the gross concrete sections, neglecting the effect of steel reinforcement, except that the transformed area of the steel shall be included for columns, arches or other compressive members.

(9) The moment of inertia of the entire superstructure sections, except railings or any curbs or sidewalks not placed monolithically with the superstructure before the falsework is released, and the moment of inertia of the full cross section of the pier or bent shall be used to determine the elastic properties of the various spans and supports.

(10) The depth of girder or slab to be used in computing moment of inertia at the centerline of support shall be obtained by extending the slope of the intrados of the member to the centerline.

(11) Rigid frames shall be considered free to sway longitudinally due to the application of vertical dead loads and vertically applied live loads, except when the structure is restrained from movement by external forces.

(12) The assumption of no moment restraint at the base of column shall be used in the analysis of rigid frames (superstructures) unless the base is known to be fully fixed. When a pinned end condition is assumed for the analysis of the superstructure, the base of column, footing and piling shall be designed to resist the moment resulting from an assumed restraint varying from zero to full fixity. The degree of restraint shall be determined by the type of footing and the character of the foundation material.

(13) Piers or bents constructed integrally with footings placed on a skew exceeding 10° shall be considered fixed at the top of footing.

1.5.3 — SPAN LENGTHS

The effective span lengths of slabs shall be as specified in Article 1.3.2.

The effective span length of freely supported beams shall not exceed the clear span plus the depth of beam.

For the analysis of all rigid frames, the span lengths shall be taken as the distance between the centers of bearings at the top of the footings.

The span length of continuous or restrained floor slabs and beams shall be the clear distance between faces of support. Where fillets making an angle of 45 degrees or more with the axis of a continuous or restrained slab are built monolithic with the slab and support, the span shall be measured from the section where the combined depth of the slab and fillet is at least one and one-half times the thickness of slab. Maximum negative moments are to be considered as existing at the ends of the span, as above defined. No portion of the fillet shall be considered as adding to the effective depth of the slab.

1.5.4 — EXPANSION

In general, provision for temperature changes shall be made in all simple spans having a clear length in excess of 40 feet.

In continuous bridges, provision shall be made in the design to resist thermal stresses induced or means shall be provided for movement caused by temperature changes.

Expansion not otherwise provided for shall be provided by means of hinged columns, rockers, sliding plates or other devices.

1.5.5 — T-BEAMS

(A) Effective Flange Width

In beam and slab construction, effective and adequate bond and shear resistance shall be provided at the junction of the beam and slab. The slab may then be considered an integral part of the beam, but its assumed effective width as a T-beam flange shall not exceed the following:
   (1) One-fourth of the span length of the beam.
   (2) The distance center to center of beams.
   (3) Twelve times the least thickness of the slab plus the width of the girder stem.

For beams having a flange on one side only, the effective overhanging flange width shall not exceed one-twelfth of the span length of the beam, nor six times the thickness of the slab, nor one-half the clear distance to the next beam.

(B) Shear

The flange shall not be considered as effective in computing the shear and diagonal tension resistance of T-beams, except in the determination of the value of j.

The horizontal shearing unit stress at the juncture of the flange and the monolithic fillet joining it to the girder stem shall not exceed that given in Article 1.5.1(C), Shear, Beams with web reinforcement.

(C) Isolated Beams

Isolated beams, in which the T-form is used only for the purpose of providing additional compression area, shall have a flange thickness of not less than one-half the width of the web, and a total flange width of not more than 4 times the width of web.

(D) Diaphragms

For T-beam spans over 40 feet in length diaphragms or spreaders shall be placed between the beams at the middle or at the third points.

(E) Construction Joints

When a construction joint is required between the slab and the stem of the beam, the shear-keys shall be designed in accordance with allowable stresses given in Article 1.5.1(C).

1.5.6 — REINFORCEMENT

(A) Spacing

The clear distance between parallel bars shall not be more than 18″ adjacent to concrete surfaces nor less than 1½ times the nominal diameter of the bars, 1½ times the maximum size of the coarse aggregate, nor 1½″.

The clear distance between bars shall also apply to the clear distance between a contact splice and adjacent splices or bars.

The maximum spacing of bars carrying stress, in a slab or wall, shall be 1½ times the thickness of slab or wall.

(B) Covering

The minimum covering measured from the surface of the concrete to the face of any reinforcing bar, shall be not less than 2 inches, except as follows:

Top of slab	1½ inches
Bottom of slab	1 inch
Stirrups and ties in T-beams	1½ inches
Stirrups and ties at outside faces of box girders	1½ inches
Stirrups and ties at inside faces of box girders	1 inch
Footings in contact with the ground	3 inches
Concrete exposed to sea water	4 inches
Concrete piles exposed to sea water	3 inches

Special consideration should be given to increasing the minimum cover in areas where chlorides or other corrosive substances are present in the soil. However, if special measures are taken to protect the steel from the corrosive substance, the special covering may be reduced but not to less than that specified in paragraph 1.

(C) Splicing

(1) Except as provided herein, all welding shall conform to AWS D12.1, "Recommended Practices for Welding Reinforcing Steel, Metal Inserts and Connections in Reinforced Concrete Construction."

(2) Lapped splices in reinforcement shall not be used for bar sizes larger than #11.

Tensile reinforcement shall preferably not be spliced at points of maximum stress. The spliced bar shall develop the computed stress at the splice point without exceeding ¾ of the permissible bond values given in Article 1.5.1(D). However, the length of lap for deformed bars shall not be less than 24 and 36 bar diameters for Grade 40 and Grade 60, respectively, nor less than 12".

(3) Where lapped splices are used in reinforcement in which the critical design stress is compressive, the minimum amount of lap shall be: with concrete having a strength of 3,000 psi or more, the length of lap for deformed bars shall be 20 bar and 24 bar diameters for Grade 40 and Grade 60, respectively, but not less than 12". When the specified concrete strengths are less than 3,000 psi, the amount of lap shall be ⅓ greater than the values given above.

Welded splices or other positive connections may be used instead of lapped splices. Where the bar size exceeds #11, welded splices or other positive connections shall be used. In bars required for compression only, the compressive stress may be transmitted by bearing of square-cut ends held in concentric contact by a suitably welded sleeve or mechanical device.

(4) An approved welded splice is one in which the bars are butted and welded so as to develop in tension at least 90% of the minimum tensile strength of the reinforcing bar. Approved positive connections for bars designed to carry critical tension or compression shall be equivalent in strength to an approved welded splice.

(D) End Anchorage and Hooks

End anchorage may be a straight extension of the bar or a standard hook. The term 'standard hook' as used herein shall mean either:

(1) A semi-circular turn plus an extension of at least four bar diameters but not less than 2½" at the free end of the bar, or

(2) A 90 degree turn plus an extension of at least 12 bar diameters at the free end of the bar, or

(3) For stirrup and tie anchorage only, either a 90 degree or 135 degree turn plus an extension of at least 6 bar diameters but not less than 2½" at the free end of the bar.

When bends are made at points of stress in the bar, an adequate radius of bend shall be provided to prevent crushing of concrete.

Standard hooks in tension may be considered as developing 10,000 psi in the bars or may be considered as extension of the bars at appropriate bond stresses.

Adequate anchorage shall be provided for the tension reinforcement in all flexural members to which the formula in Article 1.5.6 (J)(1) does not apply, such as sloped, stepped or tapered footings, brackets, or beams in which the tension reinforcement is not parallel to the compression face.

Hooks shall not be considered effective in adding to the compressive resistance of bars. Any mechanical device capable of developing the strength of the bar without damage to the concrete may be used in lieu of hooks or extensions.

(E) Extension of Reinforcement

(1) To provide for contingencies arising from unanticipated distribution of loads, yielding of supports, shifting of points of inflection, or other lack of agreement with assumed conditions governing the design of elastic structures, the reinforcement shall be extended at the supports and at other points between the supports as indicated in (2) to (5) below. These paragraphs relate to ordinary anchorage and are the minimum requirements under which normal working stresses for bond or shear are permitted.

(2) Negative tensile reinforcement at the supported end of a restrained or cantilever beam or member of a rigid frame shall be extended in or through the supporting member in such a manner as to develop the maximum tension in the bar with a bond stress not exceeding the normal working stress provided in Article 1.5.1(D).

(3) Between the supports of continuous or simple beams, every reinforcement bar shall be extended at least 15 diameters but not less than $\frac{1}{20}$ of the span length, beyond the point at which computations indicate it is no longer needed to resist stress.

(4) In simple beams and freely supported ends of continuous beams, at least $\frac{1}{3}$ of the positive reinforcement shall extend beyond the face of the supports a distance sufficient to develop $\frac{1}{2}$ the allowable stress in the bars.

(5) In restrained or continuous beams at least $\frac{1}{4}$ of the positive reinforcement shall extend beyond the face of the supports and the remainder treated as provided in (3).

(6) Dowels and bars carrying little or no theoretical stress should be embedded at least ten bar diameters from the construction joint.

(F) Structural Steel Shapes

When structural steel shapes are used for reinforcement, mechanical bond shall be provided which will effectively bond the member to the surrounding concrete mass.

(G) Interim Reinforcement for T-beams and Box Girders

When the floor slab or flange of a continuous or cantilevered T-beam or box girder is placed after the concrete in the stem has taken its set, at least 10% of the negative moment reinforcing steel shall be placed full length in the top of the beam stem to prevent cracks from falsework settlement or deflection. In lieu of the above requirement two #8 bars full length of the girders may be used.

(H) Reinforcement for Temperature and Shrinkage

Not less than $\frac{1}{8}$ square inch of reinforcement per foot shall be placed in each direction of all concrete surfaces to resist the formation of temperature and shrinkage cracks. The maximum spacing of bars shall be 18-inches. This reinforcement is not required if the surface is covered by at least 1½ feet of earth.

(I) Bundled Reinforcement

Groups of parallel reinforcing bars bundled in contact to act as a unit shall be limited to four #9 or smaller, three #11, two #14 or two #18 in any one bundle. Stirrups or ties shall enclose the bundle. Bars in bundles shall preferably be the same size.

Where spacing limitations are based on bar size, a unit of bundled bars shall be treated as a single bar of equivalent area. Bars in a bundle shall terminate at different points with at least 40 bar diameters stagger unless all of the bars end in a support.

When making bond stress calculations the external perimeter of the bundle is equal to the following:

2 Bar Bundle: 2 times perimeter of the individual bar
3 Bar Bundle: 2½ times perimeter of the individual bar
4 Bar Bundle: 3 times perimeter of the individual bar

When bundled bars are used in tied columns and as compression reinforcement in beams, the spacing of ties or hoops shall be ¾ that specified in Articles 1.5.7 and 1.5.9(D)(2).

(J) Bond Stress in Flexural Members

(1) In flexural members in which the tension reinforcement is parallel to the compression face, the flexural bond stress at any cross section shall be computed by

$$u = \frac{V}{\Sigma o \, jd}$$

Bent-up bars that are not more than d/3 from the level of the main longitudinal reinforcement may be included. Critical sections occur at the face of the support, at each point where tension bars terminate within a span, and at the point of inflection.

(2) To prevent bond failure or splitting, the calculated tension or compression in any bar at any section must be developed on each side of that section by proper embedment length, end anchorage, or for tension only, hooks. Anchorage or development bond stress, u, shall be computed as the bar forces divided by the product of Σo times the embedment length.

(3) The bond stress, u, computed as in (1) or (2) shall not exceed the limits given in 1.5.1(D), except that flexural bond stress need not be considered in compression, nor in those cases of tension where anchorage bond is less than 0.8 of the permissible.

1.5.7 — COMPRESSION REINFORCEMENT IN BEAMS

Compression reinforcement in girders and beams shall be secured against buckling by ties or stirrups adequately anchored in the concrete and spaced not more than 16 bar diameters apart. Where compression reinforcement is used, its effectiveness in resisting bending may be taken as twice the value indicated from the calculations assuming a straight-line relation between stress and strain and the modular relation of stress in steel to stress in concrete given in Article 1.5.2(4). However, in no case should a stress in compression reinforcement be greater than that allowed in Article 1.5.1(D).

1.5.8 — WEB REINFORCEMENT

(A) General

When the allowable unit shearing stress for concrete is exceeded, web reinforcement shall be provided by one of the following methods:

(1) Longitudinal bars bent up in series or in a single plane.
(2) Vertical stirrups.
(3) Combination of bent-up bars and vertical stirrups.

When any of the above methods of reinforcement are used, the concrete may be assumed to carry external vertical shear not to exceed .03 f'_c (maximum, 90 pounds per square inch) the remainder of shear being carried by the web reinforcement.

The webs of T-beams and box girders shall be reinforced with stirrups in all cases.

(B) Calculation of Shear

Diagonal tension and shear in reinforced concrete beams shall be calculated by the following formulas:

Shearing unit stress, as a measure of diagonal tension:

$$v = \frac{V}{bjd}$$

Stress in vertical web reinforcement:

$$f_v = \frac{V's}{A_v jd}$$

When a series of web bars or bent-up longitudinal bars are used, the web reinforcement shall be designed by the following formula:

$$A_v = \frac{V's}{f_v jd\ (\sin\alpha + \cos\alpha)}$$

When web reinforcement consists of bars bent up in a single plane so as to reinforce all sections of the beam which require reinforcement, the bent-up bars shall be designed by the following formula:

$$A = \frac{V'}{f_v \sin\alpha}$$

(C) Bent-up Bars

Bent-up bars used as web reinforcement may be bent at any angle between 20 and 45 degrees with the longitudinal reinforcement. The radius of bend shall not be less than 4 diameters of the bar.

The spacing of bent-up bars shall be measured at the neutral axis and in the direction of the longitudinal axis of the beam. This spacing shall not exceed three-fourths the effective depth of the beam. The first bar from the support shall cross the mid-depth of the beam at a distance from the face of the support, measured parallel to the longitudinal axis of the beam, not greater than one-half the effective depth.

(D) Vertical Stirrups

Where stirrups are required to carry shear, the maximum spacing of vertical stirrups shall be limited to ½ the depth of the beam, and where not required to carry shear, the maximum spacing shall be limited to ¾ the depth of the beam. The first stirrup shall be placed at a distance from the face of the support not greater than one-fourth of the effective depth of the beam.

(E) Anchorage

(1) The stress in a stirrup or other web reinforcement shall not exceed the capacity of its anchorage in the upper or lower one-half of the effective depth of the beam.

(2) Web reinforcement which is provided by bending into an inclined position one or more bars of the main tensile reinforcement where not required for resistance to positive or negative bending, may be considered completely anchored by continuity with the main tensile reinforcement, or by embedment of the requisite length in the upper or lower half of the beam, provided at least ½ of such embedment is as close to the upper or lower surface of the beam as the requirements of fire and rust protection allow. A hook placed close to the upper or lower surface of the beam may be substituted for a portion of such embedment.

(3) Stirrups shall be anchored at both ends by one of the following methods, or by a combination thereof:

- (a) Rigid attachment, as by welding, to the main longitudinal reinforcement. All welding shall conform to AWS D12.1.
- (b) Bending around and closely in contact with a bar of the longitudinal reinforcement, in the form of a U-stirrup or hook.
- (c) A hook placed as close to the upper or lower surface of the beam as the requirements of fire and rust protection will allow. In estimating the capacity of this anchorage the stress developed by bond between the midheight of the beam and the center of bending of the hook may be added to the capacity of the hook.
- (d) An adequate length of embedment in the upper or lower ½ of the effective depth of the beam, whether straight or bent.

1.5.9 — COLUMNS *

(A) General

The provisions of Section 5, Concrete Design, shall apply in the design of columns unless specifically modified by this article.

In the design of columns the unsupported length shall be defined as the clear distance between struts, cross beams, footings or other types of adequate restraint to lateral movement. Where a bracing member has haunches at its junction to a column, the unsupported column length shall be measured from the junction of the haunch with the column, provided that the face of the haunch makes an angle with the face of the column of at least 45 degrees. Struts or cross beams joining columns at angles greater than 30 degrees from the plane of symmetry of the column shall not be considered as adequate support. The least lateral dimension of a column shall be taken as: (1) for rectangular columns, the over-all thickness along a principal axis; (2) for spirally reinforced columns, the overall diameter including the encasement of the spirals; (3) for "T"-shaped columns, the width or depth of the T.

In a column which, for architectural or other reasons, has a larger cross section than required by the load carried, the minimum amount of longitudinal steel hereinafter specified may be reduced provided that in no case shall less longitudinal steel be used than that required by the minimum column designed with one per cent of longitudinal steel.

The notations used in this article are as follows:

A_g = over-all or gross cross-sectional area of a spirally reinforced or tied pier, pedestal or column in square inches

A_c = cross-sectional area of core of spirally reinforced columns measured to the outside diameter of the spiral, square inches

A_s = cross-sectional area of longitudinal steel

$A = A_g + (n-1) A_s$, effective area of column

$C = \dfrac{f_a}{0.40 f'_c}$, a factor used in the design of members subjected to combined axial and bending stresses

d = least lateral dimension of column, inches

e = eccentricity of resultant load on a column, measured from a gravity axis

$f_a = \dfrac{0.225 f'_c + f_s p}{1+(n-1)p}$ for spiral columns and 0.8 that amount for tied columns.

f'_c = crushing strength of 6" x 12" concrete cylinders at age of 28 days, psi.

f_e = maximum allowable compressive stress in members subjected to combined axial and bending stress, psi.

---

* This article, covering the design of reinforced concrete columns, follows in general the recommendations of the Joint Committee on Standard Specifications for Concrete and Reinforced Concrete and of the 1951 ACI Building Code Requirements for Reinforced Concrete.

f_s = Allowable working stress in flexural members [see Article 1.5.1(D)], psi.

f'_s = yield stress of spiral reinforcement (for steel grades not having a definite yield point, the stress causing a 0.2 per cent plastic set), psi.

$K = \dfrac{t^2}{2r^2}$, a factor used in the design of members subjected to combined axial and bending stresses

L = unsupported length of column, inches

n = ratio of modulus of elasticity of steel to that of concrete

P_e = a load eccentrically applied

P_p = total load on pier or pedestal, pounds

P_s = total load on spirally reinforced column, pounds

P_{sl} = total load on spirally reinforced long column, pounds

P_t = total load on tied column, pounds

P_{tl} = total load on tied long column, pounds

p = ratio of longitudinal steel area to gross column area

p' = ratio of volume of spiral reinforcement to core volume

r = radius of gyration of section (transformed section) in the direction of eccentricity or bending

t = over-all depth of column in the direction of eccentricity or bending

(B) Piers and Pedestals

The ratio of the unsupported lengths of unreinforced concrete piers or pedestals to their least dimension shall not exceed 3. The total load on any unreinforced concrete pier or pedestal shall not exceed that given by the following formula:

$$P_p = 0.25 A_g f'_c \tag{1}$$

(C) Spirally Reinforced Columns

(1) Longitudinal Reinforcement

Longitudinal reinforcement shall be placed within the area contained by the spiral reinforcement. The ratio between the area of longitudinal reinforcement and the gross area of the column, including the encasement outside the spiral reinforcement, shall be not less than 0.01 nor more than 0.08. There shall be a minimum of six longitudinal bars evenly spaced around the periphery of the column core. The diameter of bars shall be not less than five-eighths inch. For columns with a circular spirally reinforced core having excessive size or other outside shapes, the gross area to be used in determining percentage of reinforcement shall be a circle with a diameter equal to the minimum core required for structural design plus the specified outside cover.

(2) Spiral Reinforcement

Spiral reinforcement shall consist of uniform spirals held firmly in position by attachment to the longitudinal reinforcement.

Spiral reinforcement may be plain or deformed reinforcing bars or cold drawn wire conforming to AASHO M32 (ASTM A82). Splices in spiral bars should be avoided if practical or, if necessary, shall be made by welding or by a lap of 1½ turns. The pitch of spirals shall not exceed ⅙ of the core diameter. The clear distance between individual turns of the spiral shall not exceed 3 inches or be less than 1⅜ inches or 1½ times the maximum size aggregate used. Spiral reinforcement shall extend from the footing or other support to the level of the lowest horizontal reinforcement of members supported by the column.

The ratio of the volume of the spiral reinforcement to the volume of core of the column, out to out of spirals, shall be not less than:

$$p' = 0.45 \left(\frac{A_g}{A_c} - 1 \right) \frac{f'_c}{f'_s} \qquad (2)$$

The yield strength for design assumption, f'_s, shall not be taken higher than 60,000 p.s.i.

(3) Allowable Load — Short Columns

The provisions of this subarticle shall apply only to columns having ratios of unsupported height to least lateral dimension of not more than 10. The total axial load on a column shall not exceed that given by the following formula:

$$P_s = 0.225 \, f'_c \, A_g + A_s f_s \qquad (3)$$

(4) Long Columns

The total axial load on a column having a ratio of unsupported height to least lateral dimension greater than 10, but not greater than 20, shall be not greater than given by the following formula:

$$P_{sl} = P_s (1.3 - 0.03 \, L/d) \qquad (4)$$

If the L/d ratio of columns exceeds 20, the column shall be investigated for elastic stability.

(D) Tied Columns

(1) Longitudinal Reinforcement

The longitudinal reinforcement shall consist of at least four bars, and, when only four bars are used, they shall be placed at the corners of the section. Bars shall be placed at each intersection of column faces. The bars shall be not less than five-eighths inch in diameter. The ratio of the total cross-sectional area of the bars to the total cross-sectional area of the column shall be not less than 0.01 nor more than 0.04.

(2) Hoops and Lateral Ties

Hoops shall surround the longitudinal reinforcement. They shall be not less than one-fourth inch in diameter and shall be spaced not more than 12 inches apart except that this spacing may be increased in the case of pier shafts or columns having a larger cross section than required by conditions of loading. Adequate auxiliary ties shall be provided to support intermediate longitudinal bars whose distance from any tied bar exceeds 2 feet.

(3) Allowable Load — Short Columns

The provisions of this subarticle shall apply only to columns having ratios of unsupported height to least lateral dimension of not more than 10. The total axial load on a column shall be not greater than 0.8 of that given by equation (3), which results in

$$P_t = 0.8\ (0.225 f'_c A_g + A_s f_s) \tag{5}$$

(4) Long Columns

The total axial load on a column having a ratio of unsupported height to least lateral dimension greater than 10 but not greater than 20 shall be not greater than given by the following formula:

$$P_{tl} = P_t (1.3 - 0.03\ L/d) \tag{6}$$

If the L/d ratio exceeds 20, the column shall be investigated for elastic stability.

(E) Bending Moments in Columns

When beams or slabs are connected to columns, the moments induced in the columns by such beams or slabs shall be provided for in the column design.

(F) Combined Axial and Bending Stress

(1) Longitudinal Reinforcement

The limiting steel ratio of 0.04 provided in Article 1.5.9(D)(1) may be increased to 0.08 for tied columns designed to withstand combined axial and bending stresses, provided that the amount of steel spliced by lapping in any 3-foot length of column shall not exceed a steel ratio of 0.04. The size of the column shall be not less than that required by axial load alone.

(2) Ratio e/t less than 0.5

A reinforced concrete column which is symmetrical about two mutually perpendicular planes through its axis and which is subject to an axial load, P_e, combined with bending in one of the planes of symmetry shall be designed on the basis of uncracked sections provided the ratio of eccentricity to depth, e/t, is not greater than 0.5 in the plane of bending. In this case the combined fiber stress in compression is given by the following formula:

$$f_c = \frac{P_e}{A_g} \frac{\left[1 \pm \dfrac{Ke}{t}\right]}{[1+(n-1)p]} \tag{7}*$$

---

* For approximate or trial design, K may be taken as 8 for a circular spiral column and as 5 for a rectangular, tied or spiral column. The assumed value of K shall be checked for the adopted section.

The column may be designed for an equivalent axial load P_s or P_t as given by the following formula:

$$P = P_e\left(1 + C\frac{Ke}{t}\right) \qquad (8)*$$

The maximum allowable compressive stress in the concrete, f_c, in columns subjected to combined axial and bending stress as described above shall not exceed that given by the following formula:

$$f_c = f_a \frac{1 + \dfrac{Ke}{t}}{1 + \dfrac{CKe}{t}} \qquad (9)*$$

where $f_a = \dfrac{0.225 f'_c + f_s p}{1 + (n-1)p}$ for spiral columns and 0.8 that amount for tied columns.

In the case of square or rectangular columns subject to bending in both planes of symmetry, the column shall be designed on the basis of uncracked sections only when the sum of the e/t ratios about both axes does not exceed 0.5. In this case formulas (7), (8), and (9) may be used by substituting for Ke/t the sum of the Ke/t ratios in both planes of bending.

(3) Ratio e/t greater than 0.5

Reinforced concrete columns in which the e/t ratio is greater than 0.5 in the case of bending in one plane or in which the sum of the e/t ratios is greater than 0.5 in the case of bending in both planes of symmetry, shall be designed on the basis of the recognized theory for cracked sections, based on the assumption that no tension is resisted by the concrete.

In such cases the modular ratio, n, for the compressive reinforcement may be assumed as twice the value given in Article 1.5.2(4); however, the stress in the compressive reinforcement when calculated on this basis, shall not be greater than the allowable stress in tension. (See Appendix D for a method of determining the location and direction of the neutral axis.) When designed on the basis of the cracked section theory, the column shall be so proportioned that the maximum combined compressive stress in the concrete does not exceed $0.4 f'_c$. For such cases the tensile stress in the reinforcing steel shall also be investigated.

1.5.10 — CONCRETE ARCHES

(A) Shape of Arch Rings

Arch rings shall be selected as to shape in such manner that the axis of the ring shall conform, as nearly as practicable, to either the equilibrium polygon for full dead load or to the equilibrium polygon for full dead plus one-half live load over the full span, whichever produces the smallest bending stresses under combined loads.

(B) Spandrel Walls

When the spandrel walls or filled spandrel arches exceed 8 feet in height above the extrados they shall be designed as vertical slabs supported by transverse diaphragm walls or deep counterforts. Vertical cantilever walls over 8 feet in height, or counterforts having a back slope of less than 45 degrees with the vertical, shall not be used, on account of the excessive and indeterminate stresses set up in the arch ring by torsion.

(C) Expansion Joints

Vertical expansion joints shall be placed in the spandrel walls of arches to provide for movement due to temperature change and arch deflection. These joints shall be placed at the ends of spans and at intermediate points, generally not more than 50 feet apart.

(D) Reinforcement

Arch ribs in reinforced concrete construction shall be reinforced with a complete double line of longitudinal reinforcement consisting of an intradosal system and an extradosal system connected by a series of stirrups or tie-rods.

For barrel arches, a system of transverse reinforcement, thoroughly anchored to the longitudinal reinforcement, shall be used in both intrados and extrados. The transverse reinforcement shall be proportioned to resist the bending stresses due to any overturning action of the spandrel wall.

For rib arches, hoops or tie bars shall be used in connection with the longitudinal rib reinforcement, as in the case of reinforced concrete columns.

(E) Waterproofing

Preferably, the top of the arch ring and the interior faces of the spandrel walls of all filled spandrel arches shall be waterproofed with a membrane waterproofing constructed in accordance with the requirements specified in Division II, Section 17.

(F) Drainage of Spandrel Fill

The fills of filled spandrel arches shall be effectively drained by a system of tile drains or French drains laid along the intersection of the spandrel walls and arch rings and discharging through suitable outlets in the piers and abutments. The location and details of the drainage outlets shall be such as to eliminate, as far as possible, the discoloration by drainage water of the exposed masonry faces.

1.5.11 — VIADUCT BENTS AND TOWERS

When concrete columns are used in viaduct construction, bents and towers shall be effectively braced by means of longitudinal and transverse struts. For height greater than 40 feet, both longitudinal and transverse cross or diagonal bracing, preferably, shall be used and the footings for the columns forming a single bent shall be thoroughly tied together.

1.5.12 — BOX GIRDERS

(A) Effective Compression Flange Width

In girder and flange construction, consisting of a girder stem with top and bottom slab, effective and adequate bond and shear resistance shall be provided at the junction of the girder and slab. The slab may then be considered an integral part of the girder, but its effective width as a girder flange shall not exceed the following:

 (1) One fourth of the span length of the girder.
 (2) The distance center to center of girders.
 (3) Twelve times the least thickness of the slab plus the width of the girder stem.

For girders having flanges on one side only, the effective overhanging flange width shall not exceed the following:

 (1) One-twelfth of the span length of the girder.
 (2) One-half of the clear distance to the next girder.
 (3) Six times the least thickness of the slab.

(B) Flange Thickness

(1) Top Flange

The minimum thickness of the top flange shall be determined by Article 1.3.2(C) Case A, and the maximum allowable unit stresses as specified in Article 1.5.12(C) & (D), but in no case shall be less than 6 inches.

(2) Bottom Flange

The thickness of the bottom flange shall be such that the maximum allowable unit stresses as specified in Article 1.5.12(C) & (D) are not exceeded. It also shall be at least $\frac{1}{16}$ of the clear span between girders, except that it shall not be less than 5½ inches, but need not be thicker than the top flange unless required for stress considerations.

(C) Flexure

(1) Parallel to Girder

The compressive unit stress in the extreme fiber of concrete in both girder stem and flange shall not exceed that given in Article 1.5.1(C).

(2) Normal to Girder

The compressive unit stress in the extreme fiber of concrete in the girder flange shall not exceed that given in Article 1.5.1(C).

(D) Shear

The flange shall not be considered as effective in computing the shear and diagonal tension resistance of girder stems, except in the determination of the value of j.

The horizontal shearing unit stress at the junction of the flange and the monolithic fillet joining it to the girder stem shall not exceed that given in Article 1.5.1(C), Shear, Beams with web reinforcement.

Changes in girder stem thickness shall be tapered for a minimum distance of 12 times the difference in stem thickness.

(E) Reinforcement

The unit stress in steel for both girder stem and flange shall not exceed that given in Article 1.5.1(D).

(F) Flange Reinforcement

(1) Bottom Flange Reinforcement Parallel to Girders

Minimum reinforcement of 0.4% of the flange section shall be placed in the slab. A single layer of bars may be centered in the slab. Bar spacing shall not exceed 18 inches. These bars may be stopped whenever they lap with any main girder reinforcement located in the approximate center of the flange.

(2) Bottom Flange Reinforcement Normal to Girder

Minimum reinforcement of 0.5% of the flange section shall be placed in the slab, distributed over both surfaces. Bar spacing shall not exceed 18 inches.

Reinforcement provided as above for the minimum flange thickness at other points may be used in areas thickened at supports in accordance with Art. 1.5.12(B)(2) & (C).

All transverse reinforcement in the bottom flange shall be extended to the exterior face of the outside girders in each group and anchored with standard 90 degree bends.

(3) Top Flange Reinforcement

A minimum of ⅓ of the bottom layer of the transverse reinforcement in the top flange shall be extended to the exterior face of the outside girder in each group and shall be anchored with standard 90 degree bends or, if the flange extends beyond the last girder, extended beyond the girder face at least a bond length.

(G) Diaphragms

Diaphragms or spreaders shall be placed between the girders at intervals not to exceed 60 feet. Diaphragm spacing for curved girders shall be given special consideration.

(H) Flanges Supporting Pipes and Conduits

Flanges supporting both vehicle live load and pipes or conduits shall be designed using unit stresses set forth in Article 1.5.1.

Flanges supporting only dead load of structure and pipes or conduits shall be designed in the direction normal to the girder using unit stresses not exceeding 75 per cent of those set forth in Article 1.5.1.

1.5.13 — BEARINGS

Bearing devices for concrete structures shall be designed in accordance with Articles 1.7.49 through 1.7.56, or Section 12, Elastomeric Bearings.

LOAD FACTOR DESIGN

1.5.14—GENERAL

(A) Application

These specifications are intended for use in the design of simple and continuous structures of moderate (to 200′) span length. Large or unusual structures may require special study and detailed consideration of effects that can otherwise be neglected or assigned arbitrary values in the design of structures to which these specifications are intended to apply.

(B) Other Specifications

All applicable provisions of the AASHO Specifications shall apply unless specifically modified herein.

1.5.15—NOTATION

(A) Loads and Forces

B = Buoyancy
CF = Centrifugal force
D = Dead load
EQ = Earthquake
F = Longitudinal force
I = Live load impact
ICE = Ice pressure
L = Live load
LF = Longitudinal force from live load
M = Moment to be used for design of compression member
M_b = Column moment capacity under balanced conditions
M_{cr} = Moment required to crack a concrete section
M_{max} = Maximum dead load moment for section under consideration
M_u = Moment capacity of the section $\geq$ applied design load moment at a section
M_{uo} = Theoretical moment strength of a section
M_{ux} = Moment capacity in the direction of the x axis
M_{uy} = Moment capacity in the direction of the y axis
M_x = Design bending moment component in the direction of the x axis

M_y = Design bending moment component in the direction of the y axis

M_1 = Value of smaller end moment on compression member calculated from a conventional elastic frame analysis, positive if member is bent in single curvature, negative if bent in double curvature

(B) Dimensions and Constants

A_b = Loaded area

A_b' = Maximum area of the portion of the supporting surface that is geometrically similar to and concentric with the loaded area

A_g = Gross area of column section

A_s = Area of tension reinforcement

A_s' = Area of compression reinforcement

A_{sf} = Area of reinforcement to develop compressive strength of overhanging flanges in I- and T-sections

A_{st} = Total area of longitudinal reinforcement = $A_s + A_s'$ = total vertical reinforcement in columns

A_v = Area of shear reinforcement within a distance s

a = Depth of equivalent rectangular stress block = $k_1 c$

a_s = Area of an individual bar, sq. in.

b = Width of compression face of flexural member, or member subject to flexure

b' = Width of web in I- and T-sections. In tapered webs, the average width or 1.2 times the minimum width, whichever is smaller

b_o = Periphery of critical section for slabs and footings

c = Distance from extreme compression fiber to neutral axis

c_b = Distance from extreme compression fiber to neutral axis for balanced conditions

C_m = a factor relating the actual moment diagram to an equivalent uniform moment diagram

D = Nominal diameter of bars; also, overall diameter of circular section

d = Distance from extreme compression fiber to centroid of tension reinforcement

d' = Distance from extreme compression fiber to centroid of compression reinforcement

E_c = Modulus of elasticity of concrete

E_s = Modulus of elasticity of steel

e = Eccentricity of design load parallel to axis measured from the centroid of the section. It may be calculated by conventional methods of frame analysis

F = Moment magnification factor

f'_c = Specified compression strength of concrete

f_h = Tensile stress developed by a standard hook, psi

f_r = Modulus of rupture of concrete

f_y = Specified yield strength of reinforcement

h = Unsupported length of compression member
I_{cr} = Moment of inertia of the transformed cracked section
I_{eff} = Effective moment of inertia for computation of deflection
I_g = Moment of inertia of gross concrete section about the centroidal axis, neglecting the reinforcement
K = Constant for standard hook
k = Effective length factor in design of slender columns
k_1 = 0.85 for strengths, f'_c, up to 4000 psi, and shall be reduced at a rate of 0.05 for each 1000 psi of strength in excess of 4000 psi
L_a = Additional embedment length at support or at point of inflection, in.
L_d = Development length, in.
L_e = Equivalent embedment length, in.
n = E_s/E_c
p = A_s/bd
p' = A'_s/bd
p_b = Reinforcement ratio producing balanced conditions
p_f = $A_{sf}/b'd$
p_w = $A_s/b'd$
q = $A_s f_y / bd f'_c = p f_y / f'_c$
R_m = Ratio of maximum design dead load moment to maximum design total load moment, always positive
r = Radius of gyration of the concrete gross section in the direction of bending
r_b = Ratio of area of bars cut off to total area of bars at the section
s = Shear reinforcement spacing in a direction parallel to the longitudinal reinforcement
t = Flange thickness in I- and T-sections; also overall depth of section
t_{min} = Recommended minimum thickness for constant depth members
y_t = Distance from centroidal axis of gross section, neglecting the reinforcement, to extreme fiber in tension
ε_u = Maximum usable strain at the extreme concrete compression fiber, assumed equal to 0.003
ε_y = Yield strain of reinforcement corresponding to the yield strength, f_y
ϕ = Capacity modification factor

1.5.16—MATERIALS PROPERTIES

(A) Concrete

(1) The design strength, f'_c of the concrete shall be specified and the specified strength shall be indicated on the plans. The specified strength of the concrete shall be a basis for acceptance, and each class

of concrete shall be represented by a sufficient number of tests.[1] For structures designed in accordance with these specifications, the average of any three consecutive strength tests of the laboratory-cured specimens representing each class of concrete (at least two specimens shall be made for each test) shall be equal to or greater than the specified strength, f'_c, and not more than 10 percent of the strength tests shall have values less than the specified design strength, but no test shall show an average strength less than 85 percent of the specified compressive strength f'_c.

(2) The modulus of elasticity, E_c, for concrete may be taken as ($w^{1.5} \times 33 \sqrt{f'_c}$) in psi, for values of w between 90 and 155 lb. per cu ft. For normal weight concrete, E_c may be considered as $57,000 \sqrt{f'_c}$.

(B) Reinforcement

(1) Reinforcing bars shall conform to one of the following specifications, except that yield strength shall correspond to that determined by tests on full sized bars and that reinforcing bars with a specified yield strength, f_y, exceeding 60,000 psi are not permitted under these specifications.

> (a) "Specifications for Deformed Billet-Steel Bars for Concrete Reinforcement" (AASHO M31, ASTM A 615). If #14 or #18 bars meeting these specifications are to be bent, they shall also be capable of being bent 90 deg. at a minimum temperature of 60 F. around a ten-bar-diameter pin without cracking transverse to the axis of the bar.
> (b) "Specifications for Rail-Steel Deformed Bars for Concrete Reinforcement" (AASHO M42). If bars meeting these specifications are to be bent, they shall also meet the bending requirements of AASHO M31, ASTM A615 for Grade 60.
> (c) "Specifications for Axle-Steel Deformed Bars for Concrete Reinforcement" (AASHO M53, ASTM A617).

(2) The modulus of elasticity of steel reinforcement, E_s, may be taken as 29,000,000 psi.

---

[1] In the Construction Specifications, or Special Provisions, there shall be included the requirements for the minimum number of tests and specimens for each test at a given age of the concrete, the maximum number of cubic yards of structural concrete for each test, and the least number of tests for each day's concreting. The age for strength tests shall be 28 days, or where specified, the earlier age at which the concrete is to receive its full load or maximum stress. Strength Control Procedures shall be preferably in accordance with ACI 214-65, "Recommended Practice for Evaluation of Compression Tests Results of Field Concrete." Refer also the following specifications: "Method of Sampling Fresh Concrete" (AASHO T141, ASTM C 172); "Method of Making and Curing Concrete Compressive and Flexural Test Specimens in the Field" (AASHO T23, ASTM C31); "Compressive Strength of Molded Concrete Cylinders" (AASHO T22, ASTM C39).

1.5.17—LOADS AND LOAD FACTOR EQUATIONS

(A) Loads

The forces in the structure shall be determined by considering the elastic behavior of the structure under loads specified in Section 2, Loads, Articles 1.2.4 and 1.2.16 excepted.

(B) Load Factor Equations

The following Load Groups represent various combinations of loads and forces to which a structure may be subjected. Each part of such structure, or the foundation on which it rests, shall be proportioned for all combinations of such of these forces as are applicable to the particular site or type.

The maximum section required shall be used.

Group I = 1.30 [D + 5/3 (L + I)]

For all loadings less than H20, provision shall be made for an infrequent heavy load by applying Group IA loading, with the live load assumed to occupy a single lane without concurrent loading in any other lane.

Group IA = 1.30 [D + 2.2 (L + I)]

Group II = 1.30 [D + W + F + SF + B + S + T]

When earthquake loading is taken into account, Group II loading shall be used substituting EQ for W. When ice pressure is taken into account, Group II loading shall be used substituting ICE for SF.

Group III = 1.30 [D + (L + I) + CF + 0.3W + WL + F + LF]

1.5.18—STRENGTH PROVISIONS

(A) Assumptions

(1) The strength design of members for flexure and axial loads shall be based on the assumptions given in this section, and on satisfaction of the applicable conditions of equilibrium and compatibility of strains.

(2) Strain in the reinforcing steel and concrete shall be assumed directly proportional to the distance from the neutral axis.

(3) The maximum usable strain at the extreme concrete compression fiber shall be assumed equal to 0.003.

(4) Stress in reinforcement below the specified yield strength, f_y, for the grade of steel used shall be taken as E_s times the steel strain. For strains greater than that corresponding to f_y, the stress in the reinforcement shall be considered independent of strain and equal to f_y.

(5) Tensile strength of the concrete shall be neglected in flexural calculations of reinforced concrete.

(6) The relationship between the concrete compressive stress

distribution and the concrete strain may be assumed to be a rectangle, trapezoid, parabola, or any other shape which results in prediction of strength in substantial agreement with the results of comprehensive tests.

(7) The requirements of Article 1.5.18(A)(6) may be considered satisfied by an equivalent rectangular concrete stress distribution which is defined as follows: A concrete stress of 0.85 f'_c shall be assumed uniformly distributed over an equivalent compression zone bounded by the edges of the cross section and a straight line located parallel to the neutral axis at a distance $a = k_1 c$ from the fiber of maximum compressive strain. The distance c from the fiber of maximum strain to the neutral axis is measured in a direction perpendicular to that axis. The fraction k_1 shall be taken as 0.85 for strengths, f'_c, up to 4000 psi and shall be reduced continuously at a rate of 0.05 for each 1000 psi of strength in excess of 4000 psi.

(8) Balanced conditions exist at a cross section when the tension reinforcement reaches its specified yield strength, f_y, just as the concrete in compression reaches its assumed ultimate strain of 0.003.

1.5.19—CAPACITY MODIFICATION FACTORS

(A) The usable load capacities of the members shall be the calculated capacities of the members modified according to the provisions of this Article.

(B) The computed theoretical capacity shall be modified by a capacity modification factor ϕ as follows:

For flexure .... $\phi = 0.90$
For shear .... $\phi = 0.85$
For spirally reinforced compression members .... $\phi = 0.75$
For tied compression members .... $\phi = 0.70$
For bearing on concrete .... $\phi = 0.70$

Development lengths specified in Article 1.5.29 do not require a ϕ factor.

1.5.20—FLEXURE

(A) Rectangular sections with tension reinforcement only

For rectangular or flanged sections in which the neutral axis lies within the flange, the moment capacity shall be assumed as:

$$M_u = \phi[A_s f_y d (1 - 0.6q)] \quad (5\text{-}1)$$

$$= \phi[A_s f_y (d - \frac{a}{2})] \quad (5\text{-}2)$$

where

$$q = p \frac{f_y}{f'_c} \quad (5\text{-}3)$$

and

$$a = \frac{A_s f_y}{0.85 f'_c b} \quad (5\text{-}4)$$

The reinforcement ratio, p, shall not exceed 0.50 of the ratio, p_b, which produces balanced conditions at ultimate stage given by:

$$p_b = \frac{0.85\, k_1 f'_c}{f_y} \times \frac{87{,}000}{87{,}000 + f_y} \quad (5\text{-}5)$$

(B) I- and T-sections

(1) When the flange thickness equals or exceeds the depth to the neutral axis, a/k_1, the section may be designed by Equation (5-1), with, a, computed as for a rectangular beam with a width equal to the overall flange width given by Article 1.5.5(A).

(2) When the flange thickness is less than a/k_1 the design moment M shall not exceed that given by the moment capacity of the section assumed as

$$M_u = \phi [(A_s - A_{sf}) f_y (d - \frac{a}{2}) + A_{sf} f_y (d - 0.5t)] \quad (5\text{-}6)$$

where

$$A_{sf} = 0.85 (b - b') \frac{t f'_c}{f_y} \quad (5\text{-}7)$$

and

$$a = \frac{(A_s - A_{sf})\, f_y}{0.85 f'_c b'} \quad (5\text{-}8)$$

The reinforcement ratio, p_w shall not exceed 0.50 of the quantity $(p_b + p_f)$, where p_b is given by equation (5-5).

(C) Rectangular sections with compression reinforcement

The moment capacity of rectangular sections, or flanged sections in which the neutral axis lies within the flange, with compression reinforcement shall be assumed as:

$$M_u = \phi [(A_s - A'_s) f_y (d - \frac{a}{2}) + A'_s f_y (d - d')] \quad (5\text{-}9)$$

where

$$a = \frac{(A_s - A'_s)\, f_y}{0.85 f'_c b} \quad (5\text{-}10)$$

and the following condition shall exist

$$\frac{(A_s - A'_s)}{bd} \geq 0.85\, k_1 \frac{f'_c d'}{f_y d} \frac{87{,}000}{87{,}000 - f_y} \quad (5\text{-}11)$$

When the value of

$$\frac{(A_s - A'_s)}{bd}$$

is less than the value given by Equation (5-11), so that the compression steel stress is less than the yield strength, f_y, or when effects of compression steel are neglected, the calculated moment capacity shall not exceed that given by Equations (5-1) and (5-2), except when a general analysis is made on the basis of the assumptions given in Article 1.5.18(A). The quantity

$$\frac{(A_s - A_s')}{bd} = (p - p')$$

shall not exceed 0.50 of the value of p_b, given by Equation (5-5).

(D) Other cross sections

(1) For other cross sections and for cases of nonsymmetrical bending, the moment capacity, $M_u = \phi M_{uo}$, shall be computed by a general analysis based on the assumptions given in Article 1.5.18(A).

(2) The amount of tension reinforcement shall be so limited that the steel ratio, p, does not exceed 50 percent of that corresponding to balanced conditions as defined by Article 1.5.20(A).

(3) The moment capacity of the reinforced section, when cracked, shall be at least 1.5 times the moment which produces cracking of the transformed, uncracked section. This requirement, which limits the minimum tension reinforcement to be provided in the section shall apply to all sections of Article 1.5.20. The modulus of rupture of the concrete shall be used for calculating the resisting moment of the uncracked section.

1.5.21—SHEAR

(A) Shear stress

(1) The nominal design shear stress in reinforced concrete members shall be computed by:

$$v_u = \frac{V_u}{bd} \qquad (5\text{-}12)$$

For design, the maximum design shear V_u shall be considered as that at the section a distance, d, from the face of the support. Wherever applicable, effects of torsion shall be added and effects of inclined flexural compression in variable depth members shall be included.

(2) For beams of I- and T-sections, b' shall be substituted for b in Equation (5-12).

(3) The shear stress capacity of the concrete, v_{uc}, shall not exceed $2\phi\sqrt{f'_c}$ at a distance, d, from the face of the support.[3] If the

---

[3] More detailed calculation of the allowable shear stresses should be made for members subject to axial tension or compression. For these conditions and for members of lightweight concrete, refer to the ACI Building Code for concrete shear capacity formulas.

reinforcement ratio p is less than 1.2 percent then the shear stress capacity of the concrete shall be governed by

$$v_{uc} = (0.8 + 100p)\,\phi\sqrt{f'_c}$$

The design shear stress at sections between the face of the support and the section at a distance d therefrom, shall not be considered critical.

(B) Shear reinforcement

(1) Wherever the value of the design shear stress, v_u, computed by Equation (5-12) plus effects of torsion, exceeds the shear stress capacity, v_{uc}, permitted by Article 1.5.21(A)(3), shear reinforcement shall be provided to carry the excess. Such shear reinforcement shall also be provided for a distance equal to the depth, d, of the member beyond the point theoretically required. Shear reinforcement between the face of the support and the section at a distance, d, therefrom shall be the same as required at the section.

(2) When shear reinforcement perpendicular to the longitudinal axis is used, the required area of shear reinforcement shall be computed by:

$$A_v = \frac{(v_u - v_{uc})\,bs}{\phi f_y} \qquad (5\text{-}13)$$

(C) Stress restrictions

(1) The design yield strength for shear reinforcement shall not exceed 60,000 psi.

(2) The shear stress $v_u = v_{uc} + v_{us}$ shall not exceed $10\phi\sqrt{f'_c}$ in sections with shear reinforcement.

(3) In those areas subject to stress reversals caused by a single passage of the live load plus impact, at service load level, the range of tensile stress in the shear reinforcement shall be limited as in accordance with Article 1.5.25(B). The shear stress capacity of the concrete, v_{uc}, shall be zero and the design shear, v_u, shall not exceed

$$10\phi\sqrt{\frac{f'_c}{2}}$$

(D) Shear reinforcement restrictions

(1) Where shear reinforcement is required and is placed perpendicular to the axis of the member, it shall be spaced not further apart than 0.50d but not more than 24 in. Inclined stirrups and bent bars shall be so spaced that every 45 degree line, extending toward the reaction from the mid-depth of the member, 0.50d, to the longitudinal tension bars, shall be crossed by at least one line of shear reinforcement. When the design shear stress, v_u, exceeds $6\phi\sqrt{f'_c}$ the maximum spacings given above shall be reduced by one-half.

(2) Where shear reinforcement is required, its area, A_v, shall not

be less than 0.15 percent of the area computed as the product of the width of the web and the spacing of the shear reinforcement along the longitudinal axis of the member. (i.e., $A_v \geq 0.0015b's$).

(E) Shear stress in slabs and footings

(1) The shear capacity of slabs and footings in the vicinity of concentrated loads or concentrated reactions is governed by the more severe of two conditions:

(a) The slab or footing acting essentially as a wide beam, with a critical section extending in a plane across the entire width and located at a distance d from the face of the concentrated load or reaction area. For this condition the slab or footing shall be designed in accordance with Article 1.5.21(A).

(b) Two-way action for the slab or footing, with a critical section perpendicular to the plane of the slab and located so that its periphery is a minimum and approaches no closer than d/2 to the periphery of the concentrated load or reaction area. For this condition the slab or footing shall be designed as specified in the remainder of this section.

(2) The periphery shear stress shall be computed by

$$v_u = \frac{V_u}{b_o d} \quad (5\text{-}14)$$

in which V_u and b_o are taken at the critical section specified in Article 1.5.21(E)(1)(b). The periphery shear stress, v_u, shall not exceed the shear stress capacity of the concrete $v_{uc} = 4\phi\sqrt{f'_c}$ unless shear reinforcement is provided in accordance with Article 1.5.21(E)(3), in which case, v_u, shall not exceed $6\phi\sqrt{f'_c}$.

(3) When v_u exceeds $v_{uc} = 4\phi\sqrt{f'_c}$ shear reinforcement shall be provided in accordance with Articles 1.5.21(B) to 1.5.21(D), except that the design yield strength, f_y, for the shear reinforcement shall be 50 percent of that prescribed in Article 1.5.16(B). Shear reinforcement consisting of bars, rods or wires shall not be considered effective in members having an effective depth of less than 10 inches.

1.5.22—COLUMNS

(A) General

(1) All columns shall be designed to resist the combined bending and axial loads that result from the various combinations of loads and forces given in Article 1.5.17. All members subjected to a compression load shall be designed for the eccentricity, e, corresponding to the maximum moment that can accompany this loading condition, but not less than 1-inch, or 0.05t for spirally reinforced compression members, or 0.10t for tied compression members, about either principal axis.

(2) The area of longitudinal reinforcement preferably shall not be less than 1 percent, nor more than 8 percent of the gross concrete area of the column section. In a column which, for any reason, has a larger cross-section than required by the loads and moments determined in accordance with the provisions of Article 1.5.17, the minimum amount of longitudinal steel specified above may be reduced provided that in no case shall less longitudinal steel be used than that required by the minimum sized column necessary to support the loads and moments defined above, designed with one percent of longitudinal steel.

(B) Column Section Capacities

(1) Concentric Loading

The axial load capacity of a column section subjected to pure compression, P_o, is:

$$P_o = \phi [0.85 \, f'_c \, (A_g - A_{st}) + f_y \, A_{st}] \tag{5-15}$$

The capacity modification factor, ϕ, shall be the appropriate value for tied or spiral columns given in Article 1.5.19. Concentric loading is a hypothetical loading condition since columns shall be designed for eccentricities at least as large as those given in Article 1.5.22(A)(1).

(2) Pure Flexure

The assumptions given in Article 1.5.18(A), or the equations for flexure given in Article 1.5.20, may be used to determine the capacity of the column section under the hypothetical loading condition of pure flexure. The section capacity is multiplied by the capacity modification factor, ϕ, for flexure.

(3) Combined Axial Load and Flexure

The determination of the column cross-section capacity shall be based on compatibility of stress and strain using the assumptions enumerated in Article 1.5.18(A). The axial load capacity and the moment capacity thus obtained for the section shall be multiplied by the appropriate capacity modification factor, ϕ. The value of ϕ may be increased linearly from the value for columns to the value for flexure as the axial design load, P_u, decreases from $0.10 f'_c A_g$ to zero.

(4) Balanced Conditions

Balanced conditions are defined in Article 1.5.18(A)(8). For balanced conditions, the axial load capacity, P_b, and the corresponding moment capacity, M_b, shall be computed with the coefficient ϕ

for compression members using the assumptions of Article 1.5.18(A) and assuming the neutral axis located at:

$$c_b = \left(\frac{e_u}{e_u + e_y}\right) d \qquad (5\text{-}16)$$

(5) Biaxial Loading

In lieu of making the general column section analysis described above for the case of non-circular columns subjected to bending in the direction of both principal axes, the following approximate expressions may be used:

$$P_{uxy} = \frac{1}{(1/P_{ux}) + (1/P_{uy}) - (1/P_o)} \text{ when } P_u \geq 0.1 f'_c A_g \qquad (5\text{-}17)$$

or,

$$\frac{M_x}{M_{ux}} + \frac{M_y}{M_{uy}} \leq 1 \text{ when } P_u < 0.1 f'_c A_g \qquad (5\text{-}18)$$

(C) Slenderness effects in columns [4]

The influence of slender columns on behavior of the structure may be taken into account by the following approximate procedures.

(1) The unsupported length, h, of a compression member shall be taken as the clear distance between slabs, girders or other members capable of providing lateral support for the compression member. Where capitals or haunches are present, the unsupported length shall be measured to the lower extremity of the capital or haunch in the plane considered.

(2) The radius of gyration, r, may be taken equal to 0.30 times the overall dimension in the direction in which stability is being considered for rectangular compression members and 0.25 times the diameter for circular compression members. For other shapes, r may be computed for the gross concrete section.

(3) For compression members braced against sidesway, the effective length factor, k, shall be taken as 1.0, unless an analysis shows that a lower value may be used. For compression members not braced against sidesway, the effective length factor, k, shall be determined with due consideration of cracking and reinforcement on relative stiffness, and shall be greater than 1.0.

(4) For compression members braced against sidesway, the effects of slenderness may be neglected when kh/r is less than $34 - 12\, M_1/M_2$. For compression members not braced against sidesway, the effects of slenderness may be neglected when kh/r is less than 22. For all compression members with kh/r greater than 100, a more exact analysis

---

[4] See "Design of Slender Concrete Columns" by James G. MacGregor, John E. Breen, and Edward O. Pfrang, Journal of the American Concrete Institute, January 1970, pp. 6-28 for a comprehensive discussion of these provisions for designing slender columns. This article includes nomographs for determining the effective length factor, k.

than that prescribed herein shall be made. M_1 = value of smaller end moment on compression member calculated from a conventional elastic frame analysis, positive if member is bent in single curvature, negative if bent in double curvature. M_2 = value of larger end moment on compression member calculated from a conventional elastic frame analysis, always positive.

(5) Compression members shall be designed using the design axial load from a conventional frame analysis and a magnified moment M defined by

$$M = FM_2 \qquad (5\text{-}19)$$

where

$$F = \frac{C_m}{1 - P_u/\phi P_c} \geq 1.0 \qquad (5\text{-}20)$$

and

$$P_c = \frac{\pi^2 \, EI}{(kh)^2} \qquad (5\text{-}21)$$

In lieu of a more precise calculation, EI may be taken either as

$$EI = \frac{E_c I_g/5 + E_s I_s}{1 + R_m} \qquad (5\text{-}22)$$

or conservatively

$$EI = \frac{E_c I_g/2.5}{1 + R_m} \qquad (5\text{-}23)$$

where R_m is the ratio of maximum design dead load moment to maximum design total load moment.

For members braced against sidesway and without transverse loads between supports, C_m may be taken as

$$C_m = 0.6 + 0.4 \, (M_1/M_2) \qquad (5\text{-}24)$$

but not less than 0.4.

For all other cases C_m shall be taken as 1.0.

(6) When a group of columns on one level comprise a bent, or when they are connected integrally to the same superstructure, and collectively resist the sidesway of the structure, the value of F shall be computed for the column group. P_u and P_c, then shall be taken as the summation of P_u and P_c for all the columns in the group. In designing each column in the group, F shall be taken as the larger of (a) the value computed for the group as a whole, or (b) the value computed for the individual column assuming its ends to be braced against sidesway.

(7) When compression members are subject to bending about both principal axes, the moment about each axis shall be amplified by F, computed from the corresponding conditions of restraint about that axis.

(8) When design of compression members is governed by the minimum eccentricities specified in Article 1.5.22(A)(1), M_2 in Equation (5-19) shall be based on the specified minimum eccentricity, with conditions of curvature determined by either of the following:

(a) When the actual computed eccentricities are less than the specified minimum, the computed end moments may be used to evaluate the conditions of curvature.
(b) If computations show that there is no eccentricity at both ends of the member, conditions of curvature shall be based on a ratio of M_1/M_2 equal to one.

(9) In structures which are not braced against sidesway, the flexural members shall be designed for the total magnified end moments of the compression members at the joint.

1.5.23 — BEARING

(A) Bearing stresses shall not exceed $0.85\phi f'_c$, except as provided below.

(B) When the supporting surface is wider than the loaded area on all sides, the permissible bearing stress on the loaded area may be multiplied by $\sqrt{A'_b/A_b}$, but not more than 2.

(C) When the supporting surface is sloped or stepped, A'_b may be taken as the area of the lower base of the largest frustrum of a right pyramid or cone contained wholly within the support and having for its upper base the loaded area, and having side slopes of 1 vertical to 2 or more horizontal.

1.5.24 — SERVICE LOAD REQUIREMENTS

(A) Service Load Stresses

(1) For investigation of service load stresses the straight-line theory of stress and strain in flexure shall be used and the following assumptions shall be made:

(a) A section plane before bending remains plane after bending; strains vary as the distance from the neutral axis.
(b) The stress-strain relation for concrete is a straight line under service loads. Service load stresses vary as the distance from the neutral axis except for deep beams.
(c) The steel takes all the tension due to flexure.
(d) The modular ratio, $n = E_s/E_c$, may be taken as the nearest whole number (but not less than 6).

(2) In doubly reinforced beams and slabs, an effective modular ratio of $2E_s/E_c$ shall be used to transform the compression reinforcement for stress computations.

1.5.25 — FATIGUE

(A) Concrete

The range of compressive stress in the concrete caused by a single passage of live load plus impact and centrifugal force, at service load level, shall be limited to $0.5f'_c$ at points of contraflexure,[5] and at sections where stress reversals occur.

(B) Reinforcement

The range of stress in straight reinforcement caused by a single passage of live load plus impact at service load level, shall be limited to 21,000 psi.[6] Bends in primary reinforcement shall be avoided at sections having a high range of stress.

1.5.26 — FLEXURAL STRESS LIMITATIONS

(A) General

(1) The steel stress range shall comply with the fatigue provisions of Article 1.5.25(B).[6]

(2) The maximum steel stress shall be 36,000 psi[7][8]

(B) Bridges exposed to corrosive environments without a waterproof deck protection system

(1) All primary negative moment reinforcement in continuous bridges shall be increased 10 percent beyond the amount required by the provisions of Article 1.5.20.[9]

1.5.27 — DEFLECTIONS

(A) Superstructure depth recommendations

As a means of controlling long-time deflections due to creep and shrinkage, it is recommended that superstructure depths for various bridge types be not less than the value obtained from the equations in Table 1.5.27. Depths less than these may be used when specific consideration is given to limiting long-time deflections by use of compression reinforcement or some other method.

---

[5] Concrete Roadway Slabs excluded.
[6] Applicable primarily to bridge deck slabs and short span slab bridges where the dead load to total load moment ratio is less than approximately 0.25.
[7] This stress limitation is intended for 60,000 psi yield point reinforcement. For bridges designed using 40,000 psi and 50,000 psi yield point reinforcement, the maximum stresses for very long span structures where the live load stresses become insignificant would be limited by the load factor equations in Section 1.5.17(B) to about 27,700 psi for 40,000 psi yield point reinforcement and to about 34,600 psi for 50,000 psi yield point reinforcement.
[8] Applicable to bridges using 60,000 psi yield point reinforcement with spans of 150 feet or more where the dead load to total load moment ratio is equal to or greater than 0.775.
[9] This provision does not apply to reinforcement of bridge deck slabs.

TABLE 1.5.27

Recommended Minimum Thickness* for Constant Depth Members**

(t_{min} in feet)

Slabs with main reinforcement parallel and transverse to traffic S = effective span in feet	$t_{min} = 0.33 + \dfrac{S}{30}$ but no less than 0.542 feet
Tee-Beams S = actual span in feet	$t_{min} = 0.5 + \dfrac{S}{18}$
Box Girders S = actual span in feet	$t_{min} = 0.5 + \dfrac{S}{20}$

* Recommended values for continuous spans; simple spans should have about 10 percent greater thickness.

** When variable depth members are used, table values may be adjusted to account for change in relative stiffness of positive and negative moment sections.

(B) Dead load deflections at falsework removal

Unless a more comprehensive analysis is made, immediate dead load deflections upon removal of falsework shall be computed by the usual methods of formulas for elastic deflections, using the modulus of elasticity for concrete specified in Article 1.5.16(A). The effective moment of inertia shall be taken as the following, but not greater than I_g:

$$I_{eff} = (M_{cr}/M_{max})^3 I_g + [1 - (M_{cr}/M_{max})^3] I_{cr} \qquad (5\text{-}25)$$

where

$$M_{cr} = f_r I_g / y_t$$

$$f_r = 7.5 \sqrt{f'_c}$$

M_{max} = Maximum dead load moment for section under consideration

For continuous spans, the effective moment of inertia may be taken as the average of the values obtained from Equation (5-25) for the critical positive and negative moment sections.

(C) Long-time deflections caused by dead loads, creep and shrinkage

Unless a more comprehensive analysis is made, for purposes of determining falsework camber, the dead load deflection computed in (B) may be multiplied by a factor chosen from Table 1.5.27A. The long-time deflections, thus calculated, might be expected to occur over a period of about three years.

TABLE 1.5.27A

	$A'_s = 0$	$A'_s = \frac{1}{2}A_s$	$A'_s = A_s$
Climate of high humidity	2.5	1.8	1.5
Climate of average humidity	3.0	2.2	1.8
Climate of low humidity	3.5	2.5	2.0

1.5.28 — OVERLOAD

Structures proportioned by this specification will sustain without damage the following overload:

Members designed for Group I loading $= D + 5/3\,(L + I)$
Members designed for Group IA loading $= D + 2.2\,(L + I)$

1.5.29 — DEVELOPMENT OF REINFORCEMENT

(A) General

(1) The calculated tension or compression in the reinforcement at each section shall be developed on each side of that section by embedment length or end anchorage or a combination thereof. For bars in tension, hooks may be used in developing the bars.

(2) Tension reinforcement may be anchored by bending it across the web and making it continuous with the reinforcement on the opposite face of the member, or anchoring it there.

(3) The critical sections for development of reinforcement in flexural members are at points of maximum stress and at points within the span where adjacent reinforcement terminates, or is bent. The provisions of Article 1.5.29(B)(2) must also be satisfied.

(4) Reinforcement shall extend beyond the point at which it is no longer required to resist flexure for a distance equal to the effective depth of the member or 12 bar diameters, whichever is greater, except at supports of simple spans and at the free end of cantilevers.

(5) Continuing reinforcement shall have an embedment length not less than the development length, L_d, beyond the point where bent or terminated tension reinforcement is no longer required to resist flexure.

(6) Flexural reinforcement shall not be terminated in a tension zone unless one of the following conditions is satisfied:

    (a) The shear at the cutoff point does not exceed two-thirds that permitted, including the shear strength of furnished web reinforcement.
    (b) Stirrup area in excess of that required for shear and torsion is provided along each terminated bar over a distance from the termination point equal to three-fourths the effective depth of the member. The excess

stirrups shall be proportioned such that their $(A_v/b's)f_y$ is not less than 60 psi. The resulting spacing, s, shall not exceed $d/8r_b$ where r_b is the ratio of the area of bars cut off to the total area of bars at the section.

(c) For #11 and smaller bars, the continuing bars provide double the area required for flexure at the cutoff point and the shear does not exceed three-fourths that permitted.

(B) Positive moment reinforcement

(1) At least one-third the positive moment reinforcement in simple members and one-fourth the positive moment reinforcement in continuous members shall extend along the same face of the member into the support, and in beams at least 6 in.

(2) When a flexural member is part of the primary lateral load resisting system, the required positive reinforcement of Article 1.5.29 (B)(1) extended into the support shall be anchored to develop its yield stress in tension at the face of the support.

(3) At simple supports and at points of inflection, positive moment tension reinforcement shall be limited to a diameter such that L_d computed for f_y by Article 1.5.29(E) does not exceed:

$$\frac{M_{u0}}{V_u} + L_a$$

M_{u0} is the computed flexural strength assuming all reinforcement at the section to be stressed to f_y. V_u is the maximum applied shear at the section. L_a at a support shall be the sum of the embedment length beyond the center of the support and the equivalent length of any furnished hook or mechanical anchorage. L_a at a point of inflection shall be limited to the effective depth of the member or 12D, whichever is greater. The value M_{u0}/V_u in the development length limitation may be increased 30 percent when the ends of the reinforcement are confined by a compressive reaction.

(C) Negative moment reinforcement

(1) Tension reinforcement in a continuous, restrained, or cantilever member, or in any member of a rigid frame, shall be anchored in or through the supporting member by embedment length, hooks, or mechanical anchorage.

(2) Negative moment reinforcement shall have an embedment length into the span as required by Articles 1.5.29(A)(1) and 1.5.29 (A)(4).

(3) At least one-third the total reinforcement provided for negative moment at the support shall have an embedment length beyond the point of inflection not less than the effective depth of the member, 12D, or one-sixteenth of the clear span, whichever is greater.

(D) Special members

Adequate end anchorage shall be provided for tension reinforcement in flexural members where reinforcement stress is not directly proportional to moment, such as: sloped, stepped, or tapered footings; brackets; deep beams; or members in which the tension reinforcement is not parallel to the compression face.

(E) Development length of deformed bars in tension

The development length, L_d in inches, of deformed bars in tension shall be computed as the product of the basic development length of (1) and the applicable modification factor or factors of (2) and (3), but L_d shall be not less than 12 in.

(1) The basic development length in inches shall be:

For #11 or smaller bars .... $0.04 a_s f_y / \sqrt{f'_c}$
but not less than .... $0.0004 D f_y$
For #14 bars .... $0.085 f_y / \sqrt{f'_c}$
For #18 bars .... $0.11 f_y / \sqrt{f'_c}$
For deformed wire .... $0.03 D f_y / \sqrt{f'_c}$

(2) For top reinforcement [10] the basic development length shall be multiplied by a factor of 1.4.

(3) The basic development length, modified by the appropriate requirement of (2) may be multiplied by the applicable factor or factors for :

Reinforcement being developed in a length under consideration and spaced laterally at least 6 in. on center and at least 3 in. from the side of the member 0.8

Reinforcement in a flexural member in excess of that required .... (A_s required/A_s provided)

Bars enclosed within a spiral which is not less than ¼ in. diameter and not more than 4 in. pitch 0.75

(F) Development length of deformed bars in compression

(1) The development length L_d for bars in compression shall be computed as $0.02 f_y D / \sqrt{f'_c}$ but shall not be less than $0.0003 f_y D$ or 8 in. Where excess bar area is provided, the L_d length may be reduced by the ratio of required area to area provided. The development length may be reduced 25 percent when the reinforcement is enclosed by spirals not less than ¼ in. in diameter and not more than 4 in. pitch.

---

[10] Top reinforcement is horizontal reinforcement so placed that more than 12 in. of concrete is cast in the member below the bar.

(G) Development length of bundled bars

The development length of each bar of bundled bars shall be that for the individual bar, increased by 20 percent for a three-bar bundle, and 33 percent for a four-bar bundle.

(H) Standard hooks in tension

(1) Standard hooks shall be considered to develop a tensile stress in bar reinforcement $f_h = K\sqrt{f'_c}$ where K is not greater than the values in Table 1.5.29.

TABLE 1.5.29

Bar Size	$f_y = 40$ ksi	$f_y = 60$ ksi	
		Top Bars	Bottom Bars
#3 to #5	360	540	540
#6	360	450	540
#7 to #9	360	360	540
#10	360	360	480
#11	360	360	420
#14	330	330	330
#18	220	220	220

(2) An equivalent embedment length L_e shall be computed using the provisions of Article 1.5.29(E)(1) by substituting f_h for f_y and L_e for L_d.

(3) Hooks shall not be considered effective in adding to the compressive resistance of reinforcement.

(I) Combination development length

Development length L_d may consist of a combination of the equivalent embedment length of a hook or mechanical anchorage plus additional embedment length of the reinforcement.

(J) Mechanical anchorage

Any mechanical device capable of developing the strength of the reinforcement without damage to the concrete may be used as anchorage.

(K) Anchorage of shear reinforcement

(1) Shear reinforcement shall be carried as close to the compression and tension surfaces of the member as cover requirements and the proximity of other steel will permit, and in any case the end of single leg, simple U-, or multiple U-stirrup, shall be anchored by one of the following means:
    (a) A standard hook plus an effective embedment of 0.5 L_d. The effective embedment of a stirrup leg shall be taken as the distance between the middepth of the member, d/2, and the start of the hook (point of tangency).

(b) Embedment above or below the middepth, d/2, of the beam on the compression side for a full development length L_d, but not less than 24 bar diameters.
(c) Bending around the longitudinal reinforcement through at least 180 deg. Hooking or bending stirrups around the longitudinal reinforcement shall be considered effective anchorage only when the stirrups make an angle of at least 45 deg. with deformed longitudinal bars.
(d) Between the anchored ends, each bend in the continuous portion of a transverse simple U- or multiple U-stirrup shall enclose a longitudinal bar.
(e) Pairs of U-stirrups or ties so placed as to form a closed unit shall be considered properly spliced when the laps are $1.7L_d$. In members at least 18 in. deep, such splices having $a_s f_y$ not more than 9000 lb per leg may be considered adequate if the legs extend the full available depth of the member.

Section 6 — PRESTRESSED CONCRETE

1.6.1 — GENERAL

The specifications of this section are intended for design of prestressed concrete bridge members. Members designed as reinforced concrete, except for a percentage of tensile steel stressed to improve service behavior, shall conform to the applicable specifications of Section 5.

Exceptionally long span or unusual structures require detailed consideration of effects which under this Section may have been assigned arbitrary values.

1.6.2 — NOTATION

A_s = area of non-prestressed tension reinforcement.
A'_s = area of compression reinforcement.
A^*_s = area of prestressing steel.
A_{sf} = steel area required to develop the ultimate compressive strength of the overhanging portions of the flange.
A_{sr} = steel area required to develop the ultimate compressive strength of the web of a flanged section.
A_v = area of web reinforcement.
b = width of flange of flanged member or width of rectangular member.
b' = width of a web of a flanged member.
CR_c = loss of prestress due to creep of concrete.
CR_s = loss of prestress due to relaxation of prestressing steel.
CR_{sp} = loss of prestress due to relaxation of post-tensioning steel.
D = effect of dead load.
D = nominal diameter of prestressing steel.

d = distance from extreme compressive fiber to centroid of the prestressing force.
ES = Elastic shortening loss.
e = base of Naperian logarithms.
f_{cd} = average concrete compressive stress at the c.g. of the prestressing steel under full dead load.
f_{cr} = average concrete stress at the c.g. of the prestressing steel at time of release.
f'_c = compressive strength of concrete at 28 days.
f'_{ci} = compressive strength of concrete at time of initial prestress.
Δf_s = total prestress loss, excluding friction.
f_{se} = effective steel prestress after losses.
f^*_{su} = average stress in prestressing steel at ultimate load.
f'_s = ultimate strength of prestressing steel.
f_{sy} = yield strength of non-prestressed conventional reinforcement in tension.
f'_y = yield strength of non-prestressed conventional reinforcement in compression.
f^*_y = yield point stress of prestressing steel.
I = moment of inertia about the centroid of the cross section.
I = impact load.
j = ratio of distance between centroid of compression and centroid of tension to the depth d.
K = friction wobble coefficient per foot of prestressing steel.
L = effect of design live load.
L = length of prestressing steel element from jack end to point x.
M_u = ultimate flexural strength.
p = A_s/bd, ratio of non-prestressed tension reinforcement.
$p^* = A^*_s$/bd, ratio of prestressing steel.
$p' = A'_s$/bd, ratio of compression reinforcement.
Q = statical moment of cross sectional area, above or below the level being investigated for shear, about the centroid.
SH = concrete shrinkage loss.
s = longitudinal spacing of the web reinforcement.
t = average thickness of the flange of a flanged member.
T_o = steel stress at jacking end.
T_x = steel stress at any point x.
v = ultimate horizontal shear stress.
V_c = shear carried by concrete.
V_u = shear due to ultimate load and effect of prestressing.
μ = friction curvature coefficient.
α = total angular change of prestressing steel profile in radians from jacking end to point x.

1.6.3 — DESIGN THEORY

Members shall meet the ultimate strength and allowable stress requirements as specified.

Design shall be based on ultimate strength and behavior at service conditions for all load stages that may be critical during the life of the structure from the time of prestressing.

1.6.4 — BASIC ASSUMPTIONS

The following assumptions are made for design purposes:
(1) Strains vary linearly over the depth of the member throughout the entire load range.
(2) Before cracking, stress is linearly proportional to strain.
(3) After cracking, tension in the concrete is neglected.

1.6.5 — LOAD FACTORS

Load factors are multiples of the design load applied to the structure to ensure its safety. The computed ultimate capacity shall not be less than the largest value obtained from formulas 6.1, 6.2, 6.3 and 6.4. Members subject to combinations of loads and forces shall be designed for the combined effect.

$$\text{Group I} = \frac{1.30}{\phi} \times \left[D + \frac{5}{3}(L + I) \right] \qquad (6\text{-}1)$$

For all loadings less than H2O, provision shall be made for an infrequent heavy load by applying Group IA loading, with the live load assumed to occupy a single lane without concurrent loading in any other lane.

$$\text{Group IA} = \frac{1.30}{\phi} \times [D + 2.2\ (L+I)] \qquad (6\text{-}2)$$

$$\text{Group II} = \frac{1.30}{\phi} \times [D + W + F + SF + B + S + T] \qquad (6\text{-}3)$$

When earthquake loading is taken into account, Group II loading shall be used substituting EQ for W. When ice pressure is taken into account, Group II loading shall be used substituting ICE for SF.

$$\text{Group III} = \frac{1.30}{\phi} \times [D + (L+I) + CF + 0.3W + WL + F + LF] \qquad (6\text{-}4)$$

Except for the ϕ factors listed below, the symbols in the above formulas represent the moments, shears or forces caused by the loads and the effects described in Article 1.2.22.

ϕ = Factor on section strength =

1.0 for factory produced precast prestressed concrete members
0.95 for post-tensioned cast-in-place concrete members
0.90 for shear

1.6.6 — ALLOWABLE STRESSES

The design of precast prestressed members ordinarily shall be based on $f'_c = 5000$ psi. An increase to 6000 psi is permissible where, in the

Engineer's judgment, it is reasonable to expect that this strength will be obtained consistently. Still higher concrete strengths may be considered on an individual area basis. In such cases, the Engineer shall satisfy himself completely that the controls over materials and fabrication procedures will provide the required strengths. The provisions of this Section are equally applicable to prestressed concrete structures or components designed with lower concrete strengths.

(A) Prestressing steel

Temporary stress before loss due to creep and
shrinkage .... $0.70f'_s$
Stress at service load * after losses .... $0.80f^*_y$

(Overstressing to $0.80f'_s$ for short periods of time may be permitted provided the stress, after transfer to concrete in pretensioning or seating of anchorage in post-tensioning, does not exceed $0.70f'_s$).

(B) Concrete

(1) Temporary stresses before losses due to creep and shrinkage:

Compression

    Pretensioned members .... $0.60f'_{ci}$
    Post-tensioned members .... $0.55f'_{ci}$

Tension

    Precompressed tensile zone ..... No temporary allowable stresses are specified. See Article 1.6.6(B)(2) for allowable stresses after losses.

    Other Areas

        In tension areas with no bonded reinforcement .... 200 psi or $3\sqrt{f'_{ci}}$
Where the calculated tensile stress exceeds this value, bonded reinforcement shall be provided to resist the total tension force in the concrete computed on the assumption of an uncracked section. The maximum tensile stress shall not exceed .... $7.5\sqrt{f'_{ci}}$

(2) Stress at service load after losses have occurred:

Compression .... $0.40f'_c$
Tension in the precompressed tensile zone
  (a) For members with bonded reinforcement .... $6\sqrt{f'_c}$
     For severe corrosive exposure conditions, such as coastal areas .... $3\sqrt{f'_c}$
  (b) For members wihout bonded reinforcement 0

---

* Service load consists of all loads contained in Article 1.2.1 but does not include overload provisions.

Tension in other areas is limited by the allowable temporary stresses specified in Article 1.6.6(B)(1).

(3) Cracking Stress *

Modulus of rupture from tests or if not available:
For normal weight concrete ....7.5$\sqrt{f'_c}$
For sand-lightweight concrete ....6.3$\sqrt{f'_c}$
For all other lightweight concrete ....5.5$\sqrt{f'_c}$

(4) Anchorage bearing stress:

Post-tensioned anchorage at service load............3000 psi
(but not to exceed 0.9 f'_{ci})

1.6.7 — LOSS OF PRESTRESS

(A) Friction Losses

Friction losses in post-tensioned steel shall be based on experimentally determined wobble and curvature coefficients, and shall be verified during stressing operations. The values of coefficients assumed for design, and the acceptable ranges of jacking forces and steel elongations shall be shown on the plans. These friction losses shall be calculated as follows:

$$T_o = T_x \times e^{(KL+\mu\alpha)}$$

When $(KL+\mu\alpha)$ is not greater than 0.3, the following equation may be used:

$$T_o = T_x \times (1 + KL + \mu\alpha)$$

The following values for K and μ may be used when experimental data for the materials used are not available:

Type of Steel	Type of Duct	K	μ
Wire or ungalvanized strand	Bright Metal Sheathing	0.0020	0.30
	Galvanized Metal Sheathing	0.0015	0.25
	Greased or asphalt-coated and wrapped	0.0020	0.30
	Galvanized rigid	0.0002	0.25
High-strength bars	Bright Metal Sheathing	0.0003	0.20
	Galvanized Metal Sheathing	0.0002	0.15

Friction losses occur prior to anchoring but should be estimated for design and checked during stressing operations. Rigid ducts shall

---

* Refer to Article 1.6.10.

have sufficient strength to maintain their correct alignment without visible wobble during placement of concrete. Rigid ducts may be fabricated with either welded or interlocked seams. Galvanizing of the welded seam will not be required.

(B) Prestress Losses

Loss of prestress due to all causes, excluding friction, may be estimated from the following method. The method is based upon the use of 270 ksi, seven-wire, stress-relieved strand and normal-weight concrete. For data regarding the properties and effects of lightweight aggregates and low-relaxation tendons, refer to documented tests or see authorized suppliers.*

(1) Pretensioned

$$\Delta f_s = SH + ES + CR_c + CR_s$$

where: Δf_s = total prestress loss, excluding friction.

(a) Shrinkage

SH = concrete shrinkage loss computed using the following average values. Select the average ambient relative humidity for the geographic area.

Average Ambient Relative Humidity (percent)	SH (psi)
100-75	5,000
75-25	10,000
25- 0	15,000

(b) Elastic Shortening

$ES \cong 7\ f_{cr}$ = elastic shortening loss, where f_{cr} = average concrete stress at the center of gravity of the prestressing steel at time of release. ES may be estimated using the following average values:

Sections with Composite Deck Slab		Sections without Composite Deck Slab	
f_{cr} (psi)	ES (psi)	f_{cr} (psi)	ES (psi)
1,000	7,000	600	4,000
1,400	10,000	800	5,500
1,800	13,000	1,000	7,000

---

*Should more exact prestress losses be required, data representing the material to be used, the methods of curing, the ambient service condition and any pertinent structural detail should be determined for use in the method presented in "Deflections of Prestressed Concrete Members" reported by Subcommittee 5, ACI Committee 435, *Journal of the ACI*, Vol. 60, No. 12, December 1963.

(c) Creep of Concrete

$CR_c = 16 f_{cd}$ = loss due to creep of concrete, where f_{cd} = average concrete compressive stress at the center of gravity of the prestressing steel under full dead load. CR_c may be estimated using the following average values. These values apply to sections that are both with and without composite deck slabs.

f_{cd} (psi)	CR_c (psi)
500	8,000
800	13,000
1,200	19,000

(d) Relaxation of Prestressing Steel

$CR_s = 20,000 - 0.125 (SH + ES + CR_c)$ = loss due to relaxation of prestressing steel.

(2) Post-tensioned

$\Delta f_s = 0.8 (SH) + 0.5 (ES) + CR_c + CR_{sp}$
where $CR_{sp} \cong 20,000 - 0.125 [(0.8)(SH) + (0.5)(ES) + CR_c]$

1.6.8 — FLEXURE

Prestressed concrete members may be assumed to act as uncracked members subjected to combined axial and bending stresses within specified service loads.

In calculations of section properties, the transformed area of bonded reinforcement may be included in pretensioned members and in post-tensioned members after grouting; prior to bonding of tendons, areas of the open ducts shall be deducted.

1.6.9 — ULTIMATE FLEXURAL STRENGTH

(A) Rectangular Sections

For rectangular or flanged sections in which the neutral axis lies within the flange, the ultimate flexural strength shall be assumed as

$$M_u = A^*_s f^*_{su} d \left(1 - 0.6 \frac{p^* f^*_{su}}{f'_c} \right)$$

(B) Flanged Sections

If the neutral axis falls outside the flange (usually if the flange thickness is less than $1.4 d p^* f^*_{su} / f'_c$), the ultimate flexural strength shall be assumed as

$$M_u = A_{sr} f^*_{su} d \left(1 - 0.6 \frac{A_{sr} f^*_{su}}{b' d f'_c} \right) + 0.85 f'_c (b - b') t (d - 0.5t)$$

where

$A_{sr} = A^*_s - A_{sf}$ = the steel area required to develop the ultimate compressive strength of the web of a flanged section.

$A_{sf} = 0.85\,f'_c\,(b-b')\,t/f^*_{su}$ = steel area required to develop the ultimate compressive strength of the overhanging portions of the flange.

(C) Steel Stress

Unless the value of f^*_{su} can be more accurately known from detailed analysis, the following values may be used:

Bonded members........$f^*_{su} = f'_s \left(1 - 0.5\,\dfrac{p^* f'_s}{f'_c}\right)$

Unbonded members........$f^*_{su} = f_{se} + 15{,}000$

provided that:

(1) The stress-strain properties of the prestressing steel approximate those specified in Article 2.4.33 (J)

(2) The effective prestress after losses is not less than $0.5\,f'_s$.

1.6.10 — MAXIMUM AND MINIMUM STEEL PERCENTAGE

(A) Maximum Steel

Prestressed concrete members shall be designed so that the steel is yielding as ultimate capacity is approached. In general the reinforcement index shall be such that

$$p^* \dfrac{f^*_{su}}{f'_c} \qquad \text{for rectangular sections}$$

and

$$A_{sr} \dfrac{f^*_{su}}{b'df'_c} \qquad \text{for flanged sections}$$

does not exceed 0.30. For steel with reinforcement indices greater than this, the ultimate flexural strength shall be assumed not greater than:

$M_u = 0.25\,f'_c b d^2$ for rectangular sections, or

$M_u = 0.25\,b'd^2 f'_c + 0.85\,f'_c\,(b-b')\,t\,(d-0.5t)$ for flanged sections.

(B) Minimum Steel

The total amount of prestressed and non-prestressed reinforcement shall be adequate to develop an ultimate load in flexure at the critical section at least 1.2 times the cracking load calculated on the basis of the modulus of rupture (refer to Article 1.6.6 (B) (3).

1.6.11 — NONPRESTRESSED REINFORCEMENT

Nonprestressed reinforcement may be considered as contributing to the tensile strength of the beam at ultimate strength in an amount equal to its area times its yield point, provided that

$$\frac{pf_{sy}}{f'_c}+\frac{p^*f^*_{su}}{f'_c}-\frac{p'f'_y}{f'_c}\quad\text{does not exceed 0.3 for rectangular sections, or}$$

$$\frac{A_sf_{sy}}{b'df'_c}+\frac{A_{sr}f^*_{su}}{b'df'_c}-\frac{A'_sf'_y}{b'df'_c}\quad\text{does not exceed 0.3 for flanged sections.}$$

1.6.12 — CONTINUITY

(A) General

Continuous beams and other statically indeterminate structures shall be designed for adequate strength and satisfactory behavior. Behavior shall be determined by elastic analysis, taking into account the reactions, moments, shear and axial forces produced by prestressing, the effects of temperature, creep, shrinkage, axial deformation, restraint of attached structural elements, and foundation settlement.

(B) Cast-in-place Post-Tensioned Bridges

The effect of secondary moments due to prestressing shall be included in stress calculations at working load, but shall be neglected in calculating ultimate strength.

(C) Bridges Composed of Simple-Span Precast Prestressed Girders Made Continuous

(1) General

When structural continuity is assumed in calculating live loads plus impact and composite dead load moments, the effects of creep and shrinkage shall be considered in the design of bridges incorporating simple span precast, prestressed girders and deck slabs continuous over two or more spans.

(2) Positive Moment Connection at Piers

Provision shall be made in the design for the positive moments that may develop in the negative moment region due to the combined effects of creep and shrinkage in the girders and deck slab, and due to the effects of live load plus impact in remote spans. Shrinkage and elastic shortening of the pier shall be considered when significant.

Non-prestressed positive moment connection reinforcement at piers may be designed at a working stress of 0.6 times the yield strength but not to exceed 36 ksi.

(3) Negative Moments

Negative moment reinforcement shall be proportioned by ultimate strength design with load factors in accordance with Article 1.6.5.

The effect of initial precompression due to prestress in the girders may be neglected in the negative moment calculation of ultimate strength if the maximum precompression stress is less than $0.4f'_c$ and the continuity reinforcement, p, in the deck slab is less than 0.015; where $p = A_s/bd$.

The ultimate negative resisting moment shall be calculated using the compressive strength of the girder concrete regardless of the strength of the diaphragm concrete.

(4) Compressive Stress in Girders at Piers at Service Loads

The compressive stress in ends of girders at piers resulting from addition of the effects of prestressing and negative live load bending shall not exceed 0.60 f'_c.

(5) Shear

In continuous bridges of this type, shear reinforcement shall be designed according to Article 1.6.13.

The horizontal shear connection between the cast-in-place slab and the precast girder shall be designed in accordance with Article 1.6.14.

1.6.13 — SHEAR *

Prestressed concrete members shall be reinforced for diagonal tension stresses. Shear reinforcement shall be placed perpendicular to the axis of the member. The area of web reinforcement shall be

$$A_v = \frac{(V_u - V_c)s}{2f_{sy}jd}$$

but not less than

$$A_v = 100 \, b's/f_{sy}$$

where f_{sy} shall not exceed 60,000 psi

$V_c = 0.06 f'_c b'jd$ but not more than 180 b'jd.

Web reinforcement may consist of:

(1) Stirrups perpendicular to the axis of the member

(2) Welded wire fabric with wire located perpendicular to the axis of the member

The spacing of web reinforcement shall not exceed three-fourths the depth of the member.

The critical sections for shear in simply supported beams will usually not be near the ends of the span where the shear is a maximum, but at some point away from the ends in a region of high moment.

For the design of web reinforcement in simply supported members carrying moving loads, it is recommended that shear be investigated only in the middle half of the span length. The web reinforcement required at the quarter points should be used throughout the outer quarters of the span.

---

* The method for design of web reinforcement presented in ACI 318-71 is an acceptable alternate but web reinforcement shall not be less than $A_v = 100 \, b's/f_{sy}$.

For continuous bridges whose individual spans consist of precast prestressed girders, web reinforcement shall be designed for the full length of interior spans and for the interior three-fourths of the exterior span.

1.6.14 — COMPOSITE STRUCTURES

(A) General

Composite structures in which the deck is assumed to act integrally with the beam shall be interconnected in accordance with (B) (C) and (D) of this Article to transfer shear along contact surfaces and to prevent separation of elements.

(B) Shear Transfer

Full transfer of the ultimate horizontal shear forces may be assumed when contact surfaces are clean and intentionally roughened, minimum vertical ties are provided in accordance with (D) of this Article, all stirrups are fully anchored into all intersecting components, and the web members are designed to resist the entire vertical shear. Otherwise, ultimate horizontal shear stress shall be calculated and limited according to (C) and (D) of this article.

(C) Shear Capacity

In lieu of the requirements of (B) of this article, ultimate horizontal shear stress may be computed by the formula $v = V_u Q / I b'$. To resist the computed shear stress, the following values of shear capacity shall be assumed at the contact surfaces:

When the minimum steel tie requirements of (D) of this Article are met .... 75 psi

When the minimum steel-tie requirements of (D) of this Article are met and the contact surfaces of the present elements are clean and artificially roughened................ 300 psi

In addition to the above values; for each percent of stirrup or vertical tie reinforcement crossing the joint in excess of the minimum requirements of (D) of this Article.............. 150 psi

(D) Vertical Ties

All web reinforcement shall extend into cast-in-place decks. The minimum total area of vertical ties per linear foot of span shall be not less than the area of two No. 3 bars spaced at 12 in. Web reinforcement may be used to satisfy the vertical tie requirement. The spacing of vertical ties shall not be greater than four times the average thickness of the composite flange and in no case greater than 24 in.

(E) Shrinkage Stresses

In structures with a cast-in-place slab on precast beams, the differential shrinkage tends to cause tensile stresses in the slab and in the bottom of the beams. Because the tensile shrinkage develops over an extended time period, the effect on the beams is reduced by creep. Differential shrinkage may influence the cracking load and the beam deflection profile. When these factors are particularly significant, the effect of differential shrinkage should be added to the effect of loads.

1.6.15 — ANCHORAGE ZONES

For beams with post-tensioning tendons, end blocks shall be used to distribute the concentrated prestressing forces at the anchorage. Where all tendons are pretensioned wires or 7-wire strand, the use of end blocks will not be required. End blocks shall have sufficient area to allow the spacing of the prestressing steel as specified in Article 1.6.16. Preferably, they shall be as wide as the narrower flange of the beam. They shall have a length at least equal to three-fourths of the depth of the beam and in any case 24 in. In post-tensioned members a closely spaced grid of both vertical and horizontal bars shall be placed near the face of the end block to resist bursting stresses. Amounts of steel in the end grid should follow recommendations of the supplier of the anchorage. Where such recommendations are not available the grid shall consist of at least #3 bars on 3-in. centers in each direction placed not more than 1½ in. from the inside face of the anchor bearing plate.

Closely spaced reinforcement shall be placed both vertically and horizontally throughout the length of the end block in accordance with accepted methods of end block stress analysis.

In pretensioned beams, vertical stirrups acting at a unit stress of 20,000 psi to resist at least 4 percent of the total prestressing force shall be placed within the distance of d/4 of the end of the beam, the end stirrups to be as close to the end of the beam as practicable. For at least the distance d from the end of the beam, nominal reinforcement shall be placed to enclose the prestressing steel in the bottom flange. For box girders, transverse reinforcement shall be provided and anchored by extending the leg into the web of the girder.

1.6.16 — COVER AND SPACING OF STEEL

(A) Minimum cover

The following minimum concrete cover shall be provided for prestressing and conventional steel:

(1) Prestressing steel and main reinforcement........1½ in.

(2) Slab Reinforcement

    (a) Top of slab ....1½ in.
        (1) When de-icers are used ....2 in.
    (b) Bottom of slab ....1 in.

(3) Stirrups and ties ....1 in.

When de-icer chemicals are used, drainage details shall dispose of de-icer solutions without constant contact with the prestressed girders. Where such contact cannot be avoided, or in locations where members are exposed to salt water, salt spray or chemical vapor, additional cover should be provided.

(B) Minimum Spacing

The minimum clear spacing of prestressing steel at the ends of beams shall be as follows:

Pretensioning steel: three times the diameter of the steel or 1½ the maximum size of the concrete aggregate, whichever is greater.

Post-tensioning ducts: 1½ in. or 1½ times the maximum size of the concrete aggregate, whichever is the greater.

(C) Bundling

When post-tensioning steel is draped or deflected, post-tensioning ducts may be bundled in groups of three maximum, provided that the spacing specified in (B) is maintained in the end three ft. of the member. Where pretensioning steel is bundled, the deflection points shall be investigated for secondary stresses and all bundling done in the middle third of the beam length.

(D) Size of Ducts

Ducts for prestressing steel when bars are used shall have a minimum inside diameter of ⅜-in. larger than the diameter of the bars to be used. When wire or strand is used, the area of the duct shall be at least 2½ times as large as the area of prestressing steel in the duct.

1.6.17 — POST-TENSIONING ANCHORAGES AND COUPLERS

Anchorages, couplers, and splices for post-tensioned reinforcement shall develop the required ultimate capacity of the tendons without exceeding anticipated slip. Couplers and splices shall be placed in areas approved by the Engineer and enclosed in housing long enough to permit the necessary movements.

Anchorage, end fittings and exposed tendons shall be permanently protected against corrosion.

Anchor fittings for unbonded tendons shall be capable of transferring to the concrete a load equal to the capacity of the tendon under both static and cyclic loading conditions.

1.6.18 — EMBEDMENT OF PRESTRESSED STRAND

Three or seven-wire pretensioning strand shall be bonded beyond the critical section for a development length (in inches) not less than

$$\left(f^*_{su} - \frac{2}{3} f_{se}\right) D,$$

where

D is the nominal diameter in inches, f^*_{su} and f_{se} are kips per sq. in. and the parenthetical expression is considered to be without units.

Investigation may be limited to those cross-sections nearest each end of the member which are required to develop their full ultimate capacity.

Where strand is debonded at the end of a member, the development length required above shall be doubled.

1.6.19 — CONCRETE STRENGTH AT STRESS TRANSFER

Unless otherwise specified, stress shall not be transferred to concrete until the compressive strength of the concrete as indicated by test cylinders, cured by methods identical with the curing of the members, is at least 4000 psi for pre-tensioned members and 3500 psi for post-tensioned members.

1.6.20 — BEARINGS

Bearing devices for prestressed concrete structures shall be designed in accordance with Articles 1.7.49 through 1.7.56 and Section 12.

1.6.21 — SPAN LENGTHS

The effective span lengths of slabs shall be as specified in Article 1.3.2.

The effective span lengths of simply supported beams shall not exceed the clear span plus the depth of the beam.

The span length of continuous or restrained floor slabs and beams shall be the clear distance between faces of support. Where fillets making an angle of 45 degrees or more with the axis of a continuous or restrained slab are built monolithic with the slab and support, the span shall be measured from the section where the combined depth of the slab and the fillet is at least one and one-half times the thickness of the slab. Maximum negative moments are to be considered as existing at the ends of the span, as above defined. No portion of the fillet shall be considered as adding to the effective depth.

1.6.22 — EXPANSION AND CONTRACTION

In all bridges, provisions shall be made in the design to resist thermal stresses induced, or means shall be provided for movement caused by temperature changes.

Movements not otherwise provided for, including shortening during stressing, shall be provided for by means of hinged columns, rockers, sliding plates, elastomeric pads or other devices.

1.6.23 — T-BEAMS

In beam and slab composite construction, the junction between flange slab and web beam shall meet the requirements of Article 1.6.14, if the slab is to be considered an integral part with the beam. In cast-in-place or precast T-beams, equally effective shear resistance shall be provided at the junction of slab and beam.

(A) Effective Flange Width

In beam and slab construction, using precast, prestressed beams and composite cast-in-place slabs, effective and adequate bond and shear resistance shall be provided at the junction of the beam and slab. The slab may then be considered as an integral part of the beam, but its assumed effective width as a T-beam flange shall not exceed the following:

(1) One-fourth of the span length of the beam
(2) The distance center-to-center of the beams
(3) Twelve times the least thickness of the slab plus the width of the girder web.

For beams having a flange on one side only, the effective overhanging flange width shall not exceed one-twelfth of the span length of the beam, nor six times the thickness of the slab, nor one-half the clear distance to the next beam.

For monolithic prestressed construction, with normal slab span and girder spacing, the effective flange width is the distance center-to-center of beams. For very short spans, or where girder spacing is excessive, analytical investigations shall be made to determine the anticipated width of flange acting with the beam.

(B) Construction Joints

When a construction joint is required between slab and beam of a cast-in-place T-beam, the joint shall meet the requirements of Article 1.6.14.

(C) Diaphragms

Diaphragms of precast or cast-in-place construction using prestressed or non-prestressed reinforcement are recommended at span ends. Intermediate diaphragms are not required in spans up to 40 ft.; are recommended at mid-span for spans from 40 ft. to 80 ft.; and are recommended at span third points for spans in excess of 80 ft.

(D) Isolated Beams

Isolated beams, in which the T-form is used only for the purpose of providing additional compression area, shall have a flange thickness not less than one-half the web width. For composite construction, flange width shall not exceed 4 times the web width. For monolithic prestressed construction, flange width shall not exceed 15 times the web width and shall be adequate for all design loads.

1.6.24 — BOX GIRDERS

(A) Lateral Distribution of Loads for Bending Moment *

(1) Interior Beams

The live load bending moment for each interior beam in a spread box beam superstructure shall be determined by applying to the beam the fraction (D.F.) of the wheel load (both front and rear) determined by the following equation:

$$\text{D.F.} = \frac{2N_L}{N_B} + k\frac{S}{L}$$

where N_L = number of design traffic lanes (as defined as N in Art. 1.2.6)

N_B = number of beams ($4 \leq N_B \leq 10$)

S = beam spacing, in feet ($6.75 \leq S \leq 11.00$)

L = span length, in feet

$k = 0.07W - N_L(0.10N_L - 0.26) - 0.20N_B - 0.12$

W = roadway width between curbs, in feet (defined as W_c in Art. 1.2.6) ($32 \leq W \leq 66$)

(2) Exterior Beams

The live load bending moment in the exterior beams shall be determined by applying to the beams the reaction of the wheel loads obtained by assuming the flooring to act as a simple span (of length S) between beams, but shall not be less than $2N_L/N_B$.

(B) Effective Compression Flange Width

In girder and flange construction, consisting of a stem with top and bottom slab, effective and adequate bond and shear resistance shall be provided at the juncture of the girder and the slab. The slab may then be considered an intergral part of the girder, but its effective width as a girder flange shall not exceed the following:

(1) One fourth of the span length of the girder
(2) The distance center-to-center of girders
(3) Twelve times the least thickness of the slab plus the width of the girder web.

For girders having a flange on one side only, the effective overhanging width shall not exceed the following:

(1) One twelfth of the span length of the girder
(2) One half of the clear distance to the next girder
(3) Six times the least thickness of the slab

---

* The provisions of Article 1.2.9, Reduction in Load Intensity, where not applied in the development of the provisions presented in (1) and (2) above.

(C) Flange Thickness

(1) Top Flange

The minimum flange thickness shall be 1/16 of the clear distance between girders or 6 in., whichever is greater, except the minimum thickness may be reduced for factory produced precast elements to 5½ in.

(2) Bottom Flange

The maximum thickness of the bottom flange shall be determined by maximum allowable unit stresses as specified in Article 1.6.6 but in no case shall be less than 1/16 of the clear span between girders or 5½ in., whichever is the greater, except the minimum thickness may be reduced for factory produced precast elements to 5 in. Adequate fillets shall be provided at the intersections of all surfaces within the cell of a box girder, except at the junction of web and bottom flange where none are required.

(D) Minimum Bar Reinforcement for Cast-In-Place Post-Tensioned Box Girders

(1) Top Flange

The minimum top flange reinforcement shall be the same as for reinforced concrete box girders.

(2) Bottom Flange

The minimum bottom flange reinforcement shall be the same as for reinforced concrete box girders except the minimum reinforcement shall be 0.3 percent of the flange section.

(E) Shear

The horizontal shearing unit stress at the junction of the flange and the monolithic fillet joining it to the girder web shall not exceed $0.15\ f'_c$.

Changes in girder stem thickness shall be tapered for a minimum distance of 12 times the difference in web thickness.

(F) Diaphragms

Diaphragms or spreaders within the precast box beams shall be placed at midspan for spans up to 50 ft.; at third points for spans for 50 to 75 ft.; and at quarter points for spans over 75 ft.

Diaphragms or spreaders shall be placed between the girders at intervals not to exceed 80 ft. Diaphragm spacing for curved girders shall be given special consideration.

Section 7—STRUCTURAL STEEL DESIGN

1.7.1 — ALLOWABLE STRESSES (see Table 1.7.1)

Stresses are shown in pounds per square inch.

The modulus of elasticity of all grades of steel shall be assumed to be 29,000,000 psi and the coefficient of linear expansion 0.0000065 per degree Fahrenheit.

Table 1.7.1

AASHO Designation (ASTM Designation)	Structural Carbon Steel M183(A36)	M188(A441)	High Strength Low		
Thickness of Plates	Up to 8″ Incl. (5)	Over 4″ To 8″ Incl. (5)	M161(A242), M187(A440), M188(A441), M222(A588) (7)		
			M161(A242), M187(A440), M188(A441)		
			Over 1½″ To 4″ Incl.	Over ¾″ To 1½″ Incl.	¾″ and Under
			M222(A588)		
			Over 5″ to 8″ Incl. (5)	Over 4″ to 5″ Incl. (5)	Up to 4″ Incl.
Shapes (6)	All Groups (5)	Not Applicable	M161(A242), M187(A440), M188(A441)		
			Groups 4, 5 (5)	Group 3	Groups 1, 2
			M222(A588)		
			Not Applicable	Group 5 (5)	Groups 1, 2, 3, 4
Minimum tensile strength F_u	58,000	60,000	63,000	67,000	70,000
Minimum yield point or Minimum yield strength F_y	36,000	40,000	42,000	46,000	50,000
Axial tension in members with holes, net section, whichever is the smaller of $0.55F_y$ or $0.46F_u$	20,000	22,000	23,000	25,000	27,000
Axial tension in members without holes. Tension in extreme fiber of rolled shapes, girders and built-up sections subject to bending, net section Axial compression, gross section: stiffeners of plate girders. Compression in splice material, gross section $0.55F_y$	20,000	22,000	23,000	25,000	27,000
Compression in extreme fibers of rolled shapes, girders and built-up sections subject to bending, gross section, when compression flange is: (A) Supported laterally its full length by embedment in concrete $0.55F_y$ (B) Partially supported or is unsupported with $\frac{l}{b}$ not greater than (1) (2)............................	20,000 36	22,000	23,000 34	25,000 32	27,000 30
$F_b = 0.55F_y \left[1 - \frac{\left(\frac{l}{r'}\right)^2 F_y}{4\pi^2 E}\right]$ with $(r')^2 = \frac{b^2}{12}$	$=20,000-$ $7.5\left(\frac{l}{b}\right)^2$	....	$23,000-$ $10.2\left(\frac{l}{b}\right)^2$	$25,000-$ $12.2\left(\frac{l}{b}\right)^2$	$27,000-$ $14.4\left(\frac{l}{b}\right)^2$
Compression in concentrically loaded columns with $\frac{L'}{r}$ not greater than (3) (A) Riveted ends	130	....	125	125	125
$F_a = \frac{0.55F_y}{1.25}\left[1 - \frac{\left(\frac{.75L'}{r}\right)^2 F_y}{4\pi^2 E}\right]$ (B) Pinned ends	$=16,000-$ $0.30\left(\frac{L'}{r}\right)^2$	....	$18,000-$ $0.39\left(\frac{L'}{r}\right)^2$	$20,000-$ $0.46\left(\frac{L'}{r}\right)^2$	$22,000-$ $0.56\left(\frac{L'}{r}\right)^2$
$F_a = \frac{0.55F_y}{1.25}\left[1 - \frac{\left(\frac{.875L'}{r}\right)^2 F_y}{4\pi^2 E}\right]$	$=16,000-$ $0.38\left(\frac{L'}{r}\right)^2$	....	$18,000-$ $0.52\left(\frac{L'}{r}\right)^2$	$20,000-$ $0.62\left(\frac{L'}{r}\right)^2$	$22,000-$ $0.74\left(\frac{L'}{r}\right)^2$
Shear in girder webs, gross section. $F_v = 0.33F_y$	12,000	....	14,000	15,000	17,000
Bearing on milled stiffeners and other steel parts in contact (rivets and bolts excluded) $0.80F_y$	29,000	32,000	34,000	37,000	40,000
Stress in extreme fiber or pins (4) $0.80F_y$	29,000	....	34,000	37,000	40,000
Shear in pins $F_v = 0.40F_y$	14,000	....	17,000	18,000	20,000
Bearing on pins not subject to rotation (10) $0.80F_y$	29,000	32,000	34,000	37,000	40,000
Bearing on pins subject to rotation (such as used in rockers and hinges) $0.40F_y$	14,000	16,000	17,000	18,000	20,000
Bearing on power-driven rivets and high strength bolts (or as limited by the allowable bearing on the fasteners) $1.22F_y$	44,000	48,000	50,000	56,000	60,000

See page 114 for footnotes.

Alloy Structural Steel M223(A572) (8)						High Yield Strength Quenched & Tempered Alloy Steel (A514/A517)(9)	
Up to 4" Incl.	Up to 1½" Incl.			Up to 1" Incl.	Up to ½" Incl.	Over 2½" To 4" Incl.	Up to 2½" Incl.
	Shapes thru 426#/Ft.			Groups 1, 2	Group 1	Not applicable	
60,000	60,000	65,000	70,000	75,000	80,000	105,000	115,000
42,000	45,000	50,000	55,000	60,000	65,000	90,000	100,000
23,000	25,000	27,000	30,000	33,000	36,000	48,000	53,000
23,000	25,000	27,000	30,000	33,000	36,000	49,000	55,000
23,000	25,000	27,000	30,000	33,000	36,000	49,000	55,000
34	32	30	29	28	27	23	22
$23{,}000 - 10.2\left(\frac{l}{b}\right)^2$	$25{,}000 - 12.2\left(\frac{l}{b}\right)^2$	$27{,}000 - 14.4\left(\frac{l}{b}\right)^2$	$30{,}000 - 17.3\left(\frac{l}{b}\right)^2$	$33{,}000 - 20.7\left(\frac{l}{b}\right)^2$	$36{,}000 - 24.5\left(\frac{l}{b}\right)^2$	$49{,}000 - 47\left(\frac{l}{b}\right)^2$	$55{,}000 - 58\left(\frac{l}{b}\right)^2$
125	125	125	120	115	110	90	85
$18{,}000 - 0.39\left(\frac{L'}{r}\right)^2$	$20{,}000 - 0.46\left(\frac{L'}{r}\right)^2$	$22{,}000 - 0.56\left(\frac{L'}{r}\right)^2$	$24{,}000 - 0.65\left(\frac{L'}{r}\right)^2$	$26{,}000 - 0.77\left(\frac{L'}{r}\right)^2$	$29{,}000 - 0.93\left(\frac{L'}{r}\right)^2$	$40{,}000 - 1.8\left(\frac{L'}{r}\right)^2$	$44{,}000 - 2.2\left(\frac{L'}{r}\right)^2$
$18{,}000 - 0.52\left(\frac{L'}{r}\right)^2$	$20{,}000 - 0.62\left(\frac{L'}{r}\right)^2$	$22{,}000 - 0.74\left(\frac{L'}{r}\right)^2$	$24{,}000 - 0.88\left(\frac{L'}{r}\right)^2$	$26{,}000 - 1.04\left(\frac{L'}{r}\right)^2$	$29{,}000 - 1.26\left(\frac{L'}{r}\right)^2$	$40{,}000 - 2.4\left(\frac{L'}{r}\right)^2$	$44{,}000 - 2.9\left(\frac{L'}{r}\right)^2$
14,000	15,000	17,000	18,000	20,000	22,000	30,000	33,000
34,000	37,000	40,000	44,000	48,000	52,000	72,000	80,000
34,000	37,000	40,000	44,000	48,000	52,000	72,000	80,000
17,000	18,000	20,000	22,000	24,000	26,000	36,000	40,000
34,000	37,000	40,000	44,000	48,000	52,000	72,000	80,000
17,000	18,000	20,000	22,000	24,000	26,000	36,000	40,000
50,000	56,000	60,000	67,000	73,000	79,000	105,000	115,000

FOOTNOTES FOR TABLE 1.7.1

(1) Continuous or cantilever beams or girders may be proportioned for negative moment at interior supports for an allowable unit stress 20 percent higher than permitted by this formula but in no case exceeding allowable unit stress for compression flange supported its full length. If cover plates are used, the allowable static stress at the point of theoretical cutoff shall be as determined by the formula.

(2) l = length, in inches, of unsupported flange between lateral connections, knee braces or other points of support. For continuous beams and girders, l may be taken as the distance from interior support to point of dead load contraflexure if this distance is less than designated above.

For cantilever beam and girders, l may be taken as twice the distance from the support to the end of the cantilever, if this distance is less than designated above.

r' = radius of gyration, in inches, of the compression flange about the axis in the plane of the web.
b = flange width, in inches.

(3) Compression in concentrically loaded columns having $\frac{L'}{r}$ values not greater than shown may be computed from these approximate formulae, or from the more exact formulae given in Appendix C.

L' = length of member, in inches.
r = least radius of gyration of member, in inches.

For compression members with values of $\frac{L'}{r}$ greater than those shown or of known eccentricity, see Appendix C.

(4) See also Article 1.7.4

(5) Limited to 4″ thickness for structural members other than bearing assembly components.

(6) Groups 1 and 2 include all shapes except those in Groups 3, 4 and 5.

Group 3 includes Angles, Zees and Tees over ¾-inch in thickness and the following wide flange shapes:

Nominal Depth, In. x Nominal Width, In.	Weight Per ft., Lb.
36 × 16½	all weights
33 × 15¾	all weights
14 × 16	142 to 211 incl.
12 × 12	120 to 190 incl.

Group 4 includes wide flange shapes having nominal depth of 14 inches, nominal width of 16 inches, and weight per foot of 219 to 550 pounds, inclusive.
Group 5 includes wide flange shapes having nominal depth of 14 inches, nominal width of 16 inches, and weight per foot of 605 to 730 pounds, inclusive.
For breakdown of Groups 1 and 2 see ASTM A6.

(7) M222(A588) is applicable for welded structures, with Supplementary Requirement S1, Impact Properties, (See AASHO M222).

(8) Applicable for welded structures in Grades 42, 45 and 50 in plate thicknesses through 1½″ and in shape groups 1, 2 and 3, with Supplementary Requirement S2, Impact Properties, (See AASHO M223).

(9) Quenched and tempered alloy steel structural shapes and seamless mechanical tubing meeting all mechanical and chemical requirements of ASTM A514/A517 steel, except that the specified maximum tensile strength may be 140,000 psi for structural shapes and 145,000 psi for seamless mechanical tubing, shall be considered as ASTM A514/A517 steel.

(10) This shall apply to pins used primarily in axially loaded members, such as truss members and cable adjusting links. It shall not apply to pins used in members having rotation caused by expansion or deflection.

1.7.2 — ALLOWABLE STRESSES FOR WELD METAL

Unless otherwise specified, the yield point and ultimate strength of weld metal shall be equal to or greater than minimum specified value of the base metal. Allowable stresses on the effective areas of weld metal shall be as follows:

Butt Welds—
> The same as the base metal joined, except in the case of joining metals of different yields when the lower yield material shall govern.

Fillet welds—
> $F_v = 12,400$ psi on base metal with minimum specified yield point or strength of 36,000 psi.
>
> $F_v = 14,700$ psi on base metal with minimum specified yield point or strength between 40,000 psi and 50,000 psi inclusive.
>
> $F_v = 25,000$ psi on base metal with a minimum specified yield strength between 90,000 psi and 100,000 psi inclusive. AWS A5.1, E 70 electrodes, or equivalent weld metal, may be used for fillet welds on A514/A517 steel; for such welds $F_v = 14,700$ psi.

Plug welds—
> $F_v = 12,400$ psi for resistance to shear stresses only,

where F_v = allowable basic shear stress.

1.7.3 — FATIGUE STRESSES

The number of cycles of maximum stress to be considered in the design shall be selected from Table 1.7.3A unless traffic and loadometer surveys or other considerations indicate otherwise.

TABLE.1.7.3(A) — Stress Cycles

Case	Type Of Road	Number of cycles of maximum stress to be used when the type * of load producing maximum stress is:		
		H Loading	HS Loading	Lane Loading
I	1. Freeways 2. Expressways 3. Major Highways & Streets	2,000,000	500,000	100,000
II	Other Highways & Streets not included in Case I	500,000	100,000	100,000

* Where HS & H loadings give the same maximum stress, use the number of cycles under H loadings.

The number of maximum stress cycles from wind loads shall be 100,000 unless conditions at a specific site or a dynamic analysis of a particular structure indicate otherwise.

In addition to meeting basic design criteria, members and fasteners subject to repeated variations or reversals of stress shall be designed for fatigue.

Allowable fatigue stress, F_r, shall apply to live load in combination with dead loads, wind loads only, and wind loads in combination with dead loads. Allowable fatigue stresses shall not be increased for combinde loadings. F_r is given by the following formula, except as noted in Table 1.7.3(B):

$$F_r = \frac{k_1 f_{ro}}{1 - k_2 R} \quad (A)$$

where R = algebraic ratio of the minimum stress to the maximum stress. This ratio may be expressed also in terms of moment, shear, axial force or torque.

$$k_1 = 1.0 + \alpha \left(\frac{F_u}{58,000} - 1 \right), \text{ but not less than } 1.0 \quad (B)$$

α and k_2 = values of coefficients given in Table 1.7.3(B)

f_{ro} = value given in Table 1.7.3(B), in psi.

F_u = minimum tensile strength, in psi, given in Article 1.7.1, or other appropriate minimum ultimate strength.

F_r = allowable fatigue stress—see Figures 1.7.3(A) and 1.7.3(B) for graphical presentations of the more commonly used values.

Butt welded splices which conform to all of the following conditions shall be designed for F_r as determined for "Base metal adjacent to continuous flange-web fillet welds":

(1) The parts joined are of equal thickness.
(2) The parts joined are of equal widths or if of unequal widths they shall be transitioned in accordance with Article 1.7.19.
(3) Weld soundness is established by radiographic inspection and the requirements of such inspection are specified.
(4) The weld is finished smooth and flush with the base metal on all surfaces by grinding in the direction of applied stress, leaving surfaces free from depressions. Chipping may be used providing it is followed by such grinding.

Butt welded splices which do not conform to all of the foregoing conditions shall be designed for F_r as determined for "Base Metal adjacent to butt welds" and shaped in accordance with Article 1.7.19.

Brackets, clips, gussets, stiffeners, and other detail material shall not be welded to members or parts subjected to tensile stress unless the maximum stress at the point of attachment does not exceed F_r for "Base Metal in members adjacent to or connected by fillet or plug welds."

Curves ① through ⑥ are plotted for M183(A36) steel. For other steels multiply the curve values F_r by the appropriate K_1 value determined from formula (B) Art. 1.7.3

Curves ⑦ ⑧ & ⑨ are plotted for all steels.

Fatigue stress, F_r, shall not exceed the basic allowable stress.

CATEGORY		CYCLES	FORMULA	CURVE NUMBER	TYPE OF STRESS
A	BASE METAL	100,000	$\frac{60,000}{1-R}$	⑦	TENSION
		500,000	$\frac{36,000}{1-R}$	⑧	
		2,000,000	$\frac{24,000}{1-R}$	⑨	
B	BASE METAL adjacent to Friction Type Fastener	100,000			TENSION
		500,000	$\frac{k_1(20,500)}{1-0.55R}$	①	
		2,000,000			
C	BASE METAL adjacent to Bearing Type Fastener	100,000	$\frac{k_1(20,500)}{1-0.55R}$	①	TENSION OR COMPRESSION
		500,000	$\frac{k_1(17,200)}{1-0.62R}$	②	
		2,000,000	$\frac{k_1(15,000)}{1-0.67R}$	③	
D	WELD METAL or BASE METAL adjacent to Butt Weld	100,000	$\frac{k_1(20,500)}{1-0.55R}$	①	TENSION
		500,000	$\frac{k_1(17,200)}{1-0.62R}$	②	
		2,000,000	$\frac{k_1(15,000)}{1-0.67R}$	③	
G	WELD METAL	100,000	$\frac{k_1(12,000)}{1-0.50R}$	④	SHEAR
		500,000	$\frac{k_1(10,800)}{1-0.55R}$	⑤	
		2,000,000	$\frac{k_1(9,000)}{1-0.62R}$	⑥	

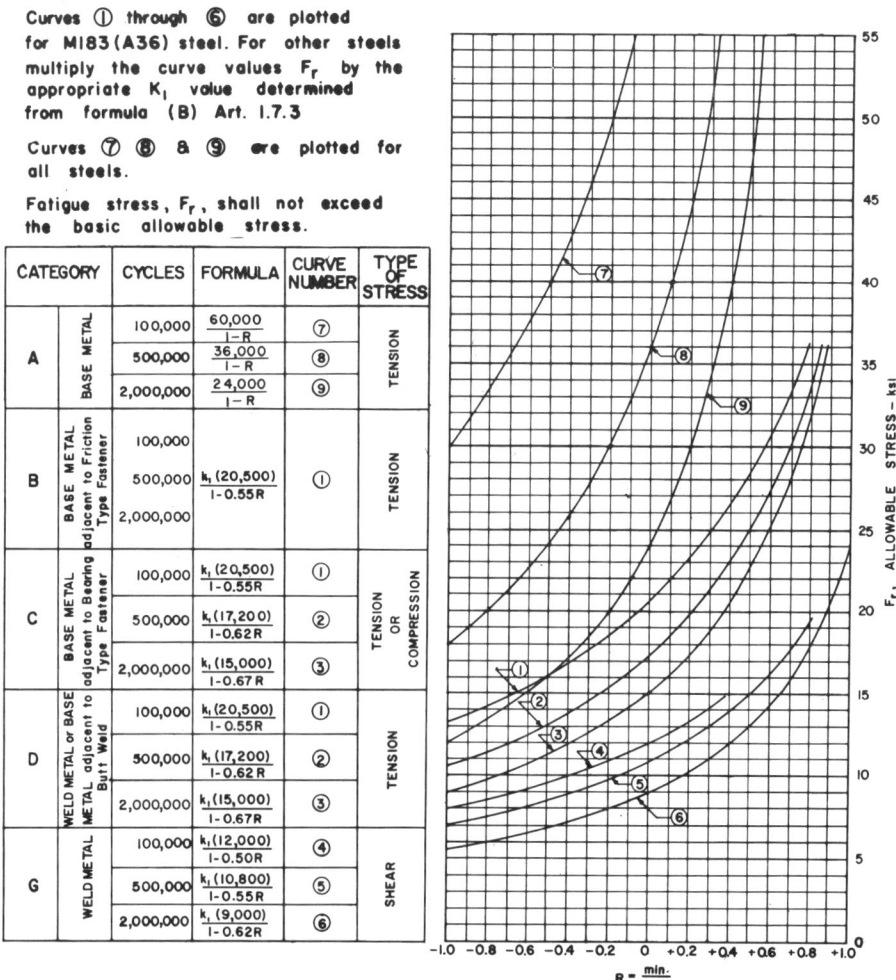

Figure 1.7.3A Fatigue Stresses

This restriction shall not apply to a compression flange at locations where stiffeners, studs, or fillet welded shear connectors only are welded.

For members with stud shear connectors attached to tension flanges the allowable stress shall not exceed F_r for "Flanges with stud shear connectors."

1.7.4 — PINS, ROLLERS AND EXPANSION ROCKERS

The effective bearing area of a pin shall be its diameter multiplied by the thickness of the material on which it bears. When parts in contact have different yield points, F_y shall be the smaller value.

FATIGUE STRESSES
Table 1.7.3B

Category	Type & Location of Material	Type of Maximum Stress	100,000 Cycles			500,000 Cycles			2,000,000 Cycles		
			f_{ro}	α	k_2	f_{ro}	α	k_2	f_{ro}	α	k_2
A	Base Metal	(4) Tension or Reversal	60,000	0	1.00	36,000	0	1.00	24,000	0	1.00
B	Base Metal adjacent to friction type fastener	(4) Tension (1) Compression	20,500 13,300	1.06 1.06	0.55 —	20,500 13,300	0.78 0.78	0.55 —	20,500 13,300	0.54 0.54	0.55 —
C	Base Metal adjacent to bearing type fastener	(4) Tension or Compression	20,500	0	0.55	17,200	0	0.62	15,000	0	0.67
D	Weld Metal or Base Metal(3) adjacent to Butt Weld	(4) Tension (1) Compression	20,500 13,300	0.65 0.65	0.55 —	17,200 10,600	0.23 0.23	0.62 —	15,000 9,000	0 0	0.67 —
E	Flanges with Stud Shear Connectors	Tension	20,500	1.06	0.55	16,500	0	0.65	11,500	0	0.75
F	Base Metal adjacent to or connected by fillet(2)(7) or plug welds	(5) Tension or Compression	21,000	0	1.00	12,500	0	1.00	8,000	0	1.00
G	Weld Metal	(4) Shear	12,000	0.78	0.50	10,800	0.36	0.55	9,000	0	0.62
H	Friction type fastener	Shear	F_v	0	0	F_v	0	0	F_v	0	0
I	Bearing type fastener	Shear	F_v	0	0.50	13,500	0	0.50	11,200	0	0.50
J	Base Metal(6) adjacent to continuous flange-web fillet weld in ASTM A514/A517 Steel	Tension or Compression	45,000	0	1.00	27,500	0	1.00	18,000	0	1.00
K	Base Metal adjacent to transverse stiffener in ASTM A514/A517 Steel	Tension (1) Compression	20,500 11,400	0.65 0.65	0.80 —	17,200 9,600	0.23 0.23	0.80 —	15,000 8,300	0 0	0.80 —

(continued on page 119)

FOOTNOTES TO TABLE 1.7.3B (continued from page 118.)

(1) Use the Formula:

$$F_r = \frac{0.55 F_y}{1 - \left(\dfrac{0.55 F_y}{k_1 f_{ro}} - 1\right) R}$$

(2) The usual continuous fillet welded flange-web connections and similar connections shall be governed by Category J, "Base Metal adjacent to continuous flange-web fillet welds."

(3) Base Metal adjacent to longitudinal butt welds and the weld metal in longitudinal butt welds shall be governed by Category J, "Base Metal adjacent to Continuous flange-web fillet welds."

(4) See graphs on figure 1.7.3A.
(5) See graphs on figure 1.7.3B.
(6) The Category G, "Weld Metal" in Table 1.7.3B does not apply to this case. Where the shear stress in the welds exceeds 15 ksi, $F_r^2 \geq F_b^2 + 3F_v^2$ in which F_b and F_v are the maximum bending and shear stresses in the weld and F_r is the allowable fatigue stress for Category J, "Base Metal Adjacent to continuous flange-web fillet welds." Intermittent fillet welds shall not be permitted.
(7) See Categories J and K for exceptions.

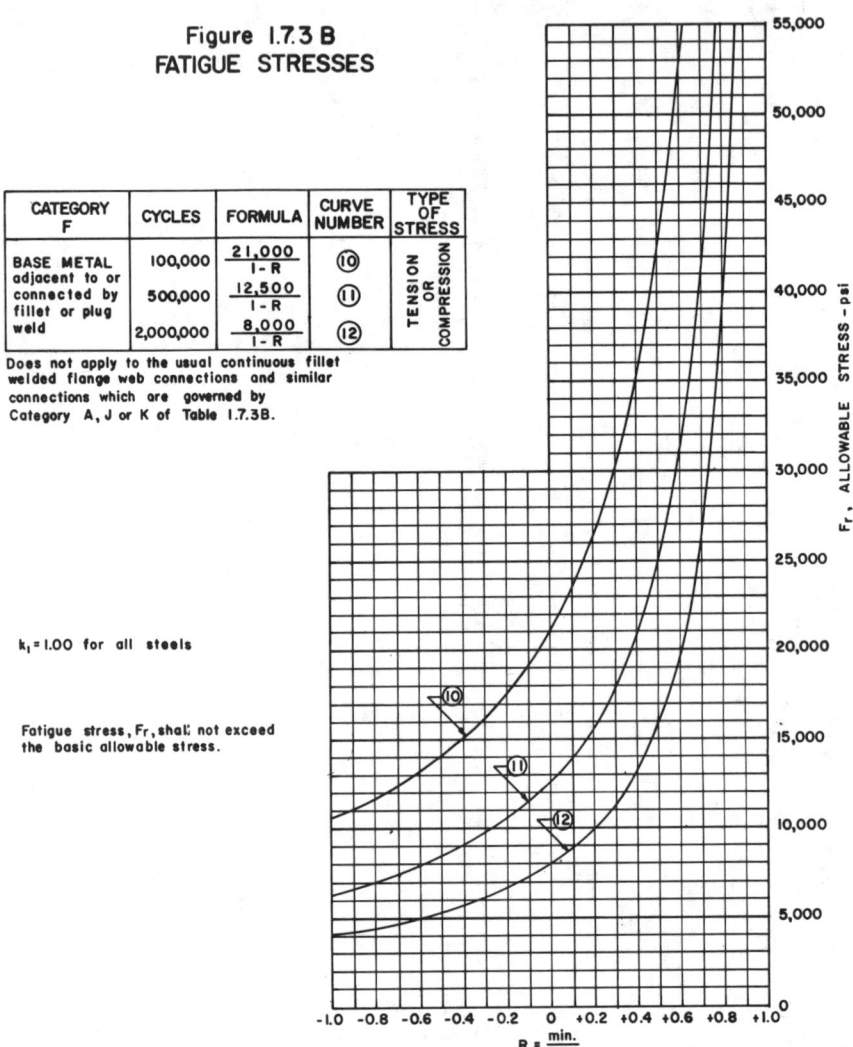

Figure 1.7.3 B
FATIGUE STRESSES

Bearing per linear inch on expansion rockers and rollers shall not exceed the values obtained by the following formulas:

Diameters up to 25 inches $p = \dfrac{F_y - 13{,}000}{20{,}000} \; 600d$

1.7.4 DESIGN

Diameters from 25 to 125 inches $p = \dfrac{F_y - 13{,}000}{20{,}000} \; 3{,}000 \; \sqrt{d}$ where:

p = allowable bearing in pounds per linear inch
d = diameter of rocker or roller in inches
F_y = minimum yield point in tension of steel in the roller or bearing plate, whichever is the smaller.

Steel may conform to one of the following designations in addition to the designations listed in Article 1.7.1:

Cold Finished Carbon Steel Bars and Shafting, AASHO, M-169 (ASTM A-108)

Carbon Steel Forgings for General Industrial Use, AASHO, M-102 (ASTM A-235)

Alloy Steel Forgings for General Industrial Use, ASTM A-237

Expansion rollers shall be not less than 4 inches in diameter

AASHO Designation with size limitations		M-169 4" in dia. or less	M-102 (To 20" in dia.)	M-102 (To 20" in dia.)	M-102 (To 10" in dia.)	None (To 20" in dia.)
ASTM Designation with Grade or Class		A-108 Grades 1016 to 1030 inc.	A-235 Class C1	A-235 Class E	A-235 Class G	**A-237 Class A
Minimum Yield Point, psi	F_y	36,000 *	33,000	37,500	50,000	50,000
Stress in Extreme Fiber, psi	0.80 F_y	29,000 *	26,000	30,000	40,000	40,000
Shear, psi	0.40 F_y	14,000 *	13,000	15,000	20,000	20,000
Bearing on pins not subject to rotation, psi ***	0.80 F_y	29,000 *	26,000	30,000	40,000	40,000
Bearing on pins subject to rotation, psi (Such as used in rockers and hinges)	0.40 F_y	14,000 *	13,000	15,000	20,000	20,000

* For design purpose only. Not a part of the A108 specifications. Supplementary material requirements should provide guarantee that material will meet these values.
** May substitute rolled material of the same properties.
*** This shall apply to pins used primarily in axially loaded members, such as truss members and cable adjusting links. It shall not apply to pins used in members having rotation caused by expansion or deflection.

1.7.5 — FASTENERS (RIVETS AND BOLTS)

In proportioning fasteners, the nominal diameter shall be used, except as otherwise noted.

The effective bearing area of a fastener shall be its diameter multiplied by the thickness of the metal on which it bears. In metal less than ⅜ inch thick, countersunk rivets, turned bolts or ribbed bolts shall not be assumed to carry stress. In metal ⅜ inch thick and over, one-half the depth of countersink shall be omitted in calculating the bearing area.

Allowable unit stresses in pounds per square inch for fasteners shall be as listed in the table below:

Type of Fastener	Tension	Bearing	Shear	
			Friction Type Connection	Bearing Type Connection
(A) *Low Carbon Steel Bolts*(1) Turned Bolts (ASTM A-307) and Ribbed Bolts	13,500 (2)	20,000		11,000
(B) *Power Driven Rivets* (rivets driven by pneumatically or electrically operated hammers are considered power driven). Structural Steel Rivet AASHO M228, grade 1 (ASTM A-502 grade 1)		40,000		13,500
Structural Steel Rivet (High Strength) AASHO M228, grade 2 (ASTM A-502 grade 2)		40,000		20,000
(C) *High Strength Bolts* High Strength Steel Bolts AASHO M-164 (ASTM A325)	36,000	40,000 (3)	13,500	20,000 (4)

High strength bolts may be substituted for Grade 1 rivets, AASHO M 228 (ASTM A-502).

All bolts except high strength bolts, shall have single self-locking nuts or double nuts.

Joints required to resist shear between their connected parts are designated as either friction type or bearing type connections. Shear connections subjected to stress reversal, or where slippage would be undesirable, shall be friction type.

Bolts in bearing type connections shall have the threads excluded from the shear planes of the contact surfaces between the connected parts. In determining whether the bolt threads are excluded from the shear planes of the contact surfaces, thread length of bolts shall be cal-

---

(1) ASTM A307 bolts shall not be used in connections subject to fatigue.
(2) Based on area at the root of thread.
(3) Does not apply to friction type connections.
(4) The allowable shear value of bolts for bearing type connections in steel with a yield point less than 42,000 psi shall be reduced by 20% when the end of the splice material is more than 24 inches from the end of the connected member, as measured along the gage line of the bolts.

culated as two thread lengths greater than the specified thread length as an allowance for thread run out.

In bearing type connections, pull-out shear in a plate should be investigated between the end of the plate and the end row of fasteners.

High strength bolts preferably shall be used for fasteners subject to computed tension or combined shear and computed tension.

For combined shear and tension in friction type joints, where applied forces reduce the total clamping force on the friction plane, the allowable unit shearing stress, f_v, for AASHO M164 (ASTM A325) high strength bolts shall not exceed the values obtained from the following equation:

$$f_v = 13{,}500 - .22 f_t$$

where f_t = tensile stress due to applied loads

When rivets or high-strength bolts in *bearing type* connections are subject to both shear and tension, the combined stress shall not exceed values obtained from the following equation:

$$s^2 + (kt)^2 = S^2$$

where s = the computed rivet or bolt unit stress in shear
 t = the computed rivet or bolt unit stress in tension
 S = the allowable rivet or bolt unit stress in shear
 k = a constant: 0.75 for rivets; 0.555 for A325 bolts with threads excluded from shear plane.

Bolted bearing type connections using high strength bolts shall be limited to members in compression and secondary members.

Where shown in the design drawings, enlarged or slotted holes may be used with high strength bolts proportioned to meet the allowable unit stresses given above except as hereinafter restricted:

1. Holes $3/16$ inch larger than bolts $7/8$ inch and less in diameter, $1/4$ inch larger than bolts 1 inch in diameter, and $5/16$ inch larger than bolts $1 1/8$ inch and greater in diameter may be used in uncoated friction type shear connections provided a hardened washer is inserted under both the head and nut.

2. Slotted holes $1/16$ inch wider than the bolt diameter and of a length more than allowed in subparagraph 1 but not more than $2 1/2$ times the bolt diameter may be used without regard to direction of loading in enclosed parts of friction-type shear connections if one-third more bolts are provided than needed to satisfy the design requirements. Only one of the inclosed parts adjacent to an individual faying surface may contain slotted holes.

3. When enlarged or slotted holes are used, the distances between edges of adjacent holes or edges of holes and edges of members

shall not be less than that permitted with conventional size holes under Art. 1.7.36.

1.7.6—CAST STEEL, DUCTILE IRON CASTINGS, MALLEABLE CASTINGS AND CAST IRON

(A) Cast Steel and Ductile Iron

For cast steel conforming to specifications for Steel Castings for Highway Bridges, AASHO M 192 (ASTM A 486), Mild-to Medium-Strength Carbon-Steel Castings for General Application AASHO M 103, (ASTM A 27) and Corrosion-Resistant Iron-Chromium-Nickel Alloy Castings for General Application, AASHO M 163 (ASTM A 296) and for Ductile Iron Castings, ASTM A 536 the following allowable stresses in pounds per square inch shall be used:

AASHO Designation	none	M192	M 192		M 163	None
ASTM Designation	A27	A486	A486		A 296	A536
Class or Grade	70-36	70	90	120	CA-15	60-40-18
Minimum yield point, F_y	36,000	60,000	95,000		65,000	40,000
Axial Tension	14,500	22,500	34,000		24,000	16,000
Tension in extreme fibers	14,500	22,500	34,000		24,000	16,000
Axial Compression, short columns	20,000	30,000	45,000		32,000	22,000
Compression in extreme fibers	20,000	30,000	45,000		32,000	22,000
Shear	9,000	13,500	21,000		14,000	10,000
Bearing, steel parts in contact	30,000	45,000	68,000		48,000	33,000
Bearing on pins not subject to rotation	26,000	40,000	60,000		43,000	28,000
Bearing on pins subject to rotation (such as used in rockers and hinges)	13,000	20,000	30,000		21,500	14,000

When in contact with castings or steel of a different yield point, the allowable unit bearing stress of the material with the lower yield point shall govern. For riveted or bolted connections, Art. 1.7.5 shall govern.

(B) Malleable Castings

For malleable castings conforming to specifications for Malleable Iron Castings, AASHO M 106, (ASTM A 47) grade No. 35018, the following allowable stresses in pounds per square inch, shall be used:

Tension 18,000
Bending in extreme fiber......... 18,000
Modulus of elasticity............ 25,000,000

(C) Cast Iron

For cast iron castings conforming to specifications for Gray Iron Castings, AASHO M 105, Class 30, the following allowable stresses in pounds per square inch, shall be used:

Bending in extreme fiber	3,000
Shear	3,000
Direct compression, short columns	12,000

1.7.7 — BRONZE OR COPPER-ALLOY

Bronze castings, AASHO M 107 (ASTM B 22) Alloys A or B or Copper-alloy Plates, AASHO M108 (ASTM B 100), shall be specified.

The allowable unit bearing stress in pounds per square inch on Bronze castings or Copper-alloy plates shall be 2,000.

1.7.8 — BEARING ON MASONRY

The allowable unit bearing stress in pounds per square inch, on the following types of masonry, shall be:

Granite	800
Sandstone and Limestone	400

Concrete:
- Bridge seats, under hinged rockers and bolsters (not subjected to high edge loading by deflecting beam, girder, or truss) 1,000
- Bridge seats, under bearing plates or non-hinged shoes (subjected to high edge loading by the direct bearing, upon the plate or shoe, of a deflecting beam or girder), average 700

The above bridge seat unit stresses will apply only where the edge of the bridge seat projects at least 3 inches (average) beyond edge of shoe or plate. Otherwise, the unit stresses permitted will be 75 percent of the above amounts.

DETAILS OF DESIGN

1.7.9 — EFFECTIVE LENGTH OF SPAN

For the calculation of stresses, span lengths shall be assumed as the distance between centers of bearings or other points of support.

1.7.10 — DEPTH RATIOS

For beams or girders the ratio of depth to length of span, preferably shall not be less than $\frac{1}{25}$.

For composite girders the ratio of the over-all depth of girder (concrete slab, plus steel girder) to the length of span preferably shall not be less than $\frac{1}{25}$, and the ratio of depth of steel girder alone to length of span shall not be less than $\frac{1}{30}$.

For trusses the ratio of depth to length of span shall not be less than $\frac{1}{10}$.

For continuous span depth ratios, the span length shall be considered as the distance between the dead load points of contraflexure.

1.7.11 — LIMITING LENGTHS OF MEMBERS

For compression members, the ratio of unsupported length to radius of gyration shall not exceed 120 for main members, or those in which the major stresses result from dead or live load, or both; and shall not exceed 140 for secondary members, or those whose primary purpose is to brace the structure against lateral or longitudinal forces, or to brace or reduce the unsupported length of other members, main or secondary.

In determining the radius of gyration for the purpose of applying the limitations of the l/r ratio, the area of any portion of a member may be neglected provided that the strength of the member as calculated without using the area thus neglected and the strength of the member as computed for the entire section with the $\frac{l}{r}$ ratio applicable thereto, both equal or exceed the computed total stress that the member must sustain.

The radius of gyration and the effective area for carrying stress of a member containing perforated cover plates shall be computed for a transverse section through the maximum width of perforation. When perforations are staggered in opposite cover plates the cross-sectional area of the member shall be considered the same as for a section having perforations in the same transverse plane.

Unsupported length shall be assumed as follows:

For the top chords of half-through trusses, the length between panel points laterally supported as indicated under Article 1.7.85; for other main members, the length between panel point intersections or centers of braced points or centers of end connections; for secondary members, the length between the centers of the end connections of such members or centers of braced points.

For tension members, except rods, eyebars, cables and plates, the ratio of unsupported length to radius of gyration shall not exceed 200 for main members, shall not exceed 240 for bracing members, and shall not exceed 140 for main members subject to a reversal of stress.

1.7.12 — DEFLECTION

The term "deflection" as used herein shall be the deflection computed in accordance with the assumption made for loading when computing the stress in the member.

Members having simple or continuous spans shall be designed so that the deflection due to live load plus impact shall not exceed $\frac{1}{800}$ of the span, except on bridges in urban areas used in part by pedestrians whereon the ratio preferably shall be $\frac{1}{1000}$.

The deflection of cantilever arms due to live load plus impact shall be limited to $\frac{1}{300}$ of the cantilever arm except for the case including pedestrian use, where the ratio preferably shall be $\frac{1}{375}$.

When spans have cross-bracing or diaphragms sufficient in depth or strength to insure lateral distribution of loads, the deflection may be computed for the standard H or HS loading, considering all beams or stringers as acting together and having equal deflection.

The moment of inertia of the gross cross-sectional area shall be used for computing the deflections of beams and girders. When the beam or girder is a part of a composite member, the live load may be considered as acting upon the composite section.

The gross area of each truss member shall be used in computing deflections of trusses. If perforated plates are used the effective area shall be the net volume divided by the length from center to center of perforations.

1.7.13 — MINIMUM THICKNESS OF METAL

Structural steel (including bracing, cross frames and all types of gusset plates), except for webs of certain rolled shapes, closed ribs in orthotropic decks, fillers and in railings, shall be not less than $5/16''$ in thickness. The web thickness of rolled beams or channels shall not be less than $0.23''$. The thickness of closed ribs in orthotropic decks shall not be less than $3/16''$.

Where the metal will be exposed to marked corrosive influences, it shall be increased in thickness or specially protected against corrosion.

It should be noted that there are other provisions in this section pertaining to thickness for fillers, segments of compression members, gusset plates, etc. As stated above fillers need not be $5/16''$ min.

For stiffeners and outstanding legs of angles, etc., refer to Article 1.7.15.

For stiffeners and other plates refer to "Plate Girders."

For compression members refer to "Trusses."

1.7.14 — EFFECTIVE AREA OF ANGLES AND TEE SECTIONS IN TENSION

The effective area of a single angle tension member, a tee section tension member, or each angle of a double angle tension member in which the shapes are connected back to back on the same side of a gusset plate, shall be assumed as the net area of the connected leg or flange plus one-half of the area of the outstanding leg.

If a double angle or tee section tension member is connected with the angles or flanges back to back on opposite sides of a gusset plate, the full net area of the shapes shall be considered as effective.

When angles connect to separate gusset plates, as in the case of a double webbed truss, and the angles are connected by stay plates located as near the gusset as practicable, or by other adequate means, the full net area of the angles shall be considered effective. If the angles are not so connected, only 80 percent of the net area shall be considered effective.

Lug angles may be considered as effective in transmitting stress, provided they are connected with at least one-third more fasteners than required by the stress to be carried by the lug angle.

1.7.15 — OUTSTANDING LEGS OF ANGLES

The widths of outstanding legs of angles in compression (except where reinforced by plates) shall not exceed the following:

In main members carrying axial stress, 12 times the thickness.
In bracing and other secondary members, 16 times the thickness.

For other limitations see Article 1.7.88.

1.7.16 — EXPANSION AND CONTRACTION

The design shall be such as to allow for total thermal movement at the rate of 1¼" in 100 feet. Provisions shall be made for changes in length of span resulting from live load stresses. In spans more than 300 feet long, allowance shall be made for expansion and contraction in the floor. The expansion end shall be secured against lateral movement.

1.7.17 — COMBINED STRESSES

All members subject to combined bending and direct stresses shall be proportioned for the maximum unit stress specified in Appendix C. When bending stresses are induced by the component of externally applied loads acting perpendicular to the axis of the member, "a" shall be assumed equal to $+1$.

1.7.18 — ECCENTRIC CONNECTIONS

Members, including bracing, preferably shall be so connected that their gravity axes will intersect in a point. Eccentric connections shall be avoided, if practicable, but if unavoidable the members shall be so proportioned that the combined fiber stresses will not exceed the allowed axial stress.

1.7.19 — SPLICES AND CONNECTIONS

Splices may be made with rivets, by high strength bolts or by the use of welding. Splices, whether in tension, compression, bending or shear, shall be designed for not less than the average of the calculated stress at the point of splice and the strength of the member at the same point but, in any event, not less than 75% of the strength of the member. Where a section changes at a splice, the strength of the smaller section is to be used for the above splice requirements. The strength of the member shall be determined by the gross section for compression members and by the net section for tension members and members primarily in bending.

As an alternate, splices of rolled flexural members may be proportioned for a shear equal to actual maximum shear multiplied by the ratio of the splice design moment and the actual moment at the splice. Web splice plates and their connections shall be designed for the portion of the design moment resisted by the web and for the moment due to the eccentricity of shear introduced by the splice connection. Flange splice plates need be designed only for the portion of the design moment not resisted by the web.

Web plates shall be spliced symmetrically by plates on each side. The splice plates for shear shall extend the full depth of the girder between flanges. In the splice there shall be not less than 2 rows of rivets or bolts on each side of the joint.

Compression members, such as columns and chords, shall have ends in close contact at riveted and bolted splices. Splices of such members, which will be fabricated and erected with close inspection and detailed with milled ends in full contact bearing at the splices, may be held in place by means of splice plates and rivets or high strength bolts proportioned for not less than 50 percent of the lower allowable stress of the sections spliced.

Tension and compression members may be spliced by means of full penetration butt welds preferably without the use of splice plates.

Splices in truss chords and columns shall be located as near to the panel points as practicable and usually on that side where the smaller stress occurs. The arrangement of plates, angles and other splice elements shall be such as to make proper provision for the stresses, both axial and bending, in the component parts of the members spliced.

For riveted and bolted flexural members, splices in flange parts shall not be used between field splices except by special permission of the Engineer. In any one flange not more than one part shall be spliced at the same cross-section. If practicable, splices shall be located at points where there is an excess of section. Riveted and bolted flange angle splices shall include two angles, one on each side of the flexural member.

In continuous spans, splices preferably shall be made at or near points of contraflexure.

Welded field splices preferably should be arranged to minimize overhead welding.

If splice plates are not in direct contact with the parts which they connect, the number of fasteners on each side of the joint shall be in excess of the number required for a direct contact splice to the extent of at least two extra transverse lines of fasteners for each intervening plate, except as provided below.

If fasteners carrying stress pass through fillers, the fillers preferably shall be extended beyond the gusset or splice material, and the extension secured by enough additional fasteners to carry the stress in the filler, which stress is to be calculated as the total stress in the member divided by the combined area of the member plus the fillers. As an alternate, the additional fasteners may be passed through the gusset or splice material without extending the filler.

If the filler is less than ¼ inch thick it shall not be extended beyond the splicing material and additional fasteners are not required. Fillers ¼ inch or more in thickness shall consist of not more than two plates, unless special permission is given by the Engineer.

In welded splices any filler ¼ in. or more in thickness shall extend beyond the edges of the splice plate and shall be welded to the part on which it is fitted with sufficient weld to transmit the splice plate stress applied at the surface of the filler as an eccentric load.

The welds joining the splice plate to the filler shall be sufficient to transmit the splice plate stress and shall be long enough to avoid overstressing the filler along the toe of the weld. Any filler less than ¼ in. thick shall have its edges made flush with the edges of the splice plate. The weld size necessary to carry the splice plate stress shall be increased by the thickness of the filler plate.

Fillers in high strength bolted friction type connections need not be extended and developed, but eccentricity of forces at short thick fillers must be considered.

Material of different widths spliced by butt welds shall have transitions conforming to Figure 1.7.19 except that for A514 and A517 steels only the transition illustrated in Fig. 1.7.19a is permitted. At butt weld splices joining material of different thicknesses there shall be a uniform slope between the offset surfaces of not more than 1 in 2½ with respect to the surface of either part.

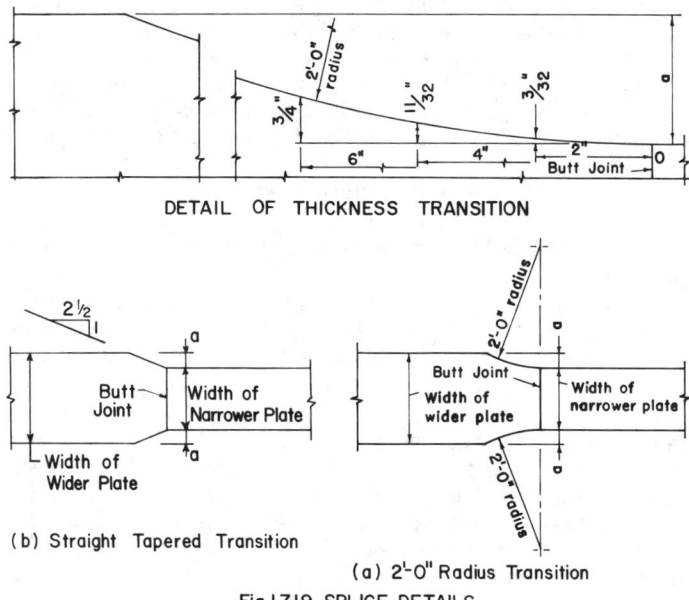

DETAIL OF THICKNESS TRANSITION

(b) Straight Tapered Transition

(a) 2'-0" Radius Transition

Fig. 1.7.19 SPLICE DETAILS

1.7.20 — STRENGTH OF CONNECTIONS

Except as otherwise provided herein, connections shall be designed for the average of the calculated stress and the strength of the member, but they shall be designed for not less than 75 percent of the strength of the member.

Connections shall be made symmetrical about the axis of the members insofar as practicable. Connections, except for lacing bars and handrails, shall contain not less than two fasteners or equivalent weld.

1.7.21 — DIAPHRAGMS, CROSS FRAMES AND LATERAL BRACING

Rolled beam and plate girder spans shall be provided with cross frames or diaphragms at each end and with intermediate cross frames or diaphragms spaced at intervals not to exceed 25 feet. Cross frames shall be as deep as practicable. Diaphragms shall be at least ⅓ and preferably ½ the girder depth. End cross frames or diaphragms shall be proportioned to adequately transmit all the lateral forces to the bearings. Special consideration shall be given to the design of cross frames used on horizontally curved steel girder bridges. These cross frames shall be designed as main members with adequate provisions for transfer of lateral forces from the girder flanges.

On spans 125 feet or longer with a concrete floor or other floor of equal rigidity, which is adequately attached to the top flanges, one plane or system of lateral bracing shall be provided near the bottom flange. Spans with timber or other non-rigid flooring shall have one system of lateral bracing near the bottom flange for spans longer than 40 feet and two systems of lateral bracing for spans 125 feet or longer. The lateral bracing shall be placed in at least one-third of the bays. Cross frames or diaphragms shall be placed in all bays.

Where beams or girders comprise the main members of through spans, such members shall be stiffened against lateral deformation by means of gusset plates or knee braces with solid webs which shall be connected to the stiffeners on the main members and the floor beams. If the unsupported length of the edge of the gusset plate (or solid web) exceeds 60 times its thickness, the plate or web shall have a stiffening plate or angles connected along its unsupported edge.

Through truss spans, deck truss spans and spandrel braced arches shall have top and bottom lateral bracing.

Bracing shall be composed of angles, other shapes or welded sections.

If a double system of bracing is used, both systems may be considered effective simultaneously if the members meet the requirements both as tension and compression members. The members shall be connected at their intersections.

The lateral bracing of compression chords, preferably shall be as deep as the chords and effectively connected to both flanges.

The smallest angle used in bracing shall be 3 by 2½ inches. There shall be not less than 2 fasteners or equivalent weld in each end connection of the angles.

1.7.22 — NUMBER OF MAIN MEMBERS ON THROUGH SPANS

Where beams, girders or trusses are used for through spans, the spans preferably shall have only two main members. Such members shall be spaced a sufficient distance apart (center to center) to be secure against overturning by the assumed lateral forces.

1.7.23 — ACCESSIBILITY OF PARTS

The accessibility of all parts of a structure for inspection, cleaning and painting shall be secured by the proper proportioning of members and the design of their details.

1.7.24 — CLOSED SECTIONS AND POCKETS

Closed sections, and pockets or depressions which will retain water, shall be avoided where practicable. Pockets shall be provided with effective drain holes or be filled with waterproofing material.

Details shall be so arranged that the destructive effects of bird life, the retention of dirt, leaves, and other foreign matter will be reduced to a minimum. Where angles are used, either singly or in pairs, they preferably shall be placed with the vertical legs extending downward.

1.7.25 — WELDING, GENERAL

Welding symbols and fabrication shall conform to the current Specifications for Welded Highway and Railway Bridges of the American Welding Society and Art. 2.10.23 except that welding of steels not covered by this specification shall conform to established welding procedures for those steels.

Material for structural members which is designed and specified to be welded shall conform to AASHO M183 (ASTM A36), AASHO M161 (ASTM A242 of a weldable grade), or AASHO M188 (ASTM A441). In addition to the steels named above and covered in Article 1.7.1, other weldable steels may be considered after the suitability and weldability of same has been thoroughly established.

1.7.26 — MINIMUM SIZE OF FILLET WELDS

The minimum fillet weld size shall be as shown in the following table. Weld size is determined by the thicker of the two parts joined unless a larger size is required by calculated stress. The weld size need not exceed the thickness of the thinner part joined.

Material Thickness of Thicker Part Joined (Inches)	Minimum Size of Fillet Weld (Inches)
To ½ inclusive	3/16
Over ½ to ¾	¼
Over ¾ to 1½	5/16
Over 1½ to 2¼	3/8
Over 2¼ to 6	½
Over 6	5/8

The minimum size seal weld shall be 3/16" fillet weld.

1.7.27 — MAXIMUM EFFECTIVE SIZE OF FILLET WELDS

The maximum size of a fillet weld that may be assumed in the design of a connection shall be such that the stresses in the adjacent base material do not exceed the values allowed in Article 1.7.1. The maximum size that may be used along edges of connected parts shall be:

(1) Along edges of material less than ¼ inch thick, the maximum size may be equal to the thickness of the material.
(2) Along edges of material ¼ inch or more in thickness, the maximum size shall be ¹⁄₁₆ inch less than the thickness of the material, unless the weld is especially designated on the drawings to be built out to obtain full throat thickness.

1.7.28 — EFFECTIVE WELD AREAS

(A) Butt Welds

The effective area shall be the effective weld length multiplied by the effective throat thickness.

(1) The effective weld length for any butt weld, square or skewed, shall be the width of the part joined, perpendicular to the direction of stress.
(2) The effective throat thickness shall be the thickness of the thinner piece of base metal joined. (No increase is permitted for weld reinforcement.)

(B) Fillet Welds

The effective area shall be the effective weld length multiplied by the effective throat thickness. (Stress in a fillet weld shall be considered as applied to this effective area, for any direction of applied load.)

(1) The effective length of straight fillet weld shall be the overall length of the full-size fillet including end returns.
(2) The effective length of a curved fillet weld shall be the length of the line generated by the centerpoint of the effective throat thickness.
(3) The effective throat thickness shall be the shortest distance from the root of the diagrammatic weld to the face.

1.7.29 — MINIMUM EFFECTIVE LENGTH OF FILLET WELDS

The minimum effective length of a fillet weld shall be four times its size and in no case less than 1½ inches.

1.7.30 — FILLET WELD END RETURNS

Fillet welds which support a tensile force that is not parallel to the axis of the weld, or which are proportioned to withstand repeated stress shall not terminate at corners of parts or members but shall be returned continuously, full size, around the corner for a length equal to twice the weld size where such return can be made in the same plane. End returns shall be indicated on design and detail drawings.

1.7.31 — LAP JOINTS

The minimum width of laps on lap joints shall be 5 times the thickness of the thinner part joined and not less than 1 inch. Lap joints joining plates or bars subjected to axial stress shall be fillet welded along the edge of both lapped parts except where the deflection of the lapped parts is sufficiently restrained to prevent opening of the joint under maximum loading.

1.7.32 — SEAL WELDS

Seal welding shall preferably be accomplished by a continuous weld combining the functions of sealing and strength, changing section only as the required strength or the requirements of minimum size fillet weld, based on material thickness, may necessitate.

1.7.33 — FILLET WELDS IN SKEWED TEE JOINTS

When joining material in skewed tee joints, fillet welds shall not be used for joints that have an included angle of less than 60 degrees.

1.7.34 — FILLET WELDS IN HOLES AND SLOTS

Fillet welds in holes or slots may be used to transmit shear in lap joints or to prevent the buckling or separation of lapped parts, and to join components of built-up members. Such fillet welds may overlap, subject to the provisions of Article 1.7.28. Fillet welds in holes or slots are not to be considered plug or slot welds.

1.7.35 — SIZE OF FASTENERS (RIVETS OR HIGH STRENGTH BOLTS)

Fasteners shall be of the size shown on the drawings, but generally shall be ¾ inch or ⅞ inch in diameter. Fasteners ⅝ inch in diameter shall not be used in members carrying calculated stress except in 2½ inch legs of angles and in flanges of sections requiring ⅝ inch fasteners.

The diameter of fasteners in angles carrying calculated stress shall not exceed one-fourth the width of the leg in which they are placed.

In angles whose size is not determined by calculated stress, ⅝ inch fasteners may be used in 2 inch legs, ¾ inch fasteners in 2½ inch legs, ⅞ inch fasteners in 3 inch legs, and 1 inch fasteners in 3½ inch legs.

Structural shapes which do not admit the use of ⅝ inch diameter fasteners shall not be used except in handrails.

1.7.36 — SPACING OF FASTENERS

The pitch of fasteners is the distance along the line of principal stress, in inches, between centers of adjacent fasteners, measured along one or more fastener lines. The gage of fasteners is the distance in inches between adjacent lines of fasteners or the distance from the back of angle or other shape to the first line of fasteners. The pitch of fasteners shall be governed by the requirements for sealing or stitch, whichever is the minimum.

The minimum distance between centers of fasteners shall be three times the diameter of the fastener but, preferably, shall not be less than the following:

For 1 inch fasteners, 3½ inches.
For ⅞ inch fasteners, 3 inches.
For ¾ inch fasteners, 2½ inches.
For ⅝ inch fasteners, 2¼ inches.

1.7.37 — MAXIMUM PITCH OF SEALING AND STITCH FASTENERS

(A) Sealing Fasteners

For sealing, the pitch on a single line adjacent to a free edge of an outside plate or shape shall not exceed 4 inches+4t or 7 inches. If there is a second line of fasteners uniformly staggered with those in the line adjacent to the free edge, at a gage "g" less than 1½ inches+4t therefrom, the staggered pitch in two such lines, considered together, shall not exceed 4 inches+4t−¾g or 7 inches−¾g but need not be less than one-half the requirement for a single line. "t"=the thickness in inches of the thinner outside plate or shape.

(B) Stitch Fasteners

In built-up members where two or more plates or shapes are in contact, stitch fasteners shall be used to insure uniform action and, in compression members, to prevent buckling. In compression members the pitch of stitch fasteners on any single line in the direction of stress shall not exceed 12t, except that, if the fasteners on adjacent lines are staggered and the gage "g" between the line under consideration and the farther adjacent line (if there are more than two lines) is less than 24t, the staggered pitch in the two lines, considered together, shall not exceed 12t or 15t−⅜g. The gage between adjacent lines of fasteners shall not exceed 24t. "t"=the thickness, in inches, of the thinner outside plate or shape. In tension members the pitch shall not exceed twice that specified for compression members and the gage shall not exceed that specified for compression members.

For pitch of fasteners in the ends of compression members, See Article 1.7.86.

1.7.38 — EDGE DISTANCE OF FASTENERS

(A) General

The minimum distance from the center of any fastener to a sheared or flame cut edge shall be:

For 1 inch fasteners, 1¾ inches.
For ⅞ inch fasteners, 1½ inches.
For ¾ inch fasteners, 1¼ inches.
For ⅝ inch fasteners, 1⅛ inches.

The minimum distance from the center of any fastener to a rolled or planed edge, except in flanges of beams and channels, shall be:

For 1 inch fasteners, 1½ inches.
For ⅞ inch fasteners, 1¼ inches.
For ¾ inch fasteners, 1⅛ inches.
For ⅝ inch fasteners, 1 inch.

In the flanges of beams and channels the distance shall be:
For 1 inch fasteners, 1¼ inches.
For ⅞ inch fasteners, 1⅛ inches.
For ¾ inch fasteners, 1 inch.
For ⅝ inch fasteners, ⅞ inch.

The maximum distance from any edge shall be eight times the thickness of the thinnest outside plate, but shall not exceed 5 inches.

(B) Special

In connections designed by bearing on the plates and having no more than two lines of fasteners parallel to the direction of the stress, the distance between the center of the nearest fastener and that end of the connected member toward which the pressure from the fastener is directed, shall not be less than the nominal shearing area of the fastener (single or double shear, as the case may be) divided by two-thirds of the plate thickness. This end distance may be proportionately less where the stress per fastener is less than that of the maximum permitted, but not less than 1.5 times the fastener diameter.

In bearing type connections having no more than two fasteners in a line parallel to the direction of stress, the distance between the center of the nearest fastener and that end of the connected member towards which the pressure from the fastener is directed shall not be less than AC/t for single shear or $2AC/t$ for double shear, where A is the nominal cross-sectional area of the fastener, t is the thickness of the connected part and C is the ratio of specified minimum tensile strength of the fastener to the specified minimum tensile strength of the connected part. This end distance may be proportionately less where the shear stress per fastener is less than that permitted in Article 1.7.5, but not less than 1½ times the fastener diameter. It need not exceed 1½ times the transverse spacing of the fasteners.

1.7.39 — LONG RIVETS

Rivets subjected to calculated stress and having a grip in excess of 4½ diameters shall be increased in number at least 1 percent for each additional 1/16 inch of grip. If the grip exceeds six times the diameter of the rivet, specially designed rivets shall be used.

1.7.40 — LINKS AND HANGERS

In pin-connected tension members other than eyebars, the net section across the pin hole shall be not less than 140 percent, and the net section back of the pin hole not less than 100 percent of the required net section of the body of the member. The ratio of the net width (through the pin hole transverse to the axis of the member) to the thickness of the segment shall not be more than 8. Flanges not bearing on the pin shall not be considered in the net section across the pin.

1.7.41 — LOCATION OF PINS

Pins shall be so located with respect to the gravity axis of the members as to reduce to a minimum the stresses due to bending.

1.7.42 — SIZE OF PINS

Pins shall be proportioned for the maximum shears and bending moments produced by the stresses in the members connected. If there are eyebars among the parts connected, the diameter of the pin shall be as specified in Article 1.7.46.

1.7.43 — PIN PLATES

When necessary for the required section or bearing area, the section at the pin holes shall be increased on each segment by plates so arranged as to reduce to a minimum the eccentricity of the segment. One plate on each side shall be as wide as the outstanding flanges will allow. At least one full width plate on each segment shall extend to the far edge of the stay plate and the others not less than 6 inches beyond the near edge. These plates shall be connected by enough rivets, bolts, or fillet and plug welds to transmit the bearing pressure, and so arranged as to distribute it uniformly over the full section.

1.7.44 — PINS AND PIN NUTS

Pins shall be of sufficient length to secure a full bearing of all parts connected upon the turned body of the pin. They shall be secured in position by hexagonal recessed nuts or by hexagonal solid nuts with washers. If the pins are bored, through rods with cap washers may be used. Pin nuts shall be malleable castings or steel. They shall be secured by cotter pins in the screw ends or else the screw ends shall be long enough to permit burring the threads.

Members shall be held against lateral movement on the pins.

1.7.45 — UPSET ENDS

Bars and rods with screw ends, where specified, shall be upset to provide a section at the root of the thread, which will exceed the net section of the body of the member by at least 15 percent.

1.7.46 — EYEBARS

Eyebars shall be of a uniform thickness without reinforcement at the pin holes. The thickness of eyebars shall be not less than $\frac{1}{8}$ of the width, nor less than $\frac{1}{2}$ inch, and not greater than 2 inches. The section of the head through the center of the pin hole shall exceed the required section of the body of the bar by at least 35 percent. The net section back of the pin hole shall not be less than 75 percent of the required net section of the body of the member. The radius of transition between the head and body of the eyebar shall be equal to or greater than the width of the head through the centerline of the pin hole. The diameter of the pin shall be not less than $\left[\frac{3}{4} + \frac{1}{4} \frac{\text{(yield point of steel)}}{100{,}000} \right]$ times the width of the body of the eyebar.

1.7.47 — PACKING OF EYEBARS

The eyebars of a set shall be symmetrical about the central plane of the truss and as nearly parallel as practicable. Bars shall be as close together as practicable and held against lateral movement, but they shall be so arranged that adjacent bars in the same panel will be separated by at least ½ inch.

Intersecting diagonal bars not far enough apart to clear each other at all times shall be clamped together at the intersection.

Steel filling rings shall be provided, if needed, to prevent lateral movement of eyebars or other members connected on the pin.

1.7.48 — FORKED ENDS

Forked ends will be permitted only where unavoidable. There shall be enough pin plates on forked ends to make the section of each jaw equal to that of the member. The pin plates shall be long enough to develop the pin plate beyond the near edge of the stay plate, but not less than the length required by Article 1.7.43.

BEARINGS

1.7.49 — FIXED BEARINGS

Fixed ends shall be firmly anchored. Bearings for spans less than 50 feet need have no provision for deflection. Spans of 50 feet or greater shall be provided with a type of bearing employing a hinge, curved bearing plates, elastomeric pads, or pin arrangement for deflection purposes.

1.7.50 — EXPANSION BEARINGS

Spans of less than 50 feet may be arranged to slide upon metal plates with smooth surfaces and no provisions for deflection of the spans need be made. Spans of 50 feet and greater shall be provided with rollers, rockers or sliding plates for expansion purposes and shall also be provided with a type of bearing employing a hinge, curved bearing plates, or pin arrangement for deflection purposes.

In lieu of the above requirements elastomeric bearings may be used. See Section 12, Division I of this specification.

1.7.51 — BRONZE OR COPPER-ALLOY SLIDING EXPANSION BEARINGS

Bronze or copper-alloy sliding plates shall be chamfered at the ends. They shall be held securely in position, usually by being inset into the metal of the pedestals or sole plates. Provisions shall be made against any accumulation of dirt which will obstruct free movement of the span.

1.7.52 — ROLLERS

Expansion rollers shall be connected by substantial side bars and shall be guided by gearing or other effectual means to prevent lateral movement, skewing and creeping. The rollers and bearing plates shall be protected from dirt and water as far as practicable, and the design shall be such that water will not be retained and that the roller nests may be inspected and cleaned easily.

1.7.53 — SOLE PLATES AND MASONRY PLATES

Sole plates and masonry plates shall have a minimum thickness of ¾ inch.

For spans on inclined grades greater than 1% without hinged bearings the sole plates shall be beveled so that the bottom of the sole plate is level, unless the bottom of the sole plate is radially curved.

1.7.54 — MASONRY BEARINGS

Beams, girders or trusses on masonry shall be so supported that the bottom chords or flanges will be above the bridge seat, preferably not less than 6 inches.

1.7.55 — ANCHOR BOLTS

Trusses, girders and rolled beam spans preferably shall be securely anchored to the substructure. Anchor bolts shall be swedged or threaded to secure a satisfactory grip upon the material used to embed them in the holes.

The following are the minimum requirements for each bearing:

> For rolled beam spans the outer beams shall be anchored at each end with 2 bolts, 1″ in diameter, set 10″ in the masonry. For trusses and girders:
>
>> Spans 50 feet in length or less; 2 bolts, 1″ in diameter, set 10″ in the masonry.
>>
>> Spans 51 to 100 feet; 2 bolts, 1¼″ in diameter, set 12″ in the masonry.
>>
>> Spans 101 to 150 feet; 2 bolts, 1½″ in diameter, set 15″ in the masonry.
>>
>> Spans greater than 150 feet; 4 bolts, 1½″ in diameter, set 15″ in the masonry.

Anchor bolts shall be designed to resist uplift as specified in Article 1.2.16.

1.7.56 — PEDESTALS AND SHOES

Pedestals and shoes preferably shall be made of cast steel or structural steel. The difference in width between the top and bottom bearing surfaces shall not exceed twice the distance between them. For hinged bearings, this distance shall be measured from the center of the pin. In built-up pedestals and shoes, the web plates and angles connecting them to the base plate shall be not less than ⅝" thick. If the size of the pedestal permits, the webs shall be rigidly connected transversely. The minimum thickness of the metal in cast steel pedestals shall be 1". Pedestals and shoes shall be so designed that the load will be distributed uniformly over the entire bearing.

Webs and pin holes in the webs shall be arranged to keep any eccentricity to a minimum. The net section through the hole shall provide 140% of the net section required for the actual stress transmitted through the pedestal or shoe. Pins shall be of sufficient length to secure a full bearing. Pins shall be secured in position by appropriate nuts with washers. All portions of pedestals and shoes shall be held against lateral movement on the pins.

FLOOR SYSTEM

1.7.57 — STRINGERS

Stringers preferably shall be framed into floorbeams. Stringers supported on the top flanges of floorbeams preferably shall be continuous over two or more panels.

1.7.58 — FLOORBEAMS

Floorbeams preferably shall be at right angles to the trusses or main girders and shall be rigidly connected thereto. Floorbeam connections preferably shall be located so the lateral bracing system will engage both the floorbeam and the main supporting member. In pin-connected trusses, if the floorbeams are located below the bottom chord pins, the vertical posts shall be extended sufficiently below the pins to make a rigid connection to the floorbeam.

1.7.59 — CROSS FRAMES

In bridges with wooden floors and steel stringers, intermediate cross frames (or diaphragms) shall be placed between stringers more than 20 feet long.

1.7.60 — EXPANSION JOINTS

To provide for expansion and contraction movement, floor expansion joints shall be provided at all expansion ends of spans and at other points where they may be necessary.

Apron plates, when used, shall be designed to bridge the joint and to prevent, so far as practicable, the accumulation of roadway debris upon the bridge seats. Preferably, they shall be connected rigidly to the end floorbeam.

1.7.61 — END CONNECTIONS OF FLOORBEAMS AND STRINGERS

The end connection shall be designed for the loads specified. The end connection angles of floorbeams and stringers shall be not less than ⅜ inch in finished thickness. Except in cases of special end floorbeam details, each end connection for floorbeams and stringers shall be made with two angles. The length of these angles shall be as great as the flanges will permit. Bracket or shelf angles which may be used to furnish support during erection shall not be considered in determining the number of fasteners required to transmit end shear.

End connection details shall be designed with special care to provide clearance for making the field connection.

End connections of stringers and floorbeams preferably shall be bolted with High Strength Bolts, however, they may be riveted or welded. In the case of welded end connections, they shall be designed for the vertical loads and the end bending moment resulting from the deflection of the members.

Where timber stringers frame into steel floorbeams, shelf angles with stiffeners shall be provided to carry the whole reaction. Shelf angles shall be not less than 7/16 inch thick.

1.7.62 — END FLOORBEAMS

There shall be end floorbeams in all square-ended trusses and girder spans and preferably in skew spans. End floorbeams for truss spans preferably shall be designed to permit the use of jacks for lifting the superstructure. For this case the allowable stresses may be increased 50 percent.

End floorbeams shall be arranged to permit painting of the side of the beam adjacent to the abutment backwall.

1.7.63 — END PANEL OF SKEWED BRIDGES

In skew bridges without end floorbeams, the end panel stringers shall be secured in correct position by end struts connected to the stringers and to the main trusses or girder. The end panel lateral bracing shall be attached to the main trusses or girders and also to the end struts. Adequate provisions shall be made for the expansion movement of stringers.

1.7.64 — SIDEWALK BRACKETS

Sidewalk brackets shall be connected in such a way that the bending stresses will be transferred directly to the floorbeams.

ROLLED BEAMS

1.7.65 — ROLLED BEAMS, GENERAL

Rolled beams, including those with welded cover plates, shall be designed by the moment of inertia method. Rolled beams with riveted cover plates shall be designed on the same basis as riveted plate girders.

The compression flanges of rolled beams supporting timber floors shall not be considered to be laterally supported by the flooring unless the floor and fastenings are specially designed to provide adequate support.

1.7.66 — BEARING STIFFENERS

Suitable stiffeners shall be provided to stiffen the webs of rolled beams at bearings when the unit shear in the web adjacent to the bearing exceeds 75% of the allowable shear for girder webs. See the related provisions of Article 1.7.73.

1.7.67 — COVER PLATES

The length of any cover plate added to a rolled beam shall be not less than (2D+3) feet where (D) is the depth of the beam in feet.

The maximum thickness of a single cover plate on a flange shall not be greater than 2 times the thickness of the flange to which the cover plate is attached. The total thickness of all cover plates should not be greater than 2½ times the flange thickness. The thickness and width of a cover plate may be varied by butt welding parts of different thickness or width, with transitions conforming to the requirements of Article 1.7.19. Such plates shall be assembled and welds ground smooth before attaching to the flange. Cover plates may be either wider or narrower than the flange to which they are attached. Cover plates wider than the flange to which they are attached must be provided with transverse end welds. The end weld may be returned around the beam flange or stopped short of the flange toes.

Any partial length welded cover plate shall extend beyond the theoretical end by the terminal distance, or it shall extend to a section where the stress in the beam flange is equal to the allowable fatigue stress for "Base Metal adjacent to or connected by fillet welds," whichever is greater. The *theoretical end* of the cover plate is the section at which the stress in the flange without that cover plate equals the allowable stress exclusive of fatigue considerations. The *terminal distance* is 2 times the nominal cover plate width for cover plates not welded across their ends, and 1½ times for cover plates welded across their ends. The width at ends of tapered cover plates shall be not less than 3 inches. The weld connecting the cover plate to the flange in its terminal distance shall be continuous and of sufficient size to develop a total stress of not less than the computed stress in the cover plate at its theoretical end. All welds connecting cover plates to beam flanges shall be continuous and shall not be smaller than the minimum size permitted by Article 1.7.26.

PLATE GIRDERS

1.7.68 — PLATE GIRDERS, GENERAL

Girders shall be proportioned by the moment of inertia method. In calculating the net moment of inertia of riveted plate girders, the gravity axis of the gross section shall be used and the moment of inertia of all holes each side of the axis shall be deducted. The tensile stress shall be computed from the moment of inertia of the entire net section and the compressive stress from the moment of inertia of the entire gross section.

The compression flanges of plate girders supporting timber floors shall not be considered to be laterally supported by the flooring unless the floor and fastenings are specially designed to provide support.

1.7.69 — FLANGES

(A) Welded Girders

Each flange may comprise a series of plates joined end to end by full penetration butt welds. Changes in flange areas may be accomplished by varying the thickness or width of the flange plate, or by adding cover plates. Where plates of varying thicknesses or widths are connected, the splice shall be made in accordance with Article 1.7.19 and welds ground smooth before attaching to the web.

When cover plates are used, they shall be designed in accordance with Article 1.7.67.

The ratio of compression flange plate width to thickness shall not exceed the value determined by the formula:

$$b/t = \frac{3250}{\sqrt{f_b}}$$ but in no case shall b/t exceed 24.

Where the calculated compressive bending stress equals .55 F_y the b/t ratios for the various grades of steel shall not exceed the following:

36,000 psi, Y.P. Min.	b/t = 23
42,000 to	
46,000 psi, Y.P. Min.	b/t = 21
50,000 psi, Y.P. Min.	b/t = 20
55,000 psi, Y.P. Min.	b/t = 19
60,000 psi, Y.P. Min.	b/t = 18
65,000 psi, Y.P. Min.	b/t = 17
90,000 psi, Y.P. Min.	b/t = 15
100,000 psi, Y.P. Min.	b/t = 14

In the above b is the flange plate width, t is the thickness, and f_b is the calculated maximum compressive bending stresses (see Art. 1.7.112 for Hybrid Girders).

(B) Riveted or Bolted Girders

Flange angles shall form as large a part of the area of the flange as practicable. Side plates shall not be used except where flange angles exceeding ⅞ inch in thickness otherwise would be required.

Width of outstanding legs of flange angles in compression, except those reinforced by plates, shall not exceed the value determined by the formula:

$$b'/t = \frac{1625}{\sqrt{f_b}}$$ but in no case shall b'/t exceed 12.

Where the calculated compressive bending stress equals .55 F_y the b'/t ratios for the various grades of steel shall not exceed the following:

36,000 psi Y.P. Min.	$b'/t = 11$
42,000 to 50,000 psi Y.P. Min.	$b'/t = 10$
55,000 to 65,000 psi Y.P. Min.	$b'/t = 9$
90,000 psi Y.P. Min.	$b'/t = 7.5$
100,000 psi Y.P. Min.	$b'/t = 7$

In the above b' is the width of a flange angle, t is the thickness, and f_b is the calculated maximum compressive stress.

The gross area of the compression flange, except for composite design, shall be not less than the gross area of the tension flange.

Flange plates shall be of equal thickness, or shall decrease in thickness from the flange angles outward. No plate shall have a thickness greater than that of the flange angles.

At least one cover plate of the top flange shall extend the full length of the girder except when the flange is covered with concrete. Any cover plate which is not full length shall extend beyond the theoretical cut off point far enough to develop the capacity of the plate or shall extend to a section where the stress in the remainder of the girder flange is equal to the allowable fatigue stress, whichever is greater. The theoretical cut off point of the cover plate is the section at which the stress in the flange without that cover plate equals the allowable stress, exclusive of fatigue considerations.

The number of fasteners connecting the flange angles to the web plate shall be sufficient to develop the increment of flange stress transmitted to the flange angles, combined with any load that is applied directly to the flange.

Legs of angles 6 inches or greater in width, connected to web plates shall have two lines of fasteners. Cover plates over 14 inches wide shall have four lines of fasteners.

1.7.70 — THICKNESS OF WEB PLATES

(A) Girders Not Stiffened Longitudinally

The web plate thickness of plate girders without longitudinal stiffeners shall not be less than that determined by the formula:

$$t = \frac{D\sqrt{f_b}}{23,000}$$ (See Figure 1.7.70)

but in no case shall the thickness be less than $D/170$.

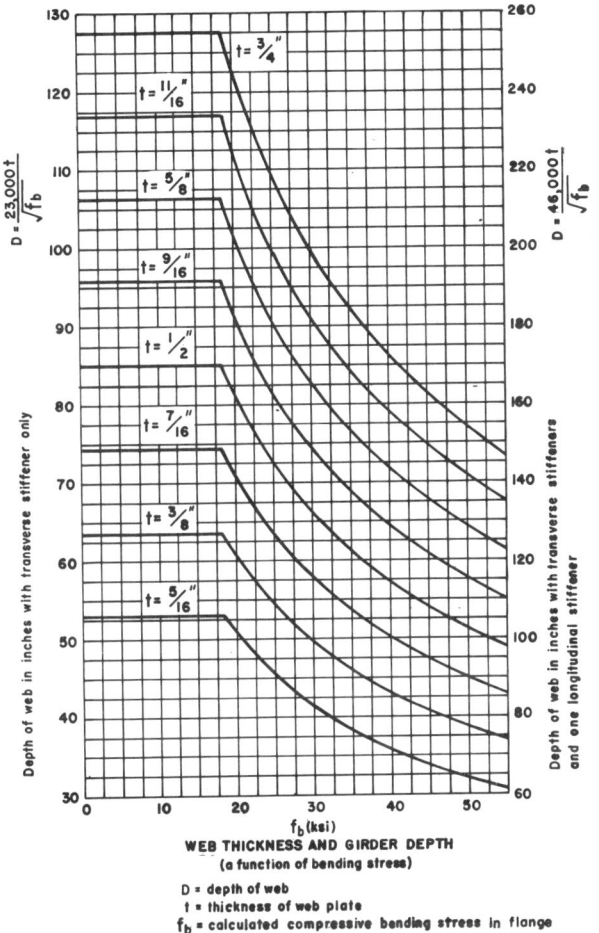

Figure 1.7.70

Where the calculated compressive bending stress in the flange equals the allowable bending stress, the thickness of the web plate, (with the web stiffened or not stiffened depending upon the requirements for transverse stiffeners), shall not be less than (where the Y.P. is for the flange material):

	36,000 psi. Y.P. Min.	D/165
	42,000 psi. Y.P. Min.	D/150
45,000 &	46,000 psi. Y.P. Min.	D/145
	50,000 psi. Y.P. Min.	D/140
	55,000 psi. Y.P. Min.	D/133
	60,000 psi. Y.P. Min.	D/127
	65,000 psi. Y.P. Min.	D/121
	90,000 psi. Y.P. Min.	D/105
	100,000 psi. Y.P. Min.	D/100

(B) Girders Stiffened Longitudinally

The web plate thickness of plate girders equipped with longitudinal stiffeners shall not be less than that determined by the formula:

$$t = \frac{D\sqrt{f_b}}{46,000} \quad \text{(See Figure 1.7.70)}$$

but in no case shall the thickness be less than D/340.

Where the calculated bending stress in the flange equals the allowable bending stress, the thickness of the web plate stiffened with transverse stiffeners in combination with one longitudinal stiffener, shall not be less than (where the Y.P. is for the flange material):

36,000 psi. Y.P. Min.	D/330
42,000 psi. Y.P. Min.	D/300
45,000 & 46,000 psi. Y.P. Min.	D/290
50,000 psi. Y.P. Min.	D/280
55,000 psi. Y.P. Min.	D/266
60,000 psi. Y.P. Min.	D/253
65,000 psi. Y.P. Min.	D/242
90,000 psi. Y.P. Min.	D/210
100,000 psi. Y.P. Min.	D/200

In the above, D (depth of web) is the clear unsupported distance, in inches, between flange components, t is the web thickness and f_b is the calculated flange bending stress.

1.7.71 — TRANSVERSE INTERMEDIATE STIFFENERS

Except as otherwise provided below, the webs of plate girders shall be stiffened at intervals not greater than the distance given by the formula:

$$d = \frac{11,000t}{\sqrt{f_v}} \quad \text{(See Figure 1.7.71A)}$$

but not greater than the clear unsupported depth of the web plate between flanges, in which:

d = the required distance between stiffeners, in inches

t = the thickness of the web plate, in inches

f_v = the average calculated unit shearing stress in the gross section of the web plate at the point considered.

The first two stiffener spaces at the simply supported ends of girders shall be one-half the value specified above.

Transverse intermediate stiffeners may be omitted if the web plate thickness is not less than the thickness determined by the formula:

$$t = \frac{D\sqrt{f_v}}{7500} \quad \text{(See Figure 1.7.71B)}$$

but in no case shall t be less than D/150.

Where the calculated shear stress equals the allowable shear stress, transverse intermediate stiffeners may be omitted if the thickness of the web is not less than:

	36,000 psi. Y.P. Min.	D/68
	42,000 psi. Y.P. Min.	D/64
45,000 &	46,000 psi. Y.P. Min.	D/60
	50,000 psi. Y.P. Min.	D/58
	55,000 psi. Y.P. Min.	D/56
	60,000 psi. Y.P. Min.	D/53
	65,000 psi. Y.P. Min.	D/51
	90,000 psi. Y.P. Min.	D/43
	100,000 psi. Y.P. Min.	D/41

Intermediate stiffeners preferably shall be made of plates for welded plate girders and shall be made of angles for riveted plate girders. They may be in pairs, one stiffener fastened on each side of the web plate, with a tight fit at the compression flange. They may however be made a single stiffener fastened to one side of the web plate. When stiffeners are used on one side only of the web plate, they shall be fastened to the compression flange.

The moment of inertia of any type of transverse stiffener shall not be less than:

$$I = \frac{d_o t^3 J}{10.92}$$

where $J = 25 \frac{D^2}{d^2} - 20$, but not less than 5.0.

In these expressions,

I = the minimum permissible moment of inertia of any type of transverse intermediate stiffener.

J = the required ratio of rigidity of one transverse stiffener to that of the web plate.

d = the required distance between stiffeners, in inches.

d_o = the actual distance between stiffeners, in inches.

D = the unsupported depth of web plate between flange components, in inches.

t = the thickness of the web plate, in inches.

When stiffeners are in pairs, the moment of inertia shall be taken about the center line of the web plate. When single stiffeners are used, the moment of inertia shall be taken about the face in contact with the web plate.

Stiffeners at points of concentrated loading shall be placed in pairs and shall be designed in accordance with Article 1.7.73.

The width of a plate or the outstanding leg of an angle intermediate stiffener shall not be less than 2 inches plus 1/30 the depth of the girder, and it shall preferably not be less than 1/4 the full width of the girder flange. The thickness of a plate or the outstanding leg of an angle intermediate stiffener shall not be less than 1/16 its width. Intermediate stiffeners may be A36 steel.

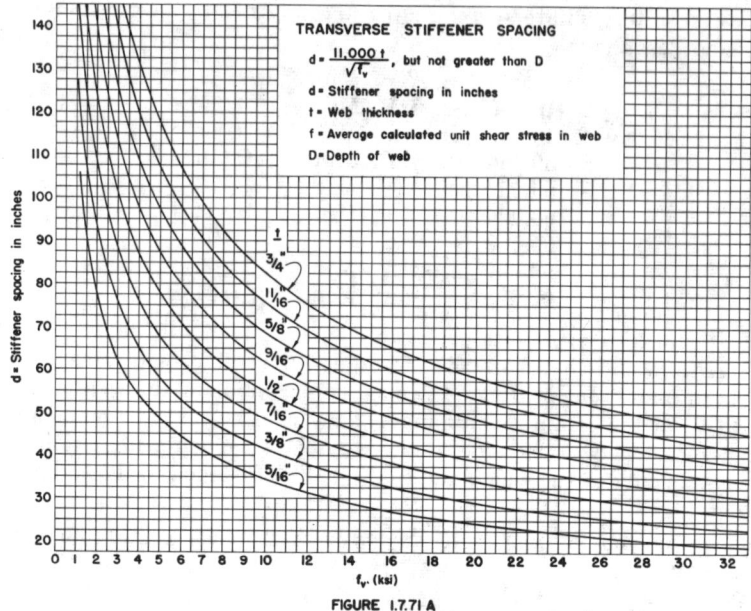

FIGURE 1.7.71 A

(See page 149 for Figure 1.7.71B)

1.7.72 — LONGITUDINAL STIFFENERS

The centerline of a plate longitudinal stiffener or the gage line of an angle longitudinal stiffener shall be D/5 from the inner surface or leg of the compression flange component. The longitudinal stiffener shall be proportioned so that:

$$I = Dt^3 \left(2.4 \frac{d_o^2}{D^2} - 0.13 \right)$$

where I = the minimum moment of inertia of the longitudinal stiffener about its edge in contact with the web plate.

D = the unsupported distance between flange components, in inches

t = the thickness of the web plate, in inches

d_o = the actual distance between transverse stiffeners, in inches

The thickness of the longitudinal stiffener shall not be less than:

$$\frac{b'\sqrt{f_b}}{2250}$$

where b' = width of stiffeners

f_b = calculated compressive bending stress in the flange

The stress in the stiffener shall not be greater than the basic allowable bending stress for the material used in the stiffener.

Longitudinal stiffeners are usually placed on one side only of the web plate. They need not be continuous and may be cut at their intersections with the transverse stiffeners.

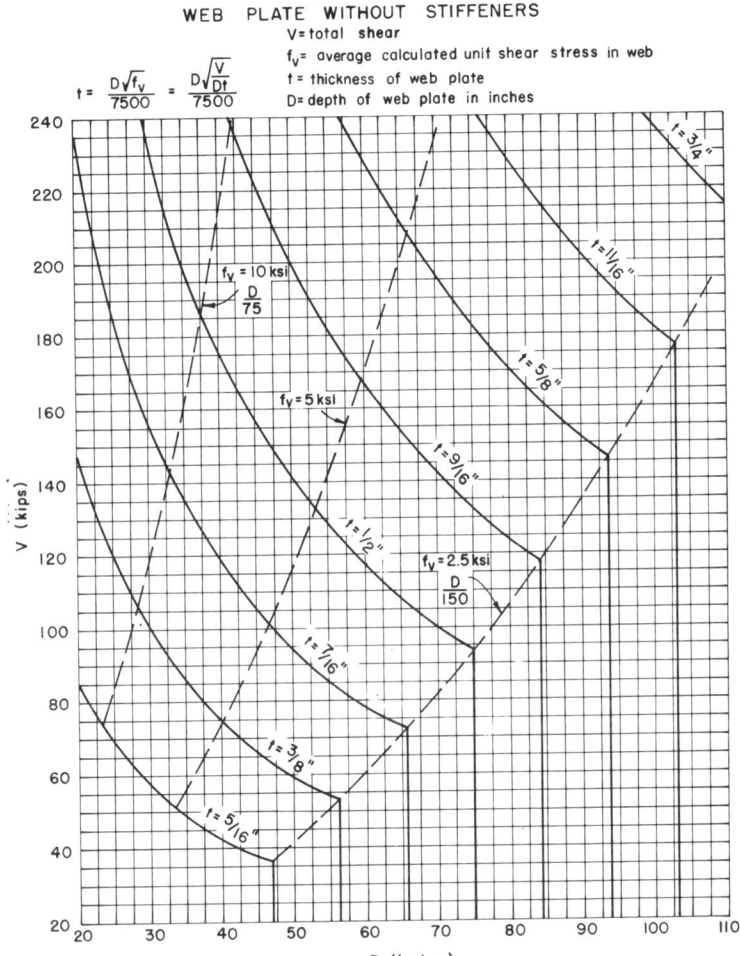

FIGURE 1.7.71 B

1.7.73 — BEARING STIFFENERS

(A) Welded Girders

Over the end bearings of welded plate girders and over the intermediate bearings of continuous welded plate girders there shall be stiffeners. They shall extend as nearly as practicable to the outer edges of the flange plates. They preferably shall be made of plates placed on both sides of the web plate. Bearing stiffeners shall be designed as columns, and their connection to the web shall be designed to transmit the entire end reaction to the bearings. For stiffeners consisting of two plates, the column section shall be assumed to comprise the two plates and a centrally located strip of the web plate whose width is equal to not more than 18 times its thickness. For stiffeners consisting of four or more plates, the column section shall be assumed to comprise the four or more plates and a centrally located strip of the web plate whose width is equal to that enclosed by the four or more plates plus a width of not more than 18 times the web plate thickness. (See Art. 1.7.113 for Hybrid Girders.) The radius of gyration shall be computed about the axis through the center line of the web plate. The stiffeners shall be ground to fit against the flange through which they receive their reaction, or attached to the flange by full penetration groove welds. Only the portions of the stiffeners outside the flange-to-web plate welds shall be considered effective in bearing. The thickness of the bearing stiffener plates shall not be less than

$$\frac{b'}{12}\sqrt{\frac{F_y}{33,000}}$$

The allowable compressive stress and the bearing pressure on the stiffeners shall not exceed the values specified in Article 1.7.1.

(B) Riveted or Bolted Girders

Over the end bearings of riveted or bolted plate girders there shall be stiffener angles, the outstanding legs of which shall extend as nearly as practicable to the outer edge on the flange angle. Bearing stiffener angles shall be proportioned for bearing on the outstanding legs of flange angles, no allowance being made for the portions of the legs being fitted to the fillets of the flange angles. Bearing stiffeners shall be arranged, and their connections to the web shall be designed to transmit the entire end reaction to the bearings. They shall not be crimped. The thickness of the bearing stiffener angles shall not be less than

$$\frac{b'}{12}\sqrt{\frac{F_y}{33,000}}$$

The allowable compressive stress and the bearing pressure on the stiffeners shall not exceed the values specified in Article 1.7.1.

1.7.74 — CAMBER

Girders should be cambered to compensate for dead load deflections and in addition thereto the camber preferably should be increased and/or decreased for the flanges to parallel the profile grade line when it is on a vertical curve.

TRUSSES

1.7.75 — TRUSSES, GENERAL

Component parts of individual truss members may be connected by welds, rivets or high strength bolts.

Preference should be given to trusses with single intersection web systems. Members shall be symmetrical about the central plane of the truss.

Trusses preferably shall have inclined end posts. Laterally unsupported hip joints shall be avoided.

Main trusses shall be spaced a sufficient distance apart center to center, to be secure against overturning by the assumed lateral forces.

For the calculation of stresses, effective depths shall be assumed as follows:

> Riveted trusses, distance between centers of gravity of the chords.
> Pin-connected trusses, distance between centers of chord pins.

1.7.76 — TRUSS MEMBERS

Chord and web truss members shall usually be made in the following shapes:

> "H" Sections, made with two side segments (composed of angles or plates) with solid web, perforated web, or web of stay plates and lacing.
>
> Channel sections, made with two angle segments, with solid web, perforated web, or web of stay plates and lacing.
>
> Single Box sections made with side channels, beams, angles and plates or side segments of plates only, connected top and bottom with perforated plates or stay plates and lacing.
>
> Single Box sections, made with side channels, beams, angles and plates or side segments of plates only, connected at top with solid cover plates and at the bottom with perforated plates or stay plates and lacing.
>
> Double Box sections, made with side channels, beams, angles and plates or side segments of plates only, connected with a conventional solid web, together with top and bottom perforated cover plates or stay plates and lacing.

If the shape of the truss permits, compression chords shall be continuous.

In chords composed of angles in channel shaped members, the vertical legs of the angles preferably shall extend downward.

If web members are subject to reversal of stress, their end connections shall not be pinned. Counters preferably shall be rigid. Adjustable counters, if used, shall have open turnbuckles, and in the design of these members an allowance of 10,000 pounds shall be made for initial stress. Only one set of diagonals in any panel shall be adjustable. Sleeve nuts and loop bars shall not be used.

1.7.77 — SECONDARY STRESSES

The design and details shall be such that secondary stresses will be as small as practicable. Secondary stresses due to truss distortion or floor-beam deflection usually need not be considered in any member, the width of which, measured parallel to the plane of distortion, is less than one-tenth of its length. If the secondary stress exceeds 4,000 pounds per square inch for tension members and 3,000 for compression members, the excess shall be treated as a primary stress. Stresses due to the flexural dead load moment of the member shall be considered as additional secondary stress.

1.7.78 — DIAPHRAGMS

There shall be diaphragms in the trusses at the end connections of floor beams.

The gusset plates engaging the pedestal pin at the end of the truss shall be connected by a diaphragm. Similarly, the webs of the pedestal shall, if practicable, be connected by a diaphragm.

There shall be a diaphragm between gusset plates engaging main members if the end tie plate is 4 feet or more from the point of intersection of the members.

1.7.79 — CAMBER

The length of the truss members shall be such that the camber will be equal to or greater than the deflection produced by the dead load.

1.7.80 — WORKING LINES AND GRAVITY AXES

Main members shall be proportioned so that their gravity axes will be as nearly as practicable in the center of the section.

In compression members of unsymmetrical section, such as chord sections formed of side segments and a cover plate, the gravity axis of the section shall coincide as nearly as practicable with the working line, except that eccentricity may be introduced to counteract dead load bending. In 2-angle bottom chord or diagonal members, the working line may be taken as the gage line nearest the back of the angle or at center of gravity for welded trusses.

1.7.81 — PORTAL AND SWAY BRACING

Through truss spans shall have portal bracing, preferably, of the 2-plane or box type, rigidly connected to the end post and the top chord flanges, and as deep as the clearance will allow. If a single plane portal is used, it shall be located, preferably, in the central transverse plane of the end posts, with diaphragms between the webs of the posts to provide for a distribution of the portal stresses. The portal bracing shall be designed to take the full end reaction of the top chord lateral system and the end posts shall be designed to transfer this reaction to the truss bearings.

Through truss spans shall have sway bracing 5 feet or more deep at each intermediate panel point. Top lateral struts shall be at least as deep as the top chord.

Deck truss spans shall have sway bracing in the plane of the end posts and at all intermediate panel points. This bracing shall extend the full depth of the trusses below the floor system. The end sway bracing shall be proportioned to carry the entire upper lateral stress to the supports through the end posts of the truss.

1.7.82 — FILLERS, DEVELOPMENT, MAXIMUM NUMBERS, ETC.

For Fillers refer to Article 1.7.19.

1.7.83 — PERFORATED COVER PLATES AND LACING BARS

The shearing force normal to the member in the planes of lacing or continuous perforated plates shall be assumed divided equally between all such parallel planes. The shearing force shall include that due to the weight of the member plus any other external force. For compression members, an additional force shall be added as obtained by the following formula:

$$V = \frac{P}{100} \left[\frac{100}{\frac{l}{r} + 10} + \frac{\frac{l}{r}}{\frac{3,300,000}{F_y}} \right]$$

In the above expression:

V = normal shearing stress in pounds.
P = allowable compressive axial load on members, in pounds.
l = length of member in inches.
r = radius of gyration of section about the axis perpendicular to plane of lacing or perforated plate in inches.
F_y = specified minimum yield point of type of steel being used.

(A) Perforated Cover Plates

When perforated cover plates are used, the following provisions shall govern their design:

(1) The ratio of length, in direction of stress, to width of perforation, shall not exceed two.

(2) The clear distance between perforations in the direction of stress, shall not be less than the distance between points of support.

(3) The clear distance between the end perforation and the end of the cover plate shall not be less than 1.25 times the distance between points of support.

(4) The point of support shall be the inner line of fasteners or fillet welds connecting the perforated plate to the flanges. For plates butt welded to the flange edge of rolled segments the point of support may be taken as the weld whenever the ratio of the outstanding flange width to flange thickness of the rolled segment is less than seven. Otherwise point of support shall be the root of the flange of the rolled segment.

(5) The periphery of the perforation at all points shall have a minimum radius of 1½ inches.

(6) For thickness of metal see Article 1.7.88.

(B) Lacing Bars

When lacing bars are used, the following provisions shall govern their design:

(1) Lacing bars of compression members shall be so spaced that the slenderness ratio of the portion of the flange in-

cluded between the lacing bar connections will be not more than 40 nor more than ⅔ of the slenderness ratio of the member.

(2) The section of the lacing bars shall be determined by the formula for axial compression in which l is taken as the distance along the bar between its connections to the main segments for single lacing, and as 70 percent of that distance for double lacing.

(3) If the distance across the member between fastener lines in the flanges is more than 15 inches, and a bar with a single fastener in the connection is used, the lacing shall be double and fastened at the intersections.

(4) The angle between the lacing bars and the axis of the member shall be approximately 45 degrees for double lacing and 60 degrees for single lacing.

(5) Lacing bars may be shapes or flat bars. For main members the minimum thickness of flat bars shall be 1/40 of the distance along the bar between its connections for single lacing and 1/60 for double lacing. For bracing members the limits shall be 1/50 for single lacing and 1/75 for double lacing.

(6) The diameter of fasteners in lacing bars shall not exceed one-third the width of the bar. There shall be at least two fasteners in each end of lacing bars connected to flanges more than 5 inches in width.

1.7.84 — GUSSET PLATES

Gusset or connection plates preferably shall be used for connecting main members, except when the members are pin-connected. The fasteners connecting each member shall be symmetrical with the axis of the member, so far as practicable, and the full development of the elements of the member shall be given consideration. The gusset plates shall be of ample thickness to resist shear, direct stress, and flexure, acting on the weakest or critical section of maximum stress.

Re-entrant cuts, except curves made for appearance, shall be avoided as far as practicable.

If the length of unsupported edge of a gusset plate exceeds the value of the expression $\dfrac{11{,}000}{\sqrt{F_y}}$ times its thickness, the edge shall be stiffened.

Listed below are the values of the expression $\dfrac{11{,}000}{\sqrt{F_y}}$ for the following grades of steel:

	36,000 psi.,	Y.P. Min.	58
	42,000 psi.,	Y.P. Min.	54
45,000 &	46,000 psi.,	Y.P. Min.	51
	50,000 psi.,	Y.P. Min.	49
	55,000 psi.,	Y.P. Min.	47
	60,000 psi.,	Y.P. Min.	45
	65,000 psi.,	Y.P. Min.	43
	90,000 psi.,	Y.P. Min.	37
	100,000 psi.,	Y.P. Min.	35

1.7.85 — HALF-THROUGH TRUSS SPANS

The vertical truss members and the floorbeams and their connections in half-through truss spans shall be proportioned to resist a lateral force of not less than 300 pounds per linear foot, applied at the top chord panel points of each truss.

The top chord shall be considered as a column with elastic lateral supports at the panel points. The critical buckling force of the column, so determined, shall exceed the maximum force from dead load, live load and impact in any panel of the top chord by not less than 50 percent.*

1.7.86 — FASTENER PITCH IN ENDS OF COMPRESSION MEMBERS

In the ends of compression members, the pitch of fasteners connecting the component parts of the member shall not exceed four times the diameter of the fastener for a length equal to 1½ times the maximum width of the member. Beyond this point, the pitch shall be increased gradually for a length equal to 1½ times the maximum width of the member until the maximum pitch is reached.

1.7.87 — NET SECTION OF RIVETED OR HIGH STRENGTH BOLTED TENSION MEMBERS

The net section of a riveted or high strength bolted tension member is the sum of the net sections of its component parts. The net section of a part is the product of the thickness of the part multiplied by its least net width.

The net width for any chain of holes extending progressively across the part shall be obtained by deducting from the gross width the sum of the diameters of all the holes in the chain and adding, for each gage space in the chain, the quantity:

$$\frac{S^2}{4g}$$

where S = pitch of any two successive holes in the chain

g = gage of the same holes

The net section of the part is obtained from the chain which gives the least net width.

For angles, the gross width shall be the sum of the widths of the legs less the thickness. The gage for holes in opposite legs shall be the sum of gages from back of angle less the thickness.

At a splice, the total stress in the member being spliced is transferred by fasteners to the splice material.

When determining the unit stress on any least net width of either splice material or member being spliced, the amount of the stress previously transferred by fasteners adjacent to the section being investigated shall be considered in determining the unit stress on the net section.

The diameter of the hole shall be taken as ⅛ inch greater than the nominal diameter of the rivet or high strength bolt, unless larger holes are permitted in accordance with Art. 1.7.5.

---

* For a discussion of columns with elastic lateral supports, refer to Timoshenko, "Theory of Elastic Stability," McGraw-Hill Book Company, First Edition, Page 122.

1.7.88 — COMPRESSION MEMBERS — THICKNESS OF METAL

Compression members shall be so designed that the main elements of the section will be connected directly to the gusset plates, pins, or other members.

The center of gravity of a built-up section shall coincide as nearly as practicable with the center of the section. Preferably, segments shall be connected by solid webs or perforated cover plates.

Plates supported on one side, outstanding legs of angles and perforated plates. For outstanding plates, the outstanding legs of angles, and perforated plates at the perforations, the b/t ratio of the plates or angle segments, when used in compression, shall not be greater than the value obtained by use of the formula:

$$b/t = \frac{1625}{\sqrt{f_a}}$$

but in no case shall b/t be greater than 12 for main members and 16 for secondary members.

(Note—b is the distance from the edge of plate or edge of perforation to the point of support.)

When the compressive stress equals the limiting factor $\frac{.55F_y}{1.25}$, the b/t ratio of the segments indicated above shall not be greater than the ratios shown for the following grades of steel:

                    36,000 psi., Y.P. Min. b/t = 12
      42,000 to 50,000 psi., Y.P. Min. b/t = 11
                    55,000 psi., Y.P. Min. b/t = 10.5
      60,000 to 65,000 psi., Y.P. Min. b/t = 10
                    90,000 psi., Y.P. Min. b/t = 8.0
               100,000 psi., Y.P. Min. b/t = 7.5

Plates supported on two edges or webs of main component segments. For members of box shape, consisting of main plates, rolled sections, or made up component segments, with cover plates, the b/t ratio of the main plates or webs of the segments, when used in compression shall not be greater than the value obtained by use of the formula:

$$b/t = \frac{4000}{\sqrt{f_a}}$$

but in no case shall b/t be greater than 45.

(Note—b is the distance between points of support for the plate and between roots of flanges for the webs of rolled segments.)

When the compressive stresses equal the limiting factor $\frac{.55F_y}{1.25}$, the b/t ratio of the plates and segments indicated above shall not be greater than the ratios shown for the following grades of steel:

                        36,000 psi., Y.P. Min. b/t = 32
                        42,000 psi., Y.P. Min. b/t = 29
          45,000 & 46,000 psi., Y.P. Min. b/t = 28
                        50,000 psi., Y.P. Min. b/t = 27
                        55,000 psi., Y.P. Min. b/t = 26
                        60,000 psi., Y.P. Min. b/t = 25
                        65,000 psi., Y.P. Min. b/t = 23
                        90,000 psi., Y.P. Min. b/t = 20
                     100,000 psi., Y.P. Min. b/t = 19

1.7.88　　　　　　　　　　　DESIGN　　　　　　　　　　　157

Solid cover plates supported on two edges or webs connecting main members or segments. For members of H or box shape consisting of solid cover plates or solid webs connecting main plates or segments, the b/t ratio of the solid cover plates or webs when used in compression shall not be greater than the value obtained by use of the formula:

$$b/t = \frac{5000}{\sqrt{f_a}}$$ but in no case shall b/t be greater than 50.

(Note—b is the unsupported distance between points of support.)

When the compressive stresses equal the limiting factor $\frac{.55F_y}{1.25}$, the b/t ratio of the cover plate and webs indicated above shall not be greater than the ratios shown for the following grades of steel:

```
              36,000 psi., Y.P. Min.    b/t=40
              42,000 psi., Y.P. Min.    b/t=37
45,000 &      46,000 psi., Y.P. Min.    b/t=35
              50,000 psi., Y.P. Min.    b/t=34
              55,000 psi., Y.P. Min.    b/t=32
              60,000 psi., Y.P. Min.    b/t=31
              65,000 psi., Y.P. Min.    b/t=29
              90,000 psi., Y.P. Min.    b/t=25
             100,000 psi., Y.P. Min.    b/t=24
```

Perforated cover plates supported on two edges. For members of box shape consisting of perforated cover plates connecting main plates or segments, the b/t ratio of the perforated cover plates when used in compression shall not be greater than the value obtained by use of the formula:

$$b/t = \frac{6000}{\sqrt{f_a}}$$ but in no case shall b/t be greater than 55.

(Note—b is the distance between points of support. Attention is directed to requirements for plate thickness at perforations, namely plate supported on one side, which also shall be satisfied.)

When the compressive stresses equal the limiting factor $\frac{.55F_y}{1.25}$, the b/t ratio of the perforated cover plates shall not be greater than the ratios shown for the following grades of steel:

```
              36,000 psi., Y.P. Min.    b/t=48
              42,000 psi., Y.P. Min.    b/t=44
45,000 &      46,000 psi., Y.P. Min.    b/t=42
              50,000 psi., Y.P. Min.    b/t=41
              55,000 psi., Y.P. Min.    b/t=39
              60,000 psi., Y.P. Min.    b/t=37
              65,000 psi., Y.P. Min.    b/t=35
              90,000 psi., Y.P. Min.    b/t=30
             100,000 psi., Y.P. Min.    b/t=29
```

In the above expressions—
f_a = the calculated compressive stress
b = is the width (defined as indicated for each expression)
t = is the plate or web thickness

The point of support shall be the inner line of fasteners or fillet welds connecting the plate to the main segment. For plates butt welded to the flange edge of rolled segments the point of support may be taken as the weld whenever the ratio of outstanding flange width to flange thickness of the rolled segment is less than seven. Otherwise point of support shall be the root of flange of rolled segment. Terminations of the butt welds are to be ground smooth.

1.7.89 — STAY PLATES

Where the open sides of compression members are not connected by perforated plates, such members shall be provided with lacing bars and shall have stay plates as near each end as practicable. Stay plates shall be provided at intermediate points where the lacing is interrupted. In main members, the length of the end stay plates between end fasteners shall be not less than 1¼ times the distance between points of support and the length of intermediate stay plates not less than ¾ of that distance. In lateral struts and other secondary members, the over-all length of end and intermediate stay plates shall be not less than ¾ of the distance between points of support.

The point of support shall be the inner line of fasteners or fillet welds connecting the stay plates to the flanges. For stay plates butt welded to the flange edge of rolled segments, the point of support may be taken as the weld whenever the ratio of outstanding flange width to flange thickness of the rolled segment is less than seven. Otherwise the point of support shall be the root of flange of rolled segment. When stay plates are butt welded to rolled segments of a member, the allowable stress in the member shall be determined in accordance with Article 1.7.3. Terminations of butt welds shall be ground smooth.

The separate segments of tension members composed of shapes may be connected by perforated plates or by stay plates or end stay plates and lacing. End stay plates shall have the same minimum length as specified for end stay plates on main compression members and intermediate stay plates shall have a minimum length of ¾ of that specified for intermediate stay plates on main compression members. The clear distance between stay plates on tension members shall not exceed 3 feet.

The thickness of stay plates shall be not less than ¹⁄₅₀ of the distance between points of support for main members, and ¹⁄₆₀ of that distance for bracing members. Stay plates shall be connected by not less than three fasteners on each side, and in members having lacing bars the last fastener in the stay plates, preferably shall also pass through the end of the adjacent bar.

RIBBED ARCHES

1.7.90 — THICKNESS OF WEB PLATES, SOLID RIB ARCHES

The thickness ratio D/t of each web plate in solid rib arches having no longitudinal stiffeners shall not be greater than the value obtained by use of the following formula:

$$D/t = \frac{7200}{\sqrt{f_a}}$$ but in no case shall D/t be greater than 60.

The thickness ratio D/t of web plates in solid rib arches equipped with longitudinal stiffeners, that is when the web is reinforced along its axis with a longitudinal stiffener of ample cross-sectional area and rigidity, shall not be greater than twice the value obtained by use of the above formula.

When the compressive stresses equal the limiting factor $\frac{.55F_y}{1.25}$ the D/t ratio of the web plates shall not be greater than the ratios shown for the following grades of steel:

		Without Longit. Stiffeners	With Longit. Stiffeners
	36,000 psi., Y.P. Min.	D/t=57	D/t=114
	42,000 psi., Y.P. Min.	D/t=53	D/t=106
45,000 &	46,000 psi., Y.P. Min.	D/t=51	D/t=102
	50,000 psi., Y.P. Min.	D/t=48	D/t=96
	55,000 psi., Y.P. Min.	D/t=46	D/t=92
	60,000 psi., Y.P. Min.	D/t=45	D/t=90
	65,000 psi., Y.P. Min.	D/t=42	D/t=84
	90,000 psi., Y.P. Min.	D/t=36	D/t=72
	100,000 psi., Y.P. Min.	D/t=34	D/t=68

BENTS AND TOWERS

1.7.91 — BENTS AND TOWERS, GENERAL

Bents, preferably shall be composed of two supporting columns, and the bents usually shall be united in pairs to form towers. The design of members for bents and towers is governed by the applicable articles under "Trusses" and "Details of Design".

1.7.92 — SINGLE BENTS

Single bents shall have hinged ends or else shall be designed to resist bending.

1.7.93 — BATTER

Bents, preferably, shall have a sufficient spread at the base to prevent uplift under the assumed lateral loadings. In general, the width of a bent at its base shall be not less than one-third of its height.

1.7.94 — BRACING

Towers shall be braced, both transversely and longitudinally, with stiff members having either welded, high strength bolted or riveted connections. The sections of members of longitudinal bracing in each panel shall not be less than those of the members in corresponding panels of the transverse bracing.

The bracing of long columns shall be designed to fix the column about both axes at or near the same point.

Column splices shall be at or close above the panel points of the bracing.

Horizontal diagonal bracing shall be placed in all towers having more than two vertical panels, at alternate intermediate panel points.

1.7.95 — BOTTOM STRUTS

The bottom struts of towers shall be strong enough to slide the movable shoes with the structure unloaded, the coefficient of friction being assumed at 0.25. Provision for expansion of the tower bracing shall be made in the column bearings.

COMPOSITE GIRDERS

1.7.96 — COMPOSITE I-GIRDERS, GENERAL

This section pertains to structures composed of steel girders with concrete slabs connected by shear connectors.

General specifications pertaining to the design of concrete and steel structures shall apply to structures utilizing composite girders where such specifications are applicable. Composite girders and slabs shall be designed and the stresses computed by the composite moment of inertia method and shall be consistent with the predetermined properties of the various materials used.

The ratio of the moduli of elasticity of steel (29,000,000 psi) to those of concrete of various design strengths shall be as follows:

f_c' = unit ultimate compressive strength of concrete as determined by cylinder tests at the age of 28 days, psi.

n = ratio of modulus of elasticity of steel to that of concrete. The value of n, as a function of the ultimate cylinder strength of concrete, shall be assumed as follows:

f_c' = 2000-2400	n = 15
= 2500-2900	= 12
= 3000-3900	= 10
= 4000-4900	= 8
= 5000 or more	= 6

The effect of creep shall be considered in the design of composite girders which have dead loads acting on the composite section. In such structures, stresses and horizontal shears produced by dead loads acting on the composite section shall be computed for "n" as given above or for this value multiplied by 3, whichever gives the higher stresses and shears.

If concrete with expansive characteristics is used, composite design should be used with caution and provision must be made in the design to accommodate the expansion.

Composite sections should preferably be proportioned so that the neutral axis lies below the top surface of the steel beam. If concrete is on the tension side of the neutral axis, it shall not be considered in computing moments of inertia or resisting moments except for deflection calculations. Mechanical anchorages shall be provided to tie the sections together and to develop stresses on the plane joining the concrete and the steel.

The steel beams, especially if not supported by intermediate falsework shall be investigated for stability during the time the concrete is in place and before it has hardened.

1.7.97 — SHEAR CONNECTORS

The mechanical means which are used at the junction of the girder and slab for the purpose of developing the shear resistance necessary to produce composite action shall conform to the specifications of the respective materials as provided in Division II. The shear connectors shall be of types which permit a thorough compaction of the concrete in order to insure that their entire surfaces are in contact with the concrete. They shall be capable of resisting both horizontal and vertical movement between the concrete and the steel.

The capacity of stud and channel shear connectors welded to the girders is given in Article 1.7.100. Channel shear connectors shall have at least 3/16 inch fillet welds placed along the heel and toe of the channel.

The clear depth of concrete cover over the tops of the shear connectors shall be not less than 2 inches. Shear connectors shall penetrate at least 2 inches above bottom of slab.

The clear distance between the edge of a girder flange and the edge of the shear connectors shall be not less than one inch.

1.7.98 — EFFECTIVE FLANGE WIDTH

In composite girder construction the assumed effective width of the slab as a T-beam flange shall not exceed the following:
(1) One-fourth of the span length of the girder.
(2) The distance center to center of girders.
(3) Twelve times the least thickness of the slab.

For girders having a flange on one side only, the effective flange width shall not exceed one-twelfth of the span length of the girder, nor six times the thickness of the slab, nor one-half the distance center to center of the next girder.

1.7.99 — STRESSES

Maximum compressive and tensile stresses in girders which are not provided with temporary supports during the placing of the permanent dead load, shall be the sum of the stresses produced by the dead loads acting on the steel girders alone and the stresses produced by the superimposed loads acting on the composite girder. When girders are provided with effective intermediate supports which are kept in place until the concrete has attained 75 percent of its required 28-day strength, the dead and live load stresses shall be computed on the basis of the composite section.

In continuous spans, the positive moment portion may be designed with composite sections as in simple spans. Shear connectors shall be provided in the negative moment portion in which the reinforcement steel embedded in the concrete is considered a part of the composite section. In case the reinforcement steel embedded in the concrete is not used in computing section properties for negative moments, shear connectors need not be provided in these portions of the spans, but additional connectors shall be placed in the region of the point of dead load contraflexure in accordance with Art. 1.7.100(A)(3). Shear connectors shall be provided in accordance with Article 1.7.100.

1.7.100 — SHEAR

(A) Horizontal Shear

The maximum pitch of shear connectors shall not exceed 24 inches, except over the interior supports of continuous beams where wider spacing may be used to avoid placing connectors at locations of high stresses in the tension flange.

Resistance to horizontal shear shall be provided by mechanical shear connectors at the junction of the concrete slab and the steel girder. The shear connectors shall be mechanical devices placed transversely across the flange of the girder spaced at regular or variable intervals. The shear connectors shall be designed for fatigue * and checked for ultimate strength.

(1) Fatigue

The range of horizontal shear shall be computed by the formula:

$$S_r = \frac{V_r Q}{I}$$

where $S_r =$ the range of horizontal shear per linear inch at the junction of the slab and girder at the point in the span under consideration.

$V_r =$ the range of shear due to live loads and impact. At any section, the range of shear shall be taken as the difference in the minimum and maximum shear envelopes (excluding dead loads).

$Q =$ The statical moment about the neutral axis of the composite section, of the transformed compressive concrete area or the area of reinforcement embedded in the concrete for negative moment.

---

* Reference is made to the paper titled "Fatigue Strength of Shear Connectors" by Roger G. Slutter and John W. Fisher in HIGHWAY RESEARCH RECORD, No. 147, published by the Highway Research Board, Washington, D.C., 1966.

I = The moment of inertia of the transformed composite girder in positive moment regions or the moment of inertia provided by the steel beam including or excluding the area of reinforcement embedded in the concrete in negative moment regions.

(In the above, the compressive concrete area is transformed into an equivalent area of steel by dividing the effective concrete flange width by the modular ratio, "n".)

The allowable range of horizontal shear, "Z_r", in pounds on an individual connector is as follows:

Channels $\quad Z_r = Bw$

Welded studs (for ratios of H/d equal to or greater than 4)

$$Z_r = \alpha d^2$$

where w = the length of a channel shear connector in inches measured in a transverse direction on the flange of a girder.

d = diameter of stud, in inches

α = 13,000 for 100,000 cycles
10,600 for 500,000 cycles
7,850 for 2,000,000 cycles

B = 4,000 for 100,000 cycles
3,000 for 500,000 cycles
2,400 for 2,000,000 cycles

H = height of stud in inches

The required pitch of shear connectors is determined by dividing the allowable range of horizontal shear of all connectors at one transverse girder crossection (ΣZ_r) by the horizontal range of shear S_r per linear inch. Over the interior supports of continuous beams the pitch may be modified to avoid placing the connectors at locations of high stresses in the tension flange provided that the total number of connectors remains unchanged.

(2) Ultimate Strength

The number of connectors so provided for fatigue shall be checked to ensure that adequate connectors are provided for ultimate strength. The number of shear connectors required between the points of maximum positive moment and the end supports or dead load points of contraflexure, and between points of maximum negative moment and the dead load points of contraflexure shall equal or exceed the number given by the formula:

$$N = \frac{P}{\phi S_u}$$

where   N = the number of connectors between points of maximum positive moment and adjacent end supports or dead load points of contraflexure, or between points of maximum negative moment and adjacent dead load points of contraflexure.

S_u = the ultimate strength of the shear connecter as given below.

ϕ = a reduction factor = 0.85.

P = force in the slab as defined hereafter as P_1, P_2, or P_3.

At points of maximum positive moment, the force in the slab is taken as the smaller value of the formulas:

$$P_1 = A_s F_y$$

or

$$P_2 = 0.85 f'_c bc$$

where A_s = total area of the steel section including coverplates.

F_y = specified minimum yield point of the steel being used.

f'_c = compressive strength of concrete at age of 28 days.

b = effective flange width given in Art. 1.7.99.

c = thickness of the concrete slab.

At points of maximum negative moment the force in the slab is taken as:

$$P_3 = A_s^r F_y^r$$

where A_s^r = total area of longitudinal reinforcing steel at the interior support within the effective flange width.

F_y^r = specified minimum yield point of the reinforcing steel.

The ultimate strength of the shear connecter is given as follows:
Channels:

$$S_u = 550 (h + t/2) \ w \ \sqrt{f'_c}$$

Welded Studs (H/d $\geq$ 4):

$$S_u = 930 \ d^2 \ \sqrt{f'_c}$$

where S_u = ultimate strength of individual shear connector, in pounds.

h = the average flange thickness of the channel flange, in inches.

t = the thickness of the web of a channel, in inches.

w = length of a channel shear connector, in inches.

f'_c = compressive strength of the concrete at 28 days, psi.

d = diameter of stud, in inches.

(3) Additional Connectors to Develop Slab Stress

The number of additional connectors required at points of contraflexure, when reinforcement steel embedded in the concrete is not used in computing section properties for negative moments, shall be computed by the formula:

$$N_c = \frac{A_r f_r}{Z_r}$$

where N_c = number of additional connectors for each beam at point of contraflexure.

A_r = total area of longitudinal slab reinforcement steel for each beam over interior support.

f_r = range of stress due to live load plus impact, in the slab reinforcement over the support (in lieu of more accurate computations, f_r may be taken as equal to 10,000 psi).

Z_r = the allowable range of horizontal shear on an individual shear connector.

The additional connectors, N_c, shall be placed adjacent to the point of dead load contraflexure within a distance equal to ⅓ the effective slab width, i.e., placed either side of this point or centered about it.

(B) Vertical Shear

The intensity of unit shearing stress in a composite girder may be determined on the basis that the web of the steel girder carries the total external shear, neglecting the effects of the steel flanges and of the concrete slab. The shear may be assumed to be uniformly distributed throughout the gross area of the web.

1.7.101 — DEFLECTION

The provisions of Article 1.7.12, in regard to deflections from live load plus impact also shall be applicable to composite girders.

When the girders are not provided with falsework or other effective intermediate support during the placing of the concrete slab, the deflection due to the weight of the slab and other permanent dead loads added before the concrete has attained 75 per cent of its required 28-day strength shall be computed on the basis of non-composite action.

1.7.102 — COMPOSITE BOX GIRDERS, GENERAL

This section pertains to the design of simple and continuous span steel-concrete composite multi-box girder bridges of moderate length. It is applicable to box girders of single cell, having width center to center of top steel flanges approximately equal to the distance center to center of adjacent top steel flanges of adjacent box girders. The cantilever overhang of the deck slab (including curbs and parapets) beyond the exterior web, shall be limited to 60 percent of the distance between the centers of adjacent top steel flanges of adjacent box girders, but in no case greater than 6 feet.

The provisions of Division I, Design, shall govern where applicable, except as specifically modified by Articles 1.7.102 through 1.7.109.

1.7.103 — LATERAL DISTRIBUTION OF LOADS FOR BENDING MOMENT

The live load bending moment for each box girder shall be determined by applying to the girder, the fraction W_L of a wheel load (both front and rear), determined by the following equation:

$$W_L = 0.1 + 1.7R + \frac{0.85}{N_w}$$

where $R = \dfrac{N_w}{\text{Number of Box Girders}}$

$N_w = W_c/12$, reduced to the nearest whole number

$W_c =$ Roadway width between curbs (in feet), or barriers if curbs are not used. R shall not be less than 0.5 nor greater than 1.5.

The provision of Article 1.2.9, Reduction in Load Intensity, shall not apply in the design of box girders when using the design load W_L given by the above equation.

1.7.104 — DESIGN OF WEB PLATES

(A) Vertical Shear

The design shear V_w for a web shall be calculated using the following equation:

$$V_w = V_v/\cos \theta$$

where $V_v =$ vertical shear

$\theta =$ angle of inclination of the web plate to the vertical

(B) Secondary Bending Stresses

Web plates may be plumb (90° to bottom of flange) or inclined. If the inclination of the web plates to a plane normal to bottom flange is no greater than 1 to 4, and the width of the bottom flange is no greater than 20 percent of the span, the transverse bending stresses resulting from distortion of the girder cross section and from vibrations of the bottom plate, need not be considered. For structures in this category transverse bending stresses due to supplementary loadings, such as utilities, shall not exceed 5,000 psi.

For structures exceeding these limits, a detailed evaluation of the transverse bending stresses due to all causes shall be made. These stresses shall be limited to a maximum stress or range of stress of 20,000 psi.

1.7.105 — DESIGN OF BOTTOM FLANGE PLATES

(A) Tension Flanges

In cases of simply supported spans, the bottom flange shall be considered completely effective in resisting bending if its width does not exceed one-fifth (⅕) the span length. If the flange plate width exceeds one-fifth (⅕) of the span, an amount equal to one-fifth (⅕) of the span only shall be considered effective.

For continuous spans, the criteria above shall be applied to the lengths between-points of contraflexure.

(B) Compression Flanges Unstiffened

Unstiffened compression flanges designed for the basic allowable stress of 0.55 F_y shall have a width to thickness ratio equal to or less than the value obtained by the use of the formula:

$$b/t = \frac{6140}{\sqrt{F_y}}$$

where $b =$ flange width between webs in inches

$t =$ flange thickness in inches

For greater b/t ratios, but not exceeding 60, the stress in an unstiffened bottom flange shall not exceed the value determined by the use of the formula:

$$f_b = 0.55 F_y - 0.224 F_y \left\{ 1 - \sin(\pi/2) \left(\frac{13{,}300 - \frac{b\sqrt{F_y}}{t}}{7160} \right) \right\}$$

For values of b/t exceeding $13{,}300/\sqrt{F_y}$, the stress in the flange shall not exceed the value given by the formula:

$$f_b = 57.6 \, (t/b)^2 \times 10^6$$

The b/t ratio preferably should not exceed 60 except in areas of low stress near points of dead load contraflexure.

Should b/t ratio exceed 45, longitudinal stiffeners should be considered.

(C) Compression Flanges Stiffened Longitudinally *

Longitudinal stiffeners shall be at equal spacings across the flange width and shall be proportioned so that the moment of inertia of each stiffener about an axis parallel to the flange and at the base of the stiffener is at least equal to:

$$I_s = \phi t^3 w$$

where $\phi = 0.07 \, k^3 n^4$ for values of n greater than 1

$\phi = 0.125 \, k^3$ for a value of $n = 1$

$w =$ width of flange between longitudinal stiffeners or distance from a web to the nearest longitudinal stiffener

---

* In solving these equations a value of k between 2 and 4 generally should be assumed.

n = number of longitudinal stiffeners

k = buckling coefficient which shall not exceed 4

For the flange, including stiffeners, to be designed for the basic allowable stress of 0.55 F_y, the ratio w/t shall not exceed the value given by the formula:

$$w/t = \frac{3070\sqrt{k}}{\sqrt{F_y}}$$

For greater values of w/t but not exceeding $6650\sqrt{k}/\sqrt{F_y}$ or 60, whichever is less, the stress in the flange, including stiffeners, shall not exceed the value determined by the formula:

$$f_b = 0.55\,F_y - 0.224\,F_y\left[1 - \sin(\pi/2)\left(\frac{6650\sqrt{k} - \frac{w\sqrt{F_y}}{t}}{3580\sqrt{k}}\right)\right]$$

For values of w/t exceeding $6650\sqrt{k}/\sqrt{F_y}$ but not exceeding 60, the stress in the flange, including stiffeners, shall not exceed the value given by the formula:

$$f_b = 14.4\,k(t/w)^2 \times 10^6$$

When longitudinal stiffeners are used, it is preferable to have at least one transverse stiffener placed near the point of dead load contraflexure. The stiffener should have a size equal to that of a longitudinal stiffener.

If the longitudinal stiffeners are placed at their maximum w/t ratio to be designed for the basic allowable design stresses of 0.55 F_y and the number of longitudinal stiffeners exceeds 2, then transverse stiffeners should be considered.

(D) Compression Flanges Stiffened Longitudinally and Transversely

The longitudinal stiffeners shall be at equal spacings across the flange width and shall be proportioned so that the moment of inertia of each stiffener about an axis parallel to the flange and at the base of the stiffener is at least equal to:

$$I_s = 8\,t^3 w$$

The transverse stiffeners shall be proportioned so that the moment of inertia of each stiffener about an axis through the centroid of the section and parallel to its bottom edge is at least equal to:

$$I_t = 0.10\,(n+1)^3\,w^3\,\frac{f_s}{E}\,\frac{A_f}{a}$$

where A_f = area of bottom flange including longitudinal stiffeners

a = spacing of transverse stiffeners

f_s = maximum longitudinal bending stress in the flange of the panels on either side of the transverse stiffener

E = modulus of elasticity of steel

For the flange, including stiffeners, to be designed for the basic allowable stress of $0.55\ F_y$, the ratio w/t for the longitudinal stiffeners shall not exceed the value given by the formula:

$$w/t = \frac{3070\ \sqrt{k_1}}{\sqrt{F_y}}$$

where $k_1 = \dfrac{[1 + (a/b)^2]^2 + 87.3}{(n+1)^2 (a/b)^2 [1 + 0.1(n+1)]}$

For greater values of w/t, but not exceeding $6650\ \sqrt{k_1} / \sqrt{f_y}$ or 60, whichever is the less, the stress in the flange, including stiffeners, shall not exceed the value determined by the formula:

$$f_b = 0.55\ F_y - 0.224\ F_y \left[1 - \sin(\pi/2) \left(\frac{6650\ \sqrt{k_1} - \dfrac{w\sqrt{F_y}}{t}}{3580\sqrt{k_1}} \right) \right]$$

For values of w/t exceeding $6650\ \sqrt{k_1} / \sqrt{F_y}$ but not exceeding 60, the stress in the flange, including stiffeners, shall not exceed the value given by the formula:

$$f_b = 14.4\ k_1\ (t/w)^2 \times 10^6$$

The maximum value of the buckling coefficient k_1, shall be 4. When k_1 has its maximum value, the transverse stiffeners shall have a spacing, a, equal to or less than 4w. If the ratio a/b exceeds 3, transverse stiffeners are not necessary.

The transverse stiffeners need not be connected to the flange plate but shall be connected to the webs of the box and to each longitudinal stiffener. The connection to the web shall be designed to resist the vertical force determined by the formula:

$$R_w = \frac{F_y S_s}{2b}$$

where S_s = section modulus of the transverse stiffener

The connection to each longitudinal stiffener shall be designed to resist the vertical force determined by the formula:

$$R_s = \frac{F_y S_s}{nb}$$

(E) Compression Flange Stiffeners, General

The width to thickness ratio of any outstanding element of the flange stiffeners shall not exceed the value determined by the formula:

$$\frac{b'}{t'} = \frac{2600}{\sqrt{F_y}}$$

where b' = width of any outstanding stiffener element

t' = thickness of outstanding stiffener element

Longitudinal stiffeners shall be extended to locations where the maximum stress in the flange does not exceed that allowed for base metal adjacent to or connected by fillet welds.

1.7.106 — DESIGN OF FLANGE TO WEB WELDS

The total effective thickness of the web-flange welds shall be not less than the thickness of the web. If fillet welds are used, they shall be on both sides of the connecting flange or web plate.

1.7.107 — DIAPHRAGMS

Diaphragms, cross-frames, or other means shall be provided within the box girders at each support to resist transverse rotation, displacement, and distortion.

Intermediate diaphragms or cross-frames are not required for steel box girder bridges designed in accordance with this specification.

1.7.108 — LATERAL BRACING

Generally no lateral bracing system is required between box girders. A horizontal load equal to 25 pounds per square foot acting on the area exposed in elevation shall be applied in the plane of the bottom flange. The section assumed to resist the horizontal load shall consist of the bottom flange acting as a web and 12 times the thickness of the webs acting as flanges. A lateral bracing system shall be provided if the combined stresses due to the specified horizontal force and dead load of steel and deck exceed 150 percent of the allowable design stress.

1.7.109 — ACCESS AND DRAINAGE

Consistent with climate, location, and materials, consideration shall be given to the providing of man-holes or other openings, either in the deck slab or in the steel box for form removal, inspection, maintenance, drainage, etc.

HYBRID GIRDERS

1.7.110 — HYBRID GIRDERS, GENERAL

This section pertains to the design of (1) noncomposite girders that have both flanges of the same minimum specified yield strength and a web with a lower minimum specified yield strength, (2) composite girders that have a tension flange with a higher minimum specified yield strength than the web and a compression flange with a minimum specified yield strength not less than that of the web and (3) girders that utilize an orthotropic deck as the top flange and have a web with a lower minimum specified yield strength than the bottom flange. It is applicable to both simple and continuous span girders. In noncomposite girders and in the negative moment portion of continuous span composite girders, the compression flange area shall be equal to the tension flange area or larger than the tension flange area by an amount not exceeding 15 percent. In composite girders, excluding the negative moment portion in continuous span girders, the compression flange area shall be equal to or smaller than the tension flange area. Steel girders that support the dead weight of the slab without composite action, but act compositely with the slab in supporting the live load, shall be considered to be composite girders. In either composite or noncomposite girders, the minimum specified yield strength of the web shall not be less than 35 percent of the minimum specified yield strength of the tension flange.

In girders that utilize an orthotropic deck as the top flange, the minimum specified yield strength of the web shall not be less than 35 percent of the minimum specified yield strength of the bottom flange in regions of positive bending moment and 50 percent of the minimum specified yield strength of the bottom flange in regions of negative bending moment. As used in this section, flange refers to the flange of the steel girder and excludes the slab and reinforcing bars.

The provisions of Division I, Design, shall govern where applicable, except as specifically modified by Articles 1.7.110 through 1.7.113.

1.7.111 — ALLOWABLE STRESSES

(A) Bending

The bending stress in the web may exceed the allowable stress for the web steel provided that the stress in each flange does not exceed the allowable stress from Art. 1.7.1 or 1.7.3 for the steel in that flange multiplied by the reduction factor

$$R = 1 - \frac{\beta\psi(1-\alpha)^2(3-\psi+\psi\alpha)}{6+\beta\psi(3-\psi)}$$

(See Figures 1.7.111A and 1.7.111B)

where

α = the minimum specified yield strength of the web divided by the minimum specified yield strength of the tension flange.*

β = the area of the web divided by the area of the tension flange.*

ψ = the distance from the outer edge of the tension flange* to the neutral axis (of the transformed section for composite girders) divided by the depth of the steel section.

The bending stress in the concrete slab in composite girders shall not exceed the allowable stress for the concrete multiplied by R.

(B) Shear

The shear stress in the web (the shear force divided by the web area) shall not exceed the allowable shear stress for the web steel.

(C) Fatigue

Hybrid girders shall be designed for fatigue as if they were homogeneous girders of the flange steel. The allowable fatigue stresses for web splices and for attachments to the web shall be based on the web steel, except that stiffener to web and flange-web fillet weld connections shall be based on the flange steel.

---

* Bottom flange of orthotropic deck bridges.

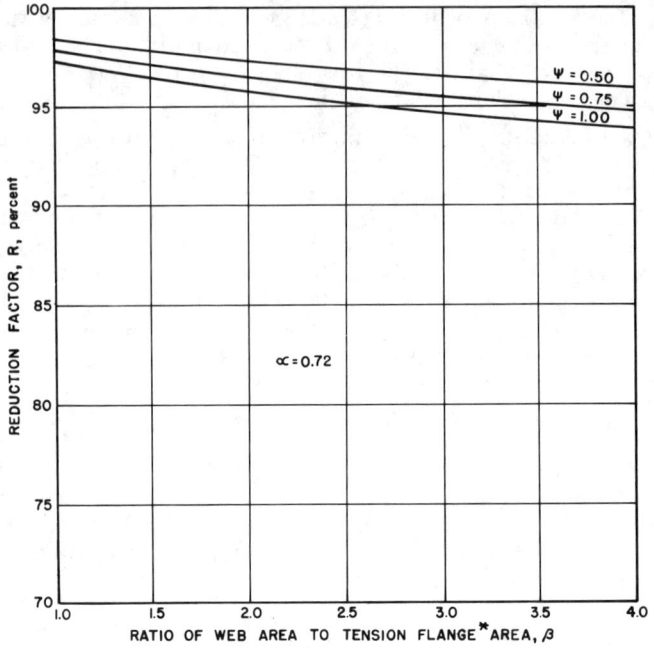

FIGURE I.7.III A

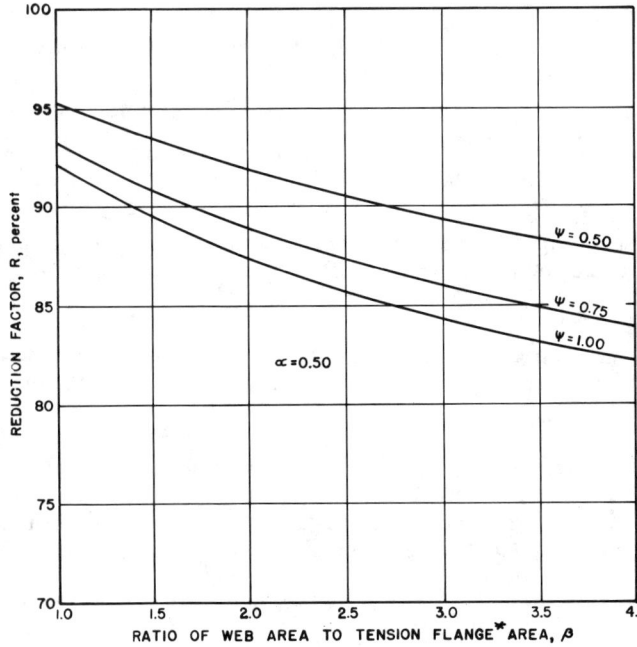

FIGURE I.7.III B

* Bottom flange of orthotropic deck bridges.

1.7.112 — PLATE THICKNESS REQUIREMENTS

In calculating the maximum width-to-thickness ratio of the flange plate according to Article 1.7.69 and the minimum thickness of the web plate according to Article 1.7.70, f_b shall be taken as the calculated bending stress in the compression flange divided by the reduction factor, R.

1.7.113 — BEARING STIFFENER REQUIREMENTS

In designing bearing stiffeners at interior supports of continuous hybrid girders for which α is less than 0.7, no part of the web shall be assumed to act in bearing.

HEAT-CURVED ROLLED BEAMS AND WELDED PLATE GIRDERS

1.7.114 — SCOPE

This section pertains to rolled beams and welded I-section plate girders heat-curved to obtain a horizontal curvature. Steels that are manufactured to a specified minimum yield point greater than 50,000 psi shall not be heat-curved.

1.7.115 — MINIMUM RADIUS OF CURVATURE

For heat-curved beams and girders, the horizontal radius of curvature measured to the centerline of the girder web shall not be less than 150 feet, and shall not be less than the larger of the values calculated (at any and all cross sections throughout the length of the girder) from the following two equations:

$$R = \frac{14bD}{\sqrt{F_y}\,\psi t}$$

$$R = \frac{7500\,b}{F_y \psi}$$

In these equations, F_y is the specified minimum yield point in ksi of steel in the girder web, ψ is the ratio of the total cross-sectional area to the cross-sectional area of both flanges, b is the widest flange width in inches, D is the clear distance between flanges in inches, t is the web thickness in inches, and R is the radius in inches.

In addition to the above requirements, the radius shall not be less than 1000 feet when the flange thickness exceeds 3 inches or the flange width exceeds 30 inches.

1.7.116 — CAMBER

To compensate for possible loss of camber of heat-curved girders in service as residual stresses dissipate, the amount of camber in inches,

Δ, at any section along the length of the girder shall be equal to:

$$\Delta = \frac{\Delta_{DL}}{\Delta_m}\left(\Delta_m + \frac{0.02L^2F_y}{EY_0}\right)$$

where Δ_{DL} is the camber in inches at any point along the span calculated by usual procedures to compensate for deflection due to dead loads or any other specified loads, Δ_m is the maximum value of Δ_{DL} in inches within the span, E is the modulus of elasticity in ksi, F_y is the specified minimum yield point in ksi of the girder flange, Y_0 is the distance from the neutral axis to the extreme outer fiber in inches (maximum distance for non-symmetrical sections), and L is the span length or distance between points of dead-load contraflexure in inches.*

LOAD FACTOR DESIGN

1.7.117 — SCOPE

Load Factor design is an alternate method for design of simple and continuous beam and girder structures of moderate length. It is a method of proportioning structural members for multiples of the design loads. To insure serviceability and durability, consideration is given to the control of permanent deformations under overloads, to the fatigue characteristics under service loadings and to the control of live load deflections under service loadings.

1.7.118 — NOTATION

A = area of cross section (in.^{2})
A_f = area of one flange of beam or girder (in.^{2})
A_s = total area of steel section including cover plates (in.^{2})
A_s = gross effective area of column cross section (in.^{2})
A_w = area of web of beam (in.^{2})
b' = width of projecting flange element (in.)
b' = width of outstanding stiffener element (in.)
D = dead load
D = distance center to center of box girder flange plates (in.)
d = depth of member (in.)
d_b = depth of beam
d_c = depth of column
d_o = distance between tranverse stiffeners (in.)
d_w = depth of steel web of a composite section (in.)
E = modulus of elasticity (29,000,000 psi)
F = stress (psi)
F_{cr} = buckling stress (psi)

---

* Part of the camber loss is attributable to construction loads and will occur during construction of the bridge; total camber loss will be complete after several months of in-service loads. Therefore, a portion of the camber increase (approximately 50 percent) should be included in the bridge profile. Camber losses of this nature (but, generally, smaller in magnitude) are also known to occur in straight beams and girders.

F_y = specified minimum yield point or yield strength of the type of steel being used (psi)
f'_c = specified 28-day compressive strength of concrete (psi)
I = impact
I = moment of inertia (in.^{4})
L_c = length of a compression member (in.)
L_b = distance between points of bracing of compression flange (in.)
L = live load
M, M_1, M_2 = moment on a cross section (in.-lb)
M_u = maximum moment capacity (in.-lb)
P = axial compression on the member (lb)
P_u = maximum axial compression capacity (lb)
r = radius of gyration (in.)
r_y = radius of gyration with respect to Y-Y axis (in.)
S = section modulus (in.^{3})
t = flange thickness (in.)
t = thickness of thinnest part connected by bolts (in.)
t_w = web thickness (in.)
V = shear force on the cross section (lb)
V_u = maximum shear capacity (lb)
Z = Plastic Section Modulus (in.^{3})
ϕ = reduction factor

1.7.119 — LOADS

Service live loads are vehicles which may operate on a highway legally without special load permit.

For design purposes, the service loads are taken as the dead, live and impact loadings described in Section 1.2 (except Art. 1.2.4).

Overloads are the live loads that can be allowed on a structure on infrequent occasions without causing permanent damage. For design purposes the maximum overload is taken as $5/3(L+I)$.

The maximum loads are the loadings specified in Article 1.7.123.

1.7.120 — DESIGN THEORY

The moments, shears and other forces shall be determined by assuming elastic behavior of the structure except as modified in Article 1.7.124(A)(3).

The members shall be proportioned by the methods specified in Articles 1.7.124 through 1.7.135 so that their computed maximum strengths shall be at least equal to the total effects of design loads multiplied by their respective load factors specified in Groups I, II and III of Article 1.7.123.

Service behavior shall be investigated as specified in Articles 1.7.136 through 1.7.138.

1.7.121 — ASSUMPTIONS

(1) Strain in flexural members shall be assumed directly proportional to the distance from the neutral axis.

(2) Stress in steel below the yield strength, F_y, of the grade of steel used shall be taken as 29,000,000 psi times the steel strain. For strain greater than that corresponding to the yield strength, F_y, the stress shall be considered independent of strain and equal to the yield strength, F_y. This assumption shall apply also to the longitudinal reinforcement in the concrete floor slab in the region of negative moment when shear developers are provided to secure composite action in this region.

(3) At maximum strength the compressive stress in the concrete slab of a composite beam shall be assumed independent of strain and equal to $0.85f'_c$.

(4) Tensile strength of concrete shall be neglected in flexural calculations.

1.7.122 — DESIGN STRENGTH FOR STEEL

The design strength for steel shall be the specified minimum yield point or yield strength, F_y, of the steel used as set forth in Article 1.7.1.

1.7.123 — MAXIMUM DESIGN LOADS

The maximum moments, shears or forces to be sustained by a stress-carrying steel member shall be computed from formulas listed below. Members subject to combinations of loads and forces shall be designed for the combined effects.

$$Group\ I = 1.30\left[D + \frac{5}{3}(L+I) \right]$$

For all loadings less than $H20$, provision shall be made for an infrequent heavy load by applying Group IA loading, with the live load assumed to occupy a single lane without concurrent loading in any other lane.

$$Group\ IA = 1.30[D + 2.2(L+I)]$$
$$Group\ II = 1.30[D + W + F + SF + B + S + T]$$

When earthquake loading is taken into account, the Group II loading shall be used substituting EQ for W. When ice pressure is taken into account, the Group II loading shall be used substituting ICE for SF.

$$Group\ III = 1.30[D + L + I + CF + 0.3W + WL + F + LF]$$

The symbols in the above formulas represent the moments, shears or forces caused by the loads and effects described in Article 1.2.22.

1.7.124 — SYMMETRICAL BEAMS AND GIRDERS

(A) Compact Sections

Symmetrical I-shaped beams with high resistance to local buckling and proper bracing to resist lateral torsional buckling qualify as compact sections. Compact sections are able to form plastic hinges which rotate at near constant moment.

Rolled or fabricated I-shaped beams meeting the requirements of paragraph (1) below shall be considered compact sections and the maximum strength shall be as computed:

$$M_u = F_y Z$$

where F_y is the specified yield point of the steel being used,
Z is the plastic section modulus.*

(1) Beams designed as compact sections shall meet the following requirements: (for certain frequently used steels these requirements are listed in Table 1).

    (a) Projecting flange element

$$b'/t \leq \frac{1600}{\sqrt{F_y}}$$

where b' is the width of the projecting flange element,
t is the flange thickness.

    (b) Web thickness

$$d/t_w \leq \frac{13{,}300}{\sqrt{F_y}}$$

where d is the depth of the beam,
t_w is the web thickness.

    (c) Lateral bracing

$$L_b/r_y \leq \frac{7000}{\sqrt{F_y}} \quad \text{when } M_2 \geq 0.7 M_1$$

or

$$L_b/r_y \leq \frac{12{,}000}{\sqrt{F_y}} \quad \text{when } M_2 < 0.7 M_1$$

where L_b is the distance between points of bracing of the compression flange,
r_y is the radius of gyration with respect to the Y-Y axis,
M_1 and M_2 are the moments at the two adjacent braced points. In no case shall L_b exceed the value given in Article 1.7.124 (B)(1)(c).

---

* See Commentary of AISI Bulletin 15 for method of computing Z. Values for rolled sections are listed in the "Manual of Steel Construction," Seventh Edition, 1970, American Institute of Steel Construction.

The required lateral bracing shall be provided by braces capable of preventing lateral displacement and twisting of the main members or by embedment of the top and sides of the compression flange in concrete.

(d) Maximum axial compression

$$P \leq 0.15 F_y A$$

where A is the area of the cross section.

(e) Maximum shear force

$$V \leq 0.55 F_y d t_w$$

(2) Article 1.7.124(A) is applicable to steels with stress-strain diagrams which exhibit a yield plateau followed by a strain hardening range.

Steels such as ASTM A36, A242, A440, A441, A572 and A588 meet these requirements. The limitations set forth in Article 1.7.124(A) are given in Table 1.

TABLE 1

F_y (psi)	36,000	42,000	46,000	50,000	55,000
b'/t	8.4	7.8	7.5	7.2	6.8
d/t	70	65	62	59	57
L_b/r_y $M_2 \geq 0.7 M_1$	37	34	33	31	30
L_b/r_y $M_2 < 0.7 M_1$	63	59	56	54	51

(3) In the design of a continuous beam of compact section complying with the provisions of Article 1.7.124(A)(1), negative moments over supports determined by elastic analysis may be reduced by a maximum of 10%. Such reduction shall be accompanied by an increase in maximum positive moment in the span equal to the average decrease of the negative moments in the span. The reduction shall not apply to negative moments produced by cantilever loading.

(B) Braced Non-Compact Sections

For rolled or fabricated I-shaped beams not meeting the requirements of Article 1.7.124(A)(1) but meeting the requirements of paragraph (1) below, the maximum strength shall be computed as:

$$M_u = F_y S$$

where S is the section modulus.

(1) The above equation is applicable to beams meeting the following requirements:

(a) Projecting flange element

$$b'/t \leq 2200/\sqrt{F_y}$$

When

$M < M_u$, b'/t may be increased by the ratio $\sqrt{M_u/M}$

(b) Web thickness
$$D/t_w \leq 150$$
where D is the clear unsupported distance between flange components.

(c) Spacing of lateral bracing for compression flange
$$L_b \leq \frac{20{,}000{,}000 A_f}{F_y d}$$
where d is the depth of beam or girder, A_f is the flange area.

(d) Maximum axial compression
Axial compression shall not exceed the value given by Article 1.7.124(A)(1)(d).

(e) Maximum shear force
$$V \leq \frac{3.5 E t^3_w}{D}$$
but not more than $0.58 F_y D t_w$.

(2) The limitations set forth in paragraph (1) above are given in Table 2.

TABLE 2

F_y (psi)	36,000	42,000	46,000	50,000	55,000	90,000	100,000
b'/t	11.6	10.7	10.3	9.8	9.4	7.3	7.0
$\dfrac{L_b d}{A_f}$	556	476	435	400	364	222	200

(C) Transition

The maximum strength of members with geometric properties falling between the limits of Articles 1.7.124(A) and (B) may be computed by straight line interpolation, except that the web thickness must always satisfy Article 1.7.124(A)(1)(b).

(D) Unbraced Sections

(1) For members not meeting the lateral bracing requirement of Article 1.7.124(B)(1)(c) the maximum strength shall be computed as:
$$M_u = F_y S \left[1 - \frac{3 F_y}{4 \pi^2 E} \left(\frac{L_b}{b'} \right)^2 \right]$$

When the ratio of stresses at the two ends of the braced length, L_b, is less than 0.7, the maximum strength, M_u, as computed by the above formula may be increased 20% but not to exceed $F_y S$.

(2) In members not meeting the requirements of Article 1.7.124 (B)(1)(e) the web shall be provided with transverse stiffeners as specified in Article 1.7.124(E).

(3) Members with axial loads in excess of $0.15F_yA$ should be designed as beam-columns as specified in Article 1.7.134

(E) Transversely Stiffened Girders

(1) For girders not meeting the shear requirements of Articles 1.7.124(A)(1)(e) and 1.7.124(B)(1)(e) transverse stiffeners are required for the web. For girders with transverse stiffeners but without longitudinal stiffeners the thickness of the web shall meet the requirement:

$$D/t_w \leq \frac{36{,}500}{\sqrt{F_y}}$$

For different grades of steel this limit is:

D/t_w	F_y (psi)
192	36,000
178	42,000
170	46,000
163	50,000
156	55,000
122	90,000
115	100,000

(2) The maximum bending strength of transversely stiffened girders meeting the requirements of Article 1.7.124(E)(1) shall be computed by Articles 1.7.124(B) or 1.7.124(D)(1) as applicable subject to the requirement of Article 1.7.124(E)(4).

(3) The shear capacity of beams and girders with webs fulfilling the requirements of Article 1.7.124(E)(1) shall be computed as:

$$V_u = V_p \left[C + \frac{0.87(1-C)}{\sqrt{1+(d_o/D)^2}} \right]$$

where:

$$V_p = 0.58 F_y D t_w$$

$$C = 18{,}000\,(t_w/D) \sqrt{\frac{1+(D/d_o)^2}{F_y}} - 0.3 \leq 1.0$$

D = clear, unsupported distance between flange components.

d_o = distance between transverse stiffeners.

(4) If a girder panel is subjected to simultaneous action of shear and bending moment with the magnitude of the shear higher than $0.6V_u$, then the moment shall be limited to not more than:

$$M/M_u = 1.375 - 0.625\, V/V_u$$

(5) Transverse stiffeners shall be spaced at a distance, d_o, according to shear capacity as specified in Article 1.7.124(E)(3) but not more than 1.5D. Transverse stiffeners may be omitted in those portions of the girders where the maximum shear force is less than the value given by Article 1.7.124(B)(1)(e).

The first stiffener space at the ends of girders with simple supports shall not be greater than D nor:

$$d_o = 14{,}500\sqrt{Dt_w^3/V}$$

The width-to-thickness ratio of transverse stiffeners shall be such that

$$b'/t \leq \frac{2{,}600}{\sqrt{F_y}}$$

where b' is the projecting width of the stiffener.

The gross cross-sectional area of intermediate transverse stiffeners shall not be less than:

$$A = [0.15\, BDt_w(1-C)(V/V_u) - 18t_w^2]Y$$

where Y is the ratio of web plate yield strength to stiffener plate yield strength

$B = 1.0$ for stiffener pairs,
1.8 for single angles,
2.4 for single plates.

C is computed by Article 1.7.124(E)(3)

The moment of inertia of transverse stiffeners with reference to the mid-plane of the web shall be not less than:

$$I = d_o\, t_w^3\, J$$

where:

$J = 2.5(D/d_o)^2 - 2$, but not less than 0.5.

Transverse stiffeners need not be in bearing with the tension flange. The maximum distance between the stiffener-to-web connection and the face of the tension flange shall not be more than $4t_w$. Stiffeners provided on only one side of the web must be in bearing against but need not be attached to the compression flange.

(F) Longitudinally Stiffened Girders

(1) Longitudinal stiffeners shall be required when the web thickness is less than that specified by Article 1.7.124(E)(1) and shall be placed at a distance D/5 from the inner surface of the compression flange.

The web thickness of plate girders with transverse stiffeners and one longitudinal stiffener shall meet the requirement:

$$D/t_w \leq \frac{73{,}000}{\sqrt{F_y}}$$

For different grades of steel, this limit is:

D/t_w	F_y (psi)
385	36,000
356	42,000
340	46,000
326	50,000
311	55,000
243	90,000
231	100,000

(2) The maximum bending strength of longitudinally stiffened girders meeting the requirements of Article 1.7.124(F)(1) shall be computed by Articles 1.7.124(B) or Article 1.7.124(D)(1) as applicable, subject to the requirement of Article 1.7.124(E)(4).

(3) The shear capacity of girders with one longitudinal stiffener shall be computed by Article 1.7.124(E)(3).

The dimensions of the longitudinal stiffener shall be such that:

(a) the width-to-thickness ratio is not greater than that given by Article 1.7.124(E)(5).

(b) the rigidity of the stiffener is not less than:

$$I \geq Dt_w^3 \left[2.4\left(\frac{d_o}{D}\right)^2 - 0.13 \right]$$

(c) the radius of gyration of the stiffener is not less than:

$$r \geq \frac{d_o\sqrt{F_y}}{23,000}$$

In computing I and r values above, a centrally located web strip not more than $18t_w$ in width shall be considered as a part of the longitudinal stiffener. Transverse stiffeners for girder panels with longitudinal stiffeners shall be designed according to Article 1.7.124(E)(5) except that the depth of subpanels shall be used instead of the total panel depth, D. In addition the section modulus of the transverse stiffener shall be not less than:

$$S_t = \frac{1}{3}(D/d_o)S_\ell$$

where D is the total panel depth (clear distance between flange components) and S_ℓ is the section modulus of the longitudinal stiffener at D/5.

1.7.125 — UNSYMMETRICAL BEAMS AND GIRDERS

(A) General

For beams and girders symmetrical about the vertical axis of the cross section but unsymmetrical with respect to the horizontal centroidal axis, the provisions of Articles 1.7.124(A) through 1.7.124(D) shall be applicable except that in computing the maximum strength by Article 1.7.124(D)(1) the term b' is replaced by 0.9b'.

(B) Unsymmetrical Sections with Transverse Stiffeners

Girders with transverse stiffeners shall be designed and evaluated by the provisions of Article 1.7.124(E) except that when D_c, the clear distance between the neutral axis and the compression flange, exceeds $D/2$ the web thickness, t_w, shall meet the requirement:

$$\frac{D_c}{t_w} \leq \frac{18{,}250}{\sqrt{F_y}}$$

(C) Longitudinally Stiffened Unsymmetrical Sections

Longitudinal stiffeners shall be required on unsymmetrical sections when the web thickness is less than that specified by Articles 1.7.124(E)(1) or 1.7.125(B).

For girders with one longitudinal stiffener and transverse stiffeners, the provisions of Article 1.7.124(F) for symmetrical sections shall be applicable provided that:

(a) the longitudinal stiffener is placed $2D_c/5$ from the inner surface or the leg of the compression flange element.

(b) When D_c exceeds $D/2$, the web thickness, t_w, shall meet the requirement:

$$\frac{D_c}{t_w} \leq \frac{36{,}500}{\sqrt{F_y}}$$

1.7.126 — COMPOSITE BEAMS AND GIRDERS

Composite beams shall be so proportioned that the following criteria are satisfied:

(a) The maximum strength of any section shall not be less than the sum of the computed moments at that section multiplied by the appropriate load factors.

(b) The web of the steel section shall be designed to carry the total external shear and must satisfy the applicable provisions of Articles 1.7.124 and 1.7.125. In such application the value of D_c shall be taken as the clear distance between the neutral axis of the composite section for live loads and the compression flange.

1.7.127 — POSITIVE MOMENT SECTIONS OF COMPOSITE BEAMS AND GIRDERS

(A) Compact Sections

When the steel section satisfies the compactness requirements of Article 1.7.127(A)(2), the maximum strength shall be computed as the resultant moment of the fully plastic stress distribution acting on the section (Figure 1.7.127).

(1) The resultant moment of the fully plastic stress distribution may be computed as follows:

(a) the compressive force in the slab, C, is equal to the smallest of the values given by the following Equations:

(1) $C = 0.85\, f'_c b t_s + (A\, F_y)_c$

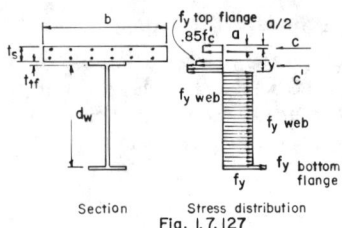

Fig. I.7.127

where b is the effective width of slab,

t_s is the slab thickness.

$(A\,F_y)_c$ is the product of the area and yield point of that part of reinforcement which lies in the compression zone of the slab.

(2) $C = (A\,F_y)_{bf} + (A\,F_y)_{tf} + (A\,F_y)_w$

where $(A\,F_y)_{bf}$ is the product of area and yield point for bottom flange of steel section (including cover plate if any),

$(A\,F_y)_{tf}$ is the product of area and yield point for top flange of steel section,

$(A\,F_y)_w$ is the product of area and yield point for web of steel section.

(3) $C = \Sigma Q_u$

where ΣQ_u is sum of ultimate strengths of shear connectors between the section under consideration and the section of zero moment.

(b) the depth of the stress block is computed from the compressive force in the slab.

$$a = \frac{C - (A\,F_y)_c}{0.85 f'_c b}$$

(c) when the compressive force in the slab is less than the value given by Equation (2) above the top portion of the steel section will be subjected to the following compressive force:

$$C' = \frac{\Sigma(A\,F_y) - C}{2}$$

(d) The location of the neutral axis within the steel section measured from the top of the steel section may be determined as follows:

for $C' < (A\,F_y)_{tf}$

$$\bar{y} = \frac{C'}{(A\,F_y)_{tf}} t_{tf}$$

for $C' \geq (A\,F_y)_{tf}$

$$\bar{y} = t_{tf} + \frac{C' - (A\,F_y)_{tf}}{(A\,F_y)_w} d_w$$

(e) the maximum strength of the section in bending is the first moment of all forces about the neutral axis, taking all forces and moment arms as positive quantities.

(2) Composite beams qualify as compact when their steel section meets the requirements of Articles 1.7.124(A)(1)(b) and 1.7.124 (A)(1)(e), and the stress-strain diagram of the steel exhibits a yield plateau followed by a strain hardening range.

(B) Non-compact Sections

When the steel section does not satisfy the compactness requirements of Article 1.7.127(A)(2) the maximum strength of the section shall be taken as the moment at first yielding.

(C) General

Maximum compressive and tensile stresses in girders which are not provided with temporary supports during the placing of dead loads shall be the sum of the stresses produced by $1.30 D_s$ acting on the steel girder alone and the stresses produced by $1.30[D_c + 5/3(L+I)]$ acting on the composite girder, where D_s and D_c are the moment caused by the dead load acting on the steel girder and composite girder, respectively.

When the girders are provided with effective intermediate supports which are kept in place until the concrete has attained 75% of its required 28-day strength, stresses are produced by the loading, $1.30[D + 5/3(L+I)]$, acting on the composite girder.

1.7.128 — NEGATIVE MOMENT SECTIONS OF COMPOSITE BEAMS AND GIRDERS

The maximum strength of beams and girders in the negative moment regions shall be computed in accordance with Articles 1.7.124 and 1.7.125 as applicable. It shall be assumed that the concrete slab does not carry tensile stresses. In cases where the slab reinforcement is continuous over interior supports, the reinforcement may be considered to act compositely with the steel section.

1.7.129 — COMPOSITE BOX GIRDERS

This section pertains to the design of simple and continuous bridges of moderate length supported by two or more single-cell composite box girders. It is applicable to box girders, having width center-to-center of top steel flanges approximately equal to the distance center-to-center of adjacent top steel flanges of adjacent box girders. The cantilever overhang of the deck slab, including curbs and parapet, shall be limited to 60 percent of the distance between the centers of adjacent top steel flanges of adjacent box girders, but in no case greater than 6 feet.

(A) Maximum Strength

The maximum strength of box girders shall be determined according to the applicable provisions of Article 1.7.126, 1.7.127 and 1.7.128. In addition, the maximum strength of the negative moment sections shall be limited by

$$M_u = F_{cr} S$$

where F_{cr} is the buckling stress of the bottom flange plate as given in Article 1.7.129(E).

(B) Lateral Distribution

The live load bending moment for each box girder shall be determined in accordance with Article 1.7.103.

(C) Web Plates

The design shear V_w for a web shall be calculated using the following equation:

$$V_w = V/\cos\theta$$

where V = one half of the total vertical shear force on one box girder,
θ = the angle of inclination of the web plate to the vertical.
The inclination of the web plates to the vertical shall not exceed 1 to 4.

(D) Tension Flanges

In the case of simply supported spans, the bottom flange shall be considered fully effective in resisting bending if its width does not exceed one-fifth the span length. If the flange plate width exceeds one-fifth of the span, only an amount equal to one-fifth of the span shall be considered effective.

For continuous spans, the requirements above shall be applied to the distance between points of contraflexure.

(E) Compression Flanges

(1) Unstiffened compression flanges designed for the yield stress, F_y, shall have a width-to-thickness ratio equal to or less than the value obtained from the formula:

$$b/t = \frac{6140}{\sqrt{F_y}}$$

where b = flange width between webs in inches,
t = flange thickness in inches.

(2) For greater b/t ratios, but not exceeding $13{,}300/\sqrt{F_y}$, the buckling stress of an unstiffened bottom flange is given by the formula:

$$F_{cr} = 0.592\, F_y \left(1 + 0.687 \sin\frac{c\pi}{2}\right)$$

in which c shall be taken as

$$c = \frac{13{,}300 - \frac{b}{t}\sqrt{F_y}}{7160}$$

(3) For values of b/t exceeding $13{,}300\sqrt{F_y}$, the buckling stress of the flange is given by the formula:

$$F_{cr} = 105\,(t/b)^2 \times 10^6$$

(4) If longitudinal stiffeners are used, they shall be equally spaced across the flange width and shall be proportioned so that the moment of inertia of each stiffener about an axis parallel to the flange and at the base of the stiffener is at least equal to:

$$I_s = \phi t^3 w$$

where $\phi = 0.07 k^3 n^4$ when n equals 2, 3, 4 or 5.

$\phi = 0.125 k^3$ when $n = 1$.

w = width of flange between longitudinal stiffeners or distance from a web to the nearest longitudinal stiffener.

n = number of longitudinal stiffeners.

k = buckling coefficient which shall not exceed 4.

For a longitudinally stiffened flange designed for the yield stress F_y, the ratio w/t shall not exceed the value given by the formula

$$w/t = \frac{3070\sqrt{k}}{\sqrt{F_y}}$$

For greater values of w/t, but not exceeding $6650\sqrt{k}/\sqrt{F_y}$, the buckling stress of the flange, including stiffeners is given by Article 1.7.129(E)(2) in which c shall be taken as:

$$c = \frac{6650\sqrt{k} - (w\sqrt{F_y}/t)}{3580\sqrt{k}}$$

For values of w/t exceeding $6650\sqrt{k}/\sqrt{F_y}$ the buckling stress of the flange, including stiffeners, is given by the formula:

$$F_{cr} = 26.2k(t/w)^2 \times 10^6$$

When longitudinal stiffeners are used, it is preferable to have at least one transverse stiffener placed near the point of dead load contraflexure. The stiffener should have a size equal to that of a longitudinal stiffener.

(5) The width-to-thickness ratio of any outstanding element of the flange stiffeners shall not exceed the value determined by the formula:

$$b'/t' = \frac{2600}{\sqrt{F_y}}$$

where b' = width of any outstanding stiffener element,

t' = thickness of outstanding stiffener element.

(F) Diaphragms

Diaphragms, cross-frames, or other means shall be provided within the box girders at each support to resist transverse rotation, displacement and distortion.

Intermediate diaphragms or cross-frames are not required for box girder bridges designed in accordance with this specification.

1.7.130 — SHEAR CONNECTORS

(A) General

The horizontal shear at the interface between the concrete slab and the steel girder shall be provided for by mechanical shear connectors throughout the simple spans and the positive moment regions of continuous spans. In the negative moment regions, shear connectors shall be provided when the reinforcement steel imbedded in the concrete is considered a part of the composite section. In case the reinforcement steel imbedded in the concrete is not considered in computing section properties of negative moment sections, shear connectors need not be provided in these portions of the span, but additional connectors shall be placed in the region of the points of dead load contraflexure as specified in Article 1.7.100(A)(3).

(B) Design of Connectors

The number of shear connectors shall be determined in accordance with Article 1.7.100(A)(2), and checked for fatigue in accordance with Article 1.7.100(A)(1) and 1.7.100(A)(3).

(C) Maximum Spacing

The maximum pitch shall not exceed 24 inches except over the interior supports of continuous beams where wider spacing may be used to avoid placing connectors at locations of high stresses in the tension flange.

1.7.131 — HYBRID GIRDERS

This section pertains to the design of (1) noncomposite beams and girders that have flanges of the same minimum specified yield strength and a web with a lower minimum specified yield strength, and (2) composite girders that have a tension flange with a higher minimum specified yield strength than the web and a compression flange with a minimum specified yield strength not less than that of the web. It is applicable to both simple and continuous girders. In noncomposite girders and in the negative moment portion of continuous composite girders, the area of the compression flange shall be equal to the area of the tension flange, or larger than the area of the tension flange by an amount not exceeding 25 percent. In composite girders, excluding the negative moment portion in continuous girders, the area of the compression flange shall be equal to or smaller than the area of the tension flange. The minimum specified yield strength of the web shall not be less than 35 percent of the minimum specified yield strength of the tension flange.

The provisions of Articles 1.7.124 through 1.7.130 shall apply to hybrid beams and girders except as modified below. In all equations of these Articles, F_y shall be taken as the minimum specified yield strength of the steel of the element under consideration.

1.7.132 — NONCOMPOSITE HYBRID GIRDERS

(A) Compact Sections

The equation of Article 1.7.124(A) for the maximum strength of compact sections shall be replaced by the expression

$$M_u = F_{yf} Z$$

where F_{yf} is the specified minimum yield strength of the flange and Z is the plastic section modulus.

In computing Z, the web thickness shall be multiplied by the ratio of the minimum specified yield strength of the web, F_{yw}, to the minimum specified yield strength F_{yf}.

(B) Braced Non-compact Sections

The equation of Article 1.7.124(B) for the maximum strength of compact sections shall be replaced by the expression

$$M_u = F_{yf} S R$$

For symmetrical sections,

$$R = \frac{12 + \beta(3\rho - \rho^3)}{12 + 2\beta}$$

where

$$\rho = F_{yw}/F_{yf}$$
$$\beta = A_w/A_f$$

For unsymmetrical sections,

$$R = 1 - \frac{\beta\psi(1-\rho)^2(3-\psi+\rho\psi)}{6+\beta\psi(3-\psi)}$$

where ψ is the distance from the outer fiber of the tension flange to the neutral axis divided by the depth of the steel section.

(C) Unbraced Noncompact Sections

The equation of Article 1.7.124(D)(1) for the maximum strength of unbraced noncompact sections shall be replaced by the expression

$$M_u = F_{yf} S \left[1 - \frac{3 F_{yf}}{4\pi^2 E} \left(\frac{L_b}{b'} \right)^2 \right] R$$

where the appropriate R is determined from (B) above.

(D) Transversely Stiffened Girders

The equation of Article 1.7.124(E)(3) for the shear capacity of transversely stiffened girders shall be replaced by the expression

$$V_u = V_p C$$

The equation for A in Article 1.7.124(E)(5) is not applicable to hybrid girders.

1.7.133 — COMPOSITE HYBRID GIRDERS

The maximum strength of the composite section shall be the moment at first yielding of the flanges times R (for unsymmetrical sections) from Article 1.7.132(B), in which ψ is the distance from the outer fiber of the tension flange to the neutral axis of the transformed section divided by the depth of the steel section.

1.7.134 — COMPRESSION MEMBERS

(A) Axial Loading

(1) Maximum Capacity

The maximum strength of concentrically loaded columns shall be computed as:

$$P_u = 0.85 \, A_s F_{cr}$$

where A_s is the gross effective area of the column cross section and F_{cr} is determined by one of the following two formulas:

$$F_{cr} = F_y \left[1 - \frac{F_y}{4\pi^2 E} \left(\frac{KL_c}{r} \right)^2 \right]$$

for $\dfrac{KL_c}{r}$ less than or equal to $\sqrt{\dfrac{2\pi^2 E}{F_y}}$

$$F_{cr} = \frac{\pi^2 E}{\left(\dfrac{KL_c}{r} \right)^2}$$

for $\dfrac{KL_c}{r}$ more than $\sqrt{\dfrac{2\pi^2 E}{F_y}}$

where

- K is effective length factor in the plane of buckling
- L_c is length of the member between points of support, in inches
- r is radius of gyration in the plane of buckling, in inches
- F_y is yield stress of the steel, in psi
- E is 29,000,000 psi
- F_{cr} is buckling stress, in psi

(2) Effective Length

The effective length factor K shall be determined as follows:

(a) For members having lateral support in both directions at its ends:
K = 0.75 for riveted, bolted or welded end connections.
K = 0.875 for pinned ends.

(b) For members having ends not fully supported laterally by diagonal bracing or an attachment to an adjacent structure, the effective length factor shall be determined by a rational procedure.*

(B) Combined Axial Load and Bending

(1) Maximum Capacity

The combined maximum axial force P and the maximum bending moment M acting on a beam-column subjected to eccentric loading shall satisfy the following equations:

$$\frac{P}{0.85 A_s F_{cr}} + \frac{MC}{M_u \left(1 - \frac{P}{A_s F_e}\right)} \leq 1.0$$

$$\frac{P}{0.85 A_s F_y} + \frac{M}{M_p} \leq 1.0$$

where

F_{cr} is buckling stress as determined by the equations of Article 1.7.134 (A) (1)

M_u is the maximum strength as determined by Articles 1.7.124 (A) (B) or (D)

$F_e = \dfrac{\pi^2 E}{\left(\dfrac{KL_c}{r}\right)^2}$ = the Euler buckling stress in the plane of bending,

C is the equivalent moment factor, as defined below.
$M_p = F_y Z$ the full plastic moment of the section,
Z is the plastic section modulus,
$\dfrac{KL_c}{r}$ is the effective slenderness ratio in the plane of bending.

(2) Equivalent Moment Factor C

If the ends of the beam-column are restrained from sidesway in the plane of bending by diagonal bracing or attachment to an adjacent laterally braced structure, then the value of equivalent moment factor, C, may be computed by the formula:

$$C = 0.6 + 0.4a, \text{ but not less than } 0.4$$

where a is the ratio of the numerically smaller to the larger end moment. The ratio a is positive when the two end moments act in an opposing sense (i.e., one acts clockwise and the other acts counterclockwise) and negative when they act in the same sense. In all cases, factor C may be taken conservatively as unity.

---

* B. G. Johnston, "Guide to Design Criteria for Metal Compression Members," John Wiley and Sons, Inc., New York, 1966.

1.7.135 — SPLICES, CONNECTIONS & DETAILS

(A) Connectors

(1) General

Connectors shall be proportioned so that their maximum strength multiplied by the reduction factor, ϕ, shall be at least equal to the effects of design loads multiplied by their respective load factors specified in Article 1.7.123. The maximum strengths multiplied by the reduction factors are listed in Table 3.

(2) Welds

The ultimate strength of weld metal in groove welds shall be equal to or greater than that of the base metal. The ultimate strength of the weld metal in fillet welds need not match the strength of the base metal. However, the welding procedure and weld metal shall be selected to insure sound welds. The effective weld area shall be taken as defined in Article 1.7.28.

(3) Bolts and Rivets

In proportioning fasteners, the nominal diameter shall be used except when a shear plane intersects the threads.

High-strength bolts preferably shall be used for fasteners subject to tension or combined shear and tension.

TABLE 3

Type of Fastener	Strength (ϕF)
Groove Weld [1]	1.00 F_y
Fillet Weld [2]	0.45 f_u
Low-Carbon Steel Bolts ASTM A307	
Tension	27 ksi
Shear [3]	25 ksi
Power-Driven Rivets ASTM A502	
Shear — Grade 1	25 ksi
Shear — Grade 2	30 ksi
High-Strength Bolts ASTM A325	
Tension [5]	76 ksi
Shear (Bearing-Type) [3,4,5]	54 ksi

(1)—F_y = yield point of connected material.
(2)—F_u = minimum strength of the welding rod metal but not greater than the tensile strength of the connected parts.
(3)—When a shear plane intersects the bolt threads, the root area shall be used.
(4)—Bearing stresses in bearing-type connections shall not exceed the tensile strength of the connected material.
(5)—For A235 bolts the tensile strength decreases for diameters greater than 7/8 in. The design value listed is for bolts up to 7/8 in. diameter. For diameters greater than 7/8 in. diameter the design value shall be computed as 0.56 F_u for tension and 0.45 F_u for shear where F_u is the ASTM minimum tensile strength of the bolt.

For combined tension and shear in bearing type connections, bolts and rivets shall be proportioned so that the shear stress does not exceed:

$$F_{vc} \leq \sqrt{F^2 - (0.6f_t)^2}$$

where F_v = shear strength of the fastener, ϕF, as given in Table 3.
f_t = tensile stress due to the applied load.

(4) Friction Joints

Friction joints shall be designed to prevent slip at the overload in accordance with Article 1.7.136(C). Maximum strength of the bolts need not be considered in the design of such joints.

(B) Connections

(1) Splices

Splices may be made with rivets, with high-strength bolts or by the use of welding. Splices, whether in tension, compression, bending or shear, shall be designed for not less than the average of the calculated stress resultant at the point of the splice and the strength of the member at the same point, but in any event not less than 75% of the maximum strength of the member. Where a section changes at a splice, the maximum strength of the splice shall be at least 75% of the smaller section spliced.

The maximum strength of the member shall be determined by the gross section for compression members. For members primarily in bending, the gross section shall be used, except that if more than 15% of each flange area is removed, that amount removed in excess of 15% shall be deducted. For tension members and splice material, the gross section shall be used unless the net section area is less than 85% of the corresponding gross area, in which case that amount removed in excess of 15% shall be deducted.

(2) Bolts Subjected to Prying Action by Connected Parts

Bolts required to support applied load by means of direct tension shall be proportioned for the sum of the external load and tension resulting from prying action produced by deformation of the connected parts. The total tension should not exceed the values given in Table 3 of Article 1.7.135.

The tension due to prying actions shall be computed as:

$$Q = \left[\frac{3b}{8a} - \frac{t^3}{20}\right] T$$

where

Q = the prying force per bolt (taken as zero when negative),
T = the direct tension per bolt due to external load,
a = distance from center of bolt to edge of plate,

b = distance from center of bolt to toe of fillet of connected part,
t = thickness of thinnest part connected, in.

(3) Rigid Connections

All rigid frame connections, the rigidity of which is essential to the continuity assumed as the basis of design, shall be capable of resisting the moments, shears, and axial loads to which they are subjected by maximum loads.

The beam web shall equal or exceed the thickness given by:

$$t_w \geq \sqrt{3}\left(\frac{M_c}{F_y d_b d_c}\right)$$

where

M_c is the column moment,
d_b the beam depth,
d_c the column depth.

When the thickness of the connection web is less than that given by the above formula, the web shall be strengthened by diagonal stiffeners or by a reinforcing plate in contact with the web over the connection area.

At joints where the flanges of one member are rigidly framed into one flange of another member, the thickness of the web (t_w) supporting the latter flange and the thickness of the latter flange (t_c) shall be checked by the formulas below. Stiffeners are required on the web of the second member opposite the compression flange of the first member when

$$t_w < \frac{A_f}{t_b + 5k}$$

and opposite the tension flange of the first member when

$$t_c < 0.4\sqrt{A_f}$$

where

t_w = thickness of web to be stiffened.
k = distance from outer face of flange to toe of web fillet of member to be stiffened,
t_b = thickness of flange delivering concentrated force,
t_c = thickness of flange of member to be stiffened,
A_f = area of flange delivering concentrated load.

1.7.136 — OVERLOAD

(A) Noncomposite Beams

For noncomposite beams the moment caused by $D + \frac{5}{3}(L+I)$ shall not exceed $0.8 F_y S$. For such beams designed for Group IA loading, the moment caused by $D + 2.2(L+I)$ shall not exceed $0.8 F_y S$. In the case of moment redistribution under the provisions of Article 1.7.124(A)(3), the above limitation shall apply to the modified moments but not to the original moments.

(B) Composite Beams

For composite beams the moment caused by $D+\frac{5}{3}(L+I)$ shall not exceed 95% of the moment at first yielding in the section. For such beams designed for Group IA loading, the moment caused by $D+2.2(L+I)$ shall not exceed 95% of the moment at first yielding in the section. In computing dead load stresses the presence or absence of temporary supports during the construction shall be considered.

(C) Friction Joints

The shear caused by the loading, $D+\frac{5}{3}(L+I)$ in friction-type high-strength bolted joints shall not exceed 21,000 psi for ASTM 325 bolts.

For combined shear and tension in friction-type joints where applied forces reduce the total clamping force on the friction plane, the maximum shear stress shall not exceed the values obtained from the following equations:

For A325 bolts

$$f_v = 21,000\,[1 - f_t/0.53F_u]$$

where F_u is the tensile strength of the bolt,

f_t is the applied tensile stress.

1.7.137 — FATIGUE

(A) General

The analysis of the probability of fatigue of steel members or connections under working loads and the allowable fatigue stresses, F_r, shall conform to Article 1.7.3, except that the limitation imposed by the basic design criteria given in Articles 1.7.1 and 1.7.2, shall not apply.

(B) Composite Construction

(1) Slab Reinforcement

When composite action is provided in the negative moment region, the range of stress in slab reinforcement shall be limited to 20,000 psi.

(2) Shear Connectors

The shear connectors shall be designed for fatigue in accordance with Article 1.7.100(A).

(C) Hybrid Beams and Girders

Hybrid girders shall be designed for fatigue in accordance with Article 1.7.111(C).

1.7.138 — DEFLECTION

The control of deflection of steel or of composite steel and concrete structures shall conform to the provision of Article 1.7.12.

ORTHOTROPIC-DECK BRIDGES

1.7.139 — ORTHOTROPIC-DECK BRIDGES, GENERAL

This section pertains to the design of steel bridges that utilize a stiffened steel plate as a deck. Usually the deck plate is stiffened by longitudinal ribs and transverse beams; effective widths of deck plate act as the top flanges of these ribs and beams. Usually the deck, including longitudinal ribs, acts as the top flange of the main box or plate girders. As used in Articles 1.7.139 through 1.7.148, the terms, rib and beam, refer to sections that include an effective width of deck plate.

The provisions of Division I, Design, shall govern where applicable, except as specifically modified by Articles 1.7.139 through 1.7.148.

An appropriate method of elastic analysis, such as the equivalent-orthotropic-slab method or the equivalent-grid method, shall be used in designing the deck. The equivalent stiffness properties shall be selected to correctly simulate the actual deck. An appropriate method of elastic analysis, such as the thin-walled-beam method, that accounts for the effects of torsional distortions of the cross-sectional shape shall be used in designing the girders of orthotropic-deck box-girder bridges. The box-girder design shall be checked for lane or truck loading arrangements that produce maximum distortional (torsional) effects.

1.7.140 — WHEEL-LOAD CONTACT AREA

The wheel loads specified in Article 1.2.5 shall be uniformly distributed to the deck plate over the rectangular area defined below:

Wheel Load, kip	Width Perpendicular to Traffic, inch	Length in Direction of Traffic, inch
8	$20+2t$	$8+2t$
12	$20+2t$	$8+2t$
16	$24+2t$	$8+2t$

In the above table, t is the thickness of the wearing surface in inches.

1.7.141 — EFFECTIVE WIDTH OF DECK PLATE

(A) Ribs and Beams

The effective width of deck plate acting as the top flange of a longitudinal rib or a transverse beam may be calculated by accepted approximate methods.*

(B) Girders

The full width of deck plate may be considered effective in acting as the top flange of the girders if the effective span of the girders is not less than: (1) 5 times the maximum distance between girder webs and (2) 10 times the maximum distance from edge of the deck to the nearest girder web. The effective span shall be taken as the actual span for simple spans and the distance between points of contraflexure for continuous spans. Alternatively, the effective width may be determined by accepted analytical methods.

The effective width of the bottom flange of a box girder shall be determined according to the provisions of Article 1.7.105(A).

1.7.142 — ALLOWABLE STRESSES

(A) Local Bending Stresses in Deck Plate

The term local bending stresses refers to the stresses caused in the deck plate as it carries a wheel load to the ribs and beams. The local transverse bending stresses caused in the deck plate by the specified wheel load plus 30 percent impact shall not exceed 30,000 psi unless a higher allowable stress is justified by a detailed fatigue analysis or by applicable fatigue-test results. For deck configurations in which the spacing of transverse beams is at least 3 times the spacing of longitudinal-rib webs, the local longitudinal and transverse bending stresses in the deck plate need not be combined with the other bending stresses covered in paragraphs (B) and (C) below.

(B) Bending Stresses in Longitudinal Ribs

The total bending stresses in longitudinal ribs due to a combination of (1) bending of the rib and, (2) bending of the girders may exceed the allowable bending stresses in Articles 1.7.1 and 1.7.3 by 25 percent. The bending stress due to each of the two individual modes shall not exceed the allowable bending stresses in Articles 1.7.1 and 1.7.3.

(C) Bending Stresses in Transverse Beams

The bending stresses in transverse beams shall not exceed the allowable bending stresses in Articles 1.7.1 and 1.7.3.

---

* Design Manual for "Orthotropic Steel Plate Deck Bridges," AISC, 1963 or "Orthotropic Bridges, Theory and Design," by M. S. Troitsky, Lincoln Arc Welding Foundation, 1967.

(D) Intersections of Ribs, Beams, and Girders

Connections between ribs and the webs of beams, holes in the webs of beams to permit passage of ribs, connections of beams to the webs of girders, and rib splices may affect the fatigue life of the bridge when they occur in regions of tensile stress. Where applicable, the number of cycles of maximum stress and the allowable fatigue stresses given in Section 1.7.3 shall be applied in designing these details; elsewhere, a rational fatigue analysis shall be made in designing the details. Connections between webs of longitudinal ribs and the deck plate shall be designed to sustain the transverse bending fatigue stresses caused in the webs by wheel loads.

1.7.143 — THICKNESS OF PLATE ELEMENTS

(A) Longitudinal Ribs and Deck Plate

Plate elements comprising longitudinal ribs, and deck-plate elements between webs of these ribs, shall meet the minimum thickness requirements of Article 1.7.88; f_a may be taken as 75 percent of the sum of the compressive stresses due to (1) bending of the rib and, (2) bending of the girder, but not less than the compressive stress due to either of these two individual bending modes.

(B) Girders and Transverse Beams

Plate elements of box girders, plate girders, and transverse beams shall meet the requirements of Articles 1.7.69, 1.7.70, 1.7.71, 1.7.72, 1.7.73, and 1.7.105.

1.7.144 — MAXIMUM SLENDERNESS OF LONGITUDINAL RIBS

The slenderness, L/r, of a longitudinal rib shall not exceed the value given by the following formula unless it can be shown by a detailed analysis that overall buckling of the deck will not occur as a result of compressive stress induced by bending of the girders:

$$\left(\frac{L}{r}\right)_{max} = 1000\sqrt{\frac{1500}{F_y} - \frac{2700F}{F_y^2}}$$

where

L = distance between transverse beams
r = radius of gyration about the horizontal centroidal axis of the rib including an effective width of deck plate
F = maximum compressive stress (in psi) in the deck plate as a result of the deck acting as the top flange of the girders; this stress shall be taken as positive
F_y = yield strength of rib material in psi

1.7.145 — DIAPHRAGMS

Diaphragms, cross frames, or other means shall be provided at each support to transmit lateral forces to the bearings and to resist transverse rotation, displacement, and distortion. Intermediate diaphragms or cross frames shall be provided at locations consistent with the analysis of the girders. The stiffness and strength of the intermediate and support diaphragms or cross frames shall be consistent with the analysis of the girders.

1.7.146 — STIFFNESS REQUIREMENTS

(A) Deflections

The deflections of ribs, beams, and girders due to live load plus impact may exceed the limitations in Article 1.7.12, but preferably shall not exceed $\frac{1}{500}$ of their span. The calculation of the deflections shall be consistent with the analysis used to calculate the stresses.

To prevent excessive deterioration of the wearing surface, the deflection of the deck plate due to the specified wheel load plus 30 percent impact preferably shall be less than $\frac{1}{300}$ of the distance between webs of ribs. The stiffening effect of the wearing surface shall not be included in calculating the deflection of the deck plate.

(B) Vibrations

The vibrational characteristics of the bridge shall be considered in arriving at a proper design.

1.7.147 — WEARING SURFACE

A suitable wearing surface shall be adequately bonded to the top of the deck plate to provide a smooth, nonskid riding surface and to protect the top of the plate against corrosion and abrasion. The wearing surface material shall provide (1) sufficient ductility to accommodate, without cracking or debonding, expansion and contraction imposed by the deck plate, (2) sufficient fatigue strength to withstand flexural cracking due to deck-plate deflections, (3) sufficient durability to resist rutting, shoving, and wearing, (4) imperviousness to water and motor-vehicle fuels and oils, and (5) resistance to deterioration from deicing salts, oils, gasolines, diesel fuels, and kerosenes.

1.7.148 — CLOSED RIBS

Closed ribs without access holes for inspection, cleaning, and painting are permitted. Such ribs shall be sealed against the entrance of moisture by continuously welding (1) the rib webs to the deck plate, (2) splices in the ribs, and (3) diaphragms, or transverse beam webs, to the ends of the ribs.

Section 8—CORRUGATED METAL AND STRUCTURAL PLATE PIPES AND PIPE-ARCHES

1.8.1 — GENERAL

The materials for the structure shall conform to the specifications set forth below, and the construction and installation shall conform to Section 23, Division II. The minimum gage or thickness shall be as determined by design in accordance with Art. 1.8.2, except that such thickness shall be increased in accordance with Art. 1.8.4 to provide for corrosion or abrasion unless there is evidence that corrosion or abrasion is not likely to occur.

Corrugated metal pipe composed of a smooth liner and corrugated shell attached integrally at seams spaced not more than 30 inches apart may be designed in accordance with Article 1.8.2 on the same basis as a standard corrugated metal pipe having the same corrugations as the shell and a weight per foot equal to the sum of the weights per foot of liner and corrugated shell. This shall be limited to a maximum pipe diameter of 84 inches, the thickness of the corrugated shell shall be at least 60% of the total thickness of shell and liner, and the specified backfill compaction shall be a minimum of 85% of standard density. Where corrosion or abrasion are anticipated, thickness of shell and liner shall be increased in accordance with Article 1.8.4 or suitable coatings shall be specified.

Corrugated metal pipe and pipe-arches may be of riveted, welded, or helical fabrication. The specifications are:

	Aluminum	Steel
Riveted	AASHO M 196	AASHO M 36 *
Continuous Welded	AASHO M 197	AASHO M 36
Spot Welded	Specification Pending	AASHO M 36
Helical Underdrain Culvert	AASHO M 197 or AASHO M 211	AASHO M 36

Structural plate pipe and pipe-arches shall be bolted. The specifications are:

	Aluminum	Steel
Bolted	AASHO M 219	AASHO M 167 (6x2 Corrugations)

Nothing included in this section shall be interpreted as prohibiting the use of new developments where usefulness can be substantiated.

1.8.2 — DESIGN

Four criteria must be considered in the structural design of a flexible buried conduit. Each considers the mutual function of the metal ring and the soil envelope surrounding it; interaction of these two materials produces a composite structure.

The criteria are:
(A) Seam Strength
(B) Handling and Installation Strength

---

* For 3×1 corrugations an equal number of $\frac{1}{2}''$ ϕ ASTM A 325 bolts may be substituted for rivets.

(C) Failure of the Conduit Wall
(D) Deflection or Flattening

(A) Seam Strength

Seam Strength must be sufficient to withstand the thrust developing from the total load supported by the conduit.

This thrust, in lbs. per lineal ft. of structure is:

$$T = (LL + DL) \times \frac{\text{Span}}{2}$$

where LL = Design Live Load, psf. See Art. 1.3.3
DL = Dead Load, psf. See Arts. 1.2.2(A) and 1.8.8
Span (or diameter), in ft.

Thrust, T, multiplied by the safety factor, (See Art. 1.8.8) should not exceed the seam strength. The strengths shown in Table 1.8.2 are recommended in the determination of fill heights. Longitudinal seams for corrugated metal pipe and pipe-arch shall develop the minimums shown in Table 1.8.2.

(B) Handling and Installation Strength

Handling and installation strength must be sufficient to withstand impact forces associated with shipping and placing of pipe. Both shop and field assembled pipe must have strength adequate to withstand compaction of the backfill without interior bracing to maintain pipe shape.

Handling rigidity is measured by a Flexibility Factor determined by the formula

$$FF = D^2/EI$$

where D = pipe diameter or maximum span, inches

E = modulus of elasticity of the pipe material, psi (see Art. 1.8.3.)

I = moment of inertia per unit length of cross section of the pipe wall, inches to the 4th power per inch.

For steel conduits, FF should generally not exceed the following values:

$2'' \times \frac{1}{2}''$ and
$2\frac{2}{3}'' \times \frac{1}{2}''$ corrugation $FF = 4.3 \times 10^{-2}$
$3'' \times 1''$ corrugation $FF = 3.3 \times 10^{-2}$
$6'' \times 2''$ corrugation $FF = 2.0 \times 10^{-2}$

For aluminum conduits, FF should generally not exceed the following values:

$2'' \times \frac{1}{2}''$ and
$2\frac{2}{3}'' \times \frac{1}{2}''$ corrugation $FF = 9.5 \times 10^{-2}$
$9'' \times 2\frac{1}{2}''$ corrugation $FF = 2.5 \times 10^{-2}$

(C) Failure of the Conduit Wall

Failure of the wall by wall crushing may occur if the wall flexibility is low (regardless of the quality of backfill). Failure of the wall

TABLE 1.8.2

Minimum Longitudinal Seam Strengths (Ultimate strength in kips per foot)

2×½ and 2⅔×½ Corrugated Steel Pipe			3×1 Corrugated Steel Pipe	
Thickness	Single Rivets	Double Rivets*	Thickness	Double Rivets
0.064	16.7	21.6	0.064	28.7
0.079	18.2	29.8	0.079	35.7
0.109	23.4	46.8	0.109	53.0
0.138	24.5	49.0	0.138	63.7
0.168	25.6	51.3	0.168	70.7

6×2 Structural Plate Steel Pipe			
Thickness	4 Bolts/ft	6 Bolts/ft	8 Bolts/ft
0.109	42.0		
0.138	62.0		
0.168	81.0		
0.188	93.0		
0.218	112.0		
0.249	132.0		
0.280	144.0	180	194

2×½ and 2⅔×½ Corrugated Aluminum Pipe			9×2½ Structural Plate Aluminum Pipe		
Thickness	Single Rivets	Double Rivets*	Thickness	Aluminum Bolts 5⅓ Bolts per ft	Steel Bolts 5⅓ Bolts per ft
0.060	9.0	14.0	0.09	22.2	
0.075	9.0	18.0	0.10	26.4	
0.105	15.6	31.5	0.125	34.8	
0.135	16.2	33.0	0.15	44.4	
0.164	16.8	34.0	0.175	52.8	
			0.20		60.0
			0.225		66.0
			0.250		72.0

* Pipes of 42" or larger diameter require double rivets.

by elastic buckling may occur if the wall flexibility is high and the backfill is compressible (poorly consolidated). Interaction of these two failure conditions, crushing and buckling, may develop in a zone between high and low wall flexibility.

It is assumed that a flexible conduit in a "soil-structure interactions system" does not fail at a specific stress defined by bending, since the conduit in yielding may transfer more of its load to the surrounding soil.

For diameters less than D, where $D = \frac{r}{k}\sqrt{\frac{24E}{f_u}}$, the ring compression stress, f_c, at which buckling becomes critical, in the interaction zone is

$$f_c = f_u - \frac{f_u^2}{48E}\left(\frac{kD}{r}\right)^2, \text{psi}.$$

For diameters greater than D, where $D = \frac{r}{k}\sqrt{\frac{24E}{f_u}}$, the ring compression stress, f_c, at which buckling becomes critical, in the elastic buckling zone is

$$f_c = \frac{12E}{\left(\frac{kD}{r}\right)^2}, \text{psi}$$

where f_u = minimum tensile strength, psi
f_c = critical stress, psi, not to exceed the yield strength
k = soil stiffness factor
D = pipe diameter or span, in.
r = radius of gyration (corrugation)
E = modulus of elasticity, psi

Design for buckling is accomplished by limiting the ring compression thrust, T, to the buckling stress multiplied by the conduit wall area per lineal foot of structure divided by the safety factor.

(D) Deflection or Flattening

The Iowa Deflection Formula provides one approach to prediction of ring deflection. It relates ring deflection to the passive side pressure resisting horizontal movement of the pipe wall and to the inherent strength of the pipe. Pipe arches need not be checked for deflection.

The Iowa Deflection Formula is:

$$X = D_1 \frac{KW_c R^3}{EI + 0.061 E'R^3}$$

where X = horizontal deflection of the pipe, in.
D_1 = deflection lag factor

K = a bedding constant (depends on bedding angle)
W_c = vertical load per unit length of pipe, lb/lin in.
R = mean radius of pipe, in.
E = modulus of elasticity of pipe, psi (see Art. 1.8.3)
I = moment of inertia per unit length of cross section of pipe wall, inches to the fourth power per inch
E' = horizontal soil modulus, psi/in.

Other methods are available for predicting ring deflection.

1.8.3 — CHEMICAL AND MECHANICAL REQUIREMENTS

(A) ALUMINUM — Corrugated Metal Pipe and Pipe-arch

Chemical — AASHO M 196 (ASTM C478) and M 197

Mechanical

Thickness, in.	Minimum Tensile Strength psi	Minimum Yield Strength psi	Minimum Elongation in 2 inches	Mod. of Elast. psi
0.051 to 0.113	31,000	24,000	4%	10×10^6
0.114 to 0.249	31,000	24,000	5%	10×10^6

(B) ALUMINUM — Structural plate pipe and pipe-arch

Chemical — AASHO M 219, Alloy 5052

Mechanical

Thickness, in.	Minimum Tensile Strength psi	Minimum Yield Strength psi	Minimum Elongation in 2 inches	Mod. of Elast. psi
0.090 to 0.175	35,500	28,000	6%	10×10^6
0.175 to 0.250	34,000	26,000	8%	10×10^6

(C) STEEL — Corrugated Metal Pipe and Pipe-arch

Chemical — AASHO M 36

Mechanical

Minimum Tensile Strength, psi	Minimum Yield Strength, psi	Minimum Elongation in 2 inches	Mod. of Elast. psi
45,000	33,000	20%	29×10^6

(D) STEEL — Structural Plate Pipe and Pipe-arch

Chemical-AASHO M 167

Mechanical

Minimum Tensile Strength, psi	Minimum Yield Strength, psi	Minimum Elongation in 2 inches	Mod. of Elast. psi
42,000	28,000	30%	29×10^6

The mechanical properties shown above are for the flat material prior to corrugating. A certificate of compliance shall be required from the manufacturer.

1.8.4 — ABRASIVE OR CORROSIVE CONDITIONS

For corrugated metal and structural plate pipes and pipe-arches having a thickness less than 0.25", the entire conduit, or bottom plates only in the case of structural plate pipe, shall be of greater thickness, or protected by other means, when required for resistance to abrasion or corrosion.

1.8.5 — RIVETS AND BOLTS

Rivets for corrugated sections and bolts for structural plate sections shall conform with the following:

Aluminum Corrugated Section:
  Rivets-Aluminum, ASTM B 316, Alloy 6053-T4

Aluminum Structural Plates:
  Bolts-Aluminum, ASTM B 211, Alloy 6061-T6
  Bolts-Steel, AASHO M 164 (ASTM A 325)

Steel Corrugated Section:
  Rivets-Steel, AASHO M 36

Steel Structural Plates:
  Bolts-Steel, AASHO M 164 (ASTM A 325)

Where end treatment requires a rigid headwall, the plates or pipe shall be anchored to the headwall with not less than ¾ inch anchor bolts at not more than 19 inch centers. Steel bolts for structural plate sections shall be torqued during installation to a minimum of 100 ft-lbs. and a maximum of 300 ft-lbs. Aluminum bolts for structural plate sections shall be torqued during installation to a minimum of 100 ft-lbs., and a maximum of 150 ft-lbs. For power driven tools, the hold-on period may vary from 2 to 5 seconds.

Bolts shall be of sufficient length to provide for a full nut.

1.8.6 — MULTIPLE STRUCTURES

Where multiple lines of pipes or pipe-arches greater than 48 inches in diameter or span are used, they shall be spaced so that adjacent sides of the pipe shall be at least one-half diameter or 3 feet apart, whichever is less, to permit adequate compaction of backfill material. For diameters up to 48 inches, the minimum spacing shall be not less than 24 inches.

1.8.7 — SLOPED ENDS — SKEWED

When the skew angle exceeds 20 degrees and the structure has the ends cut to fit the slope, the ends shall be reinforced.

1.8.8 — MAXIMUM DEPTHS OF COVER

The maximum depths of cover may be determined by use of the Iowa Deflection Formula and the following basic data. (Nothing included herein shall prohibit the use of other appropriate basic values.)

Weight of embankment—100 lbs/cu. ft.

$k = 0.44$, soil stiffness coefficient for good side fill material compacted to 85 percent of standard density based on AASHO Specification T 99 (ASTM D 698).

$E' =$ Modulus of passive soil (side fill) resistance: 700 psi.
Elongation:
    5 percent of nominal diameter
Maximum deflection:
    5 percent of nominal diameter below circular shape
Safety factors used:
    Longitudinal test seam strength = 4.0
    Pipe wall buckling   = 2.0

For pipe-arch structures placed on a stable foundation, the confining backfill must be capable of supporting a corner pressure of 2 tons per square foot. Marginally stable or compressible foundations require special investigation. Fill heights exceeding 100 feet shall be used only after a thorough investigation of the foundation material.

Section 9—STRUCTURAL PLATE ARCHES

1.9.1 — GENERAL

Structural Plate Arches shall conform to Section 8, Division I, and to the specifications set forth below, and the construction and installation shall conform to Section 23, Division II.

1.9.2 — RATIO, RISE TO SPAN

The design of single radius structural plate arches should be based on ratios of rise to span varying from 0.3 to 0.5.

1.9.3 — MINIMUM HEIGHT OF COVER

The minimum cover for design loads shall be Span/6 but not less than 12″. For construction requirements see Section 2.23.10.

1.9.4 — SCOUR CONDITIONS

Invert slabs shall be provided when scour is anticipated.

1.9.5 — MULTIPLE ARCHES

Where multiple arch spans are used, the distance between plates shall be not less than 1/10 of the longer adjoining span.

1.9.6 — SUBSTRUCTURE DESIGN

The substructure shall be designed according to specifications herein for substructures of bridges.

Section 10—TIMBER STRUCTURES

1.10.1 — ALLOWABLE STRESSES

(A) Allowable Unit Stresses for Stress-Grade Lumber

The allowable unit stresses given in Table 1.10.1 are for normal duration of loading for stress grades of sawn lumber used under continuously dry conditions as in most covered structures. For other service conditions, the following modification shall apply. For lumber used under conditions in which the moisture content of the wood is at or above the fiber saturation point, as when continuously submerged, the allowable unit stresses in Table 1.10.1 in compression parallel to the grain shall be reduced 10 percent, in compression perpendicular to the grain shall be reduced one-third, and the values for the modulus of elasticity shall be reduced one-eleventh.

(1) Use of stress grades in flexure:

Allowable unit stresses in flexure for Joist and Plank grades apply to material with the load applied to either the narrow or wide face.

Allowable unit stresses in flexure for Beam and Stringer grades apply only to material with the load applied to the narrow face.

Beam grades ordinarily are graded for use on simple spans. When used as a continuous beam the grading provisions customarily applied to the middle third of the length of simple spans shall be applied to the middle two-thirds of the length of pieces to be used over double spans and to the entire length of pieces to be used over three or more spans.

(2) Modification for condition of use for bearing perpendicular to grain:

The allowable unit stresses for compression perpendicular to the grain assume the material will be surface seasoned when installed. When used under continuously wet conditions, the tabulated values should be reduced one-third.

(B) Allowable Unit Stresses for Glued Laminated Timber

(1) The allowable unit stresses for softwood species shall be as recommended in American Institute of Timber Construction

203-70 "Standard Specifications for Structural Glued Laminated Timber of Douglas Fir, Western Larch, Southern Pine and California Redwood," and as given in Tables 1.10.1(A) and 1.10.1(B) herein. For hardwood species, the allowable unit stresses shall be as given in Table 2.10 of the "Timber Construction Manual," by the American Institute of Timber Construction, published by John Wiley & Sons, New York City, New York.

(2) The stress tables given in AITC 203-70 are divided into sections for dry-use or wet-use conditions. Allowable unit stresses for dry-use conditions are applicable when the moisture content in service is less than 16% as in most covered structures. Allowable unit stresses for wet-use conditions are applicable when the moisture content in service is 16% or more, as may occur in exterior or submerged construction, and in some structures housing wet processes or otherwise having constant high relative humidities.

(3) The stress tables in AITC 203-70 give stresses for members stressed primarily in bending (load applied perpendicular to the wide face of the lamination) and for members stressed primarily in axial tension, axial compression or loaded in bending parallel or perpendicular to the wide face of lamination. Edge joints shall be glued only in members loaded normal to the lamination edges or in members where torsion is a significant design consideration.

(4) Slope of grain, type and location of end joints, and other requirements, together with certain manufacturing requirements, must be met for these allowable unit stresses to apply. The requirements for slope of grain for softwoods are given in AITC 203-70, whereas these requirements for hardwoods are incorporated in Table 2.10 of the AITC "Timber Construction Manual." Other requirements are given in U.S. Commercial Standard 253-63.

(5) Species other than those specifically included herein may be used provided allowable unit stresses are established for them in accordance with U.S. Commercial Standard 253-63.

(C) Allowable Unit Stresses for Normal Loading Conditions

The tabulated allowable unit stresses are for normal load duration which contemplates fully stressing a member to the allowable unit stress by the application of the full design load for a duration of approximately ten years (either continuously or cumulatively). For other loading conditions, adjustments should be made as given in the following sections.

(D) Allowable Unit Stresses for Permanent Loading

When a member is fully stressed to the maximum allowable stress for long term loading conditions (greater than ten years either continuously or cumulatively), use 90 percent of the tabulated allowable unit stresses. The provisions of this paragraph do not apply to modulus of elasticity.

(E) Allowable Unit Stresses for Wind, Earthquake or Short Time Loading

When the duration of the full maximum load does not exceed the period indicated, increase the tabulated allowable unit stresses as follows: 15 per cent for 2 months duration, as for snow
25 per cent for 7 days duration
33⅓ per cent for wind or earthquake

The above increases are not cumulative. The resulting structural members shall not be smaller than required for a longer duration of loading. The provisions of this paragraph do not apply to modulus of elasticity. The increases apply to mechanical fastenings except as otherwise noted.

(F) Combined Stresses

These specifications do not cover the application of loadings which produce combined axial and bending stresses, nor the effective reductions in the tabulated stresses as a result of these loadings.

For this condition, attention is directed to Section 4, page 11, of the "Timber Construction Manual," 1966, by the American Institute of Timber Construction, published by John Wiley & Sons, New York, N.Y.

1.10.2 — FORMULAS FOR THE COMPUTATION OF STRESSES IN TIMBER

In calculating live load stresses in timber, impact shall be neglected. See Article 1.2.12(B).

(A) Horizontal Shear in Beams

Horizontal shear in beams shall be computed from the maximum shear occurring at a distance from the support equal to three times the depth of the beam, or at the quarter point, whichever is the lesser distance from the support. The live load used in computing horizontal shear shall be placed so as to produce maximum external shear at this distance from the support. This external live load shear shall be one-half the sum of 60 per cent of the shear from the undistributed wheel loads and of the shear from the wheel loads distributed laterally as specified for moment in Article 1.3.1. For undistributed wheel loads, one line of wheels is assumed to be carried by one beam.

The shear shall be calculated according to the following formula:

$$f_v = \frac{3V}{2bd}$$

Where
  f_v = horizontal shear stress in pounds per square inch
  b = width of beam in inches
  d = depth of beam in inches
  V = vertical shear in pounds

TABLE 1.10.1

Allowable Unit Stresses for Structural Lumber — Visually Graded

(The allowable unit stresses below are for normal loading conditions. See other provisions of Article 1.10.1 for adjustments of these tabulated allowable unit stresses)

Note: This represents only a partial listing of available species and grades. For a complete listing see the Supplement to 1971 Edition of "National Design Specification for Stress Grade Lumber and its Fastenings," NFPA

Species and commercial grade	Size classification	Allowable unit stress in pounds per square inch [1]							Grading rules agency
		Extreme fiber in bending "F_b"		Tension parallel to grain "F_t"	Horizontal shear "F_v"	Compression perpendicular to grain "$F_{c\perp}$"	Compression parallel to grain "F_c"	Modulus of elasticity "E"	
		Engineered uses (single)	Repetitive-member uses						
CALIFORNIA REDWOOD (Surfaced dry. Used at 19% max. m.c.)									
Clear Heart Structural	4" and less thick any width	2300		1550	145	425	2150	1,400,000	Redwood Inspection Service
Clear Structural		2300		1550	145	425	2150	1,400,000	
Select Structural	4" and less thick and wide	2050		1200	100	425	1500	1,400,000	
No. 1		1700		1000	100	425	1250	1,400,000	
No. 2		1400		800	80	425	1000	1,300,000	
No. 3		800		450	80	425	600	1,100,000	
Select Structural	4" and less thick 6" to 12" wide	1750		1200	100	425	1450	1,400,000	
No. 1		1500		1000	100	425	1250	1,400,000	
No. 2		1200		800	80	425	1000	1,300,000	
No. 3		700		450	80	425	600	1,100,000	
DOUGLAS FIR-LARCH (Surfaced dry or surfaced green. Used at 19% max. m.c.)									
Dense Select Structural	2" to 4" thick 2" to 4" wide	2450		1400	95	455	1850	1,900,000	West Coast Lumber Inspection Bureau and Western Wood Products Association (see footnotes 2 through 9)
Select Structural		2100		1200	95	385	1600	1,800,000	
Dense No. 1		2050		1200	95	455	1450	1,900,000	
No. 1		1750		1050	95	385	1250	1,800,000	
Dense No. 2		1700		1000	95	455	1150	1,700,000	
No. 2		1450		850	95	385	1000	1,700,000	
No. 3		800		475	95	385	600	1,500,000	
Dense Select Structural	2" to 4" thick 6" and wider	2100		1400	95	455	1650	1,900,000	
Select Structural		1800		1200	95	385	1400	1,800,000	
Dense No. 1		1800		1200	95	455	1450	1,900,000	
No. 1		1500		1000	95	385	1250	1,800,000	
Dense No. 2		1450		950	95	385	1250	1,700,000	
No. 2		1250		825	95	385	1050	1,700,000	
No. 3		750		475	95	385	675	1,500,000	

Table No. 1.10.1 (cont'd)

Species and commercial grade	Size classification	Allowable unit stress in pounds per square inch [1]							Modulus of elasticity "E"	Grading rules agency
		Extreme fiber in bending "F_b"		Tension parallel to grain "F_t"	Horizontal shear "F_v"	Compression perpendicular to grain "$F_{c\perp}$"	Compression parallel to grain "F_c"			
		Engineered uses (single)	Repetitive-member uses							
Dense Select Structural	Beams and Stringers	1900	...	1100	85	455	1300		1,700,000	West Coast Lumber Inspection Bureau
Select Structural		1600	...	950	85	385	1100		1,600,000	
Dense No. 1		1550	...	775	85	455	1100		1,700,000	
No. 1		1350	...	675	85	385	925		1,600,000	
Dense Select Structural	Posts and Timbers	1750	...	1150	85	455	1400		1,700,000	(see footnotes 2 through 9)
Select Structural		1500	...	1000	85	385	1200		1,600,000	
Dense No. 1		1400	...	950	85	455	1200		1,700,000	
No. 1		1200	...	825	85	385	1000		1,600,000	
Select Dex	Decking	1750	2000	...	...	385	...		1,800,000	
Commercial Dex		1450	1650	...	...	385	...		1,700,000	
Dense Select Structural	Beams and Stringers	1900	...	1250	85	455	1300		1,700,000	Western Wood Products Association
Select Structural		1600	...	1050	85	385	1100		1,600,000	
Dense No. 1		1550	...	1050	85	455	1100		1,700,000	
No. 1		1350	...	900	85	385	925		1,600,000	
Dense Select Structural	Posts and Timbers	1750	...	1150	85	455	1350		1,700,000	(see footnotes 2 through 11)
Select Structural		1500	...	1000	85	385	1150		1,600,000	
Dense No. 1		1400	...	950	85	455	1200		1,700,000	
No. 1		1200	...	825	85	385	1000		1,600,000	
Selected Decking	Decking	...	2000	...	...	...	...		1,800,000	
Commercial Decking		...	1650	...	...	...	...		1,700,000	
Selected Decking	Decking	...	2150	...	...	(Stresses apply at 15% moisture content)	...		1,900,000	
Commercial Decking		...	1800	...	...		...		1,700,000	

EASTERN HEMLOCK — TAMARACK (Surfaced dry or surfaced green. Used at 19% max. m.c.)

Species and commercial grade	Size classification	Engineered uses (single)	Repetitive-member uses	Tension parallel to grain	Horizontal shear	Compression perpendicular to grain	Compression parallel to grain		Modulus of elasticity	Grading rules agency
Select Structural	2" to 4" thick 2" to 4" wide	1800	1050	...	85	365	1350		1,300,000	Northeastern Lumber Manufacturer Association or Northern Hardwood and Pine Manufacturers Association (see footnotes 2 through 9)
No. 1		1500	900	...	85	365	1050		1,300,000	
No. 2		1250	725	...	85	365	850		1,100,000	
No. 3		700	400	...	85	365	525		1,000,000	
Select Structural	2" to 4" thick 6" and wider	1550	1050	...	85	365	1200		1,300,000	
No. 1		1300	875	...	85	365	1050		1,300,000	
No. 2		1050	700	...	85	365	900		1,100,000	
No. 3		625	400	...	85	365	575		1,000,000	

Table No. 1.10.1 (cont'd)

Species and commercial grade	Size classification	Allowable unit stress in pounds per square inch [1]							Grading rules agency
		Extreme fiber in bending "F_b"		Tension parallel to grain "F_t"	Horizontal shear "F_v"	Compression perpendicular to grain "$F_{c\perp}$"	Compression parallel to grain "F_c"	Modulus of elasticity "E"	
		Engineered uses (single)	Repetitive-member uses						
Select Structural	Beams and Stringers	1400		925	80	365	950	1,200,000	
No. 1		1150		775	80	365	800	1,200,000	
Select Structural	Posts and Timbers	1300		875	80	365	1000	1,200,000	
No. 1		1050		700	80	365	875	1,200,000	
Select Commercial	Decking	1500	1700	...	...	...	...	1,300,000	NeLMA
		1250	1450	...	...	...	...	1,100,000	
EASTERN SPRUCE (Surfaced dry or surfaced green. Used at 19% max. m.c.)									
Select Structural	2" to 4" thick 2" to 4" wide	1500	1800	875	65	255	1150	1,400,000	Northeastern Lumber Manufacturer Association or Northern Hardwood and Pine Manufacturers Association
No. 1		1300	1500	750	65	255	900	1,400,000	
No. 2		1050	1200	625	65	255	700	1,200,000	
No. 3		575	675	325	65	255	425	1,100,000	
Select Structural	2" to 4" thick 6" and wider	1300		875	65	255	1000	1,400,000	
No. 1		1100		750	65	255	900	1,400,000	
No. 2		900		600	65	255	750	1,200,000	
No. 3		525		325	65	255	475	1,100,000	
Select Commercial	Decking	1250	1450	...	...	...	...	1,400,000	
		1050	1200	...	...	...	...	1,200,000	
ENGLEMANN SPRUCE (Englemann Spruce-Lodgepole Pine) (Surfaced dry or surfaced green. Used at 19% max. m.c.)									
Selected Decking	Decking		1300					1,200,000	Western Wood Products Association
Commercial Decking			1100					1,100,000	
Selected Decking	Decking		1400	...	...	(Stresses apply at 15% moisture content)		1,300,000	
Commercial Decking			1150	...	...			1,200,000	
HEM-FIR (Surfaced dry or surfaced green. Used at 19% max. m.c.)									
Select Structural	2" to 4" thick 2" to 4" wide	1650		975	75	245	1300	1,500,000	West Coast Lumber Inspection Bureau and Western Wood Products Association (see footnotes 2 through 9)
No. 1		1400		825	75	245	1000	1,500,000	
No. 2		1150		675	75	245	800	1,400,000	
No. 3		625		375	75	245	500	1,200,000	
Select Structural	2" to 4" thick 6" and wider	1400		950	75	245	1150	1,500,000	
No. 1		1200		800	75	245	1000	1,500,000	
No. 2		1000		650	75	245	850	1,400,000	
No. 3		575		375	75	245	550	1,200,000	

1.10.2   DESIGN   213

Table No. 1.10.1 (cont'd)

Species and commercial grade	Size classification	Allowable unit stress in pounds per square inch [1]							Modulus of elasticity "E"	Grading rules agency
		Extreme fiber in bending "Fb"		Tension parallel to grain "Ft"	Horizontal shear "Fv"	Compression perpendicular to grain "Fc⊥"	Compression parallel to grain "Fc"			
		Engineered uses (single)	Repetitive-member uses							
Select Structural No. 1	Beams and Stringers	1250 1000	...	750 525	70 70	245 245	900 750	1,400,000 1,400,000	West Coast Lumber Inspection Bur.
Select Structural No. 1	Posts and Timbers	1200 975	...	800 650	70 70	245 245	950 850	1,400,000 1,400,000	(see footnotes 2 through 9)
Select Dex Commercial Dex	Decking	1400 1150	1600 1300	...	...	245 245	...	1,500,000 1,400,000	
Select Structural No. 1	Beams and Stringers	1250 1050	...	850 700	70 70	245 245	900 775	1,400,000 1,400,000	Western Wood Products Association
Select Structural No. 1	Posts and Timbers	1200 975	...	800 650	70 70	245 245	950 850	1,400,000 1,400,000	(see footnotes 2 through 11)
Selected Decking Commercial Decking	Decking	...	1600 1300	...	...	(Stresses apply at 15% moisture content)	...	1,500,000 1,400,000	
Selected Decking Commercial Decking	Decking	...	1750 1450	...	...		...	1,600,000 1,500,000	
IDAHO WHITE PINE (Surfaced dry or surfaced green. Used at 19% max. m.c.)										
Selected Decking Commercial Decking	Decking	...	1400 1150	...	...		...	1,400,000 1,300,000	Western Wood Products Association
Selected Decking Commercial Decking	Decking	...	1500 1250	...	...	(Stresses apply at 15% moisture content)	...	1,500,000 1,300,000	
LODGEPOLE PINE (Surfaced dry or surfaced green. Used at 19% max. m.c.)										
Selected Decking Commercial Decking	Decking	...	1450 1200	...	...		...	1,300,000 1,200,000	Western Wood Products Association
Selected Decking Commercial Decking	Decking	...	1550 1300	...	...	(Stresses apply at 15% moisture content)	...	1,400,000 1,200,000	
NORTHERN PINE (Surfaced dry or surfaced green. Used at 19% max. m.c.)										
Select Structural No. 1 No. 2 No. 3	2" to 4" thick 6" and wider	1400 1200 950 575	1600 1400 1100 650	950 800 650 375	70 70 70 70	280 280 280 280	1100 975 825 525	1,400,000 1,400,000 1,300,000 1,100,000	Northeastern Lumber Manufacturers Association and Northern Hardwood and Pine Manufacturers Association
Select Structural No. 1	Beams and Stringers	1250 1050	...	850 700	65 65	280 280	800 725	1,300,000 1,300,000	

Table No. 1.10.1 (cont'd)

Species and commercial grade	Size classification	Allowable unit stress in pounds per square inch [1]						Modulus of elasticity "E"	Grading rules agency
		Extreme fiber in bending "Fb"		Tension parallel to grain "Ft"	Horizontal shear "Fv"	Compression perpendicular to grain "Fc⊥"	Compression parallel to grain "Fc"		
		Engineered uses (single)	Repetitive-member uses						
Select Structural	Posts and Timbers	1150	...	800	65	280	900	1,300,000	(see footnotes 2 through 9)
No. 1		950	...	650	65	280	800	1,300,000	
Select	Decking	1350	1550	...	...	...	...	1,400,000	NeLMA
Commercial		1150	1300	...	...	...	...	1,300,000	
PONDEROSA PINE — SUGAR PINE (Ponderosa Pine-Lodgepole Pine) (Surfaced dry or surfaced green. Used at 19% max. m.c.)									Western Wood Products Association
Selected Decking	Decking	...	1350	...	...	(Stresses apply at 15% moisture content)	...	1,200,000	
Commercial Decking		...	1150	...	...		...	1,100,000	
Selected Decking	Decking	...	1450	...	...		...	1,300,000	
Commercial Decking		...	1250	...	...		...	1,100,000	
RED PINE (Surfaced dry or surfaced green. Used at 19% max. m.c.)									National Lumber Grades Author. (A Canadian agency. See footnotes 2 through 8 and 12)
Select Structural	2" to 4" thick 6" and wider	1200	1350	800	70	280	900	1,300,000	
No. 1		1100	1150	675	70	280	825	1,300,000	
No. 2		825	950	550	70	280	675	1,200,000	
No. 3		500	550	325	70	280	425	1,000,000	
Select Structural	Beams and Stringers	1050	...	625	65	280	725	1,100,000	
No. 1 Structural		875	...	450	65	280	600	1,100,000	
Select Structural	Posts and Timbers	1000	...	675	65	280	775	1,100,000	
No. 1 Structural		800	...	550	65	280	675	1,100,000	
Select	Wall and Roof Plank	1150	1350	...	...	280	...	1,300,000	
Commercial		975	1100	...	...	280	...	1,300,000	
SITKA SPRUCE (Surfaced dry or surfaced green. Used at 19% max. m.c.)									West Coast Lumber Inspection Bur.
Select Dex	Decking	1300	1500	...	...	280	...	1,500,000	
Commercial Dex		1100	1250	...	...	280	...	1,300,000	
SOUTHERN PINE (Surfaced dry. Used at 19% max. m.c.)									Southern Pine Inspection Bureau
Selected Structural	2" to 4" thick 2" to 4" wide	2100	...	1250	90	405	1600	1,800,000	
Dense Select Structural		2450	...	1450	90	475	1850	1,900,000	
No. 1		1750	...	1000	90	405	1250	1,800,000	
No. 1 Dense		2050	...	1200	90	475	1450	1,900,000	
No. 2		1250	...	725	75	345	850	1,400,000	
No. 2 Medium Grain		1450	...	850	90	405	1000	1,600,000	

Table No. 1.10.1 (cont'd)

Species and commercial grade	Size classification	Allowable unit stress in pounds per square inch [1]							Grading rules agency
		Extreme fiber in bending "F_b"		Tension parallel to grain "F_t"	Horizontal shear "F_v"	Compression perpendicular to grain "$F_c \perp$"	Compression parallel to grain "F_c"	Modulus of elasticity "E"	
		Engineered uses (single)	Repetitive-member uses						
No. 2 Dense	2" to 4" thick	1700		1000	90	475	1150	1,700,000	
No. 3	2" to 4" wide	825		475	75	345	600	1,400,000	
No. 3 Dense		950		550	90	475	700	1,500,000	
Select Structural		1800		1200	90	405	1400	1,800,000	
Dense Select Structural		2100		1400	90	475	1650	1,900,000	
No. 1		1500		1000	90	405	1250	1,800,000	
No. 1 Dense		1800		1200	90	475	1450	1,900,000	
No. 2	2" to 4" thick	1050		700	75	345	900	1,400,000	
No. 2 Medium grain	6" and wider	1250		825	90	405	1050	1,400,000	
No. 2 Dense		1450		975	90	475	1250	1,700,000	
No. 3		725		475	75	345	650	1,400,000	
No. 3 Dense		850		575	90	475	750	1,500,000	
Dense Std. Factory		2000		1200	90	475	1450	1,900,000	Southern Pine Inspection Bureau
No. 1 Factory		1400		825	90	405	1000	1,600,000	
No. 1 Dense Factory	2" to 4" thick	1650		975	90	475	1150	1,700,000	
No. 2 Factory	2" to 4" wide	1400		825	90	405	1000	1,600,000	
No. 2 Dense Factory		1650		975	90	475	1150	1,700,000	
Dense Std. Factory		1750		1200	90	475	1450	1,900,000	
No. 1 Factory		1250		825	90	405	1050	1,600,000	
No. 1 Dense Factory	2" to 4" thick	1450		975	90	475	1250	1,700,000	
No. 2 Factory	6" and wider	1250		825	90	405	1050	1,600,000	
No. 2 Dense Factory		1450		975	90	475	1250	1,700,000	
Dense Structural 86	2" to 4" thick	2750		1850	150	475	2050	1,900,000	
Dense Structural 72		2300		1550	125	475	1700	1,900,000	
WESTERN CEDARS (Surfaced dry or surfaced green. Used at 19% max. m.c.)									
Select Dex	Decking	1200	1400			295		1,100,000	West Coast Lumber Inspection Bur.
Commercial Dex	Decking	1050	1200			295		1,000,000	
Selected Decking	Decking		1400			(Stresses apply at 15% moisture content)		1,100,000	Western Wood Products Association
Commercial Decking			1200					1,000,000	
Selected Decking	Decking		1500					1,100,000	
Commercial Decking			1250					1,000,000	

FOOTNOTES FOR TABLE 1.10.1

[1] The allowable unit stresses shown for selected species and commercial grades. For stresses for other species and commercial grades not shown, the designer is referred to the grading rules of the appropriate grading rules agency.

[2] The recommended design values shown in Table 1.10.1 are applicable to lumber that will be used under dry conditions such as in most covered structures. For 2" to 4" thick lumber the DRY surfaced size should be used. In calculating design values, the natural gain in strength and stiffness that occurs as lumber dries has been taken into consideration as well as the reduction in size that occurs when unseasoned lumber shrinks. The gain in load carrying capacity due to increased strength and stiffness resulting from drying more than offsets the design effect of size reductions due to shrinkage. For 5" and thicker lumber, the surfaced sizes also may be used because design values have been adjusted to compensate for any loss in size by shrinkage which may occur.

[3] Values for "F_b," "F_t," and "F_c" for the grades of Construction and Standard apply only to 4" widths.

[4] The values in Table 1.10.1 are based on edgewise use. For dimension 2" to 4" in thickness, when used flatwise, the recommended design values for fiber stress in bending may be multiplied by the following factors:

Width	Thickness		
	2"	3"	4"
2" to 4"	1.10	1.04	1.00
6" and wider	1.22	1.16	1.11

[5] When 2" and 4" thick lumber is manufactured at a maximum moisture content of 15 percent and used in a condition where the moisture content does not exceed 15 percent, the design values shown in Table 1.10.1 may be multiplied by the following factors:

Extreme fiber in bending "F_b"	Tension parallel to grain "F_t"	Horizontal shear "F_v"	Compression perpendicular to grain "$F_c \perp$"	Compression parallel to grain "F_c"	Modulus of Elasticity "E"
1.08	1.08	1.05	1.00	1.17	1.05

[6] When 2" to 4" thick lumber is designed for use where the moisture content will exceed 19 percent for an extended period of time, the values shown in Table 1.10.1 should be multiplied by the following factors:

Extreme fiber in bending "F_b"	Tension parallel to grain "F_t"	Horizontal shear "F_v"	Compression perpendicular to grain "$F_c \perp$"	Compression parallel to grain "F_c"	Modulus of Elasticity "E"
0.86	0.84	0.97	0.67	0.70	0.97

[7] When lumber 5" and thicker is designed for use where the moisture content will exceed 19 percent for an extended period of time, the values shown in Table 1.10.1 should be multiplied by the following factors:

Extreme fiber in bending "F_b"	Tension parallel to grain "F_t"	Horizontal shear "F_v"	Compression perpendicular to grain "$F_c \perp$"	Compression parallel to grain "F_c"	Modulus of Elasticity "E"
1.00	1.00	1.00	0.67	0.91	1.00

[8] The tabulated horizontal shear values shown herein are based on the conservative assumption of the most severe checks, shakes or splits possible, as if a plane were split full length. When lumber 4" and thinner is manufactured unseasoned the tabulated values should be multiplied by a factor of 0.92.

FOOTNOTES FOR TABLE 1.10.1 (cont'd)

Specific horizontal shear values for any grade and species of lumber may be established by use of the following tables when the length of split or check is known:

When length of split is:	Multiply tabulated "F_v" value by: (Nominal 2" Lumber)
No split	2.00
½ × wide face	1.67
¾ × wide face	1.50
1 × wide face	1.33
1½ × wide face or more	1.00

When length of split on wide face is:	Multiply tabulated "F_v" value by: (3" and Thicker Lumber)
No split	2.00
½ × narrow face	1.67
1 × narrow face	1.33
1½ × narrow face or more	1.00

[9] Stress rated boards of nominal 1", 1¼" and 1½" thickness, 2" and wider, are permitted the recommended design values shown for Select Structural, No. 1, No. 2 and No. 3 grades as shown in 2" to 4" thick, 2" to 4" wide and 2" to 4" thick, 6" and wider categories when graded in accordance with those grade requirements.

[10] For species combinations shown in parentheses, the lowest design values for any species in the combination are tabulated.

[11] When "MC15" Decking is used where the moisture content will exceed 15 percent for an extended period of time, the design values tabulated to apply at 15 percent moisture content should be multiplied by the following factors: Extreme Fiber in Bending "F_b"-0.79; Modulus of Elasticity "E"-0.92.

[12] National Lumber Grades Authority is the Canadian rules-writing agency responsible for preparation, maintenance and dissemination of a uniform softwood lumber grading rule for all Canadian species.

(B) Secondary Stresses in Curved Glued Laminated Members

(1) Curvature Factor

For the curved portion of members, the allowable stress in bending shall be modified by multiplication by the following curvature factor:

$$1 - 2000(t/R)^2$$

in which
  t = thickness of lamination in inches
  R = radius of curvature of a lamination in inches
and t/R should not exceed 1/125 for Douglas fir, larch and California redwood and 1/100 for Southern pine.

No curvature factor shall be applied to stress in the straight portion of an assembly regardless of curvature elsewhere.

(2) Radial Tension or Compression

The radial stress, f_r, induced by a bending moment in a curved member, shall be defined by the following equation:

$$f_r = 3M/2Rbd$$

where
  M = bending moment in inch pounds
  R = radius of curvature at centerline of member in inches

TABLE 1.10.1A

Allowable Unit Stresses for Structural Glued Laminated Timber, Members Stressed Principally in Bending, Loaded Perpendicular to the Wide Face of the Laminations [1][2][3]

(Stresses shown below are for normal conditions of loading. See other provisions of Article 1.10.1 for adjustments of these tabulated allowable unit stresses.)

(1) Douglas Fir and Western Larch

Combination Symbol	Number of Laminations	Allowable unit stresses					
		Extreme Fiber in Bending F_b [4][5]	Tension Parallel to Grain F_t	Compression Parallel to Grain F_c	Compression ⊥ to Grain		Horizontal Shear F_v
					Tension Face $F_{c\perp}$	Compression Face $F_{c\perp}$	
		DRY CONDITIONS OF USE E = 1,800,000 psi					
22F	4-10	2200	1600	1500	410	410	165
	4-10	2200	1600	1500	450	385	165
	11-20	2200	1600	1500	450	385	165
	21-30	2200	1600	1500	450	385	165
	31-40	2200	1600	1500	450	385	165
	41 or more	2200	1600	1500	450	385	165
24F	4-10	2400	1600	1500	450	385	165
	11-20	2400	1600	1500	450	385	165
	21-25	2400	1600	1500	450	385	165
	26-35	2400	1600	1500	450	385	165
	36-40	2400	1600	1500	450	385	165
	41 or more	2400	1600	1500	450	385	165
26F	4-8	2600	1600	1500	450	410	165
	9-20	2600	1600	1500	450	410	165
	21-25	2600	1600	1500	450	410	165
	26-30	2600	1600	1500	450	410	165
	31-34	2600	1600	1500	450	410	165
	35-40	2600	1600	1500	450	410	165
	41 or more	2600	1600	1500	450	410	165

Note: The 26F combination may not be readily available and the designer should check on availability prior to specifying. The 22F and 24F combinations are generally available from all laminators.

TABLE 1.10.1A (cont'd)

(1) Douglas Fir and Western Larch

Combination Symbol	Number of Laminations	Allowable unit stresses					
		Extreme Fiber in Bending F_b[4][5]	Tension Parallel to Grain F_t	Compression Parallel to Grain F_c	Compression $\perp$ to Grain		Horizontal Shear F_v
					Tension Face $F_c \perp$	Compression Face $F_c \perp$	
WET CONDITIONS OF USE E = 1,600,000 psi								
22F	4-10	1600	1300	1100	275	275	145
	4-10	1600	1300	1100	305	260	145
	11-20	1600	1300	1100	305	260	145
	21-30	1600	1300	1100	305	260	145
	31-40	1600	1300	1100	305	260	145
	41 or more	1600	1300	1100	305	260	145
24F	4-10	1800	1300	1100	305	260	145
	11-20	1800	1300	1100	305	260	145
	21-25	1800	1300	1100	305	260	145
	26-35	1800	1300	1100	305	260	145
	36-40	1800	1300	1100	305	260	145
	41 or more	1800	1300	1100	305	260	145
Note: The 26F combination may not be readily available and the designer should check on availability prior to specifying. The 22F and 24F combinations are generally available from all laminators.								
26F	4-8	2000	1300	1100	305	275	145
	9-20	2000	1300	1100	305	275	145
	21-25	2000	1300	1100	305	275	145
	26-30	2000	1300	1100	305	275	145
	31-34	2000	1300	1100	305	275	145
	35-40	2000	1300	1100	305	275	145
	41 or more	2000	1300	1100	305	275	145

TABLE 1.10.1A (cont'd)

(2) Southern Pine

Combination Symbol		Number of Laminations	Extreme Fiber in Bending F_b [4,5,6]	Tension Parallel to Grain F_t	Compression Parallel to Grain F_c	Compression Perpendicular to Grain $F_{c\perp}$	Horizontal Shear F_v
			Allowable Unit Stresses				
			DRY CONDITIONS OF USE E = 1,800,000 psi				
18F	1	4 or more	1800	1600	1500	385	200
	2	12 or more	1800	1600	1500	385	200
20F	1	10 or more [9]	2000	1600	1500	385	200
	2	10 or more	2000	1600	1500	385	200
22F	1	6 or more [9]	2200	1600	1500	450	200
	2	14 or more	2200	1600	1500	385	200
	3	18 or more	2200	1600	1500	385	200
24F	1	4 or more	2400	1600	1500	385	200
	2	12 or more	2400	1600	1500	450	200
	3	9 or more	2400	1600	1500	385	200

Note: The 26F combination may not be readily available and the designer should check on availability prior to specifying. Other combinations listed are generally available from all laminators.

26F	1	9 or more [7,8]	2600	1600	1500	385	200
	2	14 or more	2600	1600	1500	450	200
	3	13 or more	2600	1600	1500	450	200

			WET CONDITIONS OF USE E = 1,600,000 psi				
18F	1	4 or more	1400	1300	1100	260	175
	2	12 or more	1400	1300	1100	260	175
20F	1	10 or more [9]	1600	1300	1100	260	175
	2	10 or more	1600	1300	1100	260	175
22F	1	6 or more [9]	1800	1300	1100	300	175
	2	14 or more	1800	1300	1100	260	175
	3	18 or more	1700	1300	1100	260	175
24F	1	4 or more	1900	1300	1100	260	175
	2	12 or more	2000	1300	1100	300	175
	3	9 or more	1900	1300	1100	260	175

Note: The 26F combination may not be readily available and the designer should check on availability prior to specifying. Other combinations listed are generally available from all laminators.

26F	1	9 or more [7,8]	2000	1300	1100	260	175
	2	14 or more	2000	1300	1100	300	175
	3	13 or more	2100	1300	1100	300	175

FOOTNOTES FOR TABLE 1.10.1A

[1] The tabulated stresses in this table are primarily applicable to members stressed in bending due to a load applied perpendicular to the wide face of the laminations. For combinations and stresses applicable to members loaded primarily axially or parallel to the wide face of the laminations, see Table 1.10.1B.

[2] The tabulated bending stresses are applicable to members 12 inches or less in depth. For members greater than 12 inches in depth, the requirements of Article 1.10.2 on Size Factor apply.

[3] The tabulated combinations are applicable to arches, compression members, tension members and also bending members less than 16¼ inches in depth. For bending members 16¼ inches or more in depth, footnotes 4 and 5 apply.

[4] The grading restrictions as contained in AITC 301-22, 301-24 and 301-26 tension lamination requirements shall be followed for the outermost tension laminations representing 5% of the total depth of glued laminated bending members 16¼ inches or more in depth. For all conditions of use, AITC 301-22 is applicable to combination 22F, AITC 301-24 is applicable to combination 24F and AITC 301-26 is applicable to combination 26F. See Appendix "A" of AITC 203-70 for details of these tension lamination requirements.

[5] In addition to other requirements, the tension laminations as described in AITC 301-22, 301-24 and 301-26 are required to be dense.

[6] The next inner 5% of the outermost tension laminations are to be No. 1 Dense for the same conditions as indicated by footnote number 4.

[7] For fewer than (9) laminations, add one No. 1 lamination to each outer zone.

[8] For combination 26F(1), six or fewer laminations, the allowable unit stresses for tension parallel to grain and compression parallel to grain can be increased to 1800 psi and 1600 psi respectively for the dry condition of use and to 1500 psi and 1200 psi respectively for the wet condition of use.

[9] Where fewer laminations are required, a combination with a higher allowable unit stress can be selected.

b = width of cross section in inches
d = depth of cross section in inches

When M is in the direction tending to decrease curvature (increase the radius), the stress is tension across the grain. For this condition, the tension stress across the grain shall be limited to 1/3 the allowable unit stress in horizontal shear for Southern pine and California redwood for all load conditions and for Douglas fir and larch for wind or earthquake loadings. The limit shall be 15 psi for Douglas fir and larch for other types of load. These values are subject to modifications for duration of load. If these values are exceeded, mechanical reinforcing shall be used and shall be sufficient to resist all radial tension stresses.

When M is in the direction tending to increase curvature (decrease the radius), the stress is compression across the grain and shall be limited to the allowable unit stress in compression perpendicular to the grain for all species included herein.

(C) Compression or Bearing Perpendicular to Grain

The allowable unit stresses for compression perpendicular to the grain apply to bearings of any length at the ends of the beam, and to all bearings 6 inches or more in length at any other location. When calculating the bearing area at the ends of beams, no allowance shall be made for the fact that, as the beam bends, the pressure upon the inner edge of the bearing is greater than at the end of the beam.

TABLE 1.10.1B

Allowable Unit Stresses for Structural Glued Laminated Timber, Members Stressed Principally in Axial Tension or Axial Compression, or a combination of Axial Loading Plus Bending Parallel to or Perpendicular to the Wide Face of the Laminations.[1] (Stresses shown below are for normal conditions of loading. See other provisions of Article 1.10.1 for adjustments of these tabulated allowable unit stresses.)

Combination Symbol	Number of Laminations	Tension Parallel to Grain F_t	Compression Parallel to Grain F_c	Extreme Fiber in Bending F When Loaded		Compression Perpendicular to Grain[3] $F_c \perp$	Horizontal Shear F_v When Loaded	
				Parallel to Wide Face[3]	Perpendicular to Wide Face[2,4]		Parallel to Wide Face[3]	Perpendicular to Wide Face[4]
(1) Douglas Fir and Western Larch			DRY CONDITIONS OF USE $E = 1{,}800{,}000$ psi					
1	All	1200	1500	900	1200	385	145	165
2	All	1800	1800	1500	1800	385	145	165
3	All	2200	2100	1900	2200	450	145	165
4	All	2400	2000	2100	2400	410	145	165
5	All	2600	2200	2300	2600	450	145	165
			WET CONDITIONS OF USE $E = 1{,}600{,}000$ psi					
1	All	950	1100	750	950	260	120	145
2	All	1400	1300	1100	1400	260	120	145
3	All	1800	1500	1450	1800	305	120	145
4	All	1900	1450	1500	1900	275	120	145
5	All	2000	1600	1600	2000	305	120	145
(2) Southern Pine			DRY CONDITIONS OF USE $E = 1{,}800{,}000$ psi					
1	All	1600	1400	950	1100	385	165	200
2	All	2200	1900	1700	1800	385	165	200
3	All	2600	2200	2000	2100	450	165	200
4	All	2400	2100	1950	2400	385	165	200
5	All	2600	2200	2300	2600	450	165	200
			WET CONDITIONS OF USE $E = 1{,}600{,}000$ psi					
1	All	1300	1000	750	850	260	145	175
2	All	1800	1400	1350	1450	260	145	175
3	All	2100	1600	1600	1700	300	145	175
4	All	1900	1500	1550	1950	260	145	175
5	All	2100	1600	1850	2100	300	145	175

FOOTNOTES FOR TABLE 1.10.1B

[1] The tabulated stresses in this table are primarily applicable to members loaded axially or parallel to the wide face of the laminations. For combinations and stresses applicable to members stressed principally in bending due to a load applied perpendicular to the wide face of the laminations, see Table 1.10.1A.

[2] It is not intended that these combinations be used for deep bending members, but if bending members 16¼ inches or deeper are used, the applicable AITC tension lamination requirements must be followed.

[3] The tabulated stresses are applicable to members containing three (3) or more laminations.

[4] The tabulated stresses are applicable to members containing four (4) or more laminations.

For bearings of less than 6 inches in length and not nearer than 3 inches to the end of a member, the maximum allowable load per square inch is obtained by multiplying the allowable unit stresses in compression perpendicular to grain by the following factor:

$$\frac{L+\tfrac{3}{8}}{L}$$

in which L is the length of bearing in inches measured along the grain of the wood.

The multiplying factors for indicated lengths of bearing on such small areas as plates and washers become:

Length of bearing in inches	½	1	1½	2	3	4	6 or more
Factor	1.75	1.38	1.25	1.19	1.13	1.10	1.00

In using the preceding formula and table for round washers or bearing areas, use a length equal to the diameter.

(D) Simple Solid Column Design

These formulas for simple solid columns are based on pin-end conditions but shall be applied also to square-end conditions.

Allowable unit stresses in pounds per square inch of cross-sectional area of simple solid columns shall be determined by the following formula, but such unit stresses shall not exceed the tabular values for compression parallel to grain, F_c, as provided in Article 1.10.1 and adjusted in accordance with the applicable provisions of Article 1.10.1:

$$F'_c = \frac{\pi^2 E}{2.727(\ell/r)^2} = \frac{3.619E}{(\ell/r)^2}$$

where F'_c = allowable unit stress in compression parallel to grain, in psi, adjusted for l/d ratio
E = modulus of elasticity, psi
ℓ = unsupported overall length, in inches, between points of lateral support of simple columns
r = least radius of gyration of section

For columns of square or rectangular cross-section, this formula becomes:

$$F'_c = \frac{0.30E}{(\ell/d)^2}$$

where d = dimension of least side of simple solid column in inches

For simple solid columns, the ℓ/d ratio may not exceed 50.

The values of F'_c as determined from the formulae listed are subject to adjustment for duration of load as given in Articles 1.10.1(D) and 1.10.1(E).

(E) Spaced Column Design

Spaced columns are formed of two or more individual members with their longitudinal axes parallel, separated at the ends and middle points of their length by blocking and joined at the ends by timber connectors capable of developing the required shear resistance. To obtain spaced column action, end blocks with connectors and spacer blocks are required when the individual members of a spaced column assembly have an l/d ratio greater than

$$\sqrt{\frac{0.30E}{F_c}}$$

For an assembly of members having a lesser l/d ratio, the individual members are designed as simple solid columns. Spaced columns are classified as to fixity, i.e., condition "a" or condition "b," which introduces a multiplying factor applicable in the design of its individual members. (See Figure 1.10.2).

For individual members of a spaced column, ℓ/d shall not exceed 80, nor shall ℓ_2/d exceed 40. (See Figure 1.10.2).

The individual members in a spaced column are considered to act together to carry the total column load. Each member is designed separately on the basis of its l/d ratio.

A greater l/d ratio than allowed for simple solid columns is permitted because of the end fixity developed by the connectors and end blocks. This fixity is effective only in the thickness direction. The l/d ratio in the direction of width is subject to the provisions for simple solid columns.

When a single spacer block is located within the middle tenth of the column length (1), connectors are not required for this block. If there are two or more spacer blocks, connectors are required and the distance between two adjacent blocks shall not exceed one-half the distance between centers of connectors in the end blocks.

For spaced columns used as compression members of a truss, a panel point which is stayed laterally shall be considered as the end of the spaced column, and the portion of the web members, between the individual pieces making up a spaced column, may be considered as the end blocks.

If there are two or more connectors in a contact face, the center of gravity of the group shall be used in measuring the distance from connectors in the end block to the end of the column for determining fixity condition "a" or "b." (Figure 1.10.2).

Thickness of spacer and end blocks shall not be less than that of individual members of the spaced column, nor shall thickness, width, and length of spacer and end blocks be less than required for connectors of a size and number capable of carrying the load computed in accordance with Article 1.10.2(I). Blocks thicker than a side member do not appreciably increase load capacity.

The total allowable load for a spaced column is the sum of the allowable loads for each of its individual members. Allowable unit stresses shall be determined as follows, but the maximum unit stress shall not exceed the values for compression parallel to grain "F_c" in Table 1.10.1, or as tabulated in the reference listed in Article 1.10.1 (B) (1), and as adjusted in accordance with provisions of Article 1.10.1, nor shall the load exceed that permitted by the following provisions.

The net section shall be determined by subtracting, from the full cross-sectional area of the timber, the projected area of that portion of the connector groove within the members and that portion of the bolt hole not within the connector groove located at the critical plane. (See Table 2.20.1 for typical dimensions for Timber Connectors). Where connectors are staggered, adjacent connectors, with parallel-to-grain spacing equal to or less than one connector diameter, shall be considered as occurring at the same critical section.

In tension and compression members the required net area, in square inches, shall be determined by dividing the total load transferred through the critical section by the allowable tension stress for tension members, or by the allowable compression parallel to grain stress for compression members, for the species and grade of lumber used.

For condition "a," the allowable unit stress for individual members of a spaced column, in which the connectors in end blocks are placed at a distance not exceeding l/20 from the ends, shall be determined by the formula:

$$F_c = \frac{0.75E}{(l/d)^2}$$

For condition "b," the allowable load for the individual members of a spaced column in which the connectors in end blocks are placed a distance of l/20 to l/10 from the ends and the blocks extend to the ends of the column shall be determined by the formula:

$$F_c = \frac{0.90E}{(l/d)^2}$$

The total load capacity determined by the foregoing procedure should be checked against the sum of the load capacities of the individual members taken as simple solid columns without regard to fixity, using their greater d and the l between the lateral supports which provide restraint in a direction parallel to the greater d.

The values for F_c, as above determined, are subject to the duration of load adjustments as provided in Article 1.10.1.

(F) Safe Load on Round Columns

The safe load on a round column shall not exceed that permitted for a square column of the same cross-sectional area, or as determined by the formula:

$$F'_c = \frac{3.619E}{(\ell/r)^2}$$

The values for F'_c determined by the formula may not exceed the values for compression parallel to the grain, F'_c, adjusted for service conditions and duration of load in accordance with Article 1.10.1. The allowable unit stress values as determined from the formula are subject to the duration of loading adjustments as given in Articles 1.10.1 (D) and 1.10.1(E).

In determining the least diminsion, d, for tapered columns, the diameter of a round column or the least dimension of a rectangular column, tapered at one or both ends, is taken as the sum of the minimum diameter or least dimension and one-third the difference between the minimum and maximum diameters or lesser and greater dimensions.

(G) Notched Beams

Beams notched upward in the bearing face on supports shall be limited to maximum end load R as determined by the formula:

$$R = \frac{2bd^2 F_v}{3h}$$

R = maximum end load

F_v = allowable unit horizontal shear stress

b = breadth of beam

d = depth of beam above the notch

h = total depth of beam

(H) Bearing on Inclined Surfaces

$$N = \frac{PQ}{P \sin^2 \Theta + Q \cos^2 \Theta}$$

N = unit bearing on an inclined surface

P = unit stress in compression parallel to the grain

Q = unit stress in compression perpendicular to the grain

Θ = angle in degrees between the direction of load and the direction of grain

(I) Timber Connectors

Timber connectors shall consist of devices to be used at surfaces of contact in bolted timber joints, to increase the strength or shear resistance of wood-to-wood or wood-to-steel connections.

Allowable loads, spacing of connectors, edge and end distance, bolt and washer sizes and other details of design shall be those recommended or approved by the "Design Manual for Timber Connector Construction," 1970, Timber Engineering Company; or the allowable loads may be determined by actual tests of full size joints for each condition of connector used in accordance with standard procedure.

(J) Size Factor

When the depth of a rectangular beam exceeds 12 inches, the tabulated unit stress in bending F_b, shall be reduced by multiplying the tabulated stress by the size factor, C_F, as determined from the following relationship

$$C_F = (12/d)^{1/9}$$

where C_F = size factor
d = depth of member in inches

The size factor relationship as given above is applicable to a bending member satisfying the following basic assumptions: (a) simply supported beam, (b) uniformly distributed load, and (c) span to depth ratio (ℓ/d) of 21. This factor can thus be applied with reasonable accuracy to most commonly encountered design situations. Where greater accuracy is desired for other sizes and conditions of loading, the percentage changes given in the following table may be applied directly to the size factor calculated for the basic conditions as previously stated. Straight line interpolation may be used for other ℓ/d ratios.

Span to Depth Ratio ℓ/d	% Change	Loading Condition for Simply Supported Beams	% Change
14	+2.3	Center Point	+7.8
24	−1.6	Third Point	−3.2

For more detailed analysis of the size factor and its application to the design of bending members, the designer is referred to the AITC "Timber Construction Manual." The reduction in bending stresses for deep members based on the size effect factor is only applicable to glued laminated members.

(K) Lateral Stability

(1) The tabulated allowable unit bending stresses given under Article 1.10.1 are applicable to members which are adequately braced. When deep, slender members not adequately braced are used, a reduction to the allowable unit bending stresses must be applied based on a computation of the slenderness factor of the member. In the check of lateral stability, the slenderness factor is computed by the relationship:

$$C_s = \sqrt{\frac{\ell_e d}{b^2}}$$

where C_s = slenderness factor
ℓ_e = effective length of beam, in. (see table below)
d = depth of beam, in.
b = breadth of beam, in.

EFFECTIVE LENGTH OF GLUED LAMINATED BEAMS

Type of Beam Span and Nature of Load	Value of Effective Length, ℓ_e [1]
Single span beam, load concentrated at center	$1.61\,\ell$
Single span beam, uniformly distributed load	$1.92\,\ell$
Single span beam, equal end moments	$1.84\,\ell$
Cantilever beam, load concentrated at unsupported end	$1.69\,\ell$
Cantilever beam, uniformly distributed load	$1.06\,\ell$
Single span or cantilever beam, any load (conservative value)	$1.92\,\ell$

[1] Where ℓ = unsupported length.

   (2) Beams with Various Lateral Support Conditions

   (a) Without lateral support. When the depth of a beam does not exceed its breadth, no lateral support is required and the allowable unit stress is determined by applying the appropriate provisions of Articles 1.10.1 and 1.10.2.

   (b) With lateral support. If lateral movement of the compression flange is prevented by a continuous support, there is no danger of lateral buckling, and the allowable stresses require no reduction based on a slenderness ratio concept. Also, there is no need to limit the depth-breadth ratio to 5 or 6.

   (c) When the depth of a beam exceeds the breadth, bracing must be provided at the points of bearing, and it must be so arranged as to prevent rotation of the beam at those points in a plane perpendicular to its longitudinal axis. The allowable stresses are calculated by the formulae contained in the following paragraphs for short, intermediate and long beams.

(3) The allowable unit stresses are determined from the following equations:

   (a) Short beams. When the slenderness factor, C_s, does not exceed 10, the tabular allowable unit stress in bending, F_b, adjusted in accordance with the applicable provisions of Articles 1.10.1(D), 1.10.1(E), 1.10.2(B) and 1.10.2(J) is used for design.

   (b) Intermediate beams. When the slenderness factor, C_s, is greater than 10 but does not exceed C_k, a unit stress in bending based on slenderness considerations, F_b', is calculated by the formula:

$$F_b' = F_b \left[1 - 1/3 \left(\frac{C_s}{C_k} \right)^4 \right]$$

where F_b = tabular allowable unit stress in bending, psi

$$C_k = \sqrt{\frac{3E}{5F_b}}$$

E = modulus of elasticity, psi

(c) Long Beams. When the slenderness factor, C_s, is greater than C_k, but less than 50, the unit stress in bending is calculated by the formula:

$$F_b' = \frac{0.40E}{(C_s)^2}$$

Note: In no case shall C_s be greater than 50.

For both intermediate and long beams, the allowable unit stress for design based on slenderness considerations is obtained by adjusting F_b' in accordance with the applicable provisions of Articles 1.10.1(D), 1.10.1(E) and 1.10.2(B).

Regardless of the slenderness classification into which a beam may be categorized, in no case shall the allowable unit stress in bending used for design exceed the value as obtained by adjusting the tabular allowable unit stress based on the applicable provisions of Articles 1.10.1(D), 1.10.1(E), 1.10.2(B) and 1.10.2(J).

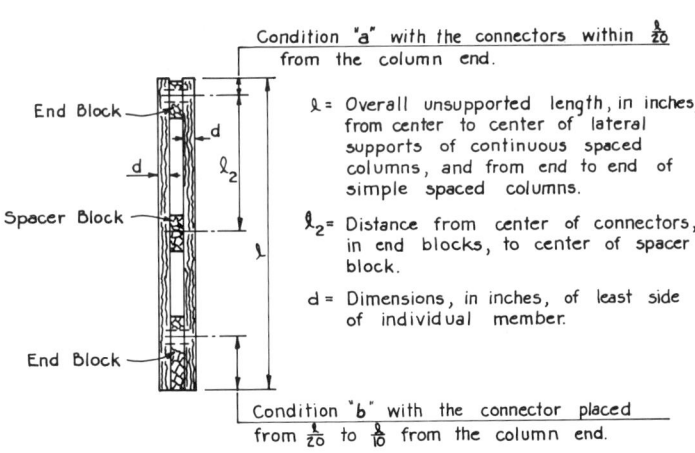

SPACED COLUMN, CONNECTOR JOINED

FIGURE 1.10.2

1.10.3 — GENERAL

All wood used in timber structures shall be preservatively treated as provided in Division II, Section 21, unless otherwise specified.

1.10.4 — BOLTS

Bolts shall be spaced center to center not closer than 4 times the bolt diameter. The distance from the center of a bolt to the end of any timber shall be not less than 7 times the bolt diameter if loaded in tension parallel to grain, nor 4 times the bolt diameter if loaded in compression parallel to the grain or in tension or compression perpendicular to the grain. For parallel to grain loading in tension or compression, the distance from any edge of the timber to the center of the nearest bolt shall be at least 1½ times the bolt diameter, except that for ℓ/d ratios more than 6, use one-half the distance between rows of bolts. For perpendicular to grain loading, the edge distance toward which the load is acting shall be at least 4 times the bolt diameter and the edge distance on the opposite edge shall be at least 1½ bolt diameters.

1.10.5 — WASHERS

A washer shall be used under all bolt heads and nuts which would otherwise come in contact with wood. Either cast or plate washers may be used and they shall be designed to prevent excessive crushing of the wood when the bolts are tightened. For bolts or rods in tension, washers shall be sufficient size to develop the tension stress in the bolt or rod without exceeding the allowable unit stress in compression perpendicular to grain for the species and grade of lumber used.

A standard circular washer shall be used under the heads of all lag screws.

1.10.6 — HARDWARE FOR SEACOAST STRUCTURES

The hardware for structures on the seacoast shall be galvanized or cadmium plated.

1.10.7 — COLUMNS AND POSTS

No column shall have an unsupported length greater than 50 times its least dimension.

The strength of built-up columns composed of two or more sticks bolted together, either with or without packing blocks, shall be considered as equal to the combined strength of the single sticks each considered as an independent column.

The strength of connector-joined spaced columns shall be determined as provided in Article 1.10.2(I).

1.10.8 — PILE AND FRAMED BENTS

(A) Pile Bents

Pile bents generally shall not exceed 40 feet in height. Pile bents over 10 feet high shall be sway-braced transversely with diagonal braces on each side of the bent, and shall be adequately braced longitudinally. In general, pile bents shall contain not less than four piles each and the outside piles, preferably, shall be battered. The piles shall be designed for safe bearing and for column action.

(B) Framed Bents

Framed bents may be supported on piles, concrete pedestals or mud sills. All bents shall be sway-braced transversely and adequate provision shall be made for longitudinal bracing. In general, framed bents shall contain not less than four posts each and the outside posts of the bent shall be battered. The posts shall be designed as columns.

(C) Sills and Mud Sills

When possible, sills shall be located clear of all earth so that there may be a free circulation of air around them. Sills shall be fastened to mud sills or piles with drift bolts of not less than ¾-inch diameter and extending into the mud sills or piles at least 6 inches. Sills shall be fastened to pedestals with dowels of not less than ¾-inch diameter, set in the pedestals and extending into the sills at least 6 inches.

Posts shall be fastened to sills by dowels of not less than ¾-inch diameter, extending at least 6 inches into the posts and sills, or by drift bolts of not less than ¾-inch diameter driven diagonally through the base of the posts and extending at least 9 inches into the sill. Posts shall be fastened to pedestals with dowels of not less than ¾-inch diameter and extending into the posts at least 6 inches.

(D) Caps

Timber caps shall be not less in size than 10 by 10 inches. They shall be fastened with drift bolts of not less than ¾-inch diameter, extending at least 9 inches into the piles or posts.

(E) Bracing

Single-story bracing shall not exceed 20 feet in height. The minimum size of transverse sway braces shall be 3 by 8 inches. All bracing shall be bolted through the piles, posts or caps at the ends; at intermediate intersections, it may be bolted or spiked. In all cases, spikes shall be provided in addition to bolts. The bolts used shall be not less than ⅝-inch diameter.

(F) Pile Bent Abutments

Pile bent abutments shall be adequately braced or anchored to resist earth pressure. Bulkhead plank shall be not less than 3 inches thick. It shall be fastened to the piles with spikes, the length of which shall be at least 3 inches greater than the thickness of the plank.

1.10.9 — TRUSSES

(A) Joints and Splices

Joints shall be detailed to shed water to the maximum degree practicable. Joints and splices shall be designed to develop the computed stresses in the members connected and, preferably, to develop the full strength of the members. Posts or struts bearing against the sides of timber members, preferably, shall be provided with metal end bearings. Joints involving end bearing on inclined surfaces shall be avoided, preferance being given to square-cut ends of timbers bearing against blocks.

Bearing surfaces of castings connecting timber members shall be milled to provide smooth, even surfaces permitting accurate fitting and complete contact of the wood and metal bearing surfaces. Rolled plates, bars and shapes used in chord splice plates, or other parts bearing upon wood surfaces, shall be true and even. The wood surfaces taking bearing upon metal parts shall be not less than ⅝ inch in width. Bolts engaging castings and structural parts shall hold them rigidly in position so that bending on the parts in contact will be reduced to a minimum. The joint details at truss panel points shall provide definite lines of action and shall be simple and as susceptible as possible of definite strength analysis. When inclined bolts are used to connect end posts or web members with chord members, they shall be placed approximately at an angle of not more than 60 degrees with the latter and when used in conjunction with joint castings, the holes in one of the connected members shall be bored ⅛ inch larger than the nominal diameter of the bolts. No daps in chords for butt blocks shall be less than ¾ inch deep.

Splices for tension members shall be designed to reduce to a minimum the effects of cross shrinkage of the timber. Neither steel splice plates of the batten type nor shear pin splices shall be used when the timbers to be spliced are more than 8 inches thick, since the shrinkage will permit the joint to become loose. Shear pin joints shall be used only with fully seasoned timber.

(B) Floor Beams

Floor beams shall be sized at bearing points. In floor beams composed of two or more timbers, the timbers preferably shall be separated by at least 2 inches for air circulation. Floor beams shall be connected to the main truss members by means of rods or structural shapes.

(C) Hangers

Hangers generally shall be rods having upset ends with a suitably designed washer or bearing plate at each end. Upset ends shall conform to the requirements specified for Structural Steel Design, Division I.

(D) Eyebars and Counters

The requirements specified for Structural Steel Design, Division I, for counters, eyebars and eyebar packing shall apply to such members when used in timber trusses.

(E) Bracing

Timber trusses shall be provided with a rigid system of laterals in the plane of the loaded chord. When the details will permit, this lateral bracing shall be securely fastened to all longitudinal stringers. Lateral bracing, preferably rigid, in the plane of the unloaded chord, and rigid portal and sway-bracing shall be provided in all trusses having sufficient headroom. Outrigger brackets connected to extensions of the floor beams shall be used for bracing through-trussses having headroom insufficient for a top lateral system.

(F) Camber

Camber, in addition to that required to provide for dead load and shrinkage, shall be provided in timber trusses in sufficient amount to give the structure a good appearance.

1.10.10 — FLOORS AND RAILINGS

(A) Stringers

Stringers shall be of sufficient length to take bearing over the full width of caps or floor beams, except outside stringers which may have butt joints. Preferably, they shall be of two panel lengths placed with staggered joints. The lapped ends of untreated stringers shall be separated at least ½ inch for air circulation. Stringers shall be secured to caps or floor beams.

(B) Bridging

Stringers shall be braced by cross bridging in each panel. The bridging shall be not less in size than 2 by 4 inches.

(C) Nailing Strips

When timber floors are supported by steel joists, the joists shall be provided with nailing strips which shall be bolted either to the top flanges or the webs.

When nailing strips are bolted to the flanges, they shall be used on all joists. They shall be not less than 4 inches deep and shall be wider than the supporting flange. They shall be secured with ⅝-inch bolts through the flanges, spaced not more than 4 feet apart and not more than 18 inches from the ends of the strips.

Nailing strips bolted to the webs shall be not less than 4 inches thick and shall be fastened with bolts spaced not farther apart than 5 feet. They shall be held clear of the flanges by blocks between the web and strip, and bolted through the web with ⅝-inch bolts spaced not more than 4 feet apart and not more than 18 inches from the ends of the strips.

(D) Flooring

Roadway floor plank shall have a nominal thickness of not less than 3 inches. Sidewalk floor plank shall have a nominal thickness of not less than 2 inches.

The minimum size of material used for laminated or strip floors shall be 2 by 4 inches.

(E) Retaining Pieces

Retaining pieces, where required, shall be not less than 6 inches in width. In general, they shall be secured in place by ⅝-inch bolts at 3-foot intervals and spiked at 1-foot intervals.

(F) Wheel Guards

Wheel guards having a cross section of not less than 4 by 6 inches shall be provided on each side of the roadway. The guard timbers shall be in lengths of not less than 12 feet. They shall be secured with ⅝-inch bolts at the ends and at intermediate points not more than 4 feet apart.

In strip floors or cambered floors, not provided with retaining pieces, the wheel guards shall be placed directly on the flooring with scupper holes at suitable intervals. In other floors the wheel guards shall be supported by scupper blocks not less than 4 inches thick and 1 foot long, held in place by spikes and a bolt through the wheel guard and flooring, and spaced not more than 4 feet center to center.

(G) Drainage

Adequate provision shall be made for the proper drainage of timber floors.

(H) Railings

Wood railings shall consist of not less than 2 horizontal lines of rails. Rails, rail posts and fastenings shall be designed for the loads specified in Article 1.1.9 — Railings.

Preferably, rails shall be surfaced 4 sides (S4S) and painted.

1.10.11 — FIRE STOPS

To check the spread of fire lengthwise of the structure, timber floors or trestles of any considerable length, preferably shall be provided with fire stops.

In timber floors these fire stops should be provided at intervals of not over 75 feet. They may consist of diaphragms of wood or fire-resistant material at least as thick as the flooring, located over caps or floorbeam and completely filling the opening between the joists.

In timber trestle bridges, in addition to the fire stops in the floor, fire curtains should be provided at intervals of 100 feet or more. These curtains may consist of plank or asbestos-covered metal spiked to the bents. They should extend downward from the bottom of the joists at least 5 feet and horizontally at least to the ends of the caps. A fire stop between the joists should be located over each curtain.

Section 11—LOAD CAPACITY RATING OF EXISTING BRIDGES

1.11.1 — OVERLOAD UNDER PERMIT

The unit working stresses used in determining the load-carrying capacity of each member of a structure to be crossed by a vehicle operating under a special permit shall take into account the type of material from which the member is made and the physical condition of the member. For structural elements for which plans are available and the properties of materials are known, the tensile stress produced by any such special permit load (including impact, if any) and dead load for structures in good condition shall not exceed:

1. 75 percent of the yield point of structural steel members or of the bars in the reinforced concrete members. This percentage should be reduced for high strength steels.
2. For prestressed concrete members, 90 percent of the yield point stress of the prestressing steel in the layer of tendons nearest the extreme tension fiber of the member; or 75 percent of the prestressing steel stress at the center of gravity of the tendons due to the ultimate moment, whichever governs.
3. A 33 percent increase in the allowable Design Stress of treated timber. For untreated timber, no overstress is to be permitted.

Compressive stresses shall be checked on a corresponding basis.

1.11.2 — IMPACT

If the condition of the permit does not eliminate the likelihood of impact, impact shall be considered as provided under Article 1.2.12.

1.11.3 — ADJUSTABLE LOADS

Overweight permits should not be approved for vehicles carrying loads such as gravel, cement, lumber, petroleum products, pipe or any other product, material or equipment which can be reduced in weight to the design or legal limit.

1.11.4 — STRESS ANALYSIS

For the load under consideration, the stress analysis used shall be in conformance with the provisions of Division I, except that a more reasonable distribution may be assumed for vehicles of unusual width or wheels of unusual size such as those used in the construction industry.

1.11.5 — ALLOWABLE STRESSES

No modification in allowable stresses for the appropriate material and those hereinafter given shall be made except as provided in Article 1.11.1.

1.11.5 — ALLOWABLE STRESSES

Allowable unit stresses are shown in pounds per square inch. The modulus of elasticity of all grades of steel shall be assumed to be 29,000,000 psi and the coefficient of linear expansion 0.0000065 per degree Fahrenheit.

	Carbon Steel	Silicon steel	Nickel steel
AASHO Designation (4)	M-94 (1961)	M-95 (1961)	M-96 (1961)
ASTM Designation (4)	A7 (1967)	A94 (1966)	A8 (1961)
Minimum Tensile Strength F_u	60,000		
Minimum Yield Point F_y	33,000	45,000	55,000
Axial tension, net section Tension in extreme fiber of rolled shapes, girders and built-up sections subject to bending Axial Compression, gross section: Stiffeners of plate girders Compression in splice material, gross section	$0.55 F_y$	24,000	30,000
Compression in extreme fibers of rolled shapes, girders and built-up sections, subject to bending, gross section, when compression flange is: (A) Supported laterally its full length by embedment in concrete.	$0.55 F_y$	24,000	30,000
(B) (1) Partially supported or unsupported	$18{,}000 - 0.52 \left(\frac{l}{r'}\right)^2$	$24{,}000 - 6.67 \left(\frac{l}{b}\right)^2$	$30{,}000 - 8.33 \left(\frac{l}{b}\right)^2$
with $\frac{l}{b}$ not greater than:	—	25	20
$\frac{l}{r'}$ not greater than:	132	—	—

(Axial tension row value: 18,000)

1.11.5 DESIGN

		M-94	M-95	M-96
Compression in concentrically loaded columns (2)				
Riveted ends	$\dfrac{.55\,F_y}{1.25}\left[1-\dfrac{\left(.75\dfrac{L'}{r}\right)^2}{4\pi^2 E}\right]F_y$	$15{,}000-0.25\left(\dfrac{L'}{r}\right)^2$	$20{,}000-0.46\left(\dfrac{L'}{r}\right)^2$	$24{,}000-0.66\left(\dfrac{L'}{r}\right)^2$
Pinned ends	$\dfrac{.55\,F_y}{1.25}\left[1-\dfrac{\left(.875\dfrac{L'}{r}\right)^2}{4\pi^2 E}\right]F_y$	$15{,}000-0.32\left(\dfrac{L'}{r}\right)^2$	$20{,}000-0.61\left(\dfrac{L'}{r}\right)^2$	$24{,}000-0.86\left(\dfrac{L'}{r}\right)^2$
with $\dfrac{L}{r}$ not greater than		140	130	120
Shear in girder webs, gross section............		11,000	14,000	17,500
Bearing on milled stiffeners and other steel parts in contact	$.80\,F_y$	26,000	36,000	44,000
Stress in extreme fiber of pins				
Bearing on pins not subject to rotation (3) (5)		26,000	32,000	40,000
Bearing on pins subject to rotation (3) (such as rockers and hinges)............		13,000	16,000	18,000
Shear in pins............	$.40\,F_y$	13,000	18,000	22,000
Shear in turned & ribbed bolts				
Bearing on power driven rivets and H.S. bolts			See Article 1.7.5	
Bearing on expansion rollers and rockers............			See Article 1.7.4	

FOOTNOTES FOR TABLE ON PAGE 236

Footnotes:

(1) Continuous or cantilever beams or girders may be proportioned for negative moment at interior supports for an allowable unit stress 20 percent higher than permitted by this formula but in no case exceeding allowable unit stress for compression flange supported its full length. If cover plates are used, the allowable static stress at the point of theoretical cutoff shall be as determined by the formula.

l = length, in inches, of unsupported flange between lateral connections, knee braces or other points of support. For continuous beams and girders, l may be taken as the distance from interior support to point of dead load contraflexure if this distance is less than that designated above.

For cantilever beams and girders, l may be taken as twice the distance from the support to the end of the cantilever if this distance is less than designated above.

r' = radius of gyration, in inches of a tee section comprised of the compression flange plus one-sixth of the web area, about the axis in the plane of the web:

for welded girders—$(r')^2 \cong b^2/12 \left(\dfrac{1}{1 + \dfrac{A_w}{6A_f}} \right)$

for riveted girders—$(r')^2 \cong 0.1b^2$

where b = flange width, in inches

A_w = area of the web, in.^{2}

A_f = area of the flange, in.^{2}

(2) Compression in concentrically loaded columns having $\dfrac{L'}{r}$ values not greater than shown may be computed from these approximate formulae, or from the more exact formulae given in Appendix C.

L' = length of member, in inches

r = least radius of gyration of member, in inches

For compression members with values of $\dfrac{L'}{r}$ greater than those shown or of known eccentricity, see Appendix C. The factor of safety to be used when using Appendix C is 1.76, 1.80 and 1.83 for M94, M95 and M96 steels respectively.

(3) The effective bearing area of a pin shall be its diameter multiplied by the thickness of the metal on which it bears.

(4) Number in parenthesis represents the last year these specifications were printed.

(5) This shall apply to pins used primarily in axially loaded members such as truss members and cable adjusting links. It shall not apply to pins used in members having rotation caused by expansion or deflection.

Section 12—ELASTOMERIC BEARINGS

1.12.1 — GENERAL

Elastomeric bearings shall be subject to the requirements of this section and to the sections applicable to the particular types of construction with which they are used.

The elastomers to be used shall conform to requirements given in Section 2.25 of this specification.

1.12.2 — DESIGN

Bearings may be plain (consisting of elastomer only) or laminated (consisting of layers of elastomer restrained at their interfaces by bonded laminates). Elastomer compounds of nominal 70 durometer hardness shall not be used in laminated bearings. Plain bearings generally will be

1.12.2 DESIGN

restricted by the requirements of this specification to conditions where little movement is anticipated.

The following terms will be used:

L = Length of a rectangular bearing parallel to the direction of translation

W = Width of a rectangular bearing perpendicular to the direction of translation

R = Radius of a circular bearing

t = Average thickness of a plain bearing or the thickness of any individual layer of elastomer in a laminated bearing (including the top and bottom layer)

T = Total effective elastomer thickness (summation of t's)

S = Shape factor (the area of the loaded face divided by the side area free to bulge)

$$S = \frac{LW}{2t(L+W)} \text{ for rectangular bearings}$$

$$S = \frac{R}{2t} \text{ for circular bearings}$$

The size of the elastomeric pad shall be such that both surfaces are in complete contact with the bearing areas.

The compressive strain of a plain bearing or of any individual layer of a laminated bearing is a function of the average unit compressive stress, the hardness of the elastomer and the shape factor. The compressive deflection of each layer is the product of the strain and the thickness of the layer. The total deflection of the bearing is the sum of the layer deflections. The shear strain of a bearing is a function of the temperature, hardness of the elastomer and average shearing unit stress. The shear deflection of a bearing is the product of the shear strain and the total effective elastomer thickness. These relationships may be taken from existing test reports, but for large bearings or groups of standard designs, they shall preferably be verified by tests of the particular designs involved.

Bearings shall have built in taper when nonparallel load surfaces would otherwise produce a compressive deflection of .06 T under dead load. Such taper shall be limited to ⅝" per foot.

To insure stability the following limits shall be observed:

Plain Bearings—Minimum L or R = 5T
　　　　　　　　Minimum W　　= 5 T

Laminated Bearings—Minimum L or R = 3 T
　　　　　　　　　　Minimum W　　= 2 T

The total of the positive and negative movements caused by anticipated temperature change shall not exceed .5 T.

The average unit pressure on elastomeric bearings shall not exceed 800 psi under a combination of dead load plus live load, not including impact. The average unit pressure due to dead load only shall not exceed

500 psi. When dead load plus live load uplift reduce the average pressure to less than 200 psi the bearing shall be secured against horizontal crawling preferably by positive attachment to the top surface or to the top and bottom surfaces. When secured to the top and bottom surfaces the bearing may be subject to momentary light tension.

The initial compressive deflection in a plain bearing or in any layer of a laminated bearing, under dead load plus live load, not including impact shall not exceed .07 t. The deflection can be determined from a plot showing the relationship of Shape Factor, load and the durometer hardness of the elastomer under consideration. These curves are generally available from manufacturers for their product.

Section 13 — STEEL TUNNEL LINER PLATES

1.13.1 — GENERAL

These criteria cover the design of cold formed panel steel tunnel liner plates. The minimum thickness shall be as determined by design in accordance with Articles 1.13.2, 3, 4, 5, 6 and the construction shall conform to Section 26 of Division II. The supporting capacity of a non-rigid tunnel lining such as a steel liner plate results from its ability to deflect under load so that side restraint developed by the lateral resistance of the soil constrains further deflection. Deflection thus tends to equalize radial pressures and to load the tunnel liner as a compression ring.

The load to be carried by the tunnel liner is a function of the type of soil. In a granular soil, with little or no cohesion, the load is a function of the angle of internal friction of the soil and the diameter of the tunnel being constructed. In cohesive soils such as clays and silty clays the load to be carried by the tunnel liner is dependent on the shearing strength of the soil above the roof of the tunnel.

A subsurface exploration program and appropriate soil tests should be performed at each installation before undertaking a design.

Nothing included in this section shall be interpreted as prohibiting the use of new developments where usefulness can be substantiated.

1.13.2 — LOADS

External load on a circular tunnel liner made up of tunnel liner plates may be predicted by various methods including actual tests. In cases where more precise methods of analysis are not employed, the external load P can be predicted by the following:

1. If the grouting pressure is greater than the computed external load, the external load P on the tunnel liner shall be the grouting pressure.
2. In general the external load can be computed by the formula
$P = P_1 + P_d$

Where: P = the external load on the tunnel liner;
P_l = the vertical load at the level of the top of the tunnel liner due to live loads;
P_d = the vertical load at the level of the top of the tunnel liner due to dead load.

For an H-20 load on an unsurfaced fill, values of P_l are approximately the following:

H (ft)	4	5	6	7	8	9	10
P_l (lbs per sq. ft)	375	260	190	140	110	90	75

H is the height of soil fill above the top of the tunnel liner.

Values of P_d may be calculated using Marston's formula for load or any other suitable method.

In the absence of adequate borings and soil tests, the full overfill height should be the basis for P_d in the tunnel liner plate design.

The following is one form of Marston's formula:

Where: $P_d = C_d W D$
C_d = coefficient for tunnel liner, Figure 1.13.1
W = total (moist) unit weight of soil
D = horizontal diameter or span of the tunnel
H = height of soil fill over the top of the tunnel

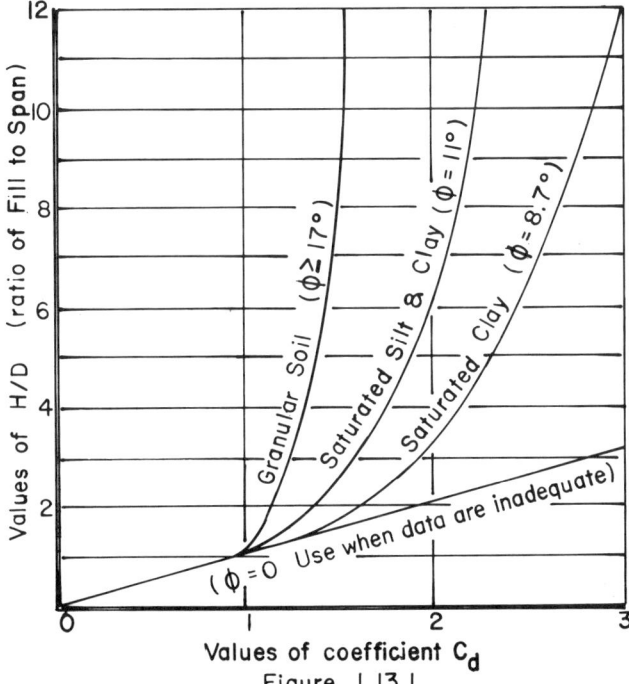

Figure 1.13.1

Diagram for Coefficient C_d for tunnels in soil (Φ = soil friction angle)

1.13.3 — DESIGN

The following criteria must be considered in the design of liner plates:

    (A) Joint strength
    (B) Handling and installation strength
    (C) Critical buckling of liner plate wall
    (D) Deflection or flattening of tunnel section.

1.13.4 — JOINT STRENGTH

Seam strength for liner plates must be sufficient to withstand the thrust developed from the total load supported by the liner plate. This thrust, T, in pounds per lineal foot is $T = PD/2$

Where P = load as defined in 1.13.2
         D = diameter or span in feet

Ultimate design longitudinal seam strengths are:

Plate thickness, inches	Ultimate strength, kips/ft	
	2 Flange	4 Flange
0.075	20.0	
0.105	30.0	26.0
0.135	47.0	43.0
0.164	55.0	50.0
0.179	62.0	54.0
0.209	87.0	67.0
0.239	92.0	81.0
0.313		115.0
0.375		119.0

Thrust T, multiplied by the safety factor, should not exceed the ultimate seam strength.

1.13.5 — HANDLING AND INSTALLATION STRENGTH

The liner plate ring shall have enough rigidity to resist the unbalanced loadings of normal construction: grouting pressures, local slough-ins and miscellaneous concentrated loads. This rigidity is measured by a Flexibility Factor determined by the formula:

$$FF = \frac{D^2}{EI}$$

Where FF = Flexibility factor
         D = Diameter, Inches
         E = Modulus of elasticity, psi
         I = Moment of inertia, inches to the 4th power per inch

For ordinary installations FF should not exceed

    2.0×10^{-2} for 2 flange
    0.9×10^{-2} for 4 flange

1.13.6 — CRITICAL BUCKLING OF LINER PLATE WALL

Wall buckling stresses are determined from the following formulae:

For diameters less than D_c, the ring compression stress at which buckling becomes critical is:

$$f_c = f_u - \left[\frac{f_u^2}{48E} \times \left(\frac{kD}{r} \right)^2 \right] \text{in psi}$$

For diameters greater than D_c:

$$f_c = \frac{12E}{\left(\dfrac{kD}{r} \right)^2} \text{in psi}$$

Where:

$$D_c = \frac{r}{k} \sqrt{\frac{24E}{f_u}} = \text{critical pipe diameter, inches}$$

f_u = minimum specified tensile strength, psi
f_c = buckling stress, psi., not to exceed minimum specified yield strength
k = soil stiffness factor, 0.22, unless more precise values are obtained
D = pipe diameter, inches
r = radius of gyration of section, inches per foot
E = modulus of elasticity, psi

Design for buckling is accomplished by limiting the ring compression thrust T to the buckling stress multiplied by the cross section area of the liner plate divided by the factor of safety:

$$T = \frac{f_c A}{FS}$$

Where T = thrust per lineal foot from Article 1.13.4
A = cross section area of liner plate, sq. in. per foot.
FS = factor of safety for buckling

1.13.7 — DEFLECTION OR FLATTENING

Deflection of a tunnel depends significantly on the amount of over excavation of the bore and is affected by delay in backpacking or inadequate backpacking. The magnitude of deflection is not primarily a function of soil modulus or the liner plate properties, so it cannot be computed with usual deflection formulae.

Where the tunnel clearances are important, the designer should oversize the structure to provide for a normal deflection. Good construction methods should result in deflections of not more than 3% of the normal diameter.

1.13.8 — CHEMICAL AND MECHANICAL REQUIREMENTS

(A) Chemical Composition

Base metal shall conform to ASTM A 569

(B) Minimum mechanical properties of flat plate before cold forming:

Tensile strength = 42,000 psi
Yield Strength = 28,000 psi
Elongation, 2 inches = 30%

1.13.9 — SECTIONAL PROPERTIES

The moment of inertia in inches to the fourth power per inch of plate width, based on the average of one ring of plates, shall conform to the following requirements:

Required Moment of Inertia	4-Flanged		2-Flanged Plate	
	Thickness	Cross Section Area	Thickness	Cross Section Area
I in.^{4}	T in.	A in.^{2}	T in.	A in.^{2}
0.031-0.035	0.105	0.134	0.075	0.096
0.036-0.040	0.105	0.134	0.105	0.135
0.041-0.045	0.105	0.134	0.105	0.135
0.046-0.050	0.135	0.172	0.135	0.174
0.051-0.055	0.135	0.172	0.135	0.174
0.056-0.060	0.164	0.209	0.135	0.174
0.061-0.065	0.164	0.209	0.164	0.213
0.066-0.070	0.164	0.209	0.164	0.213
0.071-0.075	0.179	0.227	0.164	0.213
0.076-0.080	0.209	0.264	0.179	0.233
0.081-0.085	0.209	0.264	0.179	0.233
0.086-0.090	0.250	0.313	0.209	0.272
0.091-0.095	0.250	0.313	0.209	0.272
0.096-0.100	0.250	0.313	0.209	0.272
0.101-0.105	0.313	0.388	0.239	0.312
0.106-0.110	0.313	0.388	0.239	0.312
0.111-0.115	0.313	0.388	0.239	0.312
0.116-0.120	0.313	0.388		
0.121-0.125	0.375	0.461		
0.126-0.130	0.375	0.461		
0.131-0.135	0.375	0.461		
0.136-0.140	0.375	0.461		

1.13.10 — COATINGS

Steel tunnel liner plates shall be of heavier gage or thickness or protected by coatings or other means when required for resistance to abrasion or corrosion.

1.13.11 — BOLTS

Bolts and nuts used with lapped seams shall be not less than ⅝ inch in diameter. The bolts shall conform to the specifications of ASTM A 449 for plate thicknesses equal to or greater than 0.209 inches and A 307 for plate thickness less than 0.209 inches. The nut shall conform to ASTM Designation A 563, Grade C.

Bolts and nuts used with four flanged plates shall be not less than ½ inch in diameter for plate thicknesses to and including 0.179 inches and not less than ⅝ inch in diameter for plates of greater thickness. The bolts and nuts shall be quick acting coarse thread and shall conform to ASTM Specifications A 307, Grade A.

1.13.12 — SAFETY FACTORS

Longitudinal test seam strength—3
Pipe Wall Buckling —2

Division II

CONSTRUCTION

Section 1—EXCAVATION AND FILL

2.1.1 — GENERAL

Foundation excavation shall include the removal of all material, of whatever nature, necessary for the construction of foundations and substructures in accordance with the plans or as directed by the Engineer. It shall include the furnishing of all necessary equipment and the construction of all cribs, cofferdams, caissons, unwatering, etc., which may be necessary for the execution of the work. It shall also include the subsequent removal of cofferdams and cribs and the placement of all necessary backfill as hereinafter specified. It shall also include the disposing of excavated material, which is not required for backfill, in a manner and in locations so as not to affect the carrying capacity of the channel and not to be unsightly.

Compensation for all clearing and grubbing contained within the bridge site shall, unless otherwise specified in the contract, be included in the unit price or prices bid for another item or other items. The bridge site is defined as the entire area between the right of way lines and between lines paralleling the bridge ends and passing through the longitudinal extremities of the substructure or superstructure, whichever is greater, unless a greater length is necessary for the required construction of the bridge, or unless otherwise specified in the contract.

All substructures, where practicable, shall be constructed in open excavation and, where necessary, the excavation shall be shored, braced or protected by cofferdams in accordance with approved methods. When footings can be placed in the dry without the use of cribs or cofferdams, backforms may be omitted with the approval of the Engineer and the entire excavation filled with concrete to the required elevation of the top of the footing. The additional concrete required shall be placed at the expense of the Contractor.

2.1.2 — PRESERVATION OF CHANNEL

Unless otherwise specified, no excavation shall be made outside of caissons, cribs, cofferdams, steel piling or sheeting, and the natural stream bed adjacent to the structure shall not be disturbed without permission from the Engineer. If any excavation or dredging is made at the site of the structure before caissons, cribs or cofferdams are sunk or are in place, the Contractor shall, without extra charge, after the foundation base is in place, backfill all such excavation to the original ground surface or river bed with material satisfactory to the Engineer. Material deposited within the stream area from foundation or other excavation or from the filling of cofferdams shall be removed and the stream area freed from obstruction thereby.

2.1.3 — DEPTH OF FOOTINGS

The elevation of the bottoms of footings, as shown on the plans, shall be considered as approximate only and the engineer may order, in writing, such changes in dimensions or elevation of footings as may be necessary to secure a satisfactory foundation.

2.1.4 — PREPARATION OF FOUNDATIONS FOR FOOTINGS

All rock or other hard foundation material shall be freed from all loose material, cleaned and cut to a firm surface, either level, stepped, or roughened, as may be directed by the engineer. All seams shall be cleaned out and filled with concrete, mortar or grout.

When masonry is to rest on an excavated surface other than rock, special care shall be taken not to disturb the bottom of the excavation and the final removal of the foundation material to grade shall not be made until just before the masonry is to be placed.

2.1.5 — COFFERDAMS AND CRIBS

(A) General

Cofferdams and cribs for foundation construction shall be carried to adequate depths and heights, be safely designed and constructed, and be made as water-tight as is necessary for the proper performance of the work which must be done inside them. In general, the interior dimensions of cofferdams and cribs shall be such as to give sufficient clearance for the construction of forms and the inspection of their exteriors, and to permit pumping outside of the forms. Cofferdams or cribs which are tilted or moved laterally during the process of sinking shall be righted, reset or enlarged so as to provide the necessary clearance and this shall be at the sole expense of the contractor.

When conditions are encountered which, in the opinion of the engineer, render it impracticable to unwater the foundation before placing masonry, he may require the construction of a concrete foundation seal of such dimensions as may be necessary. The foundation shall then be pumped out and the balance of the masonry placed in the dry. When weighted cribs are employed and the weight is utilized to partially overcome the hydrostatic pressure acting against the bottom of the foundation seal, special anchorage such as dowels or keys shall be provided to transfer the entire weight of the crib into the foundation seal. During the placing of a foundation seal, the elevation of the water inside the cofferdam shall be controlled to prevent any flow through the seal and if the cofferdam is to remain in place, it shall be vented or ported at low water level.

(B) Protection of Concrete

Cofferdams or cribs shall be constructed so as to protect green concrete against damage from a sudden rising of the stream and to prevent damage to the foundation by erosion. No timber or bracing shall be left in cofferdams or cribs in such a way as to extend into the substructure masonry, without written permission from the engineer.

(C) Drawings Required

For substructure work, the Contractor shall submit, upon request, drawings showing his proposed method of cofferdam construction and other details left open to his choice or not fully shown on the engineer's drawings. Such drawings shall be approved by the Engineer before construction is started on work governed by them.

(D) Removal

Unless otherwise provided, cofferdams or cribs with all sheeting and bracing shall be removed after the completion of the substructure, care being taken not to disturb or otherwise injure the finished masonry.

2.1.6 — PUMPING

Pumping from the interior of any foundation enclosure shall be done in such manner as to preclude the possibility of the movement of water through any fresh concrete. No pumping will be permitted during the placing of concrete or for a period of at least 24 hours thereafter, unless it be done from a suitable sump separated from the concrete work by a water-tight wall or other effective means.

Pumping to unwater a sealed cofferdam shall not commence until the seal has set sufficiently to withstand the hydrostatic pressure.

2.1.7 — INSPECTION

After each excavation is completed the Contractor shall notify the Engineer, and no masonry shall be placed until the Engineer has approved the depth of the excavation and the character of the foundation material.

2.1.8 — BACK-FILL

All material used for back-fill shall be of a quality acceptable to the engineer and shall be free from large or frozen lumps, wood, or other extraneous material.

All spaces excavated and not occupied by abutments, piers, or other permanent work shall be refilled with earth up to the surface of the surrounding ground, with a sufficient allowance for settlement. All back-fill shall be thoroughly compacted and, in general, its top surface shall be neatly graded.

The fill behind abutments and wing walls of all bridge structures shall be deposited in well-compacted, horizontal layers not to exceed 12 inches in thickness. The back-fill in front of such units shall be placed first to prevent the possibility of forward movement. Special precautions shall be taken to prevent any wedging action against the masonry, and the slope bounding the excavation for abutments and wing walls shall be destroyed by stepping or roughening to prevent wedge action. Jetting of the fill behind abutments and wing walls will not be permitted.

Fill placed around culverts and piers shall be deposited on both sides to approximately the same elevation at the same time.

Adequate provision shall be made for the thorough drainage of all back-filling. French drains shall be placed at weep holes.

No back-fill shall be placed against any masonry abutment, wing wall or culvert until permission shall have been given by the engineer and preferably not until the masonry has been in place 14 days, or until test cylinders show the strength to be twice the working stress used in the design.

Back-filling of sectional plate pipes and arches shall be done in accordance with Articles 2.23.5 and 2.23.6.

2.1.9 — FILLED SPANDREL ARCHES

For filled spandrel arches, the filling shall be carefully placed in such manner as to load the ring uniformly and symmetrically. The filling material shall be acceptable to the Engineer and shall be placed in horizontal layers, not to exceed 12 inches in thickness, carefully tamped and brought up simultaneously from both haunches. Wedge shaped sections of filling material against spandrels, wings or abutments will not be permitted.

2.1.10 — APPROACH EMBANKMENT

When the contract for any bridge structure requires the placement of approach embankments, they shall be constructed and paid for in accordance with the highway specifications governing this class of construction.

2.1.11 — CLASSIFICATION OF EXCAVATION

Classification, if any, of excavation will be indicated on the plans and set forth in the proposal.

2.1.12 — MEASUREMENT AND PAYMENT

Payment for foundation excavation shall include the cost of all labor, material, equipment, and other items that may be necessary or convenient to the successful completion of the excavation to the elevation of the bottom of the footings. It shall also include the cost of removing cofferdams and any surplus material which may have been thrown up during the process of excavation, and shall include the cost of back-filling in a compacted condition an amount of material equal to the amount of excavation. Any back-fill required in excess of the amount excavated shall be paid for as extra work unless a price for extra back-fill is included in the contract.

The yardage to be paid for shall be the actual number of cubic yards in original position, of material acceptably excavated in conformity with the plans or as directed by the Engineer, but no yardage shall be included in the measurement for payment which is outside of a volume bounded by vertical planes 18 inches outside of and parallel to the neat lines of the footing. The cross-sectional area measured shall not include water or other liquids, but shall include mud, muck and other similar semi-solids.

The top and bottom limits of computed volume shall be the original ground or the top of the required grading cross section, whichever is lower, and the bottom of the completed footing.

When it is necessary, in the opinion of the Engineer, to carry the foundations below the elevations shown on the plans, the excavation for the first three feet of additional depth will be included in the quantity for which payment will be made under the item Foundation Excavation. Excavation below this additional depth will be paid for as extra work, unless the Contractor is willing to accept payment at contract prices.

Section 2 — SHEET PILES

2.2.1 — GENERAL

This specification covers only sheet piling shown on the plans, or ordered by the Engineer, to be left in place so that it becomes a part of the finished structure.

2.2.2 — TIMBER SHEET PILES

The timber, unless otherwise noted, shall be preservatively treated in accordance with Division II, Section 21, and may consist of any treatable species which will satisfactorily stand driving. It shall be sawn or hewn with square corners and shall be free from worm holes, loose knots, wind shakes, decayed or unsound portions, or other defects which might impair its strength or tightness.

The piles shall be of the dimensions shown on the plans either cut from the solid material or made by building up the piles of three planks securely fastened together. The piles shall be drift sharpened at their lower ends so as to wedge the adjacent piles tightly together.

The tops of the piles shall be cut off to a straight line at the elevation indicated and shall be braced with waling strips, properly lapped and joined at all splices and corners. The wales shall preferably be in one length between corners and shall be bolted near the tops of the piles.

2.2.3 — CONCRETE SHEET PILES

Where concrete sheet piles are required, they shall be in strict accordance with the detailed design. The requirements governing the manufacture and installation of concrete sheet piling shall conform, in general, to those governing precast concrete bearing piles.

2.2.4 — STEEL SHEET PILES

Steel sheet piles shall be of the type and weight indicated on the plans or designated in the special provisions and of the material required below. The piles, when in place in the completed structure, shall be practically water-tight at the joints. Painting of steel sheet piles shall conform to Article 2.3.17.

Steel sheet piles shall conform to the requirements of AASHO M202 (ASTM A 328).

2.2.5 — MEASUREMENT AND PAYMENT

Payment for sheet piles shall include the cost of furnishing, driving and cutting off. Payment will be made on the basis of the piles driven as approved by the Engineer, except that a deduction from the payment will be made in an amount equal to the salvage value of the material cut off after driving.

Timber, concrete, and steel sheet piles will be paid for at the contract price per square foot.

Section 3—BEARING PILES

2.3.1 — MATERIALS

Steel piles shall consist of structural steel shapes of the section provided on the plans or as otherwise specified. The steel shall conform to the Specification for Structural Steel, AASHO M-183 (ASTM A-36).

2.3.2 — DESIGN AND CONDITIONS OF USE

General and Design: Refer to Division I, Section 4.
Bearing Values, Design: Refer to Article 1.4.4.
Preservative Treatment: Refer to Division II, Section 21.

2.3.3 — PREPARATION FOR DRIVING

(A) Excavation

In general, piles shall not be driven until after the excavation is complete. Any material forced up between the piles shall be removed to correct elevation without cost to the State before masonry for the foundation is placed.

(B) Caps

The heads of all concrete piles, and the heads of timber piles, when the nature of the driving is such as to unduly injure them, shall be protected by caps of approved design, preferably having a rope or other suitable cushion next to the pile head and fitting into a casting which in turn supports a timber shock block. When the area of the head of any timber pile is greater than that of the face of the hammer, a suitable cap shall be provided to distribute the blow of the hammer throughout the cross section of the pile.

The head shall be cut square and shall be shaped or chamfered to prevent splitting at its periphery.

For special types of piling, driving heads, mandrels, or other devices in accordance with the manufacturers' recommendation shall be provided so that the pile may be driven without injury.

For steel piling the heads shall be cut squarely and a driving cap shall be provided to hold the axis of the pile in line with the axis of the hammer.

(C) Collars

Collars, bands, or other devices, to protect timber piles against splitting and brooming shall be provided where necessary.

(D) Pointing

Timber piles shall be pointed where soil conditions require it. When necessary, the piles shall be shod with metal shoes of a design satisfactory to the Engineer, the points of the piles being carefully shaped to secure an even and uniform bearing on the shoes.

(E) Splicing Piles

Full length piles shall be used where practicable. In exceptional circumstances splicing of piles may be permitted. The method of splicing shall be as shown on the plans or as approved by the Engineer. When the splicing of steel piles or steel shells of special piles is done by welding, the arc method shall be given preference.

(F) Painting Steel Piles

Steel piles shall be painted as specified in Article 2.3.17.

2.3.4 — METHODS OF DRIVING

(A) General

Piles may be driven with a gravity hammer, a steam hammer, or a combination of water jets and hammer but a steam hammer is preferred. Precast concrete piles, preferably, shall be driven by means of a combination of hammer and jet.

(B) Hammers for Timber and Steel Piles

Gravity hammers for driving timber piles shall weigh not less than 2,000 pounds, preferably 3,000 pounds, and for steel piles not less than 3,000 pounds, but in no case shall the weight of the hammer be less than the combined weight of driving head and pile. The fall shall be so regulated as to avoid injury to the piles and in no case shall exceed 15 feet. When a steam hammer is used, the total energy developed by the hammer shall be not less than 6,000 foot-pounds per blow.

(C) Hammers for Concrete Piles

Unless otherwise provided, concrete piles, precast, or shells for cast-in-place piles, shall be driven with a steam hammer which shall develop an energy per blow at each full stroke of the piston, of not less than one foot-pound for each pound of weight driven. In no case shall the total energy developed by the hammer be less than 6,000 foot-pounds per blow.

If a gravity hammer is used, it shall have a weight not less than 50 per cent of the weight of the pile, but in no case less than 3,000 pounds; and the drop of the hammer shall not exceed 8 feet.

(D) Additional Equipment

In case the required penetration is not obtained by the use of a hammer complying with the above minimum requirements, the Contractor shall provide a heavier hammer, or resort to jetting at his own expense.

(E) Leads

Pile driver leads shall be constructed in such a manner as to afford freedom of movement of the hammer, and they shall be held in position by guys or stiff braces to insure support to the pile during driving. Except where piles are driven through water, the leads, preferably, shall be of sufficient length so that the use of a follower will not be necessary.

Preferably, inclined leads shall be used in driving battered piles.

(F) Followers

The driving of piling with followers shall be avoided if practicable and shall be done only under written permission of the Engineer. When followers are used, one pile from each group of 10 shall be a long pile driven without a follower, and shall be used as a test pile to determine the average bearing power of the group.

(G) Water Jets

When water jets are used, the number of jets and the volume and pressure of water at the jet nozzles shall be sufficient to freely erode the material adjacent to the pile. The plant shall have sufficient capacity to deliver at all times at least 100 pounds per square inch pressure at two ¾-inch jet nozzles. Before the desired penetration is reached, the jets shall be withdrawn and the piles shall be driven with the hammer to secure the final penetration.

(H) Accuracy of Driving

Piles shall be driven with a variation of not more than ¼ inch per foot from the vertical or from the batter shown on the plans, except that piles for trestle bents shall be so driven that the cap may be placed in its proper location without inducing excessive stresses in the piles, and foundation piles shall not be out of the position shown on the plan more than 6 inches after driving.

2.3.5 — DEFECTIVE PILES

The procedure incident to the driving of piles shall not subject them to excessive and undue abuse producing crushing and spalling of the concrete, injurious splitting, splintering and brooming of the wood or deformation of the steel. Manipulation of piles to force them into **proper position, considered by the Engineer to be excessive, will not be** permitted. Any pile damaged by reason of internal defects, or by improper driving or driven out of its proper location or driven below the **elevation fixed by the plans or by the Engineer, shall be corrected at the** contractor's expense by one of the following methods approved by the Engineer for the pile in question:

(1) The pile shall be withdrawn and replaced by a new and if necessary, a longer pile.

(2) A second pile shall be driven adjacent to the defective or low pile.

(3) The pile shall be spliced or built up as otherwise provided herein or a sufficient portion of the footing extended to properly embed the pile. Timber piles shall not be spliced without specific **permission of the Engineer. All piles pushed up by the driving of** adjacent piles or by any other cause shall be driven down again.

2.3.6 — DETERMINATION OF BEARING VALUES (See also Article 1.4.4)

(A) Loading Tests

When required, the size and number of piles shall be determined by actual loading tests. In general, these tests shall consist of the application of a test load placed upon a suitable platform supported by the pile, with suitable apparatus for accurately measuring the test load and the settlement of the pile under each increment of load.

In lieu thereof hydraulic jacks with suitable yokes and pressure gauges may be used.

The safe allowable load shall be considered as 50 per cent of that load which, after a continuous application of 48 hours, produces a permanent settlement not greater than ¼ inch measured at the top of the pile. This maximum settlement shall not be increased by a continuous application of the test load for a period of 60 hours or longer.

At least one pile for each group of 100 piles preferably should be tested.

(B) Timber Pile Formulas

When not driven to practical refusal, the bearing values of piles preferably shall be determined by load tests as specified above. In the absence of loading tests or substantiated adequate pile formulas, the safe bearing values for timber piles shall be determined by the following formulas:

$$P = \frac{2WH}{S+1.0} \text{ for gravity hammers,}$$

$$P = \frac{2WH}{S+0.1} \text{ for single-acting steam hammers,}$$

$$P = \frac{2H(W+Ap)}{S+0.1} \text{ for double-acting steam hammers,}$$

where P = safe bearing capacity in pounds,
  W = weight, in pounds, of striking parts of hammer,
  H = height of fall in feet,
  A = **area of piston in square inches,**
  p = steam pressure in pounds per square inch at the hammer,
  S = the average penetration in inches per blow for the last 5 to 10 blows for gravity hammers and the last 10 to 20 blows for steam hammers.

The above formulas are applicable only when—
  The hammer has a free fall.
  The head of the pile is not broomed or crushed.
  The penetration is reasonably quick and uniform.
  There is no sensible bounce after the blow.
  A follower is not used.

Twice the height of the bounce shall be deducted from "H" to determine its value in the formula.

Unless otherwise ordered by the Engineer timber piling shall be driven to the bearing value given on the plans or in the supplemental specifications. If bearing values are not given, timber piling shall be driven to a minimum value of twenty tons.

In case water jets are used in connection with the driving, the **bearing capacity shall be determined by the above formulas from the** results of driving after the jets have been withdrawn, or a load test may be applied.

(C) Concrete and Steel Piles

When not driven to practical refusal the bearing value for concrete and steel piles preferably shall be determined by means of loading tests **specified above. In the absence of loading tests, their safe bearing** values may be approximated by substantiated adequate pile formulas or those specified for timber piles. However, the character of the soil penetrated, conditions of driving, distribution, size, length and weight of the piles, shells or cores driven, and the computed load per pile shall be given due consideration in determining their probable safe bearing value.

2.3.7 — TEST PILES

When required, the Contractor shall drive test piles of a **length and at the location designated by the Engineer. These piles shall be of** greater length than the length assumed in the design in order to provide for any variation in soil conditions.

2.3.8 — ORDER LISTS FOR PILING

The Contractor shall furnish piles in accordance with an **itemized list, which shall be furnished by the Engineer, showing the number and** length of all piles.

In determining lengths of piles for ordering and for footage to be included in the contract, the lengths given in the order list shall be based on the lengths which are assumed to remain in the completed structure. **The Contractor shall, at his own expense, increase the lengths** given to provide for fresh heading and for such additional length as may be necessary to suit the contractor's method of operation.

2.3.9 — STORAGE AND HANDLING OF TIMBER PILES

The method of storing and handling shall be such as to avoid injury to the piles. Special care shall be taken to avoid breaking the surface of treated piles and cant-hooks, dogs or pike-poles shall not be used. Cuts or breaks in the surface of treated piling shall be given three brush coats of hot creosote oil of approved quality and hot creosote oil shall be poured into all bolt holes.

2.3.10 — CUTTING OFF TIMBER PILES

The tops of all piling shall be sawed to a true plane, as shown on the plans, and at the elevation fixed by the Engineer. Piles which support timber caps or grillage shall be sawed to conform to the plane of the bottom of the superimposed structure. In general, the length of pile above the elevation of cut-off shall be sufficient to permit the complete removal of all material injured by driving, but piles driven to very nearly the cut-off elevation shall be carefully adzed or otherwise freed from all "broomed," splintered or otherwise injured material.

2.3.11 — CUTTING OFF STEEL OR STEEL SHELL PILES

Piles shall be cut off at the required elevation. If capping is required the connection shall be made according to details shown on the plans.

2.3.12 — CAPPING TIMBER PILES

After cut-off, the heads of timber piles shall be protected as specified in Article 2.20.7.

2.3.13 — MANUFACTURE OF PRECAST CONCRETE PILES

(A) General

Piles shall be constructed in accordance with details shown on the plans.

(B) Class of Concrete

Class A or A(AE) concrete shall be used for precast concrete piles.

(C) Form Work

Forms for precast concrete piles shall conform to the general requirements for concrete form work as provided in Art. 2.4.19. Forms shall be accessible for tamping and consolidation of the concrete. Under good weather curing conditions side forms may be removed at any time not less than 24 hours after placing the concrete, but the entire pile shall remain supported for at least seven days and shall not be subjected to any handling stress until the concrete has set for at least 21 days and for a longer period in cold weather, the additional time to be determined by the Engineer.

(D) Reinforcement

Reinforcement shall be placed in accordance with details shown on the plans.

(E) Casting

The piles may be cast in either a horizontal or a vertical position. Special care shall be taken to place the concrete so as to produce satisfactory bond with the reinforcement and avoid the formation of "stone pockets," honeycomb or other such defects.

To secure uniformity and remove surplus water, the concrete in each pile shall be placed continuously and shall be compacted by vibrating or by other means acceptable to the Engineer. The forms shall be overfilled, the surplus concrete screeded off, and the top surfaces finished to a uniform, even texture similar to that produced by the forms.

(F) Finish

Trestle piling exposed to view shall be finished above the ground line in accordance with the provisions governing the finishing of concrete columns. Foundation piling, that portion of the trestle piling which will be below the ground surface, and piles for use in sea water or alkali soils shall not be finished except by pointing as above set forth.

(G) Curing

Concrete piles shall be cured as provided elsewhere in these specifications for concrete. As soon as the piles have set sufficiently they shall be removed from the forms and piled in a curing pile separated from each other by wood spacing blocks. No pile shall be driven until it has set for at least 21 days and, in cold weather, for a longer period as determined by the Engineer. Concrete piles for use in sea water or alkali soils shall be cured for not less than 30 days before being used.

2.3.14 — STORAGE AND HANDLING OF PRECAST CONCRETE PILES

Removal of forms, curing, storing, transporting and handling precast concrete piles shall be done in such a manner as to avoid excessive bending stresses, cracking, spalling or other injurious results. The method of handling shall be such as will not induce stresses exceeding those specified in Section 1.5 when computed according to Article 1.4.5G.

Piles to be used in sea water or in alkali soils shall be handled so as to avoid surface abrasions or other injuries exposing the interior concrete.

2.3.15 — MANUFACTURE OF CAST-IN-PLACE CONCRETE PILES

(A) General

Piles shall be constructed in accordance with details shown on the plans.

(B) Inspection of Metal Shells

At all times prior to the placing of concrete in the driven shells, the Contractor shall have available a suitable light for the inspection of each shell throughout its entire length. Any improperly driven, broken or otherwise defective shell shall be corrected to the satisfaction of the Engineer, by removal and replacement, or the driving of an additional pile, at no extra cost to the State.

(C) Class of Concrete

Class A or A(AE) concrete shall be used for cast-in-place piles.

(D) Reinforcement

Reinforcement shall be placed in accordance with the plans or special provisions.

(E) Placing Concrete

No concrete shall be placed until all driving within a radius of 15 feet has been completed, nor until all the shells for any one bent have been completely driven. If this cannot be done, all driving within the above limits shall be discontinued until the concrete in the last pile cast has set at least seven days.

Concrete shall be placed as specified for piles precast in the vertical position. Accumulations of water in the shells shall be removed before the concrete is placed.

2.3.16 — EXTENSIONS OR "BUILD-UPS"

Extensions, splices or "build-ups" on concrete piles, when necessary, shall be made as follows:

After the driving is completed, the concrete at the end of the pile shall be cut away, leaving the reinforcement steel exposed for a length of 40 diameters. The final cut of the concrete shall be perpendicular to the axis of the pile. Reinforcement similar to that used in the pile shall be securely fastened to the projecting steel and the necessary form work shall be placed, care being taken to prevent leakage along the pile. The concrete shall be of the same quality as that used in the pile. Just prior to placing concrete, the top of the pile shall be thoroughly wetted and covered with a thin coating of neat cement, retempered mortar or other suitable bonding material. The forms shall remain in place not less than seven days and shall then be carefully removed and the entire exposed surface of the pile finished as previously specified.

2.3.17 — PAINTING STEEL PILES AND STEEL PILE SHELLS

Unless otherwise provided, when steel piles or steel pile shells extend above the ground surface or water surface they shall be protected by three coats of paint as specified for Painting Metal Structures in Section 2.14. This protection shall extend from an elevation 2 feet below the water or ground surface to the top of the exposed steel.

2.3.18 — MEASUREMENT AND PAYMENT

(A) General

Piling, whether timber, concrete or steel, will be paid for according to Method A or B as designated in the contract.

(B) Method A

For furnishing and driving piles at the contract price per linear foot.

(1) Cutoff

The total cutoff of piling shall be paid for at the prices set forth by the State in the special provisions for those of the following items incorporated in the work:

Cutoff, untreated timber piles, per linear foot	$_____
Cutoff, treated timber piles, per linear foot	$_____
Cutoff, precast concrete piles, per linear foot	$_____
Cutoff, steel shells for piles, per linear foot	$_____
Cutoff, steel piles, per pound	$_____

(2) Furnishing and Driving

The number of linear feet to be paid for shall be the actual length of piles remaining in the completed structure and the number of linear feet of cutoff to be paid for shall be the actual number of linear feet cut off, except that no allowance will be made for lengths in excess of those ordered by the Engineer, and except that if the Contractor casts concrete piles full length of the reinforcement bars to facilitate driving, no payment will be made for that portion where concrete must be removed in order that bars may project as shown on the plans. If paid for as "cutoff," cutoff material (if the cutoff is in excess of 3 feet in length) shall become the property of the State. Cutoff material 3 feet or less in length, and other cutoff material which, in the opinion of the Engineer, is not worth salvaging shall be disposed of to the satisfaction of the **Engineer.**

(3) Payment for Furnishing and Driving Piles

Payment for furnishing and driving piles shall include the material and work specified under "Payment for Furnishing Piles" and "Payment for Driving Piles"—Method B.

(C) Method B

For furnishing piles at the contract price per linear foot. For driving piles at the contract price per linear foot.

(1) Furnishing

The number of linear feet of timber, precast concrete or steel piles to be paid for shall be the total ordered length of piles which are driven and which have been furnished in accordance with the lengths designated by the Engineer, except that if the Contractor casts concrete piles full length of the reinforcement bars to facilitate driving, no payment will be made for that portion where concrete must be removed in order that bars may project as shown on the plans. Cutoff material 3 feet or less in length and other cutoff material which, in the opinion of the Engineer, is not worth salvaging, shall be disposed of to the satisfaction of **the Engineer.**

The number of linear feet of cast-in-place piles to be paid for shall be the actual number of linear feet of piles remaining in the completed structure. The length measured shall include both the steel shell and the reinforced concrete extension as measured from the point of the tip of the pile to the bottom of the cap or bottom of the footing, as the case may be.

(2) Driving

The number of linear feet to be paid for shall be the total number of linear feet of piling remaining in the completed structure. For driving cast-in-place piles the length measured shall include both the steel shell and the reinforced concrete extensions as measured from the point of the tip of the pile to the bottom of the cap or bottom of the footing, as the case may be.

(3) Payment for Furnishing Piles

Payment for furnishing piles shall include full compensation for furnishing the piling and all material required therefor ready for placement, including all material necessary for extensions and build-ups and for completion of the pile, and for all labor, tools, hauling, equipment, handling, treatment and all work incidental to the construction of the piling prior to driving or construction of build-ups and extensions. Payment shall also include (a) reinforcement in concrete piles required to extend beyond the end of the pile for connections; (b) the fitting and attaching of steel shoes when they are specified for timber piles; (c) the furnishing and attachment of brackets, lugs, core stoppers and cap plates on steel piling.

(4) Payment for Driving Piles

Payment for driving piles per linear foot shall include full compensation for furnishing all labor, tools, materials, supplies, equipment and other necessary or incidental costs of handling, driving, cutting off piles, treatment of pile heads, constructing build-ups and extensions of concrete piles, painting of steel piles and all other incidental work connected therewith. It shall also include full compensation for all jetting, drilling, blasting, or other work necessary to obtain the required penetration or bearing values of piles.

(D) Falsework and Defective Piles

No payment will be made for the furnishing or driving of falsework piles, nor will payment be made for piles driven out of place, for defective piles, or for piles which are damaged in handling or driving.

(E) Additional Requirements

If the length of wood piles, steel piles or steel pile shells designated by the Engineer is not sufficient, the splicing, including labor, equipment and material, shall be paid for on the basis of extra work unless a contract item is provided to cover the payment.

Brackets, plates or other reinforcement on steel piles required by the Engineer in addition to those shown on the plans will be paid for as extra work.

If not covered by a contract item, metal shoes for piling, ordered by the Engineer, will be paid for at cost delivered to the site, plus 15 per cent.

No additional allowance, or adjustment, will be made in the contract prices for furnishing or driving piling because of these additional requirements.

2.3.19 — PAYMENT FOR TEST PILES

Test piles ordered by the Engineer shall be paid for as follows:
If piles are used in the structure as a result of the test, the test piles shall be paid for as in the case of other piles.

If, however, piling is not used in the structure, the test piles will be paid for as provided for extra work, due consideration being given to the cost of bringing the pile driver to the site and removing it from the work.

2.3.20 — PAYMENT FOR LOADING TESTS

Payment for loading tests shall include the cost of all material, equipment and labor incidental to making the loading test or tests as directed by the Engineer, or as specified in the special provisions. Payment shall be made on the basis of the contract price for pile loading tests, or, in the absence of such a price, shall be made on the basis of extra work.

Section 4—CONCRETE MASONRY

2.4.1 — GENERAL

Concrete masonry shall consist of portland cement, aggregates and water which shall conform to the requirements of this section and which shall be proportioned as hereinafter specified.

2.4.2 — MATERIALS

(A) Cement

(1) Portland Cement

Eight types of portland cement are recognized by these specifications, designated as follows:

Type I.	For use in general concrete construction where the special properties specified under types II, III, IV and V are not required.
Type IA.	Air-entraining portland cement for the same uses as specified under type I.
Type II.	For use in general concrete construction exposed to moderate sulfate action or where moderate heat of hydration is required.
Type IIA.	Air-entraining portland cement for the same uses as specified under type II.
Type III.	For use when high early strength is required.
Type IIIA.	Air-entraining portland cement for the same

	use as specified under type III.
Type IV.	For use when a low heat of hydration is required.
Type V.	For use when high sulfate resistance is required.

> NOTE: Attention is called to the fact that cements conforming to the requirements for types IV and V are not usually carried in stock. In advance of specifying their use purchasers or their representatives should determine whether these types of cement are, or can be, made available.

Portland cement of the type or types specified shall conform to the requirements of AASHO Specifications for Portland Cement, M 85. Unless otherwise provided, or called for in the special provisions, type I shall be furnished.

Air-entraining portland cement of the type or types specified shall conform to the above requirements. Unless otherwise provided or called for in the special provisions, type IA shall be furnished.

(2) Portland Blast-Furnace Slag Cement

Two types of portland blast-furnace slag cement are recognized by these specifications, as follows:

Type IS. Portland blast-furnace slag cement.

Type IS-A. Air-entraining portland blast-furnace slag cement.

Portland blast-furnace slag cement of the type or types specified shall conform to the requirements of AASHO Specifications for Portland Blast-Furnace Slag Cement, M 151.

(3) Masonry Cement

Two types of masonry cement are recognized by these specifications, as follows:

Type I. For use in general purpose masonry.

Type II. For use where high strength is required.

Masonry cement of the type or types specified shall conform to the requirements of AASHO Specifications for Masonry Cement, M 150 (ASTM C91).

(4) Natural Cement

Natural cement shall conform to the requirements of AASHO Specifications for Natural Cement, M 135.

(5) Sampling and Testing

Hydraulic cements shall be sampled and tested in accordance with the standard methods referred to in the applicable specifications of the American Association of State Highway Officials.

Cement may be sampled either at the mill or at the site of the work as provided in the above specification. The seals of cars containing cement which have been sampled shall not be broken except by the Engineer; otherwise additional samples shall be taken from these cars.

The Contractor shall notify the Engineer of dates of delivery so that there will be sufficient time for sampling the cement, either at the mill or upon delivery. If this is not done or if additional tests are necessary, the Contractor may be required to rehandle the cement in the storehouse for the purpose of obtaining the required samples.

(B) Water and Admixtures

(1) Quality of Water

Water for use with cement in mortar or concrete shall be subject to the approval of the Engineer. It shall not be salty or brackish and shall be reasonably clear and free from oil, acid, injurious alkali or vegetable matter.

(2) Tests for Water

When required by the Engineer the quality of the mixing water shall be determined by the Standard Method of Test for Quality of Water to be Used in Concrete, AASHO Methods of Sampling and Testing, Designation: T 26.

In sampling water for testing, care shall be taken that the containers are clean and that samples are representative.

When comparative tests are made with a water of known satisfactory quality, any indication of unsoundness, marked change in time of setting, or a reduction of more than 10 per cent in mortar strength, shall be sufficient cause for rejection of the water under test.

(3) Admixtures

Admixtures in concrete shall be used only when approved by the Engineer. Air-entraining admixtures if specified or permitted shall conform to the requirements of AASHO Standard Specifications for Air-Entraining Admixtures for Concrete, M 154 (ASTM C260).

(4) Tests for Admixtures

Air-entraining admixtures shall be tested in accordance with AASHO Standard Method of Testing Air-Entraining Admixtures for Concrete, T 157 (ASTM C 233).

(C) Fine Aggregate

(1) Fine Aggregate

All fine aggregate for concrete shall conform to the Specification for Fine Aggregate for Portland Cement Concrete, AASHO M 6. NOTE: Requirements for soundness should be stipulated in the special provisions [Refer to AASHO M 6, paragraph 4.1, 4.2 and 4.3].

(2) Sand for Mortar

Sand for mortar shall conform to the Specifications for Mortar Sand, AASHO M 45 (ASTM C 144).

(D) Coarse Aggregates

(1) Coarse Aggregates

All coarse aggregates for concrete shall conform to the Specification for Coarse Aggregate for Portland Cement Concrete, AASHO M 80 as to quality and to the Specifiction for Standard Sizes of Coarse Aggregate for Highway Construction, AASHO M 43 (ASTM D 448) as to size. NOTE: Requirements for soundness should be stipulated in the special provisions [Refer to AASHO M 80, paragraph 6.1, 6.2, 6.3].

Slag shall be used for aggregate only if its use is provided for in the special provisions.

(2) Rubble or Cyclopean Aggregate

One-man and derrick stone used in rubble or cyclopean concrete shall consist of tough, sound and durable rock. The stone shall be free from coatings, drys, seams or flaws of any character. In general, the percentage of wear shall not exceed 50 when tested in accordance with AASHO Standard Method of Test for Abrasion of Coarse Aggregate by the use of the Los Angeles Machine, T 96.

Preferably, stone shall be angular in shape and shall have a rough surface such as will thoroughly bond with the surrounding mortar.

2.4.3 — CARE AND STORAGE OF CONCRETE AGGREGATES

The handling and storage of concrete aggregates shall be such as to prevent segregation or the admixture of foreign materials. The Engineer may require that aggregates be stored on separate platforms at satisfactory locations.

When specified in the special provisions, the coarse aggregate shall be separated into two or more sizes in order to secure greater uniformity of the concrete mixture. Different sizes of aggregate shall be stored in separate stock piles sufficiently removed from each other to prevent the material at the edges of the piles from becoming intermixed.

2.4.4 — STORAGE OF CEMENT

All cement shall be stored in suitable weatherproof buildings which will protect the cement from dampness. These buildings shall be placed in locations approved by the Engineer. Provisions for storage shall be ample, and the shipments of cement as received shall be separately stored in such a manner as to provide easy access for the identification and inspection of each shipment. Storage buildings shall have a capacity for the storage of a sufficient quantity of cement to allow sampling at least 12 days before the cement is to be used. Stored cement shall meet the test requirements at any time after storage when a retest is ordered by the **Engineer.**

On small jobs, storage in the open may be permitted by written authorization from the Engineer, in which case a raised platform and ample waterproof covering shall be provided.

When required by the terms of the contract, the Contractor shall keep accurate records of the deliveries of cement and of its use in the work. Copies of these records shall be supplied to the Engineer in such form as may be required.

2.4.5 — CLASSES OF CONCRETE

Ten classes of concrete are provided for in these specifications, five classes of non-air-entrained concrete and five corresponding classes of air-entrained concrete. Each class of concrete shall be used in that part of the structure where called for on the plans or where designated by the Engineer. The classes are as follows:

Non-air-entrained concrete	Air-entrained concrete
Class A	Class A (AE)
Class B	Class B (AE)
Class C	Class C (AE)
Class X	Class X (AE)
Class Y	Class Y (AE)

The following requirements shall govern unless otherwise shown on the plans:

Class A or Class A (AE) concrete shall be used for all superstructures, except as noted below, and for reinforced substructures except where the sections are massive and lightly reinforced. The more important items of work included are slabs, beams, girders, columns, arch ribs, box culverts, reinforced abutments, retaining walls, reinforced footings, precast piles and cribbing. Concrete deposited in water shall be Class A with 10 per cent additional cement. Class A (AE) shall be used in all locations where the concrete will be exposed to severe or moderate natural weathering (alternate freezing and thawing) as well as in concrete exposed to salt water action.

Class B or Class B (AE) concrete shall be used in mass footings, pedestals, massive pier shafts, gravity walls, with none or only a small amount of reinforcement.

Class C or Class C (AE) concrete shall be used in massive unreinforced sections.

Class X or Class X (AE) concrete shall be used in massive sections, lightly reinforced where a higher grade than Class B or Class B (AE) is required.

Class Y or Class Y (AE) concrete shall be used in thin reinforced sections, for handrails except as specified for precast railing under "Railings" and for filler in steel grid floors. Class Y (AE) shall be used in all locations where the concrete will be exposed to severe or moderate weather (alternate freezing and thawing).

2.4.6 — COMPOSITION OF CONCRETE

The cement content, coarse aggregate size, consistency, air content (in the case of AE mixes) and the approximate weights of fine and coarse aggregate (saturated surface-dry basis) for each class of concrete shall be as follows:

Class of Concrete	Cement Content	Coarse Aggregate Size	Consistency (range in slump)		Air Content (range)	Approximate Weights (saturated surface-dry) of Fine and Coarse Aggregate per sack (94 lbs.) of cement			
			Vi-brated	Non-vi-brated		Rounded Coarse Aggregate		Angular Coarse Aggregate	
	Sacks per cu. yd.		Inches	Inches	Percent	Fine Lbs.	Coarse Lbs.	Fine Lbs.	Coarse Lbs.
A	6.0	1 in.-No. 4	2-4	3-5	—	220	310	235	275
A (AE)	6.0	1 in.-No. 4	2-4	3-5	4-7	200	310	215	275
B	4.5	2 in.-No. 4	1-2	2-3	—	265	485	290	440
B (AE)	4.5	2 in.-No. 4	1-2	2-3	3-6	245	485	270	440
C	3.5	2½ in.-No. 4	1-2	2-3	—	345	640	380	590
C (AE)	3.5	2½ in.-No. 4	1-2	2-3	3-6	315	640	350	590
X	5.5	2 in.-No. 4	1-2	2-3	—	205	400	230	360
X (AE)	5.5	2 in.-No. 4	1-2	2-3	3-6	185	400	205	360
Y	7.0	½ in.-No. 4	2-4	3-5	—	215	200	220	185
Y (AE)	7.0	½ in.-No. 4	2-4	3-5	6-9	200	200	205	185

The weights of fine and coarse aggregate given in the above table are based on the use of aggregates having bulk specific gravities, in a saturated surface-dry condition, of 2.65 ± 05. In the case of blast-furnace slag, the bulk specific gravity, unless determined on the particular material being used, may be assumed to be 2.25. For reasonably well-graded materials of normal physical characteristics, the use of the above indicated proportions, together with sufficient water to obtain the required consistency, will result in concrete of the specified cement content, plus or minus two percent. For aggregates having specific gravities outside the ranges indicated above, the weights shall be corrected by multiplying the weights shown in the table by the ratio of the specific gravity of the aggregate to be used and 2.65.

The relative weights of fine and coarse aggregate per sack of cement given in the above table are based on the use of a natural sand having a fineness modulus within the range of 2.70 to 2.90 and methods of placing which do not involve high frequency vibration. When sharp, angular manufactured sands or extremely coarsely graded sands are used, the relative amount of fine aggregate should be increased. For finer sands the relative amount of fine aggregate should be decreased. In general, the least amount of sand which will insure concrete of the required workability for the placing conditions involved should be used. Any change in weight of fine aggregate made by the Engineer for the purpose of adjusting workability should always be compensated for by changing the weight of coarse aggregate in the opposite direction by a corresponding amount.

When air-entrained concrete is specified, and an air-entraining admixture is to be used, the Engineer shall determine, by means of trial batches on the project, the amount of admixture required to produce an air content within the range specified. In case an air-entraining cement is to be used, the Engineer shall determine whether the cement when used in concrete in the required proportions will entrain air in the concrete within the limits specified. In case the air content, thus determined, falls outside the specified range and it is found impossible to obtain the required air content by slight adjustments in the fine-coarse aggregate ratio and/or by changes in mixing procedures, the Contractor may, if the air content is too low, use an air-entraining admixture of the same type as that used in the manufacture of the air-entraining cement, in an amount sufficient to bring the air content within the required range. If the air content is too high, the Contractor, subject to approval by the Engineer, may use a non-air-entraining cement as a replacement for a portion of the air-entraining cement in an amount sufficient to bring the air content within the required range.

2.4.7 — SAMPLING AND TESTING

Compliance with the requirements indicated in Article 2.4.6 shall be determined in accordance with the following standard methods of AASHO:

    (a) Sampling fresh concrete.......... T 141 (ASTM C172)
    (b) Cement content .... T 121 (ASTM C138)
    (c) Size of coarse aggregate.......... T 27
    (d) Consistency .... T 119 (ASTM C143)
    (e) Air content .... T 152 (ASTM C231)
    (f) Bulk specific gravity and
        absorption .... T 84
                                      T 85

Tests for strength (when required) shall be made in accordance with the following:

    (a) Molding concrete specimens in
        the field .... T 23 (ASTM C31)
    (b) Compressive strength of molded
        cylinders .... T 22 (ASTM C39)

2.4.8 — MEASUREMENT OF MATERIALS

Materials shall be measured by weighing, except as otherwise specified or where other methods are specifically authorized by the Engineer. The apparatus provided for weighing the aggregates and cement shall be suitably designed and constructed for this purpose. Each size of aggregate and the cement shall be weighed separately. The accuracy of all weighing devices shall be such that successive quantities can be measured to within 1 per cent of the desired amount. Cement in standard packages (sack) need not be weighed, but bulk cement shall be weighed. The mixing water shall be measured by volume or by weight. The water measuring device shall be susceptible of control accurate to plus or minus ½ per cent of the capacity of the tank. All measuring devices shall be subject to approval.

Where volumetric measurements are authorized by the Engineer for projects where the amount of concrete is small, the weight proportions shall be converted to equivalent volumetric proportions. In such cases, suitable allowance shall be made for variations in the moisture condition of the aggregates, including the bulking effect in the fine aggregate.

When the aggregates contain more water than the quantity necessary to produce a saturated surface-dry condition as contemplated in Article 2.4.6, representative samples shall be taken and the moisture content determined for each kind of aggregate.

When sack cement is used, the quantities of aggregates for each batch shall be exactly sufficient for one or more full sacks of cement and no batch requiring fractional sacks of cement will be permitted.

2.4.9 — MIXING CONCRETE

(A) General

Unless otherwise authorized by the Engineer, concrete shall be machine mixed at the site.

(B) Mixing at Site

Concrete shall be thoroughly mixed in a batch mixer of an approved size and type which will insure a uniform distribution of the materials throughout the mass.

The mixer shall be equipped with adequate water storage and a device for accurately measuring and automatically controlling the amount of water used in each batch. Preferably, mechanical means shall be provided for recording the number of revolutions for each batch and automatically preventing the discharge of the mixer until the materials have been mixed the specified minimum time.

The entire contents of the mixer shall be removed from the drum before materials for a succeeding batch are placed therein. No mixer having a rated capacity of less than a 1-bag batch shall be used nor shall a mixer be charged in excess of its rated capacity.

All concrete shall be mixed for a period of not less than 1½ minutes after all materials, including water, are in the mixer. During the period of mixing, the mixer shall operate at the speed for which it has been designed, but this speed shall be not less than 14 nor more than 20 revolutions per minute.

The first batch of concrete materials placed in the mixer shall contain a sufficient excess of cement, sand and water to coat the inside of the drum without reducing the required mortar content of the mix. Upon the cessation of mixing for a considerable period, the mixer shall be thoroughly cleaned.

(C) Truck Mixing

Truck mixers, unless otherwise authorized by the Engineer, shall be of the revolving drum type, watertight, and so constructed that the concrete can be mixed to insure a uniform distribution of materials throughout the mass. All solid materials for the concrete shall be accurately measured in accordance with Article 2.4.8, and charged into the drum at the proportioning plant. Except as subsequently provided, the truck mixer shall be equipped with a tank for carrying mixing water. Only the prescribed amount of water shall be placed in the tank unless the tank is equipped with a device by which the quantity of water added can be readily verified. The mixing water may be added directly to the batch, in which case a tank shall not be required. Truck mixers may be required to be provided with means by which the mixing time can be readily verified by the Engineer.

The maximum size of batch in truck mixers shall not exceed the maximum rated capacity of the mixer as stated by the manufacturer and stamped in metal on the mixer. Truck mixing shall be continued for not less than 50 revolutions after all ingredients, including the water, are in the drum. The speed shall not be less than 4 r.p.m., nor more than a speed resulting in a peripheral velocity of the drum of 225 feet per minute. Not more than 100 revolutions of mixing shall be at a speed in excess of 6 r.p.m. Mixing shall begin within 30 minutes after the cement has been added either to the water or aggregate.

When cement is charged into a mixer drum containing water or surface-wet aggregate and when the temperature is above 90° F., or when high-early strength portland cement is used, this limit shall be reduced to 15 minutes; the limitation on time between the introduction of the cement to the aggregates and the beginning of the mixing may be waived when, in the judgment of the Engineer, the aggregates are sufficiently free from moisture, so that there will be no harmful effects on the cement.

(D) Partial Mixing at the Central Plant

When a truck mixer, or an agitator provided with adequate mixing blades, is used for transportation, the mixing time at the stationary machine mixer may be reduced to 30 seconds and the mixing completed in a truck mixer or agitator. The mixing time in the truck mixer or agitator equipped with adequate mixing blades shall be as specified for truck mixing.

(E) Plant Mix

Mixing at a central plant shall conform to the requirements for mixing at the site.

(F) Time of Hauling and Placing Mixed Concrete

Concrete transported in a truck mixer, agitator, or other transportation device shall be discharged at the job and placed in its final position in the forms within 1½ hours after the introduction of the mixing water to the cement and aggregate, or the cement to the aggregate, except that in hot weather or under other conditions contributing to quick stiffening of the concrete, the maximum allowable time may be reduced by the Engineer. The maximum volume of mixed concrete transported in an agitator shall be in accordance with the specified rating.

(G) Hand Mixing

When hand mixing is authorized it shall be done on a watertight platform and in such a manner as to insure a uniform distribution of the materials throughout the mass. Mixing shall be continued until a homogeneous mixture of the required consistency is obtained.

(H) Delivery

The organization supplying concrete shall have sufficient plant capacity and transporting apparatus to insure continuous delivery at the rate required. The rate of delivery of concrete during concreting operations shall be such as to provide for the proper handling, placing and finishing of the concrete. The rate shall be such that the interval between batches shall not exceed 20 minutes. The methods of delivering and handling the concrete shall be such as will facilitate placing with the minimum of rehandling and without damage to the structure or the concrete.

(I) Retempering

The concrete shall be mixed only in such quantities as are required for immediate use and any which has developed initial set shall not be used. Concrete which has partially hardened shall not be retempered or remixed.

2.4.10 — HANDLING AND PLACING CONCRETE

(A) General

In preparation for the placing of concrete all sawdust, chips, and other construction debris and extraneous matter shall be removed from the interior of forms. Struts, stays and braces, serving temporarily to hold the forms in correct shape and alignment, pending the placing of concrete at their locations, shall be removed when the concrete placing has reached an elevation rendering their service unnecessary. These temporary members shall be entirely removed from the forms and not buried in the concrete.

No concrete shall be used which does not reach its final position in the forms within the time stipulated under Article 2.4.9(F).

Concrete shall be placed so as to avoid segregation of the materials and the displacement of the reinforcement. The use of long troughs, chutes and pipes for conveying concrete from the mixer to the forms shall be permitted only on written authorization of the Engineer. In case an inferior quality of concrete is produced by the use of such conveyors, the Engineer may order discontinuance of their use and the institution of a satisfactory method of placing.

Open troughs and chutes shall be of metal or metal lined; where steep slopes are required, the chutes shall be equipped with baffles or be in short lengths that reverse the direction of movement.

All chutes, troughs and pipes shall be kept clean and free from coatings of hardened concrete by thoroughly flushing with water after each run. Water used for flushing shall be discharged clear of the structure.

When placing operations would involve dropping the concrete more than 5 feet, it shall be deposited through sheet metal or other approved pipes. As far as practicable, the pipes shall be kept full of concrete during placing and their lower ends shall be kept buried in the newly placed concrete. After initial set of the concrete, the forms shall not be jarred and no strain shall be placed on the ends of reinforcement bars which project.

Concrete, during and immediately after depositing, shall be thoroughly compacted. The compaction shall be done by mechanical vibration subject to the fol'owing provisions:

(1) The vibration shall be internal unless special authorization of other methods is given by the Engineer or as provided herein.

(2) Vibrators shall be of a type and design approved by the **Engineer**. They shall be capable of transmitting vibration to the concrete at frequencies of not less than 4500 impulses per minute.

(3) The intensity of vibration shall be such as to visibly affect a mass of concrete of 1-inch slump over a radius of at least 18 inches.

(4) The Contractor shall provide a sufficient number of vibrators to properly compact each batch immediately after it is placed in the forms.

(5) Vibrators shall be manipulated so as to thoroughly work the concrete around the reinforcement and imbedded fixtures, and into the corners and angles of the forms.

Vibration shall be applied at the point of deposit and in the area of freshly deposited concrete. The vibrators shall be inserted and withdrawn out of the concrete slowly. The vibration shall be of sufficient duration and intensity to thoroughly compact the concrete, but shall not be continued so as to cause segregation. Vibration shall not be continued at any one point to the extent that localized areas of grout are formed.

Application of vibrators shall be at points uniformly spaced and not farther apart than twice the radius over which the vibration is visibly effective.

(6) Vibration shall not be applied directly or through the reinforcement to sections or layers of concrete which have hardened to the degree that the concrete ceases to be plastic under vibration. It shall not be used to make concrete flow in the forms over distances so great as to cause segregation, and vibrators shall not be used to transport concrete in the forms.

(7) Vibration shall be supplemented by such spading as is necessary to insure smooth surfaces and dense concrete along form surfaces and in corners and locations impossible to reach with the vibrators.

(8) The provisions of this article shall apply to the filler concrete for steel grid floor except that the vibrator shall be applied to the steel.

(9) The provisions of this article shall apply to precast piling, concrete cribbing and other precast members except that, if approved by the Engineer, the manufacturers' methods of vibrations may be used.

Concrete shall be placed in horizontal layers not more than 12 inches thick except as hereinafter provided. When less than a complete layer is placed in one operation, it shall be terminated in a vertical bulkhead. Each layer shall be placed and compacted before the preceding batch has taken initial set to prevent injury to the green concrete and avoid surfaces of separation between the batches. Each layer shall be compacted so as to avoid the formation of a construction joint with a preceding layer which has not taken initial set.

When the placing of concrete is temporarily discontinued, the concrete, after becoming firm enough to retain its form, shall be cleaned of laitance and other objectionable material to a sufficient depth to expose sound concrete. To avoid visible joints as far as possible upon exposed faces, the top surface of the concrete adjacent to the forms shall be smoothed with a trowel. Where a "feather edge" might be produced at a construction joint, as in the sloped top surface of a wing wall, an inset form shall be used to produce a blocked out portion in the preceding layer which shall produce an edge thickness of not less than 6 inches in the succeeding layer. Work shall not be discontinued within 18 inches of the top of any face, unless provision has been made for a coping less than 18 inches thick, in which case, if permitted by the Engineer, the construction joint may be made at the under side of the coping.

Immediately following the discontinuance of placing concrete all accumulations of mortar splashed upon the reinforcement steel and the surfaces of forms shall be removed. Dried mortar chips and dust shall not be puddled into the unset concrete. If the accumulations are not removed prior to the concrete becoming set, care shall be exercised not to injure or break the concrete-steel bond at and near the surface of the concrete, while cleaning the reinforcement steel.

(B) Culverts

In general, the base slab or footings of box culverts shall be placed and allowed to set before the remainder of the culvert is constructed. In this case suitable provision shall be made for bonding the sidewalls to the culvert base, preferably by means of raised longitudinal keys so constructed as to prevent, as far as possible, the percolation of water through the construction joint.

Before concrete is placed in the sidewalls, the culvert footings shall be thoroughly cleaned of all shavings, sticks, sawdust, or other extraneous material and the surface carefully chipped and roughened in accordance with the method of bonding construction joints as specified herein.

In the construction of box culverts 4 feet or less in height, the sidewalls and top slab may be constructed as a monolith. When this method of construction is used, any necessary construction joints shall be vertical and at right angles to the axis of the culvert.

In the construction of box culverts more than 4 feet in height, the concrete in the walls shall be placed and allowed to set before the top slab is placed. In this case, appropriate keys shall be left in the sidewalls for anchoring the cover slab.

Each wing wall shall be constructed, if possible, as a monolith. Construction joints, where unavoidable, shall be horizontal and so located that no joint will be visible in the exposed face of the wing wall above the ground line.

(C) Girders, Slabs and Columns

For simple spans, concrete, preferably, shall be deposited by beginning at the center of the span and working from the center toward the ends. Concrete in girders shall be deposited uniformly for the full length of the girder and brought up evenly in horizontal layers. For continuous spans, where required by design considerations, the concrete placing sequence shall be shown on the plans or in the special provisions.

Concrete in girder haunches less than 3 feet in height shall be placed at the same time as that in the girder stem, and the column or abutment tops shall be cut back to form seats for the haunches. Whenever any haunch or fillet has a vertical height of 3 feet or more, the abutment or columns, the haunch, and the girder shall be placed in three successive stages; first, up to the lower side of the haunch; second, to the lower side of the girder; and third, to completion.

For haunched continuous girders, the girder stem (including haunch) shall be poured to the top of stem. Where the size of the pour is such that it cannot be made in one pour, vertical construction joints shall preferably be located within the area of contraflexure.

Concrete in slab spans shall be placed in one continuous operation for each span unless otherwise provided.

The floors and girders of through girder superstructures shall be placed in one continuous operation unless otherwise specified, in which case special shear anchorage shall be provided to insure monolithic action between girder and floor.

Concrete in T-beam or deck girder spans may be placed in one continuous operation or may be placed in two separate operations, each of which shall be continuous; first, to the top of the girder stems, and second, to completion. In the latter case, the bond between stem and slab shall be positive and mechanical, and may be secured by means of suitable shear keys or by artificially roughening the surface of the top of the girder stem. In general, suitable keys may be formed by the use of timber blocks approximately 2 by 4 inches in cross-section and having a length 4 inches less than the width of the girder stem. These key blocks shall be spaced along the girder stems as required, but the spacing shall be not greater than 1 foot center to center. The blocks shall be beveled and oiled in such manner as to insure their ready removal, and they shall be removed as soon as the concrete has set sufficiently to retain its shape.

Concrete in box girders may be placed in two or three separate operations. In either case the bottom slab shall be poured first. Bond between the bottom slab and stem shall be positive and mechanical. If the webs are poured separately from the top slab, bond between the top slab and webs shall be secured in the same manner as for T-beams. Requirements for shear keys for T-beams shall also apply to box girders, except that keys need not be deeper than the depth to the top of bottom slab reinforcement.

Concrete in columns shall be placed in one continuous operation, unless otherwise directed. The concrete shall be allowed to set at least 12 hours before the caps are placed.

Unless otherwise permitted by the Engineer, no concrete shall be placed in the superstructure until the column forms have been stripped

sufficiently to determine the character of the concrete in the columns. The load of the superstructure shall not be allowed to come upon the bents until they have been in place at least 14 days, unless otherwise permitted by the Engineer.

(D) Arches

The concrete in arch rings shall be placed in such a manner as to load the centering uniformly.

Arch rings, preferably, shall be cast in transverse sections of such size that each section can be cast in a continuous operation. The arrangement of the sections and the sequence of placing shall be as approved by the Engineer and shall be such as to avoid the creation of initial stress in the reinforcement. The sections shall be bonded together by suitable keys or dowels. When permitted by the Engineer, arch rings may be cast in a single continuous operation.

2.4.11 — PNEUMATIC PLACING

Pneumatic placing of concrete will be permitted only if specified in the special provisions or if authorized by the Engineer. The equipment shall be so arranged that vibrations will not damage freshly placed concrete.

Where concrete is conveyed and placed by pneumatic means the equipment shall be suitable in kind and adequate in capacity for the work. The machine shall be located as close as practicable to the place of deposit. The position of the discharge end of the line shall not be more than 10 feet from the point of deposit. The discharge lines shall be horizontal or inclined upwards from the machine.

At the conclusion of placement the entire equipment shall be thoroughly cleaned.

2.4.12 — PUMPING

Placement of concrete by pumping will be permitted only if specified in the special provisions or if authorized by the Engineer. The equipment shall be so arranged that vibrations will not damage freshly placed concrete.

Where concrete is conveyed and placed by mechanically applied pressure, the equipment shall be suitable in kind and adequate in capacity for the work. The operation of the pump shall be such that a continuous stream of concrete without air pockets is produced. When pumping is completed, the concrete remaining in the pipeline, if it is to be used, shall be ejected in such a manner that there will be no contamination of the concrete or separation of the ingredients. After this operation, the entire equipment shall be thoroughly cleaned.

2.4.13 — DEPOSITING CONCRETE UNDER WATER

Concrete shall not be deposited in water except with the approval of the Engineer and under his immediate supervision; and in this case the method of placing shall be as hereinafter designated.

Concrete deposited in water shall be Class A with 10 per cent excess cement. To prevent segregation, it shall be carefully placed in a compact mass, in its final position, by means of a tremie, a bottom dump bucket or other approved method, and shall not be disturbed after being deposited. Still water shall be maintained at the point of deposit and the forms under water shall be water-tight.

For parts of structures under water, when possible, concrete seals shall be placed continuously from start to finish; the surface of the concrete shall be kept as nearly horizontal as practicable at all times. To insure thorough bonding, each succeeding layer of a seal shall be placed before the preceding layer has taken initial set.

A tremie shall consist of a tube having a diameter of not less than 10 inches, constructed in sections having flanged couplings fitted with gaskets. The tremies shall be supported so as to permit free movement of the discharge end over the entire top surface of the work and so as to permit rapid lowering when necessary to retard or stop the flow of concrete. The discharge end shall be closed at the start of work so as to prevent water entering the tube and shall be entirely sealed at all times; the tremie tube shall be kept full to the bottom of the hopper. When a batch is dumped into the hopper, the flow of concrete shall be induced by slightly raising the discharge end, always keeping it in the deposited concrete. The flow shall be continuous until the work is completed.

Depositing of concrete by the drop bottom bucket method shall conform to the following specification. The top of the bucket shall be open. The bottom doors shall open freely downward and outward when tripped. The bucket shall be completely filled and slowly lowered to avoid backwash. It shall not be dumped until it rests on the surface upon which the concrete is to be deposited and when discharged shall be withdrawn slowly until well above the concrete. The slump of concrete shall be maintained between 4 and 8 inches.

Unwatering may proceed when the concrete seal is sufficiently hard and strong. All laitance or other unsatisfactory material shall be removed from the exposed surface by scraping, chipping or other means which will not injure the surface of the concrete.

2.4.14 — CONSTRUCTION JOINTS

(A) General

Construction joints shall be made only where located on the plans or shown in the pouring schedule, unless otherwise approved by the **Engineer**.

If not detailed on the plans, or in the case of emergency, construction joints shall be placed as directed by the Engineer. Shear keys or inclined reinforcement shall be used where necessary to transmit shear or bond the two sections together.

(B) Bonding

Before depositing new concrete on or against concrete which has hardened, the forms shall be retightened. The surface of the hardened concrete shall be roughened as required by the Engineer, in a manner that will not leave loosened particles of aggregate or damaged concrete at the surface. It shall be thoroughly cleaned of foreign matter and laitance, and saturated with water. To insure an excess of mortar at the juncture of the hardened and the newly deposited concrete, the cleaned and saturated surfaces, including vertical and inclined surfaces, shall first be thoroughly covered with a coating of mortar or neat cement grout against which the new concrete shall be placed before the grout has attained its initial set.

The placing of concrete shall be carried continuously from joint to joint. The face edges of all joints which are exposed to view shall be carefully finished true to line and elevation.

2.4.15 — RUBBLE OR CYCLOPEAN CONCRETE

Rubble or cyclopean concrete shall consist of either Class B or C concrete, as specified, containing large embedded stones. It shall be used only with the approval of the Engineer in massive piers, gravity abutments, and heavy footings. The stone for this class of work may be one-man stone or derrick stone conforming to the requirements of Article 2.7.2(A).

The stone shall be carefully placed—not dropped or cast—so as to avoid injury to the forms or to the partially set adjacent masonry. Stratified stone shall be placed upon its natural bed. All stone shall be washed and saturated with water before placing.

The total volume of the stone shall not be greater than one-third of the total volume of the portion of the work in which it is placed. For walls or piers greater than 2 feet in thickness, one-man stone may be used; each stone shall be surrounded by at least 6 inches of concrete; and no stone shall be closer than 1 foot to any top surface nor any closer than 6 inches to any coping. For walls or piers greater than 4 feet in thickness, derrick stone may be used; each stone shall be surrounded by at least 1 foot of concrete; and no stone shall be closer than 2 feet to any top surface nor closer than 8 inches to any coping.

2.4.16 — CONCRETE EXPOSED TO SEA WATER

Unless otherwise specifically provided, concrete for structures exposed to sea water shall be Class A concrete as specified in Article 2.4.6. The clear distance from the face of the concrete to the nearest face of reinforcement steel shall be not less than 4 inches. The concrete shall be mixed for a period of not less than 2 minutes and the water content of the mixture shall be carefully controlled and regulated so as to produce concrete of maximum impermeability. The concrete shall be thoroughly compacted and stone pockets shall be avoided. No construction joints shall be formed between levels of extreme low water and extreme high water as determined by the Engineer. Between these levels sea water shall not come in direct contact with the concrete for a period of not less than 30 days. The original surface, as the concrete comes from the forms, shall be left undisturbed.

2.4.17 — CONCRETE EXPOSED TO ALKALI SOILS OR ALKALI WATER

Where concrete may be exposed to the action of alkaline waters or soils, special care shall be taken to place it in accordance with placing specifications herein. Whenever possible, placing shall be continuous until completion of the section or until the concrete is at least 18 inches above ground or water level. Alkaline waters or soils shall be kept from contact with the concrete during placement and for a period of at least 72 hours thereafter.

2.4.18 — FALSEWORK AND CENTERING

Unless otherwise provided, detailed plans for falsework or centering shall be supplied to the Engineer on request, but, in no case shall the Contractor be relieved of responsibility for results obtained by the use of these plans.

For designing falsework and centering, a weight of 150 pounds per cubic foot shall be assumed for green concrete. All falsework shall be designed and constructed to provide the necessary rigidity and to support the loads without appreciable settlement or deformation. The Engineer may require the Contractor to employ screw jacks or hardwood wedges to take up any settlement in the formwork either before or during the placing of concrete.

Falsework which cannot be founded on a satisfactory footing shall be supported on piling which shall be spaced, driven and removed in a manner approved by the Engineer.

Falsework shall be set to give the finished structure the camber specified or indicated on the plans.

Arch centering shall be constructed according to centering plans approved by the Engineer. Provision shall be made by means of suitable wedges, sand boxes or other device for the gradual lowering of centers, and rendering the arch self-supporting. When directed, centering shall be placed upon approved jacks in order to take up and correct any slight settlement which may occur after the placing of masonry has begun.

2.4.19 — FORMS

All forms shall be of wood or metal and shall be built mortartight and of sufficient rigidity to prevent distortion due to the pressure of the concrete and other loads incident to the construction operations. Forms shall be constructed and maintained so as to prevent warping and the opening of joints due to shrinkage of the lumber.

The forms shall be substantial and unyielding and shall be so designed that the finished concrete will conform to the proper dimensions and contours. The design of the forms shall take into account the effect of vibration of concrete as it is placed.

Forms for exposed surfaces shall be made of dressed lumber of uniform thickness, with or without a form liner of an approved type, and shall be mortartight. Forms shall be filleted at all sharp corners and shall be given a bevel or draft in the case of all projections, such as girders and copings, to insure easy removal.

Metal ties or anchorages within the forms shall be so constructed as to permit their removal to a depth of at least 2 inches from the face without injury to the concrete. In case ordinary wire ties are permitted, all wires, upon removal of the forms, shall be cut back at least ¼ inch from the face of the concrete with chisels or nippers; for green concrete, nippers are necessary. All fittings for metal ties shall be of such design that, upon their removal, the cavities which are left will be of the smallest possible size. The cavities shall be filled with cement mortar and the surface left sound, smooth, even and uniform in color.

All forms shall be set and maintained true to the line designated until the concrete is sufficiently hardened. Forms shall remain in place for periods which shall be determined as hereinafter specified. When forms appear to be unsatisfactory in any way, either before or during the placing of concrete, the Engineer shall order the work stopped until the defects have been corrected.

The shape, strength, rigidity, watertightness and surface smoothness of re-used forms shall be maintained at all times. Any warped or bulged lumber must be re-sized before being re-used. Forms which are unsatisfactory in any respect shall not be re-used.

For narrow walls and columns, where the bottom of the form is inaccessible, the lower form boards shall be left loose so that they may be removed for cleaning out extraneous material immediately before placing the concrete.

All forms shall be treated with oil or saturated with water immediately before placing the concrete. For rail members or other members with exposed faces, the forms shall be treated with an approved oil to prevent the adherence of concrete. Any material which will adhere to or discolor the concrete shall not be used.

2.4.20 — REMOVAL OF FALSEWORK, FORMS AND HOUSING

In the determination of the time for the removal of falsework, forms and housing, and the discontinuance of heating, consideration shall be given to the location and character of the structure, the weather and other conditions influencing the setting of the concrete, and the materials used in the mix.

If field operations are not controlled by beam or cylinder tests the following periods, exclusive of days when the temperature is below 40°, for removal of forms and supports may be used as a guide:

Arch centers	14 days
Centering under beams	14 days
Floor slabs	7-14 days
Walls	12-24 hrs.
Columns	1- 7 days
Sides of beams and all other parts	12-24 hrs.

If high-early strength cement is used these periods may be reduced as directed by the Engineer.

If field operations are controlled by beam or cylinder tests, the removal of forms, supports and housing, and the discontinuance of heating and curing may be begun when the strengths reach values which shall be fixed by the Engineer for the particular method of testing which is to be used. The beams or cylinders shall be cured under conditions which are not more favorable than the most unfavorable conditions for the portions of the concrete which the beams represent.

Methods of form removal likely to cause overstressing of the concrete shall not be used. In general, the forms shall be removed from the bottom upwards. Forms and their supports shall not be removed without the approval of the Engineer. Supports shall be removed in such a manner as to permit the concrete to uniformly and gradually take the stresses due to its own weight.

In general, arch centering shall be struck and the arch made self-supporting before the railing or coping is placed. This precaution is essential in order to avoid jamming of the expansion joints and variations in alignment. For filled spandrel arches, such portions of the spandrel walls shall be left for construction subsequent to the striking of centers, as may be necessary to avoid jamming of the expansion joints.

Centers shall be gradually and uniformly lowered in such a manner as to avoid injurious stresses in any part of the structure. In arch structures of two or more spans, the sequence of striking centers shall be specified or approved by the Engineer.

2.4.21 — CONCRETING IN COLD WEATHER

No concrete shall be placed when the atmospheric temperature is below 35 F without written permission of the Engineer. When directed by the Engineer, the Contractor shall enclose the structure in such a way that the concrete and air within the enclosure can be kept above 60 F for a period of seven days after placing the concrete.

If high early strength cement is used these periods may be reduced, as directed by the Engineer.

The Contractor shall supply such heating apparatus as stoves, salamanders or steam equipment and the necessary fuel. When dry heat is used, means of maintaining atmospheric moisture shall be provided. All aggregates and mixing water shall be heated to a temperature of at least 70 F but not more than 150 F; the aggregates may be heated by either steam or dry heat. If permitted by the Engineer the torch method of heating mixed concrete may be used, provided the heating apparatus shall be such as to heat the mass uniformly and avoid hot spots which will burn the materials. The temperature of the concrete shall be not less than 60 F at the time of placing in the forms. In case of extremely low temperatures, the Engineer may, at his discretion, raise the minimum limiting temperatures for water, aggregates and mixed concrete.

2.4.22 — CURING CONCRETE

Concrete surfaces exposed to conditions causing premature drying shall be protected by covering as soon as possible with canvas, straw, burlap, sand or other satisfactory material and kept moist; or if the surfaces are not covered, they shall be kept moist by flushing or sprinkling. Curing shall continue for a period of not less than seven days after placing the concrete. If high-early strength cement is used, this period may be reduced as directed by the Engineer. Other precautions to insure the development of strength shall be taken as the Engineer may direct.

2.4.23 — EXPANSION AND FIXED JOINTS AND BEARINGS

All joints shall be constructed according to details shown on the plans.

(A) Open Joints

Open joints shall be placed in the locations shown on the plans and shall be constructed by the insertion and subsequent removal of a wood strip, metal plate or other approved material. The insertion and removal of the template shall be accomplished without chipping or breaking the corners of the concrete. Reinforcement shall not extend across an open joint unless so specified on the plans.

(B) Filled Joints

Poured expansion joints shall be constructed similar to open joints. When premolded types are specified, the filler shall be in correct position when the concrete on one side of the joint is placed. When the form is removed, the concrete on the other side shall be placed. Adequate water stops of metal, rubber or plastic shall be carefully placed as shown on the plans.

(C) Premolded Expansion Joint Fillers

Non-extruding and resilient types shall conform to the Specification for Preformed Expansion Joint Fillers for Concrete of the AASHO M 153 (ASTM D 1752).

Bituminous fiber types shall conform to the Specification for Preformed Expansion Joint Fillers for Concrete, AASHO M 153 (ASTM D 1752).

Bituminous type filler shall conform to the Specification for Preformed Expansion Joint Filler for Concrete, AASHO M 33.

(D) Steel Joints

The plates, angles or other structural shapes shall be accurately shaped, at the shop, to conform to the section of the concrete floor. The fabrication and painting shall conform to the requirements of the specifications covering those items. When called for on the plans or in the special provisions, the material shall be galvanized in lieu of painting. Care shall be taken to insure that the surface in the finished plane is true and free of warping. Positive methods shall be employed in placing the joints to keep them in correct position during the placing of the concrete. The opening at expansion joints shall be that designated on the plans at normal temperature, and care shall be taken to avoid impairment of the clearance in any manner.

(E) Water Stops

Adequate water stops of metal, rubber or plastic shall be placed as shown on the plans. Where movement at the joint is provided for, the water stops shall be of a type permitting such movement without injury. They shall be spliced, welded, or soldered, to form continuous watertight joints.

(F) Sheet Copper

Sheet copper shall conform to the Specifications for Copper Sheet, Strip and Plate, AASHO M 138 (ASTM B 152).

Sheet copper shall meet the Embrittlement Test of Section 12 of M 138.

(G) Bearing Devices

Bearing plates, rockers and other expansion devices shall be constructed according to details shown on the plans. Unless set in plastic concrete or as otherwise specified, they shall be set in grout to insure uniform bearing. Bronze or copper-alloy plates shall conform to the requirements of Articles 2.11.2 (A) or 2.11.2 (B). Structural steel and painting shall conform to the specifications for those items. When called for on the plans or in the special provisions, the material shall be galvanized in lieu of painting. The rockers or other expansion devices shall be set to conform to the temperature at the time of erection.

FINISHING CONCRETE SURFACES

2.4.24 — GENERAL

Surface finishes shall be classified as follows:

Class 1. Ordinary surface finish.
Class 2. Rubbed finish.
Class 3. Tooled finish.
Class 4. Sand-blast finish.
Class 5. Wire brush, or scrubbed finish.
Class 6. Floated surface finish.

All concrete shall be given Class 1, Ordinary Surface Finish, and in addition, if further finishing is required, such other type of finish as is specified. If not otherwise specified, the following surfaces shall be given a Class 2, Rubbed Finish: The exposed faces of piers, abutments, wing walls and retaining walls; the outside faces of girders, T-beams, slabs, columns, brackets, curbs, headwalls, railings, arch rings, spandrel walls and parapets; but not on the tops and bottoms of floor slabs and sidewalks, bottoms of beams and girders, sides of interior beams and girders, backwalls above bridge seat or the underside of copings. The surface finish on piers and abutments shall include all exposed surfaces below bridge seat to 1 foot below low water elevation or 2 feet below finish ground line when such ground line is above the water surface. Wing walls shall be finished from the top to 2 feet below the finish slope lines on the outside face and shall be finished on top and for a depth of 1 foot below the top on the back sides.

Unless otherwise specified, roadway floors shall be given Class 6, Floated Surface Finish.

2.4.25 — CLASS 1, ORDINARY SURFACE FINISH

Immediately following the removal of forms, all fins and irregular projections shall be removed from all surfaces except from those which are not to be exposed or are not to be waterproofed. On all surfaces, the cavities produced by form ties and all other holes, honeycomb spots, broken corners or edges and other defects shall be thoroughly cleaned, and after having been kept saturated with water for a period of not less than three hours shall be carefully pointed and trued with a mortar of cement and fine aggregate mixed in the proportions used in the grade of the concrete being finished. Mortar used in pointing shall be not more than one hour old. The mortar patches shall be cured as specified under Article 2.4.22. All construction and expansion joints in the completed work shall be left carefully tooled and free of all mortar and concrete. The joint filler shall be left exposed for its full length with clean and true edges.

The resulting surfaces shall be true and uniform. All repaired surfaces, the appearance of which is not satisfactory to the Engineer, shall be "rubbed" as specified under Article 2.4.26.

2.4.26 — CLASS 2, RUBBED FINISH

After removal of forms, the rubbing of concrete shall be started as soon as its condition will permit. Immediately before starting this work the concrete shall be kept thoroughly saturated with water for a minimum period of three hours. Sufficient time shall have elapsed before the wetting down to allow the mortar used in the pointing of rod holes and defects to thoroughly set. Surfaces to be finished shall be rubbed with a medium coarse carborundum stone, using a small amount of mortar on its face. The mortar shall be composed of cement and fine sand mixed in proportions used in the concrete being finished. Rubbing shall be continued until all form marks, projections and irregularities have been removed, all voids filled, and a uniform surface has been obtained. The paste produced by this rubbing shall be left in place at this time.

After all concrete above the surface being treated has been cast, the final finish shall be obtained by rubbing with a fine carborundum stone and water. This rubbing shall be continued until the entire surface is of a smooth texture and uniform color.

After the final rubbing is completed and the surface has dried, it shall be rubbed with burlap to remove loose powder and shall be left free from all unsound patches, paste, powder and objectionable marks.

2.4.27 — CLASS 3, TOOLED FINISH

Finish of this character for panels and other like work may be secured by the use of a bushhammer, pick, crandall, or other approved tool. Air tools, preferably, shall be employed. No tooling shall be done until the concrete has set for at least 14 days and as much longer as may be necessary to prevent the aggregate particles from being "picked" out of the surface. The finished surface shall show a grouping of broken aggregate particles in a matrix of mortar, each aggregate particle being in slight relief.

2.4.28 — CLASS 4, SAND BLASTED FINISH

The thoroughly cured concrete surface shall be sand blasted with hard, sharp sand to produce an even fine-grained surface in which the mortar has been cut away, leaving the aggregate exposed.

2.4.29 — CLASS 5, WIRE BRUSHED OR SCRUBBED FINISH

This type of finish shall be produced by scrubbing the surface of green concrete with stiff wire or fiber brushes, using a solution of muriatic acid in the proportion of 1 part acid to 4 parts water. As soon as the forms are removed and while the concrete is yet comparatively green, the surface shall be thoroughly and evenly scrubbed as above described until the cement film or surface is completely removed and the aggregate particles are exposed, leaving an even pebbled texture presenting an appearance grading from that of fine granite to coarse conglomerate, depending upon the size and grading of aggregate used. As soon as the scrubbing has progressed sufficiently to produce the texture desired, the entire surface shall be thoroughly washed with water to which a small amount of ammonia has been added to remove all traces of the acid.

2.4.30 — CLASS 6, FLOATED SURFACE FINISH

(A) Striking Off

After the concrete is compacted as specified under Article 2.4.10 (A), the surface shall be carefully rodded and struck off with a strike board to conform to the cross section and grade shown on the plans. Proper allowance shall be made for camber, if required. The strike board may be operated longitudinally or transversely and shall be moved forward with a combined longitudinal and transverse motion, the manipulation being such that neither end is raised from the side forms during the process. A slight excess of concrete shall be kept in front of the cutting edge at all times.

(B) Floating

After striking off and consolidating as specified above, the surface shall be made uniform by longitudinal or transverse floating, or both. Longitudinal floating will be required except in places where this method is not feasible.

(C) Longitudinal Floating

The longitudinal float, operated from foot bridges, shall be worked with a sawing motion while held in a floating position parallel to the road centerline and passing gradually from one side of the pavement to the other. The float shall then be moved forward one-half of its length and the above operation repeated. Machine floating which produces equivalent results may be substituted for the above hand method.

(D) Transverse Floating

The transverse float shall be operated across the pavement by starting at the edge and slowly moving to the center and back again to the edge. The float shall then be moved forward one-half of its length and the above operations repeated. Care shall be taken to preserve the crown and cross section of the pavement.

(E) Straightedging

After the longitudinal floating has been completed and the excess water removed, but while the concrete is still plastic, the slab surface shall be tested for trueness with a straightedge. For this purpose, the contractor shall furnish and use an accurate 10-foot straightedge swung from handles 3 feet longer than one-half the width of the slab.

The straightedge shall be held in successive positions parallel to the road centerline and in contact with the surface and the whole area gone over from one side of the slab to the other as necessary. Advancement along the deck shall be in successive stages of not more than one-half the length of the straightedge. Any depressions found shall be immediately filled with freshly mixed concrete, struck off, consolidated and refinished. High areas shall be cut down and refinished. The straightedge testing and refloating shall continue until the entire surface is found to be free from observable departures from the straightedge and the slab has the required grade and contour, until there are no deviations of more than $\frac{1}{8}$ inch under the 10-foot straightedge.

(F) Final Finishing

When the concrete has hardened sufficiently, the surface shall be given a broom finish. The broom shall be of an approved type. The strokes shall be square across the slab, from edge to edge, with adjacent strokes slightly overlapped, and shall be made by drawing the broom without tearing the concrete, but so as to produce regular corrugations not over $\frac{1}{8}$ of an inch in depth. The surface as thus finished shall be free from porous spots, irregularities, depressions and small pockets or rough spots such as may be caused by accidental disturbing, during the final brooming, of particles of coarse aggregate embedded near the surface.

2.4.31 — SIDEWALK FINISH

After the concrete has been deposited in place, it shall be compacted and the surface shall be struck off by means of a strike board and floated with a wooden or cork float. An edging tool shall be used on all edges and at all expansion joints. The surface shall not vary more than $\frac{1}{8}$ inch under a 10-foot straightedge. The surface shall have a granular or matte texture which will not be slick when wet.

Sidewalk surfaces shall be laid out in blocks with an approved grooving tool as shown on the plans or as directed by the Engineer.

2.4.32 — PNEUMATICALLY APPLIED MORTAR

(A) General

This section refers to premixed sand and cement pneumatically applied by suitable mechanism and competent operators, and to which mixture the water is added immediately previous to its expulsion from the nozzle.

(B) Proportions

The proportion of cement to sand shall be based on dry and loose volumes and shall not be less than one to four for encasement of steel members, one to three for concrete repair, nor one to four and a half for special linings.

(C) Water Content

The water content shall be maintained at a practicable minimum and not in excess of 3 gallons per sack of cement as placed.

(D) Mixing

The cement and sand shall be thoroughly mixed before being charged into the machine. The sand shall contain not less than 3 nor more than 6 per cent moisture by weight.

(E) Nozzle Velocity

The velocity of the material as it leaves the nozzle must be maintained uniform at a rate determined for the given job conditions to produce minimum rebound.

(F) Nozzle Position

The nozzle shall be held in such a position and at such distance that the stream of flowing material will impinge at approximately right angles to the surface being covered without excessive impact.

(G) Rebound Sand

Rebound or accumulated loose sand shall be removed from the surface to be covered prior to placing of the original or succeeding layers of mortar.

(H) Forms

The forms shall be structurally sufficient and of such design that rebound or accumulated loose sand can freely escape or be readily removed. Shooting strips should be used at corners, edges, and on surfaces where necessary to obtain true lines and proper thickness.

(I) Joints

The pneumatically applied mortar at the end of any day's work or similar stopping periods shall be sloped off to a thin edge. Before placing an adjacent section this sloped portion shall be thoroughly cleaned and wetted.

(J) Bond

Surfaces to which pneumatically applied mortar is to be bonded shall be thoroughly cleaned of dirt, paint, grease, organic matter and loose particles. Absorptive surfaces shall be wetted before the application of the mortar.

(K) Curing

Pneumatically applied mortar shall be so applied, protected, and cured as to prevent its temperature falling below 50 F or a loss of moisture from the surface for the periods indicated below:
  (1) Where normal portland cement is used, 7 days.
  (2) Where high-early strength portland cement is used, 3 days.

Pneumatically applied mortar shall be applied only with the permission of the Engineer if the air temperature is 50 F or less.

(L) Reinforcement

The reinforcement, when required, shall be adequate from the standpoint of structural requirements and shall consist of mesh or round bars, spaced not less than 2 inches nor more than 4 inches apart either way, and having a diameter not less than that of No. 12 wire. The area of the reinforcement shall be at least 0.2 per cent of the cross-sectional area of the mortar. The reinforcement shall be at least ¼ inch from the unexposed surface of the mortar and at least ¾ inch from the exposed surface.

2.4.33 — PRESTRESSED CONCRETE

(A) General

The construction of prestressed concrete members shall conform to the requirements of preceding articles in this section except as those requirements are modified or supplemented by the provisions which follow.

(B) Supervision

Unless specifically permitted by the Engineer, the Contractor or fabricator shall provide a technician skilled in the use of the system of prestressing to be used who shall supervise the work and give the Engineer such assistance as in his judgment may be necessary.

(C) Equipment

The Contractor or fabricator shall provide all equipment necessary for the construction and the prestressing. Prestressing shall be done with approved jacking equipment. If hydraulic jacks are used, they shall be equipped with accurately reading pressure gages. The combination of jack and gage shall be calibrated and a graph or table showing the calibration shall be furnished the Engineer. Should other types of jacks be used, calibrated proving rings or other devices shall be furnished so that the jacking forces may be accurately known.

(D) Concrete

Concrete shall be controlled, mixed and handled as specified in other articles of this section unless otherwise specified herein.

Concrete shall not be deposited in the forms until the Engineer has inspected the placing of the reinforcement, conduits, anchorages, and prestressing steel and has given his approval thereof.

The concrete shall be vibrated internally or externally, or both, as ordered by the Engineer. The vibrating shall be done with care in such a manner as to avoid displacement of reinforcing, conduits, or wires.

(E) Steam Curing

Steam curing shall be done under a suitable enclosure to contain the live steam and minimize moisture and heat losses. The initial application of the steam shall be from two to four hours after the final placement of concrete to allow the initial set of the concrete to take place. If retarders are used, the waiting period before application of the steam shall be increased to from four to six hours. The steam shall be at 100 per cent relative humidity to prevent loss of moisture and to provide excess moisture for proper hydration of the cement. Application of the steam shall not be directly on the concrete. During application of the steam, the ambient air temperature shall increase at a rate not to exceed 40 F per hour until a maximum temperature of from 140 F to 160 F is reached. The maximum temperature shall be held until the concrete has reached the desired strength.

When the desired concrete strength has been reached, steam curing may be discontinued. The members shall be detensioned immediately after termination of steam curing, while the concrete and forms are still warm. In discontinuing the steam, the ambient air temperature shall not decrease at a rate to exceed 40 F per hour until a temperature has been reached about 20 F above the temperature of the air to which the concrete will be exposed. The concrete shall not be exposed to temperatures below freezing for six days after casting.

(F) Transportation and Storage

Precast girders should be transported in an upright position, and points of support and directions of the reactions with respect to the girder should be approximately the same during transportation and storage as when the girder is in its final position. In the event that the contractor deems it expedient to transport or store precast girders in other than this position, it shall be done at his own risk.

Care shall be taken during storage, hoisting, and handling of the precast units to prevent cracking or damage. Units damaged by improper storing or handling shall be replaced by the Contractor at his expense.

(G) Pretensioning Method

The prestressing elements shall be accurately held in position and stressed by jacks. A record shall be kept of the jacking force and the elongations produced thereby. Several units may be cast in one continuous line and stressed at one time. Sufficient space shall be left between ends of units to permit access for cutting after the concrete has attained the required strength. No bond stress shall be transferred to the concrete, nor shall end anchors be released, until the concrete has attained a compressive strength as shown by standard cylinders made and cured identically with the members, of at least the minimum strength shown on the plans or in the specifications for such transfer of load. The elements shall be cut or released in such an order that lateral eccentricity of prestress will be a minimum.

(H) Post-tensioning Method

The tensioning process shall be conducted so that the tension being applied and the elongation may be measured at all times. The friction loss shall be estimated as provided in Article 1.6.7. A record shall be kept of gage pressures and elongations at all times and submitted to the Engineer for his approval. Loads shall not be applied to the concrete until it has attained strength as specified in Article 2.4.33 (G) for pretensioning method.

(I) Grouting of Bonded Steel

Post-tensioned prestressed bridge members preferably shall be of the bonded type in which the tensioned steel is installed in holes or flexible metal ducts cast in the concrete and bonded to the surrounding concrete by filling the tubes or ducts with grout. The grout shall be a mixture of cement and fine sand (passing a No. 30 sieve) in the approximate proportions of one part cement to 0.75 part sand, the exact proportion to be adjusted to form a grout having the proper consistency.

All prestressing reinforcement to be bonded shall be free of dirt, loose rust, grease, or other deleterious substances. Before grouting, the ducts shall be free of water, dirt or any other foreign substance. The ducts shall be blown out with compressed air until no water comes through the duct. For long members with draped strands an open tap at the low point of the duct may be necessary.

The grout shall be fluid (consistency of thick paint) but proportioned so that free water will not separate out of the mix. Unpolished aluminum powder may be added in an amount of one to two teaspoons per sack of cement. Commercial plasticizers used in accordance with the manufacturer's recommendation may be used provided they contain no ingredients that are corrosive to steel. Sufficient pressure shall be used in grouting to force the grout completely through the duct, care being taken that rupturing of the ducts does not occur.

(J) Prestressing Reinforcement

Prestressing reinforcement shall be high-tensile-strength steel wire, high-tensile-strength seven-wire strand or high-tensile-strength alloy bars as called for on the plans or in the special provisions.

High-tensile-strength steel wire shall conform to AASHO M 204 (ASTM A 421).

High-tensile-strength seven-wire strand shall conform to the requirements of AASHO M 203 (ASTM A 416).

High-tensile-strength alloy bars shall be stress relieved and then cold stretched to a minimum of 130,000 psi. After cold stretching, the physical properties shall be as follows:

Minimum ultimate tensile strength	145,000 psi
Minimum yield strength, measured by the 0.7 percent extension under load method shall be not less than	130,000 psi
Minimum modulus of elasticity	25,000,000 psi
Minimum elongation in 20-bar diameters after rupture	4 percent
Diameter tolerance	$+0.03'', -0.01''$

(K) Testing Prestressing Reinforcement and Anchorages

All wire, strand, or bars to be shipped to the site shall be assigned a lot number and tagged for identification purposes. Anchorage assemblies to be shipped shall be likewise identified.

All samples submitted shall be representative of the lot to be furnished and, in the case of wire or strand, shall be taken from the same master roll.

All of the materials specified for testing shall be furnished free of cost and shall be delivered in time for tests to be made well in advance of anticipated time of use.

Where the Engineer intends to require nondestructive testing of one or more parts of the structure, special specifications shall be drawn giving the required details of the work.

The vendor shall furnish for testing the following samples selected **from each lot. If ordered by the Engineer, the selection of samples shall** be made at the manufacturer's plant by the inspector.

Pretensioning method.—For pretensioned strands, one sample at least 7 feet long shall be furnished in accordance with the requirements of paragraph 7.1 of AASHO M-203.

Post-tensioning method.—The following lengths shall be furnished:

For wires requiring heading—5 feet.

For wires not requiring heading—sufficient length to make up one parallel-lay cable 5 feet long consisting of the same number of wires as the cable to be furnished.

For strand to be furnished with fittings—5 feet between near ends of fittings.

For bars to be furnished with threaded ends and nuts—5 feet between threads at ends.

Anchorage assemblies.—Two anchorage assemblies shall be furnished, complete with distribution plates of each size or type to be used, if anchorage assemblies are not attached to reinforcement samples.

2.4.34 — MEASUREMENT AND PAYMENT

The payment for concrete of the various classes shall include compensation for all equipment, tools, material, falsework, forms, bracing, labor, surface finish and all other items of expense required to complete the concrete work shown on the plans, with the exception of reinforcement steel. The payment for concrete shall include the cost of joint fillers, metal drains, expansion joints and miscellaneous metal devices unless they are covered by other items in the contract. The quantity of concrete involved in fillets, scorings and chamfers 1 square inch or less in cross-sectional area shall be neglected. Payment will be made on the basis of the actual yardage within the neat lines of the structure as shown on the plans or revised by authority of the Engineer, except that deduction shall be made as follows:

(1) The volume of structural steel, including steel piling, encased in concrete.

(2) The volume of timber piles encased in concrete, assuming the volume to be .8 cubic foot per linear foot of pile.

(3) The volume of concrete piles encased in concrete.

No deduction shall be made for the volume of concrete displaced by steel reinforcement, floor drains, or expansion joint material. If a bid is asked on handrailing, that portion of the railing above the top of the roadway curb or above the surface of the sidewalk, as the case may be, shall not be included in the yardage of concrete, but shall be paid for as handrailing. Massive pylons or posts which are to be excepted from handrailing payment shall be so noted on the plans.

Payment for pneumatically applied mortar will be made on the basis of the actual number of square feet placed and accepted. The payment for pneumatically applied mortar shall include compensation for all equipment, tools, materials, labor and incidentals necessary to complete the work and shall include metal reinforcement unless otherwise provided.

Section 5—REINFORCEMENT

2.5.1 — MATERIAL

(A) Bar Reinforcement

Bar reinforcement for concrete in sizes up to and including No. 18 shall conform to the requirements of the Specifications for Deformed Billet-Steel Bars for Concrete Reinforcement, AASHO M 31 (ASTM A 615); or Rail-Steel Bars for Concrete Reinforcement AASHO M 42 with the following modifications:

(1) The use of cold twisted bars is not permitted.

(2) Steel for all bars shall be made by the open-hearth, electric furnace or basic oxygen process, unless otherwise called for in the special provisions or on the plans.

(B) Wire and Wire Mesh

Wire shall conform to the Specification for Cold-Drawn Steel Wire for Concrete Reinforcement, of the AASHO M 32 (ASTM A 82).

Wire mesh, when used as reinforcement in concrete shall conform to the Specification for Welded Steel Wire Fabric for Concrete Reinforcement of the AASHO M 55 (ASTM A 185). The type of mesh shall be approved by the Engineer.

(C) Bar Mat Reinforcement

Bar mat reinforcement for concrete shall conform to the Specification for Fabricated Steel Bar or Rod Mats for Concrete Reinforcement of the AASHO M 54 (ASTM A 184).

(D) Structural Shapes

Structural shapes used as reinforcement in concrete shall conform to the requirements for structural steel as provided in these specifications.

2.5.2 — ORDER LISTS

Before ordering material, all order lists and bending diagrams shall be furnished by the Contractor for the approval of the Engineer, and no materials shall be ordered until such lists and bending diagrams have been approved. The approval of order lists and bending diagrams by the Engineer shall in no way relieve the Contractor of responsibility for the correctness of such lists and diagrams. Any expense incident to the revision of material furnished in accordance with such lists and diagrams to make it comply with the design drawings shall be borne by the Contractor.

2.5.3 — PROTECTION OF MATERIAL

Steel reinforcement shall be protected at all times from injury. When placed in the work, it shall be free from dirt, detrimental scale, paint, oil or other foreign substance. However, when steel has, on its surface, detrimental rust, loose scale and dust which is easily removable, it may be cleaned by a satisfactory method, if approved by the Engineer.

2.5.4 — FABRICATION

Bar reinforcement shall be bent to the shapes shown on the plans.
All bars shall be bent cold, unless otherwise permitted by the Engineer. No bars partially imbedded in concrete shall be field bent except as shown on the plans or specifically permitted by the Engineer.

The radii of bend measured on the inside of the bar for standard hooks shall be not less than the following:

Bar Size	Minimum Radii
3, 4 or 5	$2\frac{1}{2}$ bar diameters
6, 7 or 8	3 bar diameters ⎱ $2\frac{1}{2}$ bar diameters for billet steel
9, 10 or 11	4 bar diameters ⎰
14 or 18	5 bar diameters

Special fabrication is required for bends exceeding 90 degrees for #14 and #18 sizes and grades having a specified yield point of 50,000 psi.

Bends for stirrups and ties shall have radii on the inside of the bar not less than one bar diameter. Bends for all other bars shall have radii on the inside of the bar not less than the values tabulated in the preceding paragraphs.

Bar reinforcement shall be shipped in standard bundles, tagged and marked in accordance with the Code of Standard Practice of the Concrete Reinforcement Steel Institute.

2.5.5 — PLACING AND FASTENING

All steel reinforcement shall be accurately placed in the positions shown on the plans and firmly held during the placing and setting of concrete. When placed in the work it shall be free from dirt, detrimental rust, loose scale, paint, oil or other foreign material. Bars shall be tied at all intersections except where spacing is less than 1 foot in each direction, in which case alternate intersections shall be tied.

Distances from the forms shall be maintained by means of stays, blocks, ties, hangers, or other approved supports. Blocks for holding reinforcement from contact with the forms shall be precast mortar blocks of approved shape and dimensions or approved metal chairs. Metal chairs which are in contact with the exterior surface of the concrete shall be galvanized. Layers of bars shall be separated by precast mortar blocks or by other equally suitable devices. The use of pebbles, pieces of broken stone or brick, metal pipe and wooden blocks shall not be permitted. The minimum spacing of bars shall be as specified in Article 1.5.6 (A). Reinforcement in any member shall be placed and then inspected and approved by the Engineer before the placing of concrete begins. Concrete placed in violation of this provision may be rejected and removal required.

If fabric reinforcement is shipped in rolls, it shall be straightened into flat sheets before being placed.

Bundled bars shall be tied together at not more than six feet centers.

2.5.6 — SPLICING

All reinforcement shall be furnished in the full lengths indicated on the plans. Splicing of bars, except where shown on the plans, will not be permitted without the written approval of the Engineer. Splices shall be staggered as far as possible.

Unless otherwise shown on the plans, bars shall be spliced in accordance with Article 1.5.6 (C). In lapped splices, the bars shall be placed in contact and wired together in such a manner as to maintain a clearance of not less than the minimum clear distance to other bars and the minimum distance to the surface of the concrete specified in Article 1.5.6(B).

2.5.7 — LAPPING

Sheets of mesh or bar mat reinforcement shall overlap each other sufficiently to maintain a uniform strength and shall be securely fastened at the ends and edges. The edge lap shall not be less than one mesh in width.

2.5.8 — SUBSTITUTIONS

Substitution of different size bars will be permitted only with specific authorization by the Engineer. The substituted bars shall have an area equivalent to the design area, or larger.

2.5.9 — MEASUREMENT

Steel reinforcement incorporated in the concrete masonry will be measured in pounds based on the total computed weight for the sizes and lengths of bars, mesh or mats shown on the plans or authorized.

The weight of mesh will be computed from the theoretical weight of plain wire. If the weight per square foot is given on the plan, that weight shall be used.

The weight of plain bars or bar mat or of deformed bars which do not comply with AASHO M 31 (ASTM A 615) will be computed from the theoretical weight of plain round or square bars of the same nominal size as shown in the following table:

Size		1/4"	3/8"	1/2"	5/8"	3/4"	7/8"	1"	1 1/8"	1 1/4"	1 1/2"
Weight in pounds per foot	Round	0.167	0.376	0.668	1.043	1.502	2.044	2.670			
	Square			0.850					3.400	4.303	5.313	7.650

The weight of bars which comply with AASHO M 31 (ASTM A 615) will be calculated as follows:

Bar	No.3	No. 4	No. 5	No. 6	No. 7	No.8	No. 9	No.10	No. 11	No. 14	No. 18
Weight lbs. per lin. ft.	.376	.668	1.043	1.502	2.044	2.670	3.400	4.303	5.313	7.65	13.60

The weight of reinforcement used in railings shall not be included when railings are paid for on a linear foot basis. The weight of reinforcement in precast piles and other items where the reinforcement is included in the contract price for the item shall not be included.

No allowance will be made for clips, wire, separators, wire chairs, and other material used in fastening the reinforcement in place. If bars are substituted upon the contractor's request and as a result more steel is used than specified, only the amount specified shall be included.

When laps are made for splices, other than those shown on the plans, for the convenience of the Contractor, the extra steel shall not be included.

2.5.10 — PAYMENT

Payment for reinforcement as determined under measurement shall be made at the contract price per pound. Payment shall include the cost of furnishing, fabricating and placing of the reinforcement.

Section 6—ASHLAR MASONRY

2.6.1 — DESCRIPTION

Ashlar masonry shall consist of first-class cut stone masonry laid in regular courses and shall include all work in which, as distinguished from rubble masonry, the individual stones are dressed or tooled to exact dimensions.

2.6.2 — MATERIALS

(A) Ashlar Stone

Stone for ashlar masonry shall be of the kind specified on the plans or in the contract. The stone shall be tough, dense, sound and durable, resistant to weathering action, reasonably fine grained, uniform in color, and free from seams, cracks, pyrite inclusions, or other structural defects. Preferably, stone shall be from a quarry, the product of which is known to be of satisfactory quality. Stone shall be of such character that it can be wrought to such lines and surfaces, whether curved or plane, as may be required. Any stone having defects which have been repaired with cement or other materials shall be rejected.

Each bidder shall submit with his bid a 6-inch cubical block of the stone he proposes to furnish and shall designate the quarry from which it is obtained. The quality of the stone furnished shall be at least equal to that of the sample. The sample shall be squared and dressed on three sides; one side shall be smooth-finished, one side fine-finished, and one side shall be given the finish indicated on the plans for exposed surfaces of face stone. The remaining sides shall be left with quarry face.

When permitted by the Engineer, bidders may submit bids, accompanied by samples as specified above, on kinds of stone other than that specified.

The stone shall be kept free from dirt, oil or any other injurious material which may prevent the proper adhesion of the mortar or detract from the appearance of the exposed surfaces.

(B) Mortar

Mortar for laying the stone and pointing shall be composed of one part of portland cement and three parts of sand unless otherwise provided. The sand shall conform to the requirements of Article 2.4.2(C).

2.6.3 — SIZE OF STONE

The individual stones shall be large and well proportioned. They shall not be less than 12 nor more than 30 inches in thickness. The thicknesses of course, if varied, shall diminish regularly from bottom to top of wall. The size of ring stones in arches shall be as shown on the plans.

2.6.4 — SURFACE FINISHES OF STONE

For the purpose of this specification the surface finishes of stone are defined as follows:

Smooth-finished: Having a surface in which the variations from the pitch line do not exceed 1/16 inch.

Fine-finished: Having a surface in which the variations from the pitch line do not exceed 1/4 inch.

Rough-finished: Having a surface in which the variations from the pitch line do not exceed 1/2 inch.

Scabbled: Having a surface in which the variations from the pitch line do not exceed 3/4 inch.

Rock-faced: Having an irregular projecting face without indications of tool marks. The projections beyond the pitch line shall not exceed 3 inches and no part of the face shall recede back of the pitch line.

2.6.5 — DRESSING STONE

Stones shall be dressed to exact sizes and shapes before being laid and shall be cut to lie on their natural beds with top and bottom truly parallel. Hollow beds will not be permitted. The bottom bed shall be the full size of the stone and no stone shall have an overhanging top. In rock-face construction the face side of any stone shall not present an undercut contour adjacent to its bottom arris giving a top-heavy, unstable appearance when laid.

Beds of face stone shall be fine-finished for a depth of not less than 12 inches.

Vertical joints of face stone shall be fine-finished and full to the square for a depth of not less than 9 inches.

Exposed surfaces of the face stone shall be given the surface finish indicated on the plans, with edges pitched to true lines and exact batter. Chisel drafts 1½ inches wide shall be cut at all exterior corners. Face stone forming the starling or nosing of piers shall be rough-finished unless otherwise specified.

Holes for stone hooks shall not be permitted to show in exposed surfaces.

2.6.6 — STRETCHERS

Stretchers shall have a width of bed of not less than 1½ times their thickness. They shall have a length of bed not less than twice nor more than 3½ times their thickness, and not less than 3 feet.

2.6.7 — HEADERS

Headers shall be placed in each course and shall have a width of not less than 1½ times their thickness. In walls having a thickness of 4 feet or less, the headers shall extend entirely through the wall. In walls of greater thickness, the length of headers shall be not less than 2½ times their thickness when the course is 18 inches or less in height, and not less than 4 feet in courses of greater height. Headers shall bond with the core or backing not less than 12 inches. Headers shall hold in the heart of the wall the same size shown in the face and shall be spaced not further apart than 8 feet center to center. There shall be at least one header to every two stretchers.

2.6.8 — CORES AND BACKING

Cores and backing shall consist either of roughly bedded and jointed headers and stretchers, as specified above, or of Class B or C concrete, as may be specified.

When stone is used for cores or backing, at least ½ of the stone shall be of the same size and character as the face stone, and with parallel ends. No course shall be less than 8 inches thick.

Concrete used for cores and backing shall conform to the requirements specified in Section 2.4.

The headers and stretchers in walls having a thickness of 3 feet or less shall have a width or length equal to the full thickness of the wall. No backing will be allowed.

2.6.9 — MIXING MORTAR

The mortar shall be hand or machine mixed, as may be required by the Engineer. In the preparation of hand-mixed mortar, the sand and cement shall be thoroughly mixed together in a clean, tight mortar box until the mixture is of uniform color, after which clean water shall be added in such quantity as to form a stiff plastic mass. Machine-mixed mortar shall be prepared in an approved mixer and shall be mixed not less than 1½ minutes. Mortar shall be used within 45 minutes after mixing. Retempering of mortar will not be permitted.

2.6.10 — LAYING STONE

(A) General

Stone masonry shall not be constructed in freezing weather or when the stone contains frost, except by written permission of the Engineer and subject to such conditions as he may require.

(B) Face Stone

Stone shall not be dropped upon, or slid over, the wall, nor will hammering, rolling or turning of stones on the wall be allowed. They shall be carefully set without jarring the stone already laid and they shall be handled with a lewis or other appliance which will not cause disfigurement.

Each stone shall be cleaned and thoroughly saturated with water before being set and the bed which is to receive it shall be cleaned and well moistened. All stones shall be well bedded in freshly made mortar and settled in place with a suitable wooden maul before the setting of the mortar. Whenever possible, the face joints shall be properly pointed before the mortar sets. Joints which cannot be so pointed shall be prepared for pointing by raking them out to a depth of 2 inches before the mortar has set. The face surfaces of stones shall not be smeared with the mortar forced out of the joints or that used in pointing. No pinning up of stones with spalls will be permitted and no spalls will be permitted in beds.

Joints and beds shall be not less than ⅜ inch nor more than ½ inch in thickness and the thickness of the joint or bed shall be uniform throughout.

The stones in any one course shall be placed so as to form bonds of not less than 12 inches with the stones of adjoining courses. Headers shall be placed over stretchers and, in general, the headers of each course shall equally divide the spaces between the headers of adjoining courses, but no header shall be placed over a joint and no joint shall be made over a header.

(C) Stone Backing and Cores

Stone backing shall be laid in the same manner as specified above for face stone, with headers interlocking with face headers when the thickness of the wall will permit. Backing shall be laid to break joints with the face stone. Stone cores shall be laid in full mortar beds so as to bond not less than 12 inches with face and backing stone and with each other. Bed joints in cores and backing shall not exceed 1 inch and vertical joints shall not exceed 4 inches in thickness.

(D) Concrete Cores and Backing

The operations involved in the handling and placing of concrete used in cores and backing shall conform to the requirements specified in Section 2.4. However, the puddling and compacting of concrete adjacent to the ashlar masonry facing shall be done in a manner that will insure the filling of all spaces around the stones and secure full contact and efficient bond with all stone surfaces.

2.6.11 — LEVELING COURSES

Stone cores and backing shall be carried up to the approximate level of the face course before the succeeding course is started.

The construction joints produced in concrete cores or backing by the intermittent placing of concrete shall be located, in general, not less than 6 inches below the top bed of any course of masonry.

2.6.12 — RESETTING

In case any stone is moved or the joint broken, the stone shall be taken up, the mortar thoroughly cleaned from bed and joints, and the stone reset in fresh mortar.

2.6.13 — DOWELS AND CRAMPS

Where required, coping stone, stone in the wings of abutments and stone in piers shall be secured with wrought-iron cramps or dowels as indicated on the plans.

Dowel holes shall be drilled through each stone before the stone is placed and, after it is in place, such dowel holes shall be extended by drilling into the underlying course not less than 6 inches.

Cramps shall be of the shapes and dimensions shown on the plans or approved by the Engineer. They shall be inset in the stone so as to be flush with the surfaces.

Cramps and dowels shall be set in lead, care being taken to completely fill the surrounding spaces with the molten metal.

2.6.14 — COPINGS

Stones for copings of wall, pier and abutment bridge seats shall be carefully selected and fully dimensioned stones. On piers, not more than two stones shall be used to make up the entire width of coping. The copings of abutment bridge seats shall be of sufficient width to extend at least 4 inches under the backwall. Each step forming the coping of a wing wall shall be formed by a single stone which shall overlap the stone forming the step immediately below it at least 12 inches.

Tops of copings shall be given a bevel cut at least 2 inches wide, and beds, bevel cuts and tops shall be fine-finished. The vertical joints shall be smooth-finished and the copings shall be laid with joints not more than ¼ inch in thickness. The under sides of projecting copings, preferably, shall have a drip bead.

Joints in copings shall be located so as to provide not less than a 12-inch bond with the stones of the under course and so that no joint will come directly under the superstructure masonry plates.

2.6.15 — ARCHES

The number of courses and the depth of voussoirs shall be as shown on the plans. Voussoirs shall be placed in the order indicated, shall be full size throughout, dressed true to template, and shall have bond not less than the thickness of the stone. Beds and joints shall be fine-finished and mortar joints shall not exceed ¾ inch in thickness. Exposed surfaces of the intrados and arch ring shall be given the surface finish indicated on the plans.

Backing may consist of Class B concrete or of large stones shaped to fit the arch, bonded to the spandrels, and laid in full beds of mortar. The extrados and interior faces of the spandrel walls shall be given a finishing coat of 1:2½ cement mortar which shall be trowelled smooth to receive the waterproofing.

Arch centering, waterproofing, drainage and filling shall be as specified for concrete arches.

2.6.16 — POINTING

Pointing shall not be done in freezing weather nor when the stone contains frost.

Joints not pointed at the time the stone is laid shall be thoroughly wet with clean water and filled with mortar. The mortar shall be well driven into the joints and finished with an approved pointing tool. The wall shall be kept wet while pointing is being done and in hot or dry weather the pointed masonry shall be protected from the sun and kept wet for a period of at least three days after completion.

After the pointing is completed and the mortar set, the wall shall be thoroughly cleaned and left in a neat and workmanlike condition.

2.6.17 — MEASUREMENT AND PAYMENT

The quantity of stone masonry to be paid for under this item shall be the number of cubic yards measured in the completed work and the limiting dimensions shall not exceed those shown upon the plans or fixed by the Engineer. The contract price shall include all labor, tools, materials and other expense incidental to the satisfactory completion of the work.

Section 7—MORTAR RUBBLE MASONRY

2.7.1 — DESCRIPTION

Mortar rubble masonry, as here specified, shall include the classes commonly known as coursed, random and random range work and shall consist of roughly squared and dressed stone laid in cement mortar.

2.7.2 — MATERIALS

(A) Rubble Stone

Stone for mortar rubble or dry rubble masonry shall be of approved quality, sound and durable, and free from segregations, seams, cracks, and other structural defects or imperfections tending to destroy its resistance to the weather. It shall be free from rounded, worn, or weathered surfaces. All weathered stone shall be rejected.

The stone shall be kept free from dirt, oil, or any other injurious material which may prevent the proper adhesion of the mortar.

(B) Mortar

The mortar used shall conform as regards materials, proportions and mixing to the mortar specified in Article 2.6.2(B).

2.7.3 — SIZE

Individual stones shall have a thickness of not less than 8 inches and a width of not less than $1\frac{1}{2}$ times the thickness. No stones, except headers, shall have a length less than $1\frac{1}{2}$ times their width. Stones shall decrease in thickness from bottom to top of wall.

The size of ring stones for arches shall be as shown on the plans.

2.7.4 — HEADERS

Headers shall hold in the heart of the wall the same size shown in the face and shall extend not less than 12 inches into the core or backing.

They shall occupy not less than ⅕ of the face area of the wall and shall be evenly distributed. Headers in walls 2 feet or less in thickness shall extend entirely through the wall.

2.7.5 — SHAPING STONE

The stones shall be roughly squared on joints, beds and faces. Selected stone, roughly squared and pitched to line, shall be used at all angles and ends of walls. If specified, all corners or angles in exterior surfaces shall be finished with a chisel draft.

All shaping or dressing of stone shall be done before the stone is laid in the wall, and no dressing or hammering which will loosen the stone will be permitted after it is placed.

2.7.6 — LAYING STONE

Stone masonry shall not be constructed in freezing weather or when the stone contains frost, except by written permission of the Engineer and subject to such conditions as he may require.

The masonry shall be laid to line and in courses roughly leveled up. The bottom or foundation courses shall be composed of large, selected stones and all courses shall be laid with bearing beds parallel to the natural bed of the material.

Each stone shall be cleaned and thoroughly saturated with water before being set and the bed which is to receive it shall be clean and well moistened. All stones shall be well bedded in freshly made mortar. The mortar joints shall be full and the stones carefully settled in place before the mortar has set. No spalls will be permitted in the beds. Joints and beds shall have an average thickness of not more than 1 inch.

Whenever possible the face joints shall be properly pointed before the mortar becomes set. Joints which cannot be so pointed shall be prepared for pointing by raking them out to a depth of 2 inches before the mortar has set. The face surfaces of stones shall not be smeared with the mortar forced out of the joints or that used in pointing.

The vertical joints in each course shall break with those in adjoining courses at least 6 inches. In no case shall a vertical joint be so located as to occur directly above or below a header.

In case any stone is moved or the joint broken, the stone shall be taken up, the mortar thoroughly cleaned from bed and joints, and the stone reset in fresh mortar.

2.7.7 — COPINGS, BRIDGE SEATS AND BACKWALLS

Copings, bridge seats and backwalls shall be of the materials shown on the plans and when not otherwise specified shall be of Class A concrete which shall conform to the requirements of Section 2.4.

Concrete copings shall be made in sections extending the full width of the wall, not less than 12 inches in thickness, and from 5 to 10 feet long. The sections may be cast in place or precast and set in place in full mortar beds.

2.7.8 — ARCHES

The number of courses and the depth of voussoirs shall be as shown on the plans. Voussoirs shall be placed in the order indicated, shall be full size throughout and shall have bond not less than their thickness. Beds shall be roughly pointed to bring them to radial planes. Radial joints shall be in planes parallel to the transverse axis of the arch and, when measured at the intrados, shall not exceed ¾ inch in thickness. Joints perpendicular to the arch axis shall not exceed 1 inch in thickness when measured at the intrados. The intrados face shall be dressed sufficiently to permit the stone to rest properly upon the centering. Exposed faces of the arch ring shall be rock-faced with edges pitched to true lines.

The work shall be carried up symmetrically about the crown, the stone being laid in full mortar beds and the joints grouted where necessary. Pinning by the use of stone spalls will not be permitted.

Backing may consist of Class B concrete or of large stones shaped to fit the arch, bonded to the spandrels, and laid in full beds of mortar. The extrados and interior faces of the spandrel walls shall be given a finished coat of 1:2½ cement mortar which shall be trowelled smooth to receive the waterproofing.

Arch centering, waterproofing, draining and filling shall be as specified for concrete arches.

2.7.9 — POINTING

Pointing shall not be done in freezing weather or when the stone contains frost.

Joints not pointed at the time the stone is laid shall be thoroughly wet with clean water and filled with mortar. The mortar shall be well driven into the joints and finished with an approved pointing tool. The wall shall be kept wet while pointing is being done and in hot or dry weather the pointed masonry shall be protected from the sun and kept wet for a period of at least three days after completion.

After the pointing is completed and the mortar set, the wall shall be thoroughly cleaned and left in a neat and workmanlike condition.

2.7.10 — MEASUREMENT AND PAYMENT

The quantity of stone masonry to be paid for under this item shall be the number of cubic yards measured in the completed work and the limiting dimensions shall not exceed those shown upon the plans or fixed by the engineer. The contract price shall include all labor, tools, materials and other items incidental to the satisfactory completion of the work.

Concrete used in connection with rubble masonry shall be paid for as in the case of other concrete construction.

Section 8—DRY RUBBLE MASONRY

2.8.1 — DESCRIPTION

Dry rubble masonry as here specified shall include the classes commonly known as coursed, random and random range work and shall consist of roughly squared and dressed stone laid without mortar.

2.8.2 — MATERIALS

Stone for mortar rubble or dry rubble masonry shall be of approved quality, sound and durable, and free from segregations, seams, cracks, and other structural defects or imperfections tending to destroy its resistance to the weather. It shall be free from rounded, worn, or weathered surfaces. All weathered stone shall be rejected.

2.8.3 — SIZE OF STONE

The stones shall conform in size to the requirements specified in Section 2.7.

2.8.4 — HEADERS

Headers shall conform to the requirements specified in Section 2.7.

2.8.5 — SHAPING STONE

The stones shall be roughly squared on joints, beds and faces. Selected stone, roughly squared and pitched to line, shall be used at all angles and ends of walls.

2.8.6 — LAYING STONE

The masonry shall be laid to line and in courses roughly leveled up. The bottom or foundation courses shall be composed of large, selected stones and all courses shall be laid with bearing beds parallel to the natural bed of the material. Face joints shall not exceed 1 inch in width.

In laying dry rubble masonry, care shall be taken that each stone takes a firm bearing at not less than three separate points upon the underlying course. Open joints, both front and rear, shall be "chinked" with spalls fitted to take firm bearing upon their top and bottom surfaces, for the purpose of securing firm bearing throughout the length of the stone.

When required by the terms of the contract, the open joints on the rear surfaces of abutments or retaining walls shall be "slushed" thoroughly with mortar to prevent seepage of water through the joints.

2.8.7 — COPINGS, BRIDGE SEATS AND BACKWALLS

Copings, bridge seats and backwalls, when used in connection with dry rubble masonry, shall conform to the requirements specified in Section 2.7.

2.8.8 — MEASUREMENT AND PAYMENT

The quantity of stone masonry to be paid for under this item shall be the number of cubic yards measured in the completed work and the limiting dimensions shall not exceed those shown upon the plans or fixed by the Engineer. The contract price shall include all labor, tools, materials and other expense incidental to the satisfactory completion of the work.

Concrete used in connection with rubble masonry shall be paid for as in the case of other concrete construction.

Section 9—BRICK MASONRY

2.9.1 — DESCRIPTION

Brick masonry shall consist of brick laid in cement mortar and shall include such construction with building brick or ornamental brick as may be specified. Brick pavements are not included under this designation.

2.9.2 — MATERIALS

(A) Brick

Brick for masonry construction shall conform to the Specification for Building Brick (made from clay or shale) for the AASHO M 114. The grade of brick to be furnished shall be as specified in the special provisions.

The brick shall have a fine-grained, uniform, and dense structure, free from lumps of lime, laminations, cracks, checks, soluble salts, or other defects which may in any way impair their strength, durability, appearance, or usefulness for the purpose intended. Bricks shall emit a clear, metallic ring when struck with a hammer.

(B) Mortar

The mortar used shall conform, as regards materials, proportions and mixing, to the mortar specified in Article 2.6.2(B).

2.9.3 — CONSTRUCTION

The brick shall be laid in such manner as will thoroughly bond them into the mortar by means of the "shove-joint" method; "buttered" or plastered joints will not be permitted. All brick must be thoroughly saturated with water before being laid. The arrangement of headers and stretchers shall be such as will thoroughly bond the mass and, unless otherwise specified, brick work shall be of alternate headers and stretchers with consecutive courses breaking joints. Other types of bonding, as for ornamental work, shall be as specified on the plans.

All joints shall be completely filled with mortar. They shall not be less than ¼ inch and not more than ½ inch in thickness and the thickness shall be uniform throughout. All joints shall be finished properly as the work progresses and on exposed faces they shall be neatly struck, using the "weather" joint.

No spalls or bats shall be used except for shaping around irregular openings or when unavoidable to finish out a course, in which case full bricks shall be placed at the corners, the bats being placed in the interior of the course.

Piers and walls may be built of solid brick work, or may consist of a brick shell backed with concrete or other suitable material as specified on the plans. None but expert brick layers shall be employed on the work and all details of the construction shall be in accordance with the most approved practice and to the satisfaction of the Engineer.

2.9.4 — COPINGS, BRIDGE SEATS AND BACKWALLS

The tops of retaining walls, abutment wing walls and similarly exposed brick work shall be provided, in general, with either a stone or concrete coping. The underside of the coping shall have a batter or drip bead, at least 1 inch beyond the face of the brick work wall. The coping upon an abutment backwall will commonly have no projection beyond its bridge seat face. When concrete is used it shall be of Class A quality. For thin copings, mortar of the same proportions as used for laying the brick may be used to produce precast sections not less than 3 feet nor more than 5 feet in length. No coping shall be less than 4 inches thick.

Copings of piers and abutment bridge seats shall be of Ashlar stone work or of Class A concrete and shall conform to the requirements for "Ashlar Masonry" or "Concrete Masonry" as the plan may indicate. When not shown upon the plans, concrete shall be used.

2.9.5 — MEASUREMENT AND PAYMENT

The quantity of brick work to be paid for under this item shall be the number of cubic yards of brick masonry actually placed in the structure in accordance with the plans or as modified by written instructions from the Engineer. This price shall include all labor, materials and other expense incidental to the satisfactory completion of the work. Filling material for the interior of the wall, when not of brick, and concrete or mortar copings, shall be paid for on the basis of the number of cubic yards actually placed.

Section 10 — STEEL STRUCTURES

FABRICATION

2.10.1 — TYPE OF FABRICATION

These specifications apply to riveted, bolted, and welded construction.

2.10.2 — QUALITY OF WORKMANSHIP

Workmanship and finish shall be equal to the best general practice in modern bridge shops.

2.10.3 — MATERIALS
(A) Structural Steel

(1) General
Steel shall be furnished according to the following specifications. Unless otherwise specified, structural carbon steel and structural rivet steel shall be furnished.

(2) Structural Steel

(a) Carbon Steel
Unless otherwise specified, structural carbon steel for riveted, bolted or welded construction shall conform to:
Structural Steel, AASHO M183 (ASTM A36).

(b) Eyebars
Steel for eyebars shall be of a weldable grade. These grades include structural steel conforming to:

Structural Steel, AASHO M183 (ASTM A36)

High Strength Low Alloy Structural Steel, of a weldable quality, AASHO M161 (ASTM A242)

High-Strength Low Alloy Structural Manganese Vanadium Steel, AASHO M188 (ASTM A441)

High-Strength Low-Alloy Structural Steel with 50,000 psi Minimum Yield Point to 4 In. Thick, AASHO M222 (ASTM A588 with Supplementary Requirement S1 of AASHO M222 mandatory.)

(3) High Strength Low Alloy Structural Steel
High strength low alloy structural steel shall conform to:

High Strength Low Alloy Structural Steel, AASHO M161 (ASTM A242)

High-Strength Structural Steel, AASHO M187 (ASTM A440)

High-Strength Low Alloy Structural Manganese Vanadium Steel, AASHO M188 (ASTM A441)

High-Strength Low-Alloy Columbium-Vanadium Steels of Structural Quality, AASHO M223 (ASTM A572)

High-Strength Low-Alloy Structural Steel with 50,000 psi Minimum Yield Point to 4 In. Thick, AASHO M222 (ASTM A588)

(4) High Strength Low Alloy Structural Steel for Welding
High strength low alloy structural steel for welding shall conform to:

High-Strength Low Alloy Structural Manganese Vanadium Steel, AASHO M188 (ASTM A441)

High Strength Low Alloy Structural Steel, AASHO M161 (ASTM A242) of a weldable quality.

High-Strength Low-Alloy Columbium-Vanadium Steels of Structural Quality, Grades 42, 45, and 50, AASHO M223 (ASTM A572 with Supplementary Requirement S2 of AASHO M223 mandatory)

High-Strength Low-Alloy Structural Steel with 50,000 psi Minimum Yield Point to 4 In. Thick, AASHO M222 (ASTM A588 with Supplementary Requirement S1 of AASHO M222 mandatory)

(5) High Strength Structural Steel for Riveted or Bolted Construction

High strength structural steel for riveted or bolted construction shall conform to:

    High-Strength Structural Steel, AASHO M187 (ASTM A 440)

    High Strength Low Alloy Structural Steel, AASHO M161 (ASTM A242)

    High-Strength Low-Alloy Columbium-Vanadium Steels of Structural Quality, AASHO M223 (ASTM A572)

    High-Strength Low-Alloy Structural Steel with 50,000 psi Minimum Yield Point to 4 In. Thick, AASHO M222 (ASTM A588)

    High-Strength Low Alloy Structural Manganese Vanadium Steel, AASHO M188 (ASTM A441)

(6) High-Yield-Strength, Quenched and Tempered Alloy Steel Plate

High yield strength, quenched and tempered alloy steel plate shall conform to:

    High-Yield-Strength, Quenched and Tempered Alloy Steel Plate, suitable for welding, ASTM A514

    High-Strength Alloy Steel Plates, Quenched and Tempered for pressure vessels, ASTM A517

    Quenched and tempered alloy steel structural shapes and seamless mechanical tubing, meeting all of the mechanical and chemical requirements of A514/A517 steel, except that the specified maximum tensile strength may be 140,000 psi for structural shapes and 145,000 psi for seamless mechanical tubing, shall be considered as A514/A517 steel.

(7) Structural Rivet Steel

Structural rivet steel shall conform to:

    Steel Structural Rivets AASHO M 228, Grade 1 (ASTM A502, grade 1).

(8) High Strength Structural Rivet Steel

High strength structural rivet steel shall conform to:

    Steel Structural Rivets AASHO M 228, Grade 2 (ASTM A502, grade 2).

(9) High Strength Bolts

Bolts, nuts and circular washers shall conform to:

    High Strength Bolts for Structural Steel Joints, including Suitable Nuts and Plain Hardened Washers, AASHO M164 (ASTM A325). Bolts manufactured to AASHO M164 requirements are identified by marking on the top of the head with three radial lines and the symbol A325, and nuts are marked on one face with three similar circumferential markings, 120 degrees apart, or alternatively, with C, 2, D, 2H or DH. Bolt and nut dimensions shall conform to the dimensions shown in Table 2.10.3A, and to the requirements for Heavy Hexagon Structural Bolts and for Heavy Semi-Finished Hexagon Nuts given in ANSI Standard B18.2.1 and B18.2.2, except as allowed in the following paragraph.

Subject to the approval of the Engineer, high strength steel lock-pin and collar fasteners may be used as an alternate for high strength bolts or rivets as shown on the plans. The shank and head of the high strength steel lock-pin and collar fasteners shall meet the chemical composition and mechanical property requirements of AASHO M 164 (ASTM A325). Each fastener shall provide a solid shank body of sufficient diameter to provide tensile and shear strength equivalent to or greater than the bolt or rivet specified, shall have a cold forged head on one end, of type and dimensions as approved by the Engineer, a shank length suitable for material thickness fastened, locking grooves, breakneck groove and pull grooves (all annular grooves) on the opposite end. Each fastener shall provide a steel locking collar, of proper size for shank diameter used, which, by means of suitable installation tools, is cold swaged into the locking grooves forming a head for the grooved end of the fastener after the pull groove section has been removed. The steel locking collar shall be a standard product of an established manufacturer of lock-pin and collar fasteners, as approved by the Engineer.

Circular washers shall be flat and smooth and their nominal dimensions shall conform to the dimensions given in Table 2.10.3B, except that for lock-pin and collar fasteners, flat washers need not be used.

Beveled washers for American Standard beams and channels shall be square or rectangular, shall taper in thickness, and shall conform to the dimensions given in Table 2.10.3B.

Where necessary, washers may be clipped on one side to a point not closer than $7/8$ of the bolt diameter from the center of the washer.

(10) Copper Bearing Steels

When copper bearing steel is specified, the steel shall contain not less than 0.2 per cent of copper.

(11) Welded Stud Shear Connectors

(a) Shear connector studs shall conform to the requirements of Cold Finished-Carbon Steel Bars and Shafting, AASHO M 169 (ASTM A 108), cold-drawn bars, grades 1015, 1018, or 1020, either semi- or fully-killed. If flux retaining caps are used, the steel for the caps shall be of a low carbon grade suitable for welding and shall comply with Cold-Rolled Carbon Steel Strip, ASTM A 109.

TABLE 2.10.3A

Nominal Bolt Size, D	Bolt Dimensions, In Inches			Nut Dimensions, In Inches	
	Heavy Hexagon Structural Bolts			Heavy Semi-Finished Hexagon Nuts	
	Width Across Flats F	Height H	Thread length T	Width Across Flats W	Height H
1/2	7/8	5/16	1	7/8	31/64
5/8	1 1/16	25/64	1 1/4	1 1/16	39/64
3/4	1 1/4	15/32	1 3/8	1 1/4	47/64
7/8	1 7/16	35/64	1 1/2	1 7/16	55/64
1	1 5/8	39/64	1 3/4	1 5/8	63/64
1 1/8	1 13/16	11/16	2	1 13/16	1 7/64
1 1/4	2	25/32	2	2	1 7/32
1 3/8	2 3/16	27/32	2 1/4	2 3/16	1 11/32
1 1/2	2 3/8	15/16	2 1/4	2 3/8	1 15/32

TABLE 2.10.3B

WASHER DIMENSIONS[a]

Bolt Size D	Circular Washers				Square or Rectangular Beveled Washers for American Standard Beams and Channels		
	Nominal Outside Diameter[b]	Nominal Diameter of Hole	Thickness		Minimum side dimension	Mean thickness	Slope of taper in thickness
			Min.	Max.			
1/2	1 1/16	17/32	.097	.177	1 3/4	5/16	1:6
5/8	1 5/16	21/32	.122	.177	1 3/4	5/16	1:6
3/4	1 15/32	13/16	.122	.177	1 3/4	5/16	1:6
7/8	1 3/4	15/16	.136	.177	1 3/4	5/16	1:6
1	2	1 1/16	.136	.177	1 3/4	5/16	1:6
1 1/8	2 1/4	1 1/4	.136	.177	2 1/4	5/16	1:6
1 1/4	2 1/2	1 3/8	.136	.177	2 1/4	5/16	1:6
1 3/8	2 3/4	1 1/2	.136	.177	2 1/4	5/16	1:6
1 1/2	3	1 5/8	.136	.177	2 1/4	5/16	1:6
1 3/4	3 3/8	1 7/8	.178[c]	.28[c]	—	—	—
2	3 3/4	2 1/8	.178	.28	—	—	—
Over 2 to 4 Incl.	2D − 1/2	D + 1/8	.24[d]	.34[d]	—	—	—

[a] Dimensions in inches
[b] May be exceeded by 1/4 in.
[c] 3/16 in. nominal
[d] 1/4 in. nominal

(b) Tensile properties as determined by tests of bar stock after drawing or of finished studs shall conform to the following requirements:

Tensile Strength	(min.)	60,000
Yield Strength *	(min.)	50,000
Elongation	(min.)	20% in 2 inches
Reduction of area	(min.)	50%

* As determined by a 0.2% offset method.

(c) Tensile properties shall be determined in accordance with the applicable sections of ASTM A 370, Mechanical Testing of Steel Products. Tensile tests of finished studs shall be made on studs welded to test plates using a test fixture similar to that shown in Figure 2.10.23B. If fracture occurs outside of the middle half of the gage length, the test shall be repeated.

(d) Finished studs shall be of uniform quality and condition, free from injurious laps, fins, seams, cracks, twists, bends or other injurious defects. Finish shall be as produced by cold drawing, cold rolling or machining.

(e) The manufacturer shall certify that the studs as delivered are in accordance with the material requirements of this section. Certified copies of in-plant quality control test reports shall be furnished to the Engineer upon request.

(f) The Engineer may select, at the Contractor's expense, studs of each type and size used under the contract, as necessary for checking the requirements of this section.

(12) Unfilled Tubular Steel Piles

Unfilled Tubular Steel Piles shall conform to the requirements of Welded and Seamless Steel Pipe Piles ASTM Designation A252, Grade 2, with Chemical Requirements meeting ASTM Designation A53, Grade B.

(B) Steel Forgings and Steel Shafting

(1) Carbon Steel Forgings

Steel forgings shall conform to the Specifications for Carbon Steel Forgings for General Industrial Use, AASHO M 102 (ASTM 235). Class C 1 forgings shall be furnished unless otherwise specified.

(2) Cold Finished Carbon Steel Shafting

Cold finished carbon steel shafting shall conform to the specifications for Cold Finished Carbon Steel Bars and Shafting, AASHO M 169 (ASTM A 108). Grade Designation 1016-1030, inclusive, shall be furnished unless otherwise specified.

(3) Alloy Steel Forgings

Alloy steel forgings shall conform to the Specifications for Alloy Steel Forgings for General Industrial Use, ASTM A 237. Class A forging shall be furnished unless otherwise specified.

(C) Steel Castings

(1) Steel Castings for Highway Bridges

Steel castings for use in highway bridge components shall conform to Standard Specification for Steel Castings for Highway Bridges, AASHO M 192 (ASTM A 486) or Mild-to-Medium-Strength Carbon-Steel Castings for General Applications AASHO M 103 (ASTM A 27). The class 70 or grade 70-36 of steel, respectively, shall be used unless otherwise specified.

(2) Chromium Alloy-Steel Castings

Chromium alloy steel castings shall conform to the Specification for Corrosion-Resistant Iron-Chromium and Iron-Chromium-Nickel Alloy Castings for General Application, AASHO M 163 (ASTM A 296). Grade CA 15 shall be furnished unless otherwise specified.

(D) Iron Castings

(1) General

Iron castings shall be gray iron castings conforming to the Specification for Gray Iron Castings, AASHO M 105, Class No. 30 unless otherwise specified.

(2) Workmanship and Finish

Iron castings shall be true to pattern in form and dimensions, free from pouring faults, sponginess, cracks, blow holes, and other defects in positions affecting their strength and value for the service intended.

Castings shall be boldly filleted at angles and the arrises shall be sharp and perfect.

(3) Cleaning

All castings must be sandblasted or otherwise effectively cleaned of scale and sand so as to present a smooth, clean, and uniform surface.

(E) Ductile Iron Castings

(1) General

Ductile iron castings shall conform to the Specifications for Ductile Iron Castings, ASTM A 536, Grade 60-40-18 unless otherwise specified. In addition to the specified test coupons, test specimens from parts integral with the castings, such as risers, shall be tested for castings weighing more than 1000 pounds to determine that the required quality is obtained in the castings in the finished condition.

(2) Workmanship and Finish

Iron castings shall be true to pattern in form and dimensions, free from pouring faults, sponginess, cracks, blow holes, and other defects in positions, affecting their strength and value for the service intended.

Castings shall be boldly filleted at angles and the arrises shall be sharp and perfect.

(3) Cleaning

All castings must be sandblasted or otherwise effectively cleaned of scale and sand so as to present a smooth, clean, and uniform surface.

(F) Malleable Castings

(1) General

Malleable castings shall conform to the Specification for Malleable Iron Castings, AASHO M 106 (ASTM A 47); Grade No. 35018 shall be furnished unless otherwise specified.

(2) Workmanship and Finish

Malleable castings shall be true to pattern in form and dimensions, free from pouring faults, sponginess, cracks, blow holes, and other defects in positions affecting their strength and value for the service intended.

The castings shall be boldly filleted at angles and the arrises shall be sharp and perfect. The surfaces shall have a workmanlike finish.

(3) Cleaning

All castings must be sandblasted or otherwise effectively cleaned of scale and sand so as to present a smooth, clean, and uniform surface.

(G) Bronze Castings and Copper-Alloy Plates

(1) Bronze Castings

Bronze castings shall conform to Standard Specifications for Bronze Castings for Bridges and Turntables, AASHO M 107 (ASTM B 22) Alloys A or B.

(2) Copper-Alloy Plates

Copper alloy plates shall conform to Standard Specifications for Rolled Copper-Alloy Bearing and Expansion Plates and Sheets for Bridge and other Structural Uses, AASHO M 108 (ASTM B 100).

(H) Sheet Lead

Sheet lead shall conform to the requirements for Common Desilverized Lead of the Specification for Pig Lead, AASHO M 112 (ASTM B 29).

(I) Sheet Zinc

Sheet zinc shall conform to the requirements for Type II of the Specifications for Rolled Zinc, AASHO M 113 (ASTM B 69).

(J) Galvanizing

When galvanizing is shown on the plans or specified in the special provisions ferrous metal products shall be galvanized in accordance with the Specifications for Zinc (Hot-Galvanized) Coatings on Products Fabricated from Rolled, Pressed, and Forged Steel Shapes, Plates, Bars and Strip, AASHO M 111 (ASTM A 123).

(K) Canvas and Red Lead for Bedding Masonry Plates and Equivalent Bearing Areas

The canvas shall conform to the Standard Specifications for Numbered Cotton Duck and Army Duck, AASHO M 166 (ASTM D 230), and to the weight specified. The red lead paint shall conform to the specifications for paint for metals, Article 2.14.2.

(L) Preformed Fabric Pads

The preformed fabric pads shall be composed of multiple layers of 8-ounce cotton duck impregnated and bound with high-quality natural rubber or of equivalent and equally suitable materials compressed into resilient pads of uniform thickness. The number of plies shall be such as to produce the specified thickness, after compression and vulcanizing. The finished pads shall withstand compression loads perpendicular to the plane of the laminations of not less than 10,000 pounds per square inch without detrimental reduction in thickness or extrusion.

2.10.4 — STORAGE OF MATERIALS

Structural material, either plain or fabricated, shall be stored at the bridge shop above the ground upon platforms, skids, or other supports. It shall be kept free from dirt, grease and other foreign matter, and shall be protected as far as practicable from corrosion.

2.10.5 — STRAIGHTENING MATERIAL AND CURVING ROLLED BEAMS AND WELDED GIRDERS

(A) Straightening Material

Rolled material, before being laid off or worked, must be straight. If straightening is necessary, it shall be done by methods that will not injure the metal. Heat straightening of ASTM A514/A517 steel shall be done only under rigidly controlled procedures, each application subject to the approval of the Engineer. In no case shall the maximum temperature of the steel exceed 1125F. Sharp kinks and bends shall be cause for rejection of the material.

(B) Curving Rolled Beams and Welded Girders

(1) Materials

Steels that are manufactured to a yield point greater than 50,000 psi shall not be heat curved.

(2) Type of Heating

Beams and girders may be curved by either continuous or V-type heating as approved by the Engineer. For the continuous method, a strip along the edge of the top and bottom flange shall be heated simultaneously; the strip shall be of sufficient width and temperature to obtain the required curvature. For the V-type heating, the top and bottom flanges shall be heated in truncated triangular or wedge-shaped areas having their base along the flange edge and spaced at regular intervals along each flange; the spacing and temperature shall be as required to obtain the required curvature, and heating shall progress along the top and bottom flange at approximately the same rate.

For the V-type heating, the apex of the truncated triangular area applied to the inside flange surface shall terminate just before the juncture of the web and the flange is reached.* When the radius of curvature is 1000 feet or more, the apex of the truncated triangular heating pattern applied to the outside flange surface shall extend to the juncture of the flange and web. When the radius of curvature is less than 1000 feet, the apex of the truncated triangular heating pattern applied to the outside flange surface shall extend past the web for a distance equal to ⅛ of the flange or 3 inches, whichever is less. The truncated triangular pattern shall have an included angle of approximately 15 to 30 degrees, but the base of the triangle shall not exceed 10 inches. Variations in the patterns prescribed above may be made with the approval of the Engineer.

For both types of heating, the flange edges to be heated are those that will be on the inside of the horizontal curve after cooling. Heating both inside and outside flange surfaces is only mandatory when the flange thickness is 1¼ inches or greater, in which case, the two surfaces shall be heated concurrently. The maximum temperature shall be as prescribed below.

(3) Temperature

The heat-curving operation shall be conducted in such a manner that the temperature of the steel does not exceed 1150 F as measured by temperature indicating crayons or other suitable means. The girder shall not be artificially cooled until after naturally cooling to 600 F; the method of artificial cooling is subject to the approval of the Engineer.

(4) Position for Heating

The girder may be heat-curved with the web in either a vertical or a horizontal position. When curved in the vertical position, the girder must be braced or supported in such a manner that the tendency of the girder to deflect laterally during the heat-curving process will not cause the girder to overturn.

When curved in the horizontal position, the girder must be supported near its ends and at intermediate points, if required, to obtain a uniform curvature; the bending stress in the flanges due to the dead weight of the girder must not exceed the usual allowable design stress. When the girder is positioned horizontally for heating, intermediate safety catch blocks must be maintained at the midlength of the girder within 2 inches of the flanges at all

---

* To avoid unnecessary web distortion, special care shall be taken when heating the inside flange surfaces (the surfaces that intersect the web) so that heat is not applied directly to the web.

times during the heating process to guard against a sudden sag due to plastic flange buckling.

(5) Sequence of Operations

The girder shall be heat-curved in the fabrication shop before it is painted. The heat curving operation may be conducted either before or after all the required welding of transverse intermediate stiffeners is completed. However, unless provisions are made for girder shrinkage, connection plates and bearing stiffeners shall be located and attached after heat curving. If longitudinal stiffeners are required, they shall be heat-curved or oxygen-cut separately and then welded to the curved girder. When cover plates are to be attached to rolled beams, they may be attached before heat curving if the total thickness of one flange and cover plate is less than 2½ inches and the radius of curvature is greater than 1000 feet. For other rolled beams with cover plates, the beams must be heat-curved before the cover plates are attached; cover plates must be either heat curved on oxygen-cut separately and then welded to the curved beam.

(6) Camber

Girders shall be cambered before heat curving. Camber for rolled beams may be obtained by heat-cambering methods approved by the Engineer. For plate girders, the web shall be cut to the prescribed camber with suitable allowance for shrinkage due to cutting, welding, and heat curving.* However, subject to the approval of the Engineer, moderate deviations from specified camber may be corrected by a carefully supervised application of heat.

(7) Measurement of Curvature and Camber

Horizontal curvature and vertical camber shall not be measured for final acceptance before all welding and heating operations are completed and the flanges have cooled to a uniform temperature. Horizontal curvature shall be checked with the girder in the vertical position by measuring off-sets from a string line or wire attached to both flanges or by using other suitable means; camber shall be checked by adequate means.

2.10.6 — FINISH

Portions of the work exposed to view shall be finished neatly. Shearing, flame cutting and chipping shall be done carefully and accurately.

---

* The heat-curving process may tend to change the vertical camber present before heating. This effect shall be most pronounced when the top and bottom flanges are of unequal widths on a given tranverse cross section.

2.10.7 — RIVET AND BOLT HOLES

(A) Holes for Rivets, High Strength Bolts and Unfinished Bolts *

All holes for rivets or bolts shall be either punched or drilled. Material forming parts of a member composed of not more than five thicknesses of metal may be punched $\frac{1}{16}$ inch larger than the nominal diameter of the rivets or bolts whenever the thickness of the material is not greater than $\frac{3}{4}$ inch for structural steel, $\frac{5}{8}$ inch for high-strength steel or $\frac{1}{2}$ inch for quenched and tempered alloy steel, unless subpunching and reaming is required under Article 2.10.10.

When there are more than five thicknesses or when any of the main material is thicker than $\frac{3}{4}$ inch for structural steel, $\frac{5}{8}$ inch for high-strength steel, or $\frac{1}{2}$ inch for quenched and tempered alloy steel, all holes shall either be subdrilled or drilled full size.

When required under Article 2.10.10, all holes shall be either subpunched or subdrilled (subdrilled if thickness limitation governs) $\frac{3}{16}$ inch smaller and, after assembling, reamed $\frac{1}{16}$ inch larger or drilled full size to $\frac{1}{16}$ inch larger than the nominal diameter of the rivets or bolts.

When permitted by Article 1.7.5, enlarged or slotted holes are allowed with high strength bolts.

(B) Holes for Ribbed Bolts, Turned Bolts or other Approved Bearing Type Bolts

All holes for ribbed bolts, turned bolts or other approved bearing-type bolts shall be subpunched or subdrilled $\frac{3}{16}$ inch smaller than the nominal diameter of the bolt and reamed assembled or to a steel template or, after assembling, drilled from the solid at the option of the Fabricator. In any case the finished holes shall provide a driving fit as specified on the plans or in the special provisions.

2.10.8 — PUNCHED HOLES

The diameter of the die shall not exceed the diameter of the punch by more than $\frac{1}{16}$ inch. If any holes must be enlarged to admit the rivets or bolts, such holes shall be reamed. Holes must be clean cut without torn or ragged edges. Poor matching of holes will be cause for rejection.

2.10.9 — REAMED OR DRILLED HOLES

Reamed or drilled holes shall be cylindrical, perpendicular to the member and shall comply with the requirements of Article 2.10.7 as to size. Where practicable, reamers shall be directed by mechanical means. Burrs on the outside surfaces shall be removed. Poor matching of holes will be cause for rejection. Reaming and drilling shall be done with twist drills. If required by the Engineer, assembled parts shall be taken apart for removal of burrs caused by drilling. Connecting parts requiring reamed or drilled holes shall be assembled and securely held while being reamed or drilled and shall be match marked before disassembling.

---

* See Art. 2.10.19 for bolts included in designation "Unfinished Bolts."

2.10.10 — SUBPUNCHING AND REAMING OF FIELD CONNECTIONS

Unless otherwise specified in the special provisions or on the plans, holes in all field connections and field splices of main members of trusses, arches, continuous beam spans, bents, towers (each face), plate girders and rigid frames shall be subpunched (or subdrilled if subdrilling is required according to Article 2.10.7) and subsequently reamed while assembled or to a steel template, as required by Article 2.10.14. All holes for floor beam and stringer field end connections shall be subpunched and reamed to a steel template or reamed while assembled. Reaming or drilling full size of field connection holes through a steel template shall be done after the template has been located with utmost care as to position and angle and firmly bolted in place. Templates used for reaming matching members, or the opposite faces of a single member, shall be exact duplicates. Templates used for connections on like parts or members shall be so accurately located that the parts or members are duplicates and require no match-marking.

For any connection, in lieu of subpunching and reaming or subdrilling and reaming, the fabricator may, at his option, drill holes full size with all thicknesses of material assembled in proper position.

If additional sub-punching and reaming is required, it shall be specified in the special provisions or on the plans.

2.10.11 — ACCURACY OF PUNCHED AND DRILLED HOLES

All holes punched full size, subpunched, or subdrilled shall be so accurately punched that after assembling (before any reaming is done) a cylindrical pin $\frac{1}{8}$ inch smaller in diameter than the nominal size of the punched hole may be entered perpendicular to the face of the member, without drifting, in at least 75 per cent of the contiguous holes in the same plane. If the requirement is not fulfilled, the badly punched pieces will be rejected. If any hole will not pass a pin $\frac{3}{16}$ inch smaller in diameter than the nominal size of the punched hole, this will be cause for rejection.

2.10.12 — ACCURACY OF REAMED AND DRILLED HOLES

When holes are reamed or drilled, 85 per cent of the holes in any contiguous group shall, after reaming or drilling, show no offset greater than $\frac{1}{32}$ inch between adjacent thicknesses of metal.

All steel templates shall have hardened steel bushings in holes accurately dimensioned from the center lines of the connection as inscribed on the template. The center lines shall be used in locating accurately the template from the milled or scribed ends of the members.

2.10.13 — FITTING FOR RIVETING AND BOLTING

Surfaces of metal in contact shall be cleaned before assembling. The parts of a member shall be assembled, well pinned, and firmly drawn together with bolts before reaming or riveting is commenced. Assembled pieces shall be taken apart, if necessary, for the removal of burrs and shavings produced by the reaming operation. The member shall be free from twists, bends, and other deformation.

Preparatory to the shop riveting of full-sized punched material, the rivet holes, if necessary, shall be spear-reamed for the admission of the rivets. The reamed holes shall not be more than 1/16 inch larger than the nominal diameter of the rivets.

End connection angles, stiffener angles, and similar parts shall be carefully adjusted to correct position and bolted, clamped, or otherwise firmly held in place until riveted.

Parts not completely riveted in the shop shall be secured by bolts, in so far as practicable, to prevent damage in shipment and handling.

2.10.14 — SHOP ASSEMBLING

The field connections of main members of trusses, arches, continuous beam spans, bents, towers (each face), plate girders and rigid frames shall be assembled in the shop with milled ends of compression members in full bearing, and then shall have their sub-size holes reamed to specified size while the connections are assembled. Assembly shall be Full Truss or Girder Assembly unless Progressive Truss or Girder Assembly, Full Chord Assembly, Progressive Chord Assembly, or Special Complete Structure Assembly is specified in the special provisions or on the plans.

(A) Full Truss or Girder Assembly

Full Truss or Girder Assembly shall consist of assembling all members of each truss, arch rib, bent, tower face, continuous beam line, plate girder or rigid frame at one time.

(B) Progressive Truss or Girder Assembly

Progressive Truss or Girder Assembly shall consist of assembling initially for each truss, arch rib, bent, tower face, continuous beam line, plate girder, or rigid frame at least three contiguous shop sections or all members in at least three contiguous panels but not less than the number of panels associated with three contiguous chord lengths (i.e., length between field splices) and not less than 150 feet in the case of structures longer than 150 feet. At least one shop section or panel or as many panels as are associated with a chord length shall be added at the advancing end of the assembly before any member is removed from the rearward end, so that the assembled portion of the structure is never less than that specified above.

(C) Full Chord Assembly

Full Chord Assembly shall consist of assembling, with geometric angles at the joints, the full length of each chord of each truss or open spandrel arch, or each leg of each bent or tower, then reaming their field connection holes while the members are assembled and reaming the web member connections to steel templates set at geometric (not cambered) angular relation to the chord lines.

Field connection holes in web members shall be reamed to steel templates. At least one end of each web member shall be milled or shall be scribed normal to the longitudinal axis of the member and the templates at both ends of the member shall be accurately located from one of the milled ends or scribed lines.

(D) Progressive Chord Assembly

Progressive Chord Assembly shall consist of assembling contiguous chord members in the manner specified for Full Chord Assembly and in the number and length specified for Progressive Truss or Girder Assembly.

(E) Special Complete Structure Assembly

Special Complete Structure Assembly shall consist of assembling the entire structure, including the floor system. (This procedure is ordinarily needed only for complicated structures such as those having curved girders, or extreme skew in combination with severe grade or camber.)

Each assembly, including camber, alignment, accuracy of holes and fit of milled joints, shall be approved by the Engineer before reaming is commenced.

A camber diagram shall be furnished the Engineer by the Fabricator showing the camber at each panel point of each truss, arch rib, continuous beam line, plate girder or rigid frame. When the shop assembly is Full Truss or Girder Assembly or Special Complete Structure Assembly, the camber diagram shall show the camber measured in assembly. When any of the other methods of shop assembly is used, the camber diagram shall show calculated camber.

2.10.15 — DRIFTING OF HOLES

The drifting done during assembling shall be only such as to bring the parts into position, and not sufficient to enlarge the holes or distort the metal. If any holes must be enlarged to admit the rivets, they shall be reamed.

2.10.16 — MATCH-MARKING

Connecting parts assembled in the shop for the purpose of reaming holes in field connections shall be match-marked, and a diagram showing such marks shall be furnished the Engineer.

2.10.17 — RIVETS

The size of rivets called for on the plans shall be the size before heating. Rivet heads shall be of standard shape, unless otherwise specified, and of uniform size for the same diameter of rivet. They shall be full, neatly made, concentric with the rivet holes, and in full contact with the surfaces of the member.

2.10.18 — FIELD RIVETS

Sufficient field rivets shall be furnished to rivet the entire structure with an ample surplus to replace all rivets burned, lost or cut out.

2.10.19 — BOLTS AND BOLTED CONNECTIONS

The specifications of this article do not pertain to the use of high strength bolts. Bolted connections fabricated with high strength bolts shall conform to Article 2.10.20.

(A) General

Bolts shall be unfinished, turned or ribbed bolts conforming to the requirements for Grade A Bolts of Specification for Low-Carbon Steel Externally and Internally Threaded Standard Fasteners, ASTM A 307. Bolted connections shall be used only as indicated by the plans or special provisions. Bolts shall have single self-locking nuts or double nuts unless otherwise shown on the plans or in the special provisions. Beveled washers shall be used where bearing faces have a slope of more than 1:20 with respect to a plane normal to the bolt axis.

Except as otherwise provided in this article, construction shall conform to applicable specifications for riveted structures.

(B) Unfinished Bolts

Unfinished bolts shall be furnished unless other types are specified.

(C) Turned Bolts

The surface of the body of turned bolts shall meet the ANSI roughness rating value of 125. Heads and nuts shall be hexagonal with standard dimensions for bolts of the nominal size specified or the next larger nominal size. Diameter of threads shall be equal to the body of the bolt or the nominal diameter of the bolt specified. Holes for turned bolts shall be carefully reamed with bolts furnished to provide for a light driving fit. Threads shall be entirely outside of the holes. A washer shall be provided under the nut.

(D) Ribbed Bolts

The body of ribbed bolts shall be of an approved form with continuous longitudinal ribs. The diameter of the body measured on a circle through the points of the ribs shall be $5/64$ inch greater than the nominal diameter specified for the bolts.

Ribbed bolts shall be furnished with round heads conforming to ANSI B 18.5 unless otherwise specified. Nuts shall be hexagonal, either recessed or with a washer of suitable thickness. Ribbed bolts shall make a driving fit with the holes. The hardness of the ribs shall be such that the ribs do not mash down enough to pemit the bolts to turn in the holes during tightening. If for any reason the bolt twists before drawing tight, the hole shall be carefully reamed and an oversized bolt used as a replacement.

2.10.20 — CONNECTIONS USING HIGH STRENGTH BOLTS

(A) General

This article covers the assembly of structural joints using AASHO M 164 (ASTM A 325) high strength carbon steel bolts, or equivalent fasteners, tightened to a high tension. The bolts are used in holes conforming to the requirements of Articles 2.10.7, 2.10.8, and 2.10.9.

(B) Bolts, Nuts and Washers

Bolts, nuts and washers shall conform to the requirements of Article 2.10.3.(A)9.

(C) Bolted Parts

The slope of surfaces of bolted parts in contact with the bolt head and nut shall not exceed 1:20 with respect to a plane normal to the bolt axis. Bolted parts shall fit solidly together when assembled and shall not be separated by gaskets or any other interposed compressible material.

When assembled, all joint surfaces, including those adjacent to the bolt heads, nuts or washers, shall be free of scale, except tight mill scale, and shall also be free of dirt, loose scale, burrs, other foreign material and other defects that would prevent solid seating of the parts.

Contact surfaces within friction-type joints shall be free of oil, paint, lacquer or rust inhibitor, except that hot-dip galvanizing will be permitted, provided that contact surfaces are scored by wire brushing or blasting after galvanizing and prior to assembly. The wire brushing treatment shall be a light application of manual or power brushing that marks or scores the surface but removes relatively little of the zinc coating. The blasting treatment shall be a light "brush-off" treatment which will produce a dull gray appearance. However, neither treatment should be severe enough to produce any break or discontinuity in the zinc surface.

When specified on the plans galvanized high strength structural bolts shall be furnished and installed in conformance with ASTM A325-71A, in which a special lubricant is required for the nuts and additional test requirements are specified for a full size assembled joint.

(D) Installation

(1) Bolt Tension

Each fastener shall be tightened to provide, when all fasteners in the joint are tight, at least the minimum bolt tension shown in Table 2.10.20A for the size of fastener used.

Threaded bolts shall be tightened with properly calibrated wrenches or by the turn-of-nut method. If required because of bolt entering and wrench operation clearances, tightening by either procedure may be done by turning the bolt while the nut is prevented from rotating. Impact wrenches, if used, shall be of adequate capacity and sufficiently supplied with air to perform the required tightening of each bolt in approximately ten seconds.

(2) Washers

All fasteners shall have a hardened washer under the element (nut or bolt head) turned in tightening.

TABLE 2.10.20A

BOLT TENSION

Bolt Size, in inches	Minimum Bolt Tension (1) in pounds AASHO M 164 (ASTM A 325) Bolts
½	12,050
⅝	19,200
¾	28,400
⅞	39,250
1	51,500
1⅛	56,450
1¼	71,700
1⅜	85,450
1½	104,000

(1) Equal to the proof load (length measurement method) given in AASHO M 164 (ASTM A 325).

Where an outer face of the bolted parts has a slope of more than 1:20 with respect to a plane normal to the bolt axis, a smooth beveled washer shall be used to compensate for the lack of parallelism.

(3) Calibrated Wrench Tightening

When calibrated wrenches are used to provide the bolt tension specified in paragraph (D)(1) above, their setting shall be such as to induce a bolt tension 5% to 10% in excess of this value. These wrenches shall be calibrated at least once each working day by tightening, in a device capable of indicating actual bolt tension, not less than three typical bolts of each diameter from the bolts to be installed. Power wrenches shall be adjusted to stall or cut-out at the selected tension. If manual torque wrenches are used the torque indication corresponding to the calibrating tension shall be noted and used in the installation of all bolts of the tested lot. Nuts shall be turned in the tightening direction when torque is measured. When using calibrated wrenches to install several bolts in a single joint, the wrench shall be returned to "touch up" bolts previously tightened which may have been loosened by the tightening of subsequent bolts, until all are tightened to the prescribed amount.

(4) Turn-of-Nut Tightening

When the turn-of-nut method is used to provide the bolt tension specified in paragraph (D)(1), there shall first be enough bolts brought to a "snug tight" condition to insure that the parts of the joint are brought into full contact with each other. Snug tight is defined as the tightness atttained by a few impacts of an impact wrench or the full effort of a man using an ordinary spud wrench. Following this initial operation, bolts shall be placed in any remaining holes in the connection and brought to snug tightness. All bolts in the joint shall then be tightened additionally by the applicable amount of nut rotation specified in Table 2.10.20B with tightening progressing systematically from the most rigid part of the joint to its free edges. During this operation there shall be no rotation of the part not turned by the wrench.

TABLE 2.10.20B

NUT ROTATION [1] [2] FROM SNUG TIGHT CONDITION

Disposition of Outer Faces of Bolted Parts		
Both faces normal to bolt axis, or one face normal to axis and other face sloped(3) (bevel washer not used)		Both faces sloped(3) from normal to bolt axis (bevel washers not used)
Bolt length(4) not exceeding 8 diameters or 8 inches	Bolt length(4) exceeding 8 diameters or 8 inches	For all lengths of bolts
½ turn	⅔ turn	¾ turn

(1) For coarse thread heavy hexagon structural bolts of all sizes and lengths and heavy hexagon semi-finished nuts.
(2) Nut rotation is rotation relative to bolt regardless of the element (nut or bolt) being turned. Tolerance on rotation: ⅙ turn (60°) over and nothing under.
(3) Slope 1:20 maximum.
(4) Bolt length is measured from underside of head to extreme end of bolt.

(5) Lock-Pin and Collar Fasteners

The installation of lock-pin and collar fasteners shall be by methods and procedures approved by the Engineer.

(E) Inspection

(1) The Engineer shall determine that the requirements of paragraphs (2) and (3) following, are met in the work. When the calibrated wrench method of tightening is used, the Engineer shall have full opportunity to witness the calibration tests prescribed in paragraph (D) (3), above.

(2) The Engineer shall observe the installation and tightening of bolts to determine that the selected tightening procedure is properly used and shall determine that all bolts are tightened.

(3) The following inspection shall be used unless a more extensive or different inspection procedure is specified.

(a) Either the Engineer or the Contractor in the presence of the Engineer, at the Engineer's option, shall use an inspection wrench which may be either a torque wrench or a power wrench that can be accurately adjusted in accordance with the requirements of paragraph (D)(3), above.

(b) Three bolts of the same grade, size,* and condition as those under inspection shall be placed individually in a calibration device capable of indicating bolt tension. There shall be a washer under the part turned in tightening each bolt.

(c) When the *inspecting wrench* is a torque wrench, each bolt specified in paragraph (E)(3)(b) shall be tightened in the calibration device by any convenient means to the minimum tension specified for its size in paragraph (D)(1). The *inspecting wrench* then shall be applied to the tightened bolt and the torque necessary to turn the nut or head 5 degrees (approximately 1 inch at 12 inch radius) in tightening direction shall be determined. The average torque measured in the tests of three bolts shall be taken as the *job inspecting torque* to be used in the manner specified in paragraph (E)(3)(e).

(d) When the *inspecting wrench* is a power wrench it shall be adjusted so that it will tighten each bolt specified in paragraph (E)(3)(b) to a tension at least 5 but not more than 10 per cent greater than the minimum tension specified for its size in paragraph (D)(1). This setting of wrench shall be taken as the *job inspecting torque* to be used in the manner specified in paragraph (E)(3)(e).

(e) Bolts, represented by the sample prescribed in paragraph (E)(3)(b), which have been tightened in the structure shall be inspected by applying, in the tightening direction, the *inspecting wrench* and its *job inspecting torque* to 10 percent of the bolts, but not less than two bolts, selected at random in each connection. If no nut or bolt head is turned by this application of the *job inspecting torque*, the connection shall be accepted as properly tightened. If any nut or bolt head is turned by the application of the *job inspecting torque*, this torque shall be applied to all bolts in the connection, and all bolts whose nut or head is turned by the *job inspecting torque* shall be tightened and reinspected, or alternatively, the Fabricator or Erector, at his option, may retighten all of the bolts in the connection and then resubmit the connection for the specified inspection.

(4) The procedures for inspecting and testing the lock-pin and collar fasteners and their installation to assure that the required preload tension is provided shall be as approved by the Engineer.

2.10.21 — RIVETING

Rivets shall be heated uniformly to a "light cherry red color" and shall be driven while hot. Any rivet whose point is heated more than the remainder shall not be driven. When a rivet is ready for driving, it shall be

---

* Length may be any length representative of bolts used in the structure.

free from slag, scale and other adhering matter. Any rivet which, in the opinion of the Engineer, is scaled excessively, will be rejected.

All rivets that are loose, burned, badly formed, or otherwise defective shall be removed and replaced with satisfactory rivets. Any rivet whose head is defective in size or whose head is driven off center will be considered defective and shall be removed. Stitch rivets that are loosened by driving of adjacent rivets shall be removed and replaced with satisfactory rivets. Caulking, recupping or double gunning of rivet heads will not be permitted.

Shop rivets shall be driven by direct-acting rivet machines when practicable.

Approved beveled rivet sets shall be used for forming rivet heads on sloping surfaces. When the use of a direct-acting rivet machine is not practicable, pneumatic hammers of approved size shall be used. Pneumatic bucking tools will be required when, in the opinion of the Engineer, the size and length of the rivets warrant their use.

Rivets may be driven cold provided their diameter is not over ⅜ inch.

2.10.22 — PLATE CUT EDGES

(A) Edge Planing

Sheared edges of plate more than ⅝-inch in thickness and carrying calculated stress shall be planed to a depth of ¼-inch. Re-entrant corners shall be filleted to a minimum radius of ¾-inch before cutting.

(B) Visual Inspection and Repair of Plate Cut Edges

The following provisions shall apply to permissible repairs to discontinuities discovered by visual inspection of the plate-cut edges before fabrication, welding or during routine examination of welded joints by radiography or ultrasonic inspection in plates of all steels covered by these Specifications, in thicknesses up to and including 4 inches maximum. These principally include discontinuities resulting from gas pockets or blow holes and shrinkage cavities which are manifested as "laminations" or "pipe" characterized by a distinct separation of the metal which is parallel to the plane of the plate. To a lesser extent, these include discontinuities resulting from entrapped slag or refractory, or deoxidation products, which are manifested as deposits of foreign material in the steel, but only in the case where these deposits are parellel to the plane of the plate. Multiple discontinuities shall be considered continuous when located in the same plane within five percent of the plate thickness and separated by a distance less than the length of the smaller of two adjacent discontinuities.

The corrective procedures described in Table 2.10.22(B) shall not apply to discontinuities in rolled plate surfaces, which discontinuities may be corrected by the fabricator in accordance with the provisions of AASHO M160 (ASTM A6).

These provisions are not applicable to material stressed in tension in the through-thickness direction.

The limits of all internal discontinuities required to be explored, which are not explored to their full depth by other means, shall be determined by ultrasonic inspection.

Plates containing discontinuities of the types shown in Figure 2.10.22B and as described in the following Table 2.10.22(B) may be corrected by measures listed hereunder.

The length of the discontinuity is the visible long dimension on the plate-cut edge and the depth is the distance that the discontinuity extends into the plate from the cut edge.

In making any repairs, the amount of metal removed shall be the minimum necessary to remove the discontinuity or to determine that the permissible limit is not exceeded. Plate edges may be at any angle with respect to the rolling direction, but the directions of discontinuities shall be considered with respect to the directions of the plate edges. All repairs of the discontinuities made by welding shall conform to applicable provisions of AWS D1.1-72 as modified by these specifications.

Edges of plates shall be inspected and required repairs completed as early as feasible in the fabrication sequence so as to allow maximum opportunity for the fabricator to incorporate repaired plates in the least critical areas.

TABLE 2.10.22(B)

Description of Discontinuity	Repair Required
Any discontinuity 1″ in length or less	None—need not be explored
Any discontinuity over 1″ in length and ⅛″ maximum depth.	None—depth should be explored.
Any discontinuity over 1″ in length with depth over ⅛″ but not greater than ¼″.	Remove—need not weld.
Any discontinuity over 1″ in length with depth over ¼″ but not greater than 1″.	Completely remove and weld. Aggregate length of welding not over 20% of plate edge length being repaired.
Any discontinuity over 1″ in length with depth greater than 1″.	Applicable provisions, as found in FOOTNOTES below.

FOOTNOTES FOR FIG. 2.10.22B

(1) Where such a discontinuity, as (A), (B), or (C), is determined visually prior to completing the joint, the size and shape of the discontinuity shall be determined by ultrasonic inspection. The area of the discontinuity shall be determined as the area of total loss of back reflection when tested in accordance with the procedures of ASTM A435.

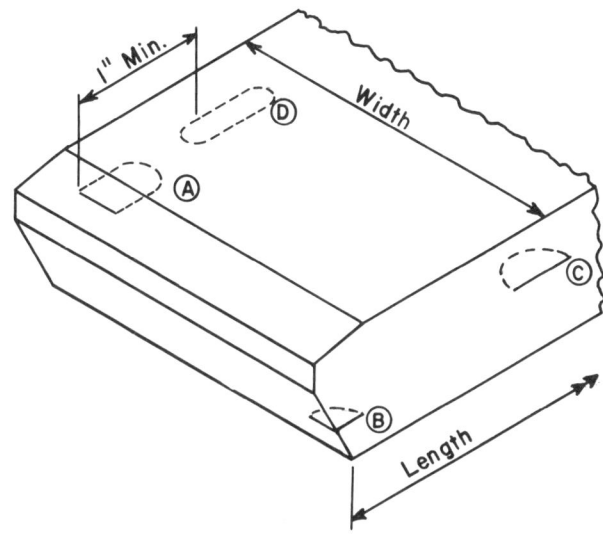

Figure 2.10.22 B

(2) For acceptance of Type (A), (B), and (D) discontinuities the area of the discontinuity (or the aggregate area of multiple discontinuities) shall not exceed 4% of plate area (length × width) except that if the width of the discontinuity, or the aggregate width of discontinuities on any transverse section, as measured perpendicular to the plate length, exceeds 20% of the plate width, the 4% plate area shall be reduced by the percentage amount of the width exceeding 20%. (For example: If discontinuity is 30% of plate width, area of discontinuity cannot exceed 3.6% of plate area.) The discontinuity on the cut edge of the plate shall be gouged out to a depth of 1" beyond its intersection of the surface by chipping, air carbon-arc gouging or grinding and blocked off by welding with the manual shielded metal-arc process in layers not to exceed $\frac{1}{8}$" in thickness.

(3) If a discontinuity, (D), not exceeding the allowable area in (2), is discovered after the joint has been completed and is determined to be 1" or more away from the face of the weld as measured on the plate surface, no repair of the discontinuity is required. If the discontinuity (D), is less than 1" away from the face of the weld, it shall be gouged out to a distance of 1" from the fusion zone of the weld by chipping, air carbon-arc gouging or grinding and blocked off by welding with the manual shielded metal-arc process for at least 4 layers not to exceed $\frac{1}{8}$" thickness per layer. The submerged arc or other process may be used for the remaining layers.

(4) If the area of the discontinuity (A), (B), or (D) exceeds the allowable in (2), the plate or subcomponent shall be rejected and replaced, or repaired at the discretion of the Engineer.

(5) The aggregate length of weld repair shall not exceed 20% of the length of the plate cut edge without approval of the Engineer.

(6) All repairs shall be in accordance with AASHO 2.10.23. Gouging of the discontinuity may be done from either plate surface or edge.

(7) For discontinuities of Type (C), if the actual net cross sectional area which would remain after removal of the discontinuity would be 98% or greater, of the net area of the plate based on nominal dimensions, it shall be unnecessary to repair by welding. Such corrections shall be faired to the plate-cut surface with a slope not exceeding 1 in 10.

(8) For discontinuities of Type (C), if the actual net cross sectional area which would remain after removal of the discontinuity would be less than 98% of the net area of the plate based on nominal dimensions, the area may be repaired by welding with the approval of the Engineer.

(9) For discontinuities of Types (A) and (B), all repair welds in A514/A517 steel shall be made with low hydrogen electrodes not exceeding $\frac{5}{32}$" in diameter. All repair welds in A514/A517 steel shall be inspected not less than 48 hours after they are completed and the butt weld shall not be made until after the repair weld has been approved by the Engineer.

(10) For discontinuities of type (C), in A514/A517 steel, no welding repairs shall be permitted.

2.10.23 — WELDS

(A) General

Welding of steel structures, when authorized in accordance with the provision of Division I, shall conform to Specifications for Welded Highway and Railway Bridges of the American Welding Society AWS D1.1-72, when using structural steel of AASHO M183 (ASTM A36), high strength low alloy structural steel of AASHO M161 (ASTM A242) of a weldable quality, AASHO M188 (ASTM A441), AASHO M223 (ASTM A572 with Supplementary Requirement S2 of AASHO M223 mandatory) Grades 42, 45 and 50 and AASHO M222 (ASTM A588 with Supplementary Requirement S1 of AASHO M222 mandatory) unless these (AASHO) Specifications conflict with the AWS Specifications, in which case these (AASHO) Specifications will govern. Welding procedures shall be qualified in accordance with Art. 9.2.3 of D1.1-72 except as otherwise provided in this Article. Reference to A514 steel in AWS D1.1-72 shall be applicable to A517 steel.

Welding procedures for ASTM A514/A517 steel shall be submitted for approval and the procedures qualified prior to the start of production welding.

No temporary or permanent welds, if not shown on the plans or permitted in the Specifications, shall be made without specific written authorization by the Engineer.

(B) Filler Metal

All electrodes for manual shielded metal-arc welding shall conform to the low-hydrogen classification requirements of the latest edition of the American Welding Society's Filler Metal Specification AWS A5.1 or AWS A5.5 and be capable of producing weld metal having an impact strength of at least 20 ft-lbs (Charpy V-Notch) at 0 F. All bare electrodes and flux used in combination for submerged arc welding, the electrode and gas shielding used in combination for gas metal-arc welding, or the electrode and shielding medium used in combination for flux cored-arc welding of steels shall conform to the requirements in the latest edition of the American Welding Society AWS A5.17, A5.18 or A5.20 capable of producing weld metal having a minimum impact strength of 20 ft-lbs (Charpy V-Notch at 0 F, or shall be capable of producing low alloy weld metal having the mechanical properties listed in paragraphs 4.12 and 4.16 of AWS D1.1-72.

Low alloy weld properties shall be determined from a multiple pass weld made in accordance with the requirements of the latest edition of the applicable Specification (AWS A5.17, A5.18, or A5.20) or the welding procedure specification. Each user shall demonstrate that each combination of electrode and flux or electrode and shielding medium will produce weld metal having the above mechanical properties until the applicable AWS Filler Metal Specification is issued; at that time the AWS Filler Metal Specification will control. The test assembly for Grades F100 and F110 shall be made using ASTM A514/A517 steel. The Engineer may accept evidence of record of a combination that has

2.10.23

been satisfactorily tested in lieu of the test required provided the same welding procedure is used.

Electrodes conforming to AWS A5.1 shall be purchased in hermetically sealed containers or shall be dried for at least two hours between 450 and 500 F before they are used. Electrodes conforming to AWS A5.5 shall be purchased in hermetically sealed containers or shall be dried one hour ± 15 minutes at a temperature of 800 F ± 25 F before being used. All electrodes for use in welding A514/A517 steel having a strength lower than that of the E100 classification shall be dried for 1 hour ± 15 minutes at a temperature of 800 F ± 25 F before being used. Electrodes shall be died prior to use if the hermetically sealed container shows evidence of damage. Immediately after removal from hermetically sealed containers or from drying ovens, electrodes shall be stored in ovens held at a temperature of at least 250 F. E70XX electrodes that are not used within four hours, E80XX within two hours, E90XX within one hour, and E100XX and E110X within one-half hour after removal from hermetically sealed containers or removal from a drying or storage oven shall be redried before use. Electrodes which have been wet shall not be used. Electrodes shall be redried no more than one time.

Flux used for submerged arc welding shall be non-hygroscopic, dry and free of contamination from dirt, mill scale, or other foreign material. All flux shall be purchased in moisture-proof packages capable of being stored under normal conditions for at least six months without such storage affecting its welding characteristics or weld properties. Flux from packages damaged in transit or handling shall be discarded or shall be dried before use at a minimum temperature of 250 F for one hour. Flux shall be placed in the dispensing system immediately upon opening a package. If flux is used from an open package or an open hopper that has been inoperative for four hours or more, the top one inch shall be discarded. Flux that has been wet shall not be used. Flux fused in welding shall not be reused.

(1) Electrodes, and flux electrodes combinations.

Base Metal (1) AASHO Designation (ASTM Designation)	Welding Process (2) (3) (4)			
	Shielded Metal-Arc	Submerged Arc	Gas Metal-Arc	Flux Cored Arc
M183 (A36)	AWS A5.1 or A5.5 E7016, 18 or 28	AWS A5.17 F61, F62, F63 or F64-EXXX	AWS A5.18 E70S-1B, 2,3,6 or E70U-1	AWS A5.20 E60T-8
M161 (A242); M188 (A441); M223 (A572) Gr. 42, 45, and 50; M222 (A588)	AWS A5.1 or A5.5 E7016, 18 or 28	AWS A5.17 F71, F72, F73 or F74-EXXX	AWS A5.18 E70S-1B, 2,3,6 or E70U-1	AWS 5.20 E70T-1, 5 or 6
A514, A517	AWS A5.5 E11018-M	Grade F110	Grade E110S	Grade E110T

(See page 330 for table notes)

Use of same type filler metal having next higher mechanical properties as listed in AWS Specification is permitted.

(1) In joints involving base metals of different yield points or strength, filler metal applicable to the lower strength base metal may be used.

(2) When welds are to be stress relieved the deposited weld metal shall not exceed 0.05 percent vanadium.

(3) See AWS Art. 4.20 for Electroslag and Electrogas weld metal requirements, Appendix C Impact Requirements mandatory.

(4) Lower strength filler metal may be used for fillet welds and partial penetration groove welds when indicated on the plans or in the special provisions.

(2) Electrodes and Flux-electrode combinations for unpainted applications. For exposed bare unpainted applications of A588 steel requiring deposited weld metal with atmospheric corrosion resistance and coloring characteristics similar to that of the base metal, the filler metal shall conform to the requirements listed in Table 4.14 of AWS D1.1-72.

In multiple-pass welds, the underlying layers may be deposited with one of the filler metals specified in the table for 2.10.23(B)(1) provided the last three layers for shielded metal-arc, gas metal-arc, or flux-cored arc welding or the last two layers for submerged-arc welding are deposited with one of the filler metals specified in the above table.

This procedure may also be used where the weld is made from one side on a backing strip which will be removed, providing the same number of layers are deposited against the backing strip with a filler metal specified in the table above.

After removing run-off tabs, the exposed weld metal of the underlying layers shall be removed by gouging to a minimum depth of ¼-inch from the edge of the base material and replaced by welding with one of the filler metals specified in the table above. The weld shall be ground flush with the edge of the material.

(C) Preheat and Interpass Temperature

With the exclusion of Electroslag and Electrogas welding, preheat and interpass temperatures shall be as follows:

MINIMUM PREHEAT AND INTERPASS TEMPERATURE [1]

Thickness of Thickest Point at Point of Welding-Inches	Welding Process	
	Shielded Metal-Arc Welding with Low Hydrogen Electrodes; Submerged Arc Welding with Carbon or Alloy Steel Wire. Neutral Flux; Gas Metal-Arc Welding; or Flux Cored Arc Welding	Submerged Arc Welding with Carbon Steel Wire, Alloy Flux
	AASHO M183 (ASTM A36); AASHO M161 (ASTM A242); AASHO M188 (ASTM A441); AASHO M223; AASHO M222 / ASTM A514/517	ASTM A514/517
To ¾, incl.	50 F / 50 F	50 F
Over ¾ to 1½, incl.	70 F / 125 F	200 F
Over 1½ to 2½, incl.	150 F / 175 F	300 F
Over 2½	225 F / 225 F	400 F

[1] Welding shall not be done when the ambient temperature is lower than 0 F. When the base metal is below the temperature listed for the welding process being used and the thickness of material being welded, it shall be preheated (except as otherwise provided) in such manner that the surface of the parts on which weld metal is being deposited are at or above the specified minimum temperature for a distance equal to the thickness of the part being welded, but not less than 3 inches, both laterally and in advance of the welding. Preheat and interpass temperatures must be sufficient to prevent crack formation. Temperature above the minimum shown may be required for highly restrained welds. For A514/A517 steel the maximum preheat and interpass temperature shall not exceed 400 F for thicknesses up to 1½ in. inclusive, and 450 F for greater thicknesses. Heat input when welding A514/A517 steel shall not exceed the steel producer's recommendation. The use of stringer beads to avoid overheating is recommended. Welding shall be carried continuously to completion or to a point that will insure freedom from cracking before the joint is allowed to cool below the minimum specified preheat and interpass temperature.

(D) Qualification of Welders, Welding Operators and Tackers

All welders, welding operators and tackers shall be qualified in accordance with the requirements of AWS D1.1-72. If a fabricating shop prequalifies its welders, welding operators and tackers according to the standard qualification procedure of the American Welding Society and certifies to the Engineer that the welder, welding operator or tacker has been prequalified within twelve months previous to the beginning of work on the subject structure, the Engineer may consider him qualified. The certificate shall state that the welder, welding operator or tacker has been doing satisfactory welding of the required type within the three month period previous to the subject work. A certification shall be submitted for each welder, welding operator or tacker, and for each project, stating the name of the welder, welding operator or tacker, the name and title of the person who conducted the examination, the kind of specimens, the position of welds, the results of the tests and the date of the examination. Such a certification of prequalification may also be accepted as proof that a welder, welding operator and tacker on field welding is qualified, if the contractor who submits it is properly staffed and equipped to conduct such an examination or if the examining and testing is done by a recognized agency which is staffed and equipped for such purpose.

In addition, evidence must be presented, satisfactory to the Engineer, that each welder, welding operator and tacker has had at least three months satisfactory experience welding A514/A517 steel. In lieu of such experience, each welder, welding operator and tacker shall be instructed in welding A514/A517 steel. The instruction course shall meet the approval of the Engineer as adequate to result in a working knowledge of the procedures for welding A514/A517 steel.

(E) Procedure Qualification

Procedure qualification of groove welds shall be in accordance with AWS D1.1-72, except for the following modifications for groove welds in A514/A517 steel:

(1) A procedure qualification is required:
   (a) with a plate thickness equal to the thickest material to be welded on the structure.
   (b) whenever there is an increase in plate thickness or whenever there is a decrease in plate thickness of 1-inch or more. Thickness is a procedure variable.
   (c) whenever the number and location of passes is changed, except a proportional reduction in the number of passes is permissible without requalification whenever the thickness reduction is less than 1-inch and the heat input does not exceed the manufacturer's recommendation for the heat input of the thickness of material being welded.
   (d) whenever the heat input is changed by more than $\pm 10\%$
   (e) whenever the arc voltage is changed by more than $\pm 7\%$
   (f) whenever the speed of travel is changed by more than $\pm 10\%$
   (g) whenever the interpass temperature is changed by more than ± 25 F.

(2) Reporting of the procedure qualification shall include specific values for all variables listed in (1) above and for those included in the AWS D1.1-72 procedure qualification.

Procedure qualification of fillet welds shall be in accordance with AWS D1.1-72 and in addition for fillet welding of A514/A517 steel over 1-inch in thickness with weld metal having minimum specified tensile strength of over 90,000 psi, the qualification test of T-joints shall be in accordance with Figure 2.10.23A.

All qualification tests of groove welds shall be radiographically tested for soundness. Radiographic testing shall apply only to the portion of the weld between the discard strips as indicated in the applicable figures of section 5 of AWS D1.1-72, except that a minimum of 6-inches of effective weld length shall be tested.

If the radiographic inspection indicates any defect or porosity exceeding the requirements of Art. 9.25 of AWS D1.1-72, the procedure test shall be considered as having failed, except that for the procedure qualification test for A514/A517 steel there shall be no defects other than allowable porosity in the qualification test weld.

(F) Inspection of Welds

(1) General

Inspection of welds in all steel other than A514/A517 may begin immediately after they are completed. Welds in A514/A517 steel shall be inspected not less than 48 hours after they are completed.

In addition to inspection as required by AWS D1.1-72 and as required below under Radiographic and Magnetic Particle Testing, all welds shall be visually examined. Procedures, technique and standards of acceptance shall be in accordance with AWS D1.1-72.

(2) Radiographic Testing

Butt welds in main members shall be tested radiographically as follows:

(a) All tension splices and all splices subject to reversal of stress, except that on girder and beam webs, only ⅙ of the web depth, beginning at the point or points of maximum tension, and 25 percent of the remainder of the web depth need be radiographed.

(b) 25% of each compression and shear splice, except that for splices in built-up members requiring less than four feet of groove weld in compression, only one joint, connecting the thickest components in each splice, need be radiographed. Maximum spacing of the radiographs shall be four times the length of the radiographs. Alternatively, 25% of the compression and shear splices selected by the Engineer, shall be radiographed. When and if unacceptable defects are found in more than 10% of the radiographs of the compression and shear splices of a member, all compression and shear splices of that and succeeding members shall be radiographed until the accumulated rejection level falls to 10% or less, at which time the radiographing shall revert to the 25% level specified above.

PROCEDURE QUALIFICATION
FILLET WELDED TEE JOINT TEST
FOR A 514/A517 STEEL IN FLANGES OVER 1" THICK AND WITH MINIMUM SPECIFIED TENSILE STRENGTH OF DEPOSITED WELD METAL OVER 90,000 psi.
MANUAL, AUTOMATIC AND SEMIAUTOMATIC

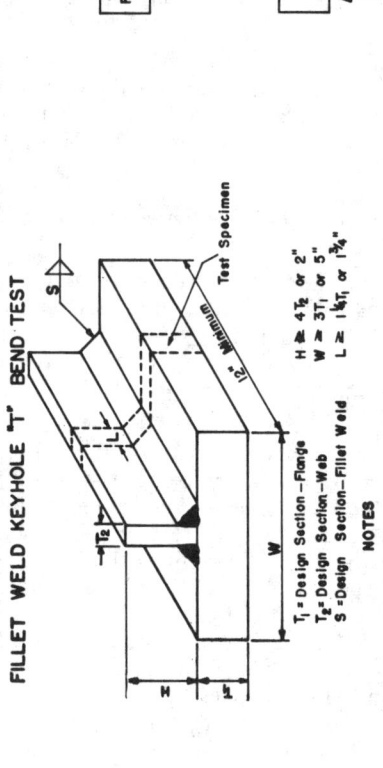

FILLET WELD KEYHOLE "T" BEND TEST

T_1 = Design Section — Flange
T_2 = Design Section — Web
S = Design Section — Fillet Weld

$H \geq 4T_2$ or 2"
$W \geq 3T_1$ or 5"
$L = 1\frac{1}{4}T_1$ or $1\frac{3}{4}$"

NOTES

I. PREPARATION OF SPECIMENS
   Test specimens may be sawed or machined (not flame cut) from a welded sample illustrated above.
   These specimens are prepared as shown above.

II. TESTING
   A. Macro Etch Test (See Procedure Qualification AWS D2.0)
   B. Bend Tests
      Specimen shall be loaded and failed as illustrated above.
   C. Hardness Tests
      Hardness Tests shall be made on a lightly etched section of Macroetched test specimens with a suitable machine.

III. TEST RESULTS REQUIRED
   A. Macro Etch Test – (See Procedure Qualification AWS D2.0)
   B. Bend Tests
      $\angle$ must not be less than 15° at failure
   C. Hardness
      Brinell Hardness of weld metal shall be within the following limits:
      1. Max. Brinell Hardness
         $\dfrac{\text{Max. Specified or Tested Tensile Strength (a) of Parent Metal}}{500} + 70$
      2. Min. Brinell Hardness
         $\dfrac{\text{Min. Specified Tensile Strength of Weld Metal}}{500}$

      This test may be performed with a Rockwell Machine and converted to Brinell Hardness using ASTM conversion chart.
         (a) Whichever is greater.

FIGURE 2.10.23 A

(c) Should a radiograph indicate rejectable defects, the areas on each side of the defect shall be radiographed to determine the extent of the defective work.

(d) Defective welds shall be radiographed after repair.

(e) The inspector shall view the making of the radiographs, examine and interpret the radiographs and the technician's reports, approve satisfactory welds, disapprove or reject unsatisfactory welds, approve satisfactory methods proposed by the contractor for repairing disapproved welds and inspect the preparation and rewelding of disapproved welds. The inspector shall record the location and findings of all radiographic inspections, together with descriptions of any repairs made.
of any repairs made.

(3) **Magnetic Particle Testing**

(a) For A514/A517 steel, fillet welds and all other welds except the butt welds radiographed in accordance with subparagraph (2) above, shall be 100% magnetic particle tested. Defective welds shall be retested after repair.

(b) The inspector shall view the making of the magnetic particle inspections, examine and interpret the magnetic particle patterns, approve satisfactory welds, disapprove or reject unsatisfactory welds, approve satisfactory methods proposed by the Contractor for repairing disapproved welds, and inspect the preparation and rewelding of disapproved welds. The inspector shall record the locations of inspected areas and defects found by magnetic particle inspection, together with a description of any repairs made.

(G) Stud Shear Connectors

(1) Stud shear connectors shall be of a design suitable for electric end welding and shall be end welded to steel beams, girders or plates with automatically timed stud welding equipment connected to a suitable power source. The type, size of diameter, and length of stud shall be as specified by the plans, specifications or special provisions. (See Figure 2.10.23B for allowable tolerances or dimensions). A maximum variation of one inch from the location shown will be accepted provided the adjacent studs are not closer than 2½ inches center to center, and clearances specified in Article 1.7.98 are maintained. Adequate provision shall be made in the fabrication of structural members to compensate for loss of camber due to welding of the shear connectors.

(2) If two or more stud welding guns are to be operated from the same power source, they shall be interlocked so that only one gun can operate at a time and so that the power source has fully recovered from making one weld before another weld is started. The power source shall be adequate for the size of stud being welded.

(3) Studs shall not be painted or galvanized. The studs shall be free from excessive rust, scale, rust pits and oil at the time of welding as determined by the Engineer. The beam surface to which the studs are welded shall be free from excessive mill scale, rust, dirt, paint, grease or any other material which might impair the quality of the weld. When necessary to obtain satisfactory welds, the areas on the beam, girder or plate to which the studs are to be welded shall be wire-brushed, peened, prick-punched or ground free of scale or rust.

(4) Welding shall not be done when the ambient temperature is below 0F, or when the surface is wet or exposed to rain or snow.

(5) When the temperature of the base metal is below 0F, preheating will be required in accordance with AWS Specifications and two studs in each 100 studs welded shall be bent 45° in addition to the first two bent as specified in paragraph (G)(13). Preheating must be to 70F and temperature of the preheated base metal shall be maintained between 32F and 70F during the welding operation.

(6) While in operation the welding gun shall be held in position without movement until the weld metal has solidified.

(7) An arc shield (ferrule) of heat-resistant ceramic or other material shall be furnished with each stud. The material shall not be detrimental to the welds or cause excessive slag and shall have sufficient strength so as not to crumble or break due to thermal or structural shock before the weld is completed.

Ferrules furnished with shop welded studs shall be removed in the shop prior to delivery, and ferrules furnished with field welded studs shall be removed before placing concrete slabs.

(8) Flux for welding shall be furnished with each stud, either attached to the end of the stud or combined with the arc shield for automatic application in the welding operation.

(9) Only qualified studs shall be used. A stud to be qualified shall have passed the tests prescribed in Paragraph (12) of this article. The arc shield used in production shall be the same as used in the qualification tests. Qualification of the studs, prior to use under the contract, shall be at the expense of the manufacturer.

(10) The Contractor shall submit to the Engineer for approval, before installation, information on the studs to be furnished as follows:
    (a) The name of the manufacturer.
    (b) A detailed description of the stud and arc shield.
    (c) A certification from the manufacturer that the stud is qualified as specified in paragraph (12).
    (d) A notarized copy of the qualification test report as certified by the testing laboratory.

(11) The studs, after welding, shall be free from any defect or substance which would interfere with their function as shear connectors.

(12) Stud Weldability Qualification Procedure
    (a) Purpose
        The purpose of this procedure is to prescribe weldability tests that will qualify a shear connector stud for welding under shop or field conditions. The tests may be performed by a university, independent laboratory or other testing agency

satisfactory to the Engineer. The agency performing the tests shall submit to the manufacturer of the stud a certified report giving procedures and results for all tests including the information listed under subparagraph (h).

(b) Duration of Qualification

A type and size of stud with arc shield, once qualified, is considered qualified until the manufacturer makes any change in the base of the stud, the flux or the arc shield which affects the welding characteristics.

(c) Preparation of Specimens

(1) Test specimens shall be prepared by welding representative studs to the center of square specimen plates, ½ to ¾ inch thick, of structural steel, AASHO M183 (ASTM A 36). At the option of the manufacturer studs may be welded to a large plate and the specimen plates cut to a size suitable for test equipment used.

(2) Studs shall be welded with power source, welding gun and control equipment as recommended by the manufacturer. Welding voltage, current and time (see subparagraph (d)) shall be measured by suitable instrumentation and recorded for each specimen. Lift and plunge shall be at the optimum setting as recommended by the manufacturer.

(d) Number of Test Specimens

(1) Thirty (30) test specimens shall be welded consecutively with optimum time held constant, but current 10% above optimum. Optimum current and time shall be the midpoint of the range normally recommended by the manufacturer for production welding.

(2) Thirty (30) test specimens shall be welded consecutively with time held constant at optimum but with current 10% below optimum.

(e) Qualification Tests

(1) Tensile tests—Ten (10) of the specimens welded in accordance with subparagraph (d)(1) and ten (10) of the specimens welded in accordance with subparagraph (d)(2) shall be subjected to a tensile test in a fixture similar to that shown in Figure 2.10.23C. A stud shall be considered as qualified for tensile tests if all test specimens have a minimum tensile strength of 60,000 psi.

(2) Bend tests—Twenty (20) of the specimens welded in accordance with subparagraph (d)(1) and twenty (20) specimens welded in accordance with subparagraph (d)(2) shall be placed in a bend testing device shown in Figure 2.10.23D and bent alternately 30 degrees in opposite directions until failure occurs. A stud shall be considered qualified for bend tests, if, on all test specimens, fracture occurs in the stud or in the plate, but not in the weld.

(f) Retests

If a failure occurs in any of the tensile or bend test

groups, that group may be retested. If failure repeats, the stud shall fail to qualify.

(g) Qualification

For a manufacturer's studs and arc shields to be qualified, each group of thirty (30) studs shall, by tests or retest, meet the requirements prescribed in subparagraph (e)(1) and (e)(2).

(h) Report of Tests

The laboratory report shall include the following:

(1) Drawings which show shapes and dimensions with tolerances of studs, arc shields and flux.

(2) A complete description of materials used in the studs and arc shields, including the quantity and analysis of the flux.

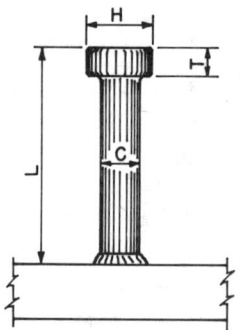

STUD SHEAR CONNECTOR[2]

DIMENSIONS AND TOLERANCES

STANDARD DIMENSIONS – INCHES			
C	L[1]	H	T
$\frac{3}{4}" +.000" \atop -.015"$	$4" +.062" \atop -.125"$	$1\frac{1}{4}" \pm \frac{1}{64}"$	$\frac{3}{8}"$ Min.
$\frac{7}{8}" +.000" \atop -.015"$	$4" +.062" \atop -.125"$	$1\frac{3}{8}" \pm \frac{1}{64}"$	$\frac{3}{8}"$ Min.

FIGURE 2.10.23B

(1) 4" Length is standard. Other lengths may be obtained by special order.

(2) Finished studs shall be of uniform quality and condition, free from injurious laps, fins, seams, cracks, twists, bends, or other injurious defects. Finish shall be as produced by cold drawing, cold rolling, or machining. Heads of shear connector or anchor studs are subject to cracks or bursts, which are names for the same thing. Cracks and bursts designate an abrupt interruption of the periphery of the head of the stud by radial separation of the metal. Such interruptions do not adversely affect the structural strength, corrosion resistance or other functional requirements of shear connector or anchor studs. A stud with cracks or bursts deeper than one-half the distance from the periphery of the head of the shank may be cause for rejection.

CONSTRUCTION

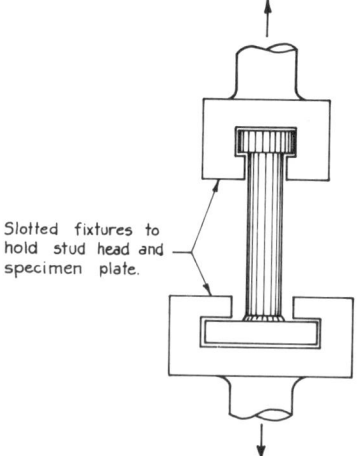

TENSILE TEST FIXTURE

FIGURE 2.10.23C

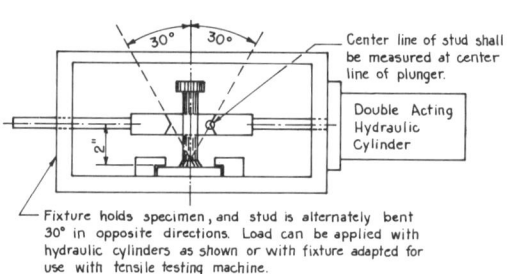

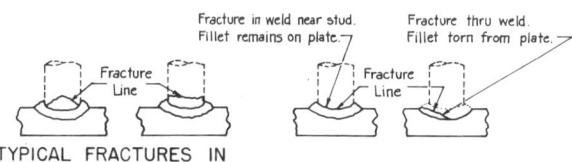

BEND TEST FIXTURE AND FAILURES

FIGURE 2.10.23D

(3) A certification that the studs and arc shields described in the report are qualified in accordance with paragraph (12)(g) of this article.

(13) The first two studs welded on each beam or girder, after being allowed to cool, shall be bent 45° by striking the stud with a hammer. If failure occurs in the weld of either stud, the procedure shall be corrected and two successive studs successfully welded and tested before any more studs are welded to the beam or girder. The Engineer shall be promptly informed of any changes in the welding procedure at any time during construction.

(14) Studs on which a full 360° weld is not obtained may, at the option of the contractor, be repaired by adding a 3/16" fillet weld in place of the lack of weld, using the shielded metal-arc process with low-hydrogen welding electrodes.

(15) If the reduction in the height of studs as they are welded becomes less than normal, welding shall be stopped immediately and not resumed until the cause has been corrected.

(16) Before welding a new stud where a defective one has been removed, the area shall be ground smooth and flush, or in the case of a pullout of metal, the pocket shall be filled with weld metal using the shielded metal-arc process with low-hydrogen welding electrodes and then ground flush. In compression areas of flanges, a new stud may be welded adjacent to the defective area in lieu of repair and replacement on existing weld area. (See paragraph (G)(1) of this article.)

(17) Inspection Requirements

(a) After the studs have been welded to the beams a visual inspection shall be made and each stud shall be given a light blow with a hammer. Any stud which does not have a complete end weld, any stud which does not emit a ringing sound when given a light blow with a hammer, any stud that has been repaired by welding or any stud which has less than normal reduction in height due to welding, shall be struck with a hammer and bent 15° from the correct axis of installation, and in the case of a defective or repaired weld the stud shall be bent 15° in the direction that will place that defective portion of the weld in the greatest tension. Studs that crack either in the weld or in the shank shall be replaced.

Studs which are to be replaced for the above reasons or because they have been otherwise rendered unacceptable may be manually welded with a full 360° 1/4" fillet weld for 3/4" studs and with a full 360° 5/16" fillet weld for 7/8" and 1" studs as specified under G(14) and G(16) above.

(b) The Engineer, at his option, may select additional studs to be subjected to the bend test specified above.

(c) The studs tested that show no sign of failure shall be left in the bent position.

(d) If during the progress of the work, inspection and testing indicate in the judgment of the Engineer that the shear connectors being obtained are not satisfactory, the Contractor will be required at his expense to make such changes in welding procedure, welding equipment and type of shear connector as necessary to secure satisfactory results.

(e) At the option of the Engineer, the Contractor may be required at any time to submit studs of the types used under the contract for check qualification by the Engineer in accordance with the procedures of 2.10.23G. In the event such studs are required for checking their qualification, payment will be made to the contractor for the studs so furnished.

2.10.24 — OXYGEN CUTTING

Steel and weld metal may be oxygen cut, provided a smooth and regular surface free from cracks and notches is secured, and provided that an accurate profile is secured by the use of a mechanical guide. Hand cutting shall be done only where approved by the Engineer.

Mill scale and extraneous material shall be removed from the torch side of A514/A517 steel plates along the lines to be flame cut, when necessary to obviate excessive notches.

All oxygen cutting shall be in accordance with AWS D1.1-72, except that occasional notches or gouges in edges of A514/A517 steel shall not be repaired by welding except under the following conditions:

(a) Cutting defects not more than $3/16$ inch deep in plate edges which will form the faces of a groove weld joint and which will subsequently be completely fused with the weld may be repaired by welding. Nonmetallic stringers or pipes opening to these edges shall be removed to a depth of $1/4$ inch below the surface by grinding or chipping and the gouge repaired by welding. Laminations opening to these edges shall be removed to a depth of $1/2$ inch below the surface by grinding and chipping and the gouge repaired by welding.

(b) Cutting defects not more than $3/16$ inch deep in plate edges which will form a fillet-welded corner joint shall be repaired by welding only on the part of the edge which will become the faying surface for the joint and the fusion zone of the fillet weld. The part of the defect outside the toe of the completed fillet weld shall be removed by machining or grinding, and faired with the surface of the cut on a bevel of 1 to 6 or less.

(c) All welding for these repairs shall be made by suitably preparing the defect, welding with low hydrogen electrodes not exceeding $5/32$ inch in diameter, observing the applicable requirements of Article 2.10.23, and grinding the completed weld smooth and flush with the adjacent surface to produce a workmanlike finish.

Oxygen cut edges of AASHO M 187 (ASTM A440) steel $1/2$ inch or greater in thickness * shall be removed to a depth of at least $1/8$ inch by

---

* AASHO M 161 (ASTM A242) steel not approved for welded construction requires this treatment of oxygen cut edges in all thicknesses and with bend tests, as applicable, conforming to AASHO M 161 (ASTM A242) for the thickness of material under consideration.

machining or grinding except that machine flame cut edges may be used without such removal if the edges are softened after cutting: (a) by heating the cut edge uniformly and progressively to a red heat, visible in ordinary shop light (1,150 to 1,250 F) to a depth of at least $\frac{1}{16}$ inch; or (b) by means of a post heating torch attached to and following the cutting torch with the tips, gas pressure, speed of travel, and the distance of the post heating torch from the kerf regulated to the thickness of the steel. Bend test specimens 1½ inches wide and of the full thickness of the material or with thickness reduced to ¾ inch in accordance with Paragraph 6.3.5 of ASTM A6 and having edges flame cut and flame softened in accordance with this article shall meet the bend test requirements specified in AASHO M 187 (ASTM A440) for the thickness of material under consideration.

Oxygen cut surfaces of members carrying calculated stress shall have their corners rounded to $\frac{1}{16}$ inch radius by grinding after cutting.

2.10.25 — FACING OF BEARING SURFACES

The surface finish of bearing and base plates and other bearing surfaces that are to come in contact with each other or with concrete shall meet the ANSI surface roughness requirements as defined in ANSI B46.1, Surface Roughness, Waviness and Lay, Part I:

Steel slabs	ANSI	2,000
Heavy plates in contact in shoes to be welded	ANSI	1,000
Milled ends of compression members, milled or ground ends of stiffeners and fillers	ANSI	500
Bridge rollers and rockers	ANSI	250
Pins and pin holes	ANSI	125
Sliding bearings	ANSI	125

2.10.26 — ABUTTING JOINTS

Abutting joints in compression members and girder flanges, and in tension members where so specified on the drawings, shall be faced and brought to an even bearing. Where joints are not faced, the opening shall not exceed ¼ inch.

2.10.27 — END CONNECTION ANGLES

Floorbeams, stringers and girders having end connection angles shall be built to exact length shown on the plans measured between the heels of the connection angles, with a permissible tolerance of plus 0 inch to minus $\frac{1}{16}$ inch. Where continuity is to be required, end connections shall be faced. The thickness of the connection angles shall not be less than ⅜ inch, nor less than that shown on the detail drawings, after facing.

2.10.28 — LACING BARS

The ends of lacing bars shall be neatly rounded unless another form is required.

2.10.29 — FINISHED MEMBERS

Finished members shall be true to line and free from twists, bends and open joints.

2.10.30 — WEB PLATES

In girders having no cover plates and not to be encased in concrete, the top edge of the web plate shall not extend above the backs of the flange angles and shall not be more than ⅛ inch below at any point. Any portion of the plate projecting beyond the angles shall be chipped flush with the backs of the angles. Web plates of girders having cover plates may be ½ inch less in width than the distance back to back of flange angles.

Splices of webs in girders without cover plates shall be sealed on top with red lead paste prior to painting.

At web splices, the clearance between the ends of the web plates shall not exceed ⅜ inch. The clearance at the top and bottom ends of the web splice plates shall not exceed ¼ inch.

2.10.31 — BENT PLATES

Unwelded, cold-bent, load-carrying, rolled-steel plates shall conform to the following:

(a) They shall be so taken from the stock plates that the bend line will be at right angles to the direction of rolling, except that cold-bent ribs for orthotropic deck bridges may be bent in the direction of rolling if permitted by the Engineer.

(b) Bending shall be such that no cracking of the plate occurs. Minimum bend radii, measured to the concave face of the metal, are shown in the following table:

THICKNESS IN INCHES

	Up to ½	Over ½ to 1	Over 1 to 1½	Over 1½ to 2½	Over 2½ to 4
All grades of structural steel in this specification	2 t	2½ t	3 t	3½ t	4t

NOTE: Low alloy steel in thicknesses over ½" may require hot bending for small radii.

Allowance for springback of A514 and A517 steels should be about 3 times that for structural carbon steel. For brake press forming, the lower die span should be at least 16 times the plate thickness. Multiple hits are advisable.

If a shorter radius is essential, the plates shall be bent hot at a temperature not greater than 1200 F, except for A514/A517 steel. If A514/A517 steel plates to be bent are heated to a temperature greater than 1125 F, they must be requenched and tempered in accordance with the producing mill's practice. Hot bent plates shall conform to requirement (a) above.

(c) Before bending, the corners of the plate shall be rounded to a radius of $\frac{1}{16}$ inch throughout the portion of the plate at which the bending is to occur.

2.10.32 — FIT OF STIFFENERS

End stiffeners of girders and stiffeners intended as supports for concentrated loads shall have full bearing (either milled, ground or, on weldable steel in compression areas of flanges, welded as shown on the plans or specified) on the flanges to which they transmit load or from which they receive load. Stiffeners not intended to support concentrated loads shall, unless shown or specified otherwise, fit sufficiently tight to exclude water after being painted. Fillers under stiffeners shall fit within $\frac{1}{4}$ inch at each end.

2.10.33 — EYEBARS

Pin holes may be flame cut at least two inches smaller in diameter than the finished pin diameter. All eyebars that are to be placed side by side in the structure shall be securely fastened together in the order that they will be placed on the pin and bored at both ends while so clamped. Eyebars shall be packed and match marked for shipment and erection. All identifying marks shall be stamped with steel stencils on the edge of one head of each member after fabrication is completed so as to be visible when the bars are nested in place on the structure. The eyebars shall be straight and free from twists and the pin holes shall be accurately located on the centerline of the bar. The inclination of any bar to the plane of the truss shall not exceed $\frac{1}{16}$ inch to a foot.

The edges of eyebars that lie between the transverse centerline of their pin holes shall be cut simultaneously with two mechanically operated torches abreast of each other, guided by a substantial template, in such a manner as to prevent distortion of the plates.

2.10.34 — ANNEALING AND STRESS RELIEVING

Structural members which are indicated in the contract to be annealed or normalized shall have finished machining, boring and straightening done subsequent to heat treatment. Normalizing and annealing (full annealing) shall be as specified in ASTM E44. The temperatures shall be maintained uniformly throughout the furnace during the heating and cooling so that the temperature at no two points on the member will differ by more than 100 F at any one time.

Members of A514/A517 steels shall not be annealed or normalized and shall be stress relieved only with the approval of the Engineer.

A record of each furnace charge shall identify the pieces in the charge and show the temperatures and schedule actually used. Proper instruments including recording pyrometers, shall be provided for determining at any time the temperatures of members in the furnace. The records of the treatment operation shall be available to and meet the approval of the Engineer. The holding temperature for stress relieving A514/A517 steel shall not exceed 1125 F.

Members, such as bridge shoes, pedestals, or other parts which are built up by welding sections of plate together shall be stress relieved in accordance with the procedure of paragraph 3.9 of AWS D1.1-72 when required by the plans, specifications or special provisions governing the contract.

2.10.35 — PINS AND ROLLERS

Pins and rollers shall be accurately turned to the dimensions shown on the drawings and shall be straight, smooth, and free from flaws. Pins and rollers more than 9 inches in diameter shall be forged and annealed. Pins and rollers 9 inches or less in diameter may be either forged and annealed or cold-finished carbon-steel shafting.

In pins larger than 9 inches in diameter, a hole not less than 2 inches in diameter shall be bored full length along the axis after the forging has been allowed to cool to a temperature below the critical range, under suitable conditions to prevent injury by too rapid cooling, and before being annealed.

2.10.36 — BORING PIN HOLES

Pin holes shall be bored true to the specified diameter, smooth and straight, at right angles with the axis of the member and parallel with each other unless otherwise required. The final surface shall be produced by a finishing cut.

The distance outside to outside of end holes in tension members and inside to inside of end holes in compression members shall not vary from that specified more than $\frac{1}{32}$ inch. Boring of holes in built-up members shall be done after the riveting is completed.

2.10.37 — PIN CLEARANCES

The diameter of the pin hole shall not exceed that of the pin by more than $\frac{1}{50}$ inch for pins 5 inches or less in diameter, or by $\frac{1}{32}$ inch for larger pins.

2.10.38 — THREADS FOR BOLTS AND PINS

Threads for all bolts and pins for structural steel construction shall conform to the Unified Standard Series UNC — ANSI B1.1, Class 2A for external threads and Class 2B for internal threads, except that pin ends having a diameter of $1\frac{3}{8}$ inches or more shall be threaded 6 threads to the inch.

2.10.39 — PILOT AND DRIVING NUTS

Two pilot nuts and two driving nuts for each size of pin shall be furnished, unless otherwise specified.

2.10.40 — NOTICE OF BEGINNING OF WORK

The Contractor shall give the Engineer ample notice of the beginning of work at the mill or in the shop, so that inspection may be provided. The term "mill" means any rolling mill or foundry where material for the work is to be manufactured. No material shall be manufactured, or work done in the shop, before the Engineer has been so notified.

2.10.41 — FACILITIES FOR INSPECTION

The Contractor shall furnish facilities for the inspection of material and workmanship in the mill and shop, and the inspectors shall be allowed free access to the necessary parts of the works.

2.10.42 — INSPECTOR'S AUTHORITY

Inspectors shall have the authority to reject any material or work which does not meet the requirements of these specifications. In case of dispute the Contractor may appeal to the Engineer, whose decision shall be final.

2.10.43 — WORKING DRAWINGS AND IDENTIFICATION OF STEEL DURING FABRICATION

(A) Working Drawings

The Contractor shall submit copies of the detailed shop drawings to the Engineer for approval. Any work done prior to the approval of these plans shall be at the contractor's risk. When material must be ordered in advance, specific approval of such an action shall be obtained by the Contractor prior to placing the order. Shop drawings for steel structures shall give full detailed dimensions and sizes of component parts of the structure and details of all miscellaneous parts, such as pins, nuts, bolts, rivets, drains, etc.

The Contractor shall expressly understand that the Engineer's approval of the working drawings submitted by the Contractor covers the requirements for "strength and detail," and the Engineer assumes no responsibility for errors in dimensions.

(B) Identification of Steels During Fabrication

(1) Identification by Contractor

The Engineer shall be furnished with complete certified mill test reports showing chemical analysis and physical tests for each heat of steel, for all members unless excepted by the Engineer. Each piece of steel to be fabricated shall be properly identified for the Engineer.

Shop drawings shall specifically identify each piece that is to be made of steel which is to be other than AASHO M 183 (ASTM A36) steel. Pieces made of different grades of steel shall not be given the same assembling or erecting mark, even though they are of identical dimensions and detail.

The contractor's system of assembly-marking individual pieces, required to be made of steel other than AASHO M 183 (ASTM A36) steel, and the issuance of cutting instructions to the shop (generally by cross-referencing of the assembly-marks shown on the shop drawings with the corresponding item covered on the mill purchase order) shall be such as to maintain identity of the mill test report number.

The Contractor may furnish from stock, material that he can identify by heat number and mill test report.

Any excess material placed in stock for later use shall be marked with the mill test report number and shall be marked with its AASHO M 160 (ASTM A6) specification identification color code (see Table 2.10.43) if any, when separated from the full-size piece furnished by the supplier.

(2) Identification of Steels During Fabrication.

During fabrication, up to the point of assembling members, each piece of steel, other than AASHO M 183 (ASTM A36) steel, shall show clearly and legibly its specification identification color code shown in Table 2.10.43.

Individually marked pieces of steel which are used in furnished size, or reduced from furnished size only by end or edge trim, that does not disturb the heat number or color code or leave any usable piece, may be used without further color coding provided that the heat number or color code remains legible.

Pieces of steel, other than AASHO M 183 (ASTM A36) steel, which are to be cut to smaller size pieces shall, before cutting, be legibly marked with the AASHO M 160 (ASTM A6) specification identification color code.

Individual pieces of steel, other than AASHO M 183 (ASTM A36) steel, which are furnished in tagged lifts or bundles shall be marked with the AASHO M 160 (ASTM A6) specification identification color code immediately upon being removed from the bundle or lift.

Pieces of steel, other than AASHO M 183 (ASTM A36) steel, which prior to assembling into members, will be subject to fabricating operations such as blast cleaning, galvanizing, heating for forming, or painting which might obliterate paint color code marking, shall be marked for grade by steel die stamping or by a substantial tag firmly attached.

The following identification color code shall be used to identify material required to meet the individual specifications listed in Table 2.10.43.

TABLE 2.10.43
IDENTIFICATION COLOR CODES

AASHO M161	(A 242)	Blue
AASHO M187	(A 440)	Brown
AASHO M188	(A 441)	Yellow
	A 514	Red
	A 517	Red and Blue
AASHO M223	(A 572)	Grade 42 — Green and White
		Grade 45 — Green and Black
		Grade 50 — Green and Yellow
		Grade 55 — Green and Brown
		Grade 60 — Green and Gray
		Grade 65 — Green and Blue
AASHO M222	(A 588)	Blue and Yellow

Other steels, except AASHO M 183 (ASTM A36) steel, not covered above, nor included in the AASHO M 160 (ASTM A6) Specification, shall have an individual color code which shall be established and on record for the Engineer.

(3) Certification of Identification.

Upon request, the Contractor shall furnish an affidavit certifying that throughout the fabrication operation he has maintained the identification of steel in accordance with this specification.

2.10.44 — WEIGHING OF MEMBERS

In case it is specified that any part of the material is to be paid for by actual weight, finished work shall be weighed in the presence of the Inspector, if practicable. In such case, the Contractor shall supply satisfactory scales and shall perform all work involved in handling and weighing the various parts.

2.10.45 — FULL SIZE TESTS

When full size tests of fabricated structural members or eyebars are required by the contract, the plans or specifications shall state the number and nature of the tests, the results to be attained and the measurements of strength, deformation or other performance that are to be made. The Contractor shall provide suitable facilities, material, supervision and labor necessary for making and recording the tests. The members tested in accordance with the contract shall be paid for in accordance with Article 2.10.64. The cost of testing including equipment, handling, supervision, labor and incidentals for making the tests shall be included in the contract price for the fabrication or fabrication and erection of structural steel, whichever is the applicable item in the contract, unless otherwise specified.

2.10.46 — MARKING AND SHIPPING

Each member shall be painted or marked with an erection mark for identification and an erection diagram shall be furnished with erection marks shown thereon.

The Contractor shall furnish to the Engineer as many copies of material orders, shipping statements and erection diagrams as the Engineer may direct. The weights of the individual members shall be shown on the statements. Members weighing more than 3 tons shall have the weights marked thereon. Structural members shall be loaded on trucks or cars in such a manner that they may be transported and unloaded at their destination without being excessively stressed, deformed or otherwise damaged.

Bolts and rivets of one length and diameter and loose nuts or washers of each size shall be packed separately. Pins, small parts and packages of bolts, rivets, washers and nuts shall be shipped in boxes, crates, kegs or barrels, but the gross weight of any package shall not exceed 300 lbs. A list and description of the contained material shall be plainly marked on the outside of each shipping container.

2.10.47 — PAINTING

Unless otherwise shown on the plans or specified, all iron and steel surfaces shall be cleaned and painted in accordance with Section 2.14 "PAINTING METAL STRUCTURES."

ERECTION

2.10.48 — ORTHOTROPIC-DECK BRIDGES

(A) Protection of Deck Plate After Sand Blasting

If sand blasting to a white metal, or an equivalent method, is used to prepare the deck plate to receive a wearing surface, a protective coating shall be applied to the plate immediately after cleaning.

(B) Dimensional Tolerance Limits

Dimensional tolerance limits for orthotropic-deck bridge members shall be applied to each completed but unloaded member and shall be as specified in the AWS Specification referred to in Article 2.10.23 (A) except as superseded hereinafter. The deviation from detailed flatness, straightness, or curvature at any point shall be the perpendicular distance from that point to a templet edge having the detailed straightness or curvature and which is in contact with the element at two other points. The term element as used herein, refers to individual panels, stiffeners, flanges, or other pieces. The templet edge may have any length not exceeding the greatest dimension of the element being examined and, for any panel, not exceeding 1.5 times the least dimension of the panel; it may be placed anywhere within the boundaries of the element. The deviation shall be measured between adjacent points of contact of the templet edge with the element; the distance between these adjacent points of contact shall be used in the formulas to establish the tolerance limits for the segment being measured whenever this distance is less than the applicable dimension of the element specified for the formula.

(1) Flatness of Panels

The term panel as used in this article means a clear area of steel plate surface bounded by stiffeners, webs, flanges, or plate edges and not further subdivided by any such elements. The provisions of this article apply to all panels in the bridge; for plates stiffened on one side only such as orthotropic deck plates or flanges of box girders, this includes the total clear width on the side without stiffeners as well as the panels between stiffeners on the side with stiffeners.

The maximum deviation from detailed flatness or curvature of a panel shall not exceed the greater of:

$$\tfrac{3}{16} \text{ inch or } \frac{D}{144\sqrt{T}} \text{ inch}$$

where

D = the least dimension in inches along the boundary of the panel and
T = the minimum thickness in inches of the plate comprising the panel

(2) Straightness of Longitudinal Stiffeners Subject to Calculated Compressive Stress, Including Orthotropic-Deck Ribs

The maximum deviation from detailed straightness or curvature in any direction perpendicular to its length of a longitudinal stiffener subject to calculated compressive stress, including each orthotropic-deck rib, shall not exceed:

$$\frac{L}{480}$$

where

L = the length of the stiffener or rib between cross members, webs, or flanges.

(3) Straightness of Transverse Web Stiffeners and Other Stiffeners not Subject to Calculated Compressive Stress

The maximum deviation from detailed straightness or curvature in any direction perpendicular to its length of a transverse web stiffener or other stiffener not subject to calculated compressive stress shall not exceed:

$$\frac{L}{240}$$

where

L = the length of the stiffener between cross members, webs, or flanges.

2.10.49 — ERECTION OF STRUCTURE

If the substructure and superstructure are built under separate contracts, the Department will provide the masonry, constructed to correct lines and elevations and properly finished, and will establish the lines and elevations required for setting the steel.

The Contractor shall erect the metalwork, remove the temporary construction, and do all work required to complete the bridge or bridges as covered by the agreement, including the removal of the old structure or structures, if stipulated, all in accordance with the plans and these specifications.

2.10.50 — PLANS

If the fabrication and erection of the superstructure are done under separate contracts, the Department will furnish detail plans for the bridge or bridges to be erected, including shop details, camber diagrams, erection diagrams, list of field rivets and bolts, and copy of shipping statements showing a list of parts and their weights.

2.10.51 — PLANT

The Contractor shall provide the falsework and all tools, machinery and appliances, including drift pins and fitting-up bolts, necessary for the expeditious handling of the work.

2.10.52 — DELIVERY OF MATERIALS

If the contract is for erection only, the Contractor shall receive the materials entering into the finished structure free of charge at the place designated and loaded or unloaded as specified. The Contractor shall unload promptly upon delivery any material delivered to the place designated, which he is required to unload. Otherwise, he shall be responsible for demurrage charges.

2.10.53 — HANDLING AND STORING MATERIALS

Material to be stored shall be placed on skids above the ground. It shall be kept clean and properly drained. Girders and beams shall be placed upright and shored. Long members, such as columns and chords, shall be supported on skids placed near enough together to prevent injury from deflection. If the contract is for erection only, the Contractor shall check the material turned over to him against the shipping lists and report promptly in writing any shortage or injury discovered. He shall be responsible for the loss of any material while in his care, or for any damage caused to it after being received by him.

2.10.54 — FALSEWORK

The falsework shall be properly designed and substantially constructed and maintained for the loads which will come upon it. The Contractor, if required, shall prepare and submit to the Engineer for approval, plans for falsework or for changes in an existing structure necessary for maintaining traffic. Approval of the contractor's plans shall not be considered as relieving the Contractor of any responsibility.

2.10.55 — METHODS AND EQUIPMENT

Before starting the work of erection, the Contractor shall inform the Engineer fully as to the method of erection he proposes to follow, and the amount and character of equipment he proposes to use, which shall be subject to the approval of the Engineer. The approval of the Engineer shall not be considered as relieving the Contractor of the responsibility for the safety of his method or equipment or from carrying out the work in full accordance with the plans and specifications. No work shall be done until such approval by the Engineer has been obtained.

2.10.56 — BEARINGS AND ANCHORAGES

Bridge bearings shall be set level, in exact position, and must have full and even bearing on the masonry.

Elastomeric bearing pads, if used, shall set directly on the concrete masonry.

Cast iron or steel, or rolled steel bearings shall be bedded on the masonry with alternate layers of red lead and canvas, or a single thickness of sheet lead or preformed fabric bearing pad.

The Contractor shall drill holes for anchor bolts and set them in portland cement grout, or pre-set them as shown on the plans or as specified.

Location of anchors and setting of rockers or rollers shall take into account any variation from mean temperature at time of setting and anticipated lengthening of bottom chord or bottom flange due to dead load after setting, the intention being that, as near as practicable, at mean temperature and under dead load the rockers and rollers shall set vertical and anchor bolts at expansion bearings will center their slots. Care shall be taken that full and free movement of the superstructure at the movable bearings is not restricted by improper setting or adjustment of bearings or anchor bolt and nuts.

Bridge bearings shall not be placed on masonry bearing areas which are irregular or improperly formed.

2.10.57 — STRAIGHTENING BENT MATERIAL AND CAMBERING

(A) Straightening Bent Material

The straightening of plates, angles, other shapes and built-up members, when permitted by the Engineer, shall be done by methods that will not produce fracture or other injury. Distorted members shall be straightened by mechanical means or, if approved by the Engineer, by the careful planned and supervised application of a limited amount of localized heat, except that heat straightening of A514/A517 steel members shall be done only under rigidly controlled procedures, each application subject to the approval of the Engineer. In no case shall the maximum temperature of the A514/A517 steel exceed 1125F, nor shall the temperature exceed 950F at the weld metal or within 6 inches of weld metal. Heat shall not be applied directly on weld metal. In all other steels, the temperature of the heated area shall not exceed 1200 F (a dull red) as controlled by temperature indicating crayons, liquids or bimetal thermometers.

Parts to be heat straightened shall be substantially free of stress and from external forces, except stresses resulting from mechanical means used in conjunction with the application of heat.

Following the straightening of a bend or buckle, the surface of the metal shall be carefully inspected for evidence of fracture.

(B) Cambering

Correction of errors in camber in welded beams and girders of A514/A517 material shall be done only under rigidly controlled procedures, each application subject to approval of the Engineer.

2.10.58 — ASSEMBLING STEEL

The parts shall be accurately assembled as shown on the plans and any match-marks shall be followed. The material shall be carefully handled so that no parts will be bent, broken or otherwise damaged. Hammering which will injure or distort the members shall not be done. Bearing surfaces and surfaces to be in permanent contact shall be cleaned before the members are assembled. Unless erected by the cantilever method, truss spans shall be erected on blocking so as to give the trusses proper camber. The blocking shall be left in place until the tension chord splices are fully riveted or bolted and all other truss connections pinned and bolted. Rivets or permanent bolts in splices of butt joints of compression members and rivets or permanent bolts in railings shall not be driven or tightened until the span has been swung. Splices and field connections shall have one half of the holes filled with bolts and cylindrical erection pins (half bolts and half pins) before riveting or bolting with high-strength bolts. Splices and connections carrying traffic during erection shall have three-fourths of the holes so filled.

Fitting-up bolts shall be of the same nominal diameter as the rivets or high-strength bolts, and cylindrical erection pins shall be $\frac{1}{32}$ inch larger.

2.10.59 — RIVETING

Pneumatic hammers shall be used for field riveting, except when the use of hand tools is permitted by the Engineer. Rivets larger than ⅞ inch in diameter shall not be driven by hand. Cup-faced dollies, fitting the head closely to insure good bearing, shall be used. Connections shall be accurately and securely fitted up before the rivets are driven. Drifting shall be only such as to draw the parts into position and not sufficient to enlarge the holes or distort the metal. Unfair holes shall be reamed or drilled.

Rivets shall be heated uniformly to a light "cherry red" color and shall be driven while hot. They shall not be overheated or burned. Rivet heads shall be full and symmetrical, concentric with the shank, and shall have full bearing all around. They shall not be smaller than the heads of the shop rivets. Rivets shall be tight and shall grip the connected parts securely together. Caulking or recupping will not be permitted. In removing rivets, the surrounding metal shall not be injured; if necessary, they shall be drilled out.

2.10.60 — PIN CONNECTIONS

Pilot and driving nuts shall be used in driving pins. They shall be furnished by the Contractor without charge. Pins shall be so driven that the members will take full bearing on them. Pin nuts shall be screwed up tight and the threads burred at the face of the nut with a pointed tool.

2.10.61 — MISFITS

The correction of minor misfits involving harmless amounts of reaming, cutting and chipping will be considered a legitimate part of the erection. However, any error in the shop fabrication or deformation resulting from handling and transportation which prevents the proper assembling and fitting up of parts by the moderate use of drift pins or by a moderate amount of reaming and slight chipping or cutting, shall be reported immediately to the inspector and his approval of the method of correction obtained. The correction shall be made in his presence. If the contract provides for complete fabrication and erection, the Contractor shall be responsible for all misfits, errors and injuries and shall make the necessary corrections and replacements. If the contract is for erection only, the inspector, with the cooperation of the Contractor, shall keep a correct record of labor and materials used and the Contractor shall render within 30 days an itemized bill for the approval of the Engineer.

2.10.62 — REMOVAL OF OLD STRUCTURE AND FALSEWORK

If stipulated in the agreement, the Contractor shall dismantle the old structure which, unless otherwise provided, shall be the property of the Department, and shall store the material in the immediate vicinity of the bridge site as the Engineer may direct. If the old structure is to be re-erected, it shall be dismantled without unnecessary damage and the parts match-marked and carefully stockpiled.

Upon completion of the erection and before final acceptance, the Contractor shall remove all falsework, excavated or useless materials, rubbish and temporary buildings, replace or renew any fences damaged and restore in an acceptable manner all property, both public and private, which may have been damaged during the prosecution of this work, and shall leave the bridge site and adjacent highway in a neat and presentable condition satisfactory to the Engineer. All excavated material or falsework placed in the stream channel during construction shall be removed by the Contractor before final acceptance.

2.10.63 — METHOD OF MEASUREMENT

The weight of the metal work to be paid for under the item of structural steel shall be computed on the following basis:

(a) Unit weights, pounds per cubic foot—

Aluminum, cast or wrought	173.0
Bronze, cast	536.0
Copper-alloy	536.0
Copper sheet	558.0
Iron, cast	445.0
Iron, mallable	470.0
Iron, wrought	487.0
Lead, sheet	707.0
Steel, rolled, cast, copper bearing, silicon, nickel and stainless	490.0
Zinc	450.0

(b) The weights of rolled shapes shall be computed on the basis of their nominal weights per foot as shown on the drawings, or listed in the hand books.

The weights of plates shall be computed on the basis of the nominal weight for their width and thickness as shown on the drawings, plus an estimated over-run computed as one-half the "Permissible Variation in Thickness and Weight" as tabulated in Specification, "General Requirements for Delivery of Rolled Steel Plates, Shapes, Steel Piling and Bars for Structural Use," AASHO M 160 (ASTM A 6).

(c) The weight of castings shall be computed from the dimensions shown on the approved shop drawings, deducting for open holes. To this weight shall be added 5 per cent allowance for fillets and over-run. Scale weights may be substituted for computed weights in the case of castings or of small complex parts for which accurate computations of weight would be difficult.

(d) The weight of temporary erection bolts, shop and field paint, boxes, crates and other containers used for shipping, and materials used for supporting members during transportation and erection, shall not be included.

(e) In computing pay weight on the basis of computed net weight the following stipulations in addition to those in paragraphs a, b, c and d above shall apply.

The weight shall be computed on the basis of the net finished dimensions of the parts as shown on the approved shop drawing, deducting for copes, cuts, clips and all open holes, except rivet and bolt holes.

The weight of all rivet heads, both field and shop, shall be included on the basis of the following weights:

Diameter of Rivet Inches	Weights per 100 heads, pounds
½	4
⅝	7
¾	12
⅞	18
1	26
1⅛	36
1¼	48

The weight of heads, nuts, single washers and threaded stick-through of all high tensile strength shop bolts shall be included on the basis of the following weights:

Diameter of Bolt Inches	Weight per 100 Bolts, pounds
½	19.7
⅝	31.7
¾	52.4
⅞	80.4
1	116.7
1⅛	165.1
1¼	212.0
1⅜	280.0
1½	340.0

The weight of weld metal shall be computed on the basis of the theoretical volume from dimensions of the welds.

(f) In computing pay weight on the basis of computed gross weight, the following stipulations in addition to those of paragraphs a, b, c and d above shall apply.

The weight shall be figured on the basis of rectangular dimensions for all plates, and ordered over-all lengths for all structural shapes; except that (A) when parts can be economically cut in multiples from material of larger dimensions, the computed weight shall be that of the material from which the parts are cut, and (B) all material shall be ordered to produce as little waste as practicable when cut and finished by modern shop methods.

No deductions from the computed weight of rolled steel shall be made for copes, clips, sheared edges, punchings, borings, milling or planning; or from the computed weight of castings to allow for drillings or borings. The weight of shop rivets shall be computed on the basis of reasonable average lengths, in accordance with the following table:

Diameter of Rivets, In.....	½	⅝	¾	⅞	1	1⅛	1¼
Weight per 100 Rivets, Lbs..	20	30	50	100	150	250	325

The weight of field rivets shall be computed on the basis of the field rivet shipping list.

The weight of weld metal shall be computed on the basis of the theoretical volume of the dimensions of the welds. To this weight shall be added 50 per cent allowance for overrun.

(g) In computing pay weight on the basis of scale weights, the pay quantity of structural steel will be the shop scale weight of the fabricated members, which shall be weighed on satisfactory scales in the presence of the inspector. If the shop paint has been applied to the completed member when weighed, 0.4 of 1 per cent of the weight of the member shall be deducted from the scale weights to compensate for weight of shop paint. The weight of field rivets shall be based on the approved shipping list. No payment will be made for any weight in excess of 1½ per cent above the computed net weight of the whole item.

2.10.64 — BASIS OF PAYMENT

The contract price for fabrication and erection of structural steel shall include all labor, materials, transportation, and shop and field painting necessary for the proper completion of the work in accordance with the contract. The contract price for fabrication without erection shall include all labor and materials necessary for the proper completion of the work in accordance with the contract.

Under contracts containing an item for structural steel, all metal parts other than metal reinforcement for concrete, such as anchor bolts and nuts, shoes, rockers, rollers, bearing and slab plates, pins and nuts, expansion dams, roadway drains and scuppers, weld metal, bolts embedded in concrete, cradles and brackets, railing and railing posts, blast plates and water stops, shall be paid for as structural steel unless otherwise stipulated.

Payment will be made on a pound-price or a lump sum basis as required by the terms of the contract, but unless stipulated otherwise, it shall be on a pound-price basis.

The payment in pound-price contracts shall be based on the computed net weight of metal in the fabricated and erected structures unless it is provided that payment shall be based on computed gross weight or scale weight.

For members comprising both carbon steel and other special steel or material, when separate unit prices are provided for same, the weight of each class of steel in each such member shall be separately computed, and paid for at the contract unit price therefor.

Full-size members which are tested in accordance with the specifications, when such tests are required by the contract, shall be paid for at the same rate as for comparable members for the structure. Members which fail to meet the contract requirements, and members rejected as a result of tests, shall not be paid for by the purchaser.

Section 11 — BRONZE OR COPPER-ALLOY BEARING AND EXPANSION PLATES

2.11.1 — GENERAL

Plates shall be of the kind of metal specified in the special provisions or as shown on the plans.

2.11.2 — MATERIALS

(A) Bronze Bearing and Expansion Plates

Bronze bearing and expansion plates shall conform to the Specification for Bronze Castings for Bridges and Turntables, AASHO M 107 (ASTM B 22). Alloy B shall be furnished unless otherwise specified.

(B) Rolled Copper-Alloy Bearings and Expansion Plates

Rolled copper-alloy bearing and expansion plates shall conform to the Specification for Rolled Copper-Alloy Bearing and Expansion Plates and Sheets for Bridge and Other structural Uses, AASHO M 108 (ASTM B 100). Alloy No. 1 shall be furnished unless otherwise specified.

(C) Metal Powder Sintered Bearings and Expansion Joints (Oil Impregnated)

Metal powder sintered bearings and expansion plates shall conform to the specifications for such material of the ASTM B 438, Grade 1, Type II or Grade 2, Type I.

2.11.3 — BRONZE PLATES

Plates shall be cast according to details shown on the plans. Sliding surfaces shall be planed parallel to the movement of the spans and polished unless detailed otherwise.

2.11.4 — COPPER-ALLOY PLATES

Plates shall be furnished according to details shown on the plans. Finishing of the rolled plates will not be required provided they have a plane, true and smooth surface.

2.11.5 — PLACING

Bearing plates shall be accurately set in correct position as shown on the plans and shall have a uniform bearing over the whole area. Provision shall be made to keep the plates in correct position as the concrete is being placed.

2.11.6 — MEASUREMENT AND PAYMENT

The weight to be paid for shall be the inspector's certified shop scale weight of the plates as placed in the structure, unless otherwise provided. If specified in the contract or permitted by the Engineer, computed weights, obtained as herein described, may be made the basis of payment.

Payment shall be made at the contract price per pound. Payment shall include the furnishing of material and all labor and incidental work that is required.

Section 12—STEEL GRID FLOORING

2.12.1 — GENERAL

Steel grid flooring shall be of the open type, or the concrete filled type as specified in the special provisions or as shown on the plans.

The floor shall meet the requirements for the design of steel grid floors Article 1.3.6. Before fabrication or construction is undertaken the Contractor shall submit complete shop and assembly details to the Engineer for approval.

2.12.2 — MATERIALS

(A) Steel

All steel shall conform to the Specification for Structural Steel of the AASHO M 183 (ASTM A 36), AASHO M 161 (ASTM A 242) of a weldable grade or AASHO M 188 (ASTM A 441). Unless the material is galvanized it shall have a copper content of 0.2 percent.

(B) Protective Treatment (Shop Coat)

Open type floors, preferably, shall be galvanized in accordance with the Specification for Zinc (Hot-Galvanized) Coatings on Products Fabricated from Rolled, Pressed and Forged Steel Shapes, Plates, Bars, and Strip, AASHO M 111 (ASTM A 123).

In lieu of galvanizing, the floor may be painted if specified in the special provisions. The paint shall be applied according to the specifications for Painting Metal Structures, except that dipping will be permitted. The paint shall be as specified for metal structures unless paint of other type is required by the special provisions.

(C) Concrete

All concrete in filled steel grid floors shall conform to the specification for concrete, Section 2.4. The concrete and the size of aggregate shall be as specified for Class Y concrete.

(D) Skid Resistance

The upper edges of all members forming the wearing surface of an open type grid flooring should be fabricated or treated to give the maximum skid resistance.

2.12.3 — ARRANGEMENT OF SECTIONS

Where the main elements are normal to center line of roadway, the units generally shall be of such length as to extend over the full width of the roadway for roadways up to 40 feet, but in every case the units shall extend over at least three panels. Where joints are required, the ends of the main floor members shall be welded at the joints over their full cross-sectional area, or otherwise connected to provide full continuity.

Where the main elements are parallel to center line of roadway, the sections shall extend over not less than three panels, and the ends of abutting units shall be welded over their full cross-sectional area, or otherwise connected to provide full continuity in accordance with the design.

2.12.4 — PROVISION FOR CAMBER

Unless otherwise provided on the plans, provision for camber shall be made as follows:

Steel units so rigid that they will not readily follow the camber required shall be cambered in the shop. To provide a bearing surface parallel to the crown of the roadway the stringers shall be canted or provided with shop-welded beveled bearing bars. If beveled bars are used they shall be placed along the center line of the stringer flange, in which case the design span length shall be governed by the width of the bearing bar instead of by the width of the stringer flange.

Longitudinal stringers shall be mill cambered or provided with bearing strips so that the completed floor after dead-load deflection shall conform to the longitudinal camber shown on the plans.

2.12.5 — FIELD ASSEMBLY

Areas of considerable size shall be assembled before the floor is welded to its supports. The main elements shall be made continuous and sections shall be connected together along their edges by welding, bolting or riveting. The connections shall meet with the approval of the Engineer. The rivets may be cold driven.

2.12.6 — CONNECTION TO SUPPORTS

The floor shall be connected to its steel supports by welding. Before any welding is done the floor shall either be loaded to make a tight joint with full bearing or it shall be clamped down. The location, length and size of the welds shall be subject to the approval of the Engineer, but in no case shall they be less than the manufacturer's standards.

The ends of all the main steel members of the slab shall be securely fastened together at the sides of the roadway for the full length of the span by means of steel plates or angles welded to the ends of the main members, or by thoroughly encasing the ends with concrete.

2.12.7 — WELDING

All shop and field welding shall be done in accordance with the current Specifications of the American Welding Society for Welding Highway and Railway Bridges unless these (AASHO) Specifications conflict with the AWS Specifications, in which case these (AASHO) Specifications shall govern.

Surfaces to be welded shall be free from paint, grease, loose scale, rust and other material that will prevent a proper weld. A thin coating of linseed oil, without pigment, need not be removed. Any clinkers or slag caused by flame cutting or other causes shall be removed before welding.

2.12.8 — REPAIRING DAMAGED GALVANIZED COATINGS

All galvanizing that has been chipped off or damaged in handling or transporting, or in welding or riveting, shall be repaired by field galvanizing by the application of a paste composed of approved zinc powder and flux with a minimum amount of water. The places to be coated shall be thoroughly cleaned, including removal of slag on welds, before the paste is applied. The surface to be coated shall first be heated with a torch to a sufficient temperature so that all metallics in the paste are melted when applied to the heated surface. Extreme care shall be taken to see that the galvanized surfaces are not damaged by the torch. The flux in the paste will cause a black substance to appear on the surface of the coated parts, and this black substance shall be removed by wiping off with waste or by the quick application of cold water.

2.12.9 — CONCRETE FILLER

Floor types with bottom flanges not in contact with each other shall be provided with bottom forms of metal or wood to retain the concrete filler without excessive leakage.

If metal form strips are used they shall fit tightly on the bottom flanges of the floor members and be placed in short lengths so as to extend only about 1 inch onto the edge of each support, but in all cases the forms shall be such as will result in adequate bearing of slab on the support.

The concrete shall be mixed, placed and cured in accordance with the requirements of Section 2.4. The concrete shall be thoroughly compacted by vibrating the steel grid floor. The vibrating device and the manner of operating it shall be subject to the approval of the Engineer.

2.12.10 — PAINTING

Flooring furnished without galvanizing, but with a shop coat of paint shall be given two field coats of paint in accordance with the requirements of Section 2.14.

If a structural steel plate is used on the bottom of a filled type floor, the bottom surface of the plate shall be painted one shop coat and two field coats of paint.

2.12.11 — MEASUREMENT AND PAYMENT

Payment for steel grid floor, open or concrete filled type, shall include the furnishing of all materials, equipment, tools, and labor necessary for the satisfactory completion of the work. Payment will be made on the basis of the number of square feet of steel grid floor complete in place, unless otherwise specified.

Section 13—RAILINGS

2.13.1 — GENERAL

Railings for bridges shall include all work constructed above the top of sidewalk, top of curb more than six inches wide or, if a sidewalk or curb is not used, above the top of the roadway. Railings for wingwalls or retaining walls shall include all work above the top of wall, or if used, top of curb or top of sidewalk. The item shall include the furnishing of all material, equipment, tools, supplies and labor necessary for the proper construction of the railing as shown on the plans or as provided in the special provisions. Payment will be as shown on the plans or in the special provisions.

2.13.2 — MATERIALS

All materials shall conform to the requirements of the applicable AASHO Material Specifications, the plans or the special provisions.

2.13.3 — LINE AND GRADE

The line and grade of the railing shall be true to that shown on the plans, and shall not follow any unevenness in the superstructure. Unless otherwise specified or shown on the plans, the handrail and curbs on bridges, whether superelevated or not, shall be vertical.

METAL RAILING

2.13.4 — CONSTRUCTION

Fabrication and erection shall be done in accordance with the requirements of Section 2.10. In the case of welded railings, all exposed joints shall be finished by grinding or filing after welding to give a neat appearing job.

Metal railings shall be carefully adjusted prior to fixing in place to insure proper matching at abutting joints, correct alignment and camber throughout their length. Holes for field connections shall be drilled with the railing in place in the structure at proper grade and alignment. Welding may be substituted for rivets in field connections with the approval of the Engineer.

2.13.5 — PAINTING

Unless otherwise specified, metal railing shall be given one shop coat of paint, and three coats of paint after erection. Painting shall conform to the requirements of Section 2.14.

CONCRETE RAILING

2.13.6 — GENERAL

In no case shall concrete railings be placed until the centering or falsework for the span has been released, rendering the span self-supporting.

2.13.7 — MATERIALS

Except as modified herein, materials shall conform to the requirements of Sections 2.4 and 2.5.

2.13.8 — RAILINGS CAST-IN-PLACE

The portion of the railing or parapet which is to be cast-in-place shall be constructed in accordance with the requirements of Section 2.4. Special care shall be exercised to secure smooth and tight-fitting forms which can be rigidly held in line and grade and removed without injury to the concrete.

Forms shall either be of single width boards or shall be lined with suitable material which shall meet with the approval of the Engineer. Form joints in plane surfaces will not be permitted.

All moldings, panel work, and bevel strips shall be constructed according to the detail plans with neatly mitered joints and all corners in the finished work shall be true, sharp and clean-cut and shall be free from cracks, spalls or other defects.

2.13.9 — PRECAST RAILS

Moist tamped mortar precast members shall be made of a mixture of cement and sand approximately in the proportions of one part of cement to two and one-half parts of sand. The sand shall be specially selected for color and grading. The sand shall be screened through a screen having $\frac{1}{8}$ inch square meshes, and all oversize particles shall be discarded. Only sufficient water shall be used in mixing to permit the immediate removal of the member from the mold.

Moist tamped mortar precast members shall be cast in mortar-tight metal or metal lined molds. The precast members shall be removed from the molds as soon as practicable and shall be kept damp for a period of at least 10 days. During this period they shall be protected from the sun and from wind. Any members that show checking or soft corners or surfaces shall be rejected. The method of storage and handling shall be such as to preserve true and even edges and corners, and any precast members which become chipped, marred, or cracked before or during the process of placing shall be rejected.

In the construction of cast-in-place railing caps and copings built in connection with precast balusters, the balusters shall be protected from staining and disfigurement during the process of placing and finishing the concrete.

2.13.10 — SURFACE FINISH

The surfaces of railings shall conform to the requirements of Section 2.4.

2.13.11 — EXPANSION JOINTS

Expansion joints shall be so constructed as to permit freedom of movement. After all other work is completed, all loose or thin shells of mortar likely to spall under movement shall be carefully removed from all expansion joints by means of a sharp chisel.

STONE AND BRICK RAILING

2.13.12 — GENERAL

Unless otherwise specified, the materials used in masonry or brick railings and parapets shall conform to, and the work shall be done in accordance with, the requirements of these specifications for the particular class of work involved. The work shall be done in accordance with the detailed plans, the workmanship shall be first class in every particular, and the finished construction shall be neat in appearance and true to line and grade.

WOOD RAILING

2.13.13 — GENERAL

Wood railings shall be constructed according to the requirements of Sections 2.20 and 2.21.

2.13.14 — MEASUREMENT AND PAYMENT

Payment for railing shall include all materials, tools, equipment, supplies, labor, and other costs necessary for the satisfactory completion of the work.

The reinforcement steel included in payment for rail shall be determined as follows: The portion of slab or beam bars which project into the handrail shall be paid for as metal reinforcement, but the portion of the handrail steel which extends into the slab or beams shall be considered as part of the handrail.

Payment will be made on the basis of the number of linear feet of railing measured along the center line of the railing. When steel railings are shown on steel structures and no separate bid is taken for railing, the railing will be paid for at the price bid per pound for structural steel.

Section 14—PAINTING METAL STRUCTURES

2.14.1 — GENERAL

The painting of metal structures shall include, unless otherwise provided in the contract, the preparation of the metal surfaces, the application, protection and drying of the paint coatings, and the supplying of all tools, tackle, scaffolding, labor and materials necessary for the entire work.

2.14.2 — MATERIAL

(A) Shop Coat (Prime Coat)

The shop or prime coat of paint for metal shall be a red lead paint and shall conform to the Specification for Red Lead Ready-Mixed Paint M 72.

Red lead pigment in the dry form or as a paste in oil shall conform to ASTM D 83. The 97% grade shall be specified for dry pigment.

In mixing for paint, raw linseed oil (ASTM D234) may be replaced with boiled linseed oil (ASTM D260) to the extent of 50% of the total oil content.

The paint, preferably, shall be factory mixed. As an alternative the pigment shall be furnished in the form of red lead paste.

(B) First Field Coat

When the finished coat of paint is specified to be aluminum, black or graphite paint, or colored green, brown or dark gray, the first field coat shall be a red lead paint as specified for the shop coat, tinted light brown as required, with lamp black in an amount not to exceed ¼ pound per gallon of linseed oil.

When the finished coat is to be white or gray, a first field coat conforming to the Specification for White and Tinted Ready-Mixed Paint (Lead and Zinc Base), AASHO M 70 may be used in lieu of red lead paint. The paint shall be tinted as directed by the Engineer.

(C) Second Field Coat (Finish Coat)

The paint to be used for the second field coat shall be as required by the special provisions or as noted on the plans. It shall conform to one of the following AASHO specifications:

(1) Foliage Green Bridge Paint, M 67.
(2) Black Bridge Paint, M 68.
(3) Aluminum Paint (Paste-Mixing Vehicle), M 69.
(4) White and Tinted Ready-Mixed Paint (Lead and Zinc Base), M 70.
(5) Red Lead (Dry and Paste-in-Oil) and Paint Made Therefrom, M 71.
(6) Red Lead Ready-Mixed Paint, M 72.

If red lead is used for the second field coat it shall be tinted with lampblack as directed by the Engineer.

2.14.3 — NUMBER OF COATS AND COLOR

All steel shall be painted one shop or prime coat, and with not less than two field coats, as specified in Article 2.14.2. The color shall be as specified or determined by the Engineer. The coats shall be sufficiently different in color to permit detection of incomplete application.

2.14.4 — MIXING OF PAINT

Paint shall be factory mixed except as provided in this section. All paint shall also be field mixed before applying in order to keep the pigments in uniform suspension.

2.14.5 — WEATHER CONDITIONS

Paint shall not be applied when the air temperature is below 40 F or when the air is misty, or when, in the opinion of the Engineer, conditions are otherwise unsatisfactory for the work. It shall not be applied upon damp or frosted surfaces.

Material painted under cover in damp or cold weather shall remain under cover until dry or until weather conditions permit its exposure in the open. Painting shall not be done when the metal is hot enough to cause the paint to blister and produce a porous paint film.

2.14.6 — APPLICATION

(A) General

Painting shall be done in a neat and workmanlike manner. Paint may be applied with hand brushes or by spraying except that aluminum paint preferably shall be applied by spraying. By either method the coating of paint applied shall be smoothly and uniformly spread so that no excess paint will collect at any point. If work done by spraying is not satisfactory to the Engineer, hand brushing will be required.

(B) Brushing

When brushes are used, the paint shall be so manipulated under the brush as to produce a smooth, uniform, even coating in close contact with the metal or with previously applied paint, and shall be worked into all corners and crevices.

(C) Spraying

Power spraying equipment shall apply the paint in a fine, even spray without the addition of any thinner. In cool weather, the paint may be warmed to reduce the viscosity for use. Such warming shall be accomplished by heating the paint containers in water or by placing them on steam radiators.

Paint when applied with spray equipment shall be immediately followed by brushing when necessary to secure uniform coverage and to eliminate wrinkling, blistering and airholes.

(D) Inaccessible Surfaces

On all surfaces which are inaccessible for paint brushes, the paint shall be applied by spraying or applied with sheepskin daubers, to insure thorough covering.

2.14.7 — REMOVAL OF PAINT

If the painting is unsatisfactory to the Engineer, the paint shall be removed and the metal thoroughly cleaned and repainted.

2.14.8 — THINNING PAINT

Paint as delivered in containers when thoroughly mixed is ready for use. If it is necessary in cool weather to thin the paint in order that it shall spread more freely, this shall be done only by heating in hot water or on steam radiators, and liquid shall not be added nor removed unless permitted by the Engineer.

2.14.9 — PAINTING GALVANIZED SURFACES

Galvanized surfaces which are required to be painted shall be treated as follows:

For the purpose of conditioning the surface of galvanized surfaces for painting, the painting shall be deferred as long as possible in order that the surface may weather.

Before painting galvanized surfaces they shall be treated as follows:

In 1 gallon of soft water dissolve 2 ounces each of copper chloride, copper nitrate, and sal ammoniac, then add 2 ounces of commercial muriatic acid. This should be done in an earthen or glass vessel, never in tin or other metal receptacle. Apply the solution with a wide flat brush to the galvanized surface, after which it will assume a dark, almost black, color, which on drying becomes a grayish film.

2.14.10 — CLEANING OF SURFACES

(A) General

Surfaces of metal to be painted shall be thoroughly cleaned, removing rust, loose mill scale, dirt, oil or grease and other foreign substances. Unless cleaning is to be done by blast cleaning, all weld areas, before cleaning is begun, shall be neutralized with a proper chemical, after which it shall be thoroughly rinsed with water.

Three methods of cleaning are provided herein. Any of these methods may be used unless otherwise specified.

(B) Method A — Hand Cleaning

The removal of rust, scale and dirt shall be done by the use of metal brushes, scrapers, chisels, hammers or other effective means. Oil and grease shall be removed by the use of a suitable effective solvent. Bristle or wood fiber brushes shall be used for removing loose dust.

(C) Method B — Blast Cleaning

All steel shall be cleaned by either the centrifugal wheel or the air blast method. The blast cleaning shall remove all mill scale and other substances down to the bare metal. Special attention shall be given to the cleaning of corners and re-entrant angles. Before painting, all metallic shot and grit or sand shall be removed from the surfaces. The cleaning shall be approved by the Engineer prior to painting. The material shall be painted before rust forms.

Blast cleaning should be performed with SAE No. S-330 shot or smaller, SAE No. G-25 grit or smaller, or dry sand passing through a 16 mesh screen, U.S. sieve series.

(D) Method C — Flame Cleaning

Unless otherwise provided in the supplemental specifications, all metal, except the inside surfaces of boxed members and other surfaces which will be inaccessible to the flame cleaning operation after the member is assembled, shall be flame cleaned in accordance with the following operations:

(1) Oil, grease and similar adherent matter shall be removed by washing with a suitable solvent. Excess solvent shall be wiped from the work before proceeding with subsequent operations.

(2) The surfaces to be painted shall be cleaned and dehydrated (freed of occluded moisture) by the passage of oxyacetylene flames which have an oxygen to acetylene ratio of at least one. The inner cones of these flames shall have a ratio of length to port diameter of at least 8 and shall be not more than 0.15 inch center to center. The oxyacetylene flames shall be traversed over the surfaces of the steel in such manner and at such speed that the surfaces are dehydrated; and dirt, rust, loose scale, scale in the form of blisters or scabs, and similar foreign matter are freed by the rapid, intense heating by the flames. The flames shall not be traversed so slowly that loose scale or other foreign matter is fused to the surface of the steel. The number, arrangement and manipulation of the flames shall be such that all parts of the surfaces to be painted are adequately cleaned and dehydrated.

(3) Promptly after the application of the flames, the surfaces of the steel shall be wire brushed, hand scraped wherever necessary, and then swept and dusted to remove all free material and foreign particles. Compressed air shall not be used for this operation.

(4) Paint shall be applied promptly after the steel has been cleaned and while the temperature of the steel is still above that of the surrounding atmosphere, so that there will be no recondensation of moisture on the cleaned surfaces.

(E) Surfaces Inaccessible After Assembly

Unless otherwise provided, the inside surfaces of boxed members and other surfaces which will be inaccessible to the flame cleaning operation after the member is assembled shall be cleaned by Method A, Hand Cleaning. If flame cleaning of such surfaces is required, it shall be so stated in the special provisions and the following will apply:

The inside surfaces of boxed members and other surfaces which will be inaccessible to the flame cleaning operation after the member is assembled, shall be cleaned as specified in paragraphs (D)(1) and (D)(2) and wire brushed but not painted before the member is boxed or assembled. After all fabrication of the member is completed, its inside surfaces shall be hand wire brushed or hand scraped wherever necessary in order to remove dirt and other foreign substances which may have accumulated after the surfaces were originally cleaned. The outside surfaces of the members shall then be cleaned and dehydrated, wire brushed, and hand scraped wherever necessary. All surfaces shall then be swept and dusted to remove free material and foreign particles and the member completely painted.

2.14.11 — SHOP PAINTING

Unless otherwise specified, steelwork shall be given one coat of approved paint after it has been accepted by the inspector and before it is shipped from the plant.

Surfaces not in contact but inaccessible after assembly or erection shall be painted three coats. The shop contact surfaces shall not be painted. Field contact surfaces shall receive a shop coat of paint, except contact surfaces of high strength bolted, friction type joints, and main splices for chords of trusses and large girder splices involving multiple thicknesses of material where a shop coat of paint would make erection difficult. Field contact surfaces not painted with the shop coat shall be given a coat of approved lacquer or other protective coating if it is expected that there will be a prolonged period of exposure before erection.

Surfaces which will be in contact with concrete shall not be painted.

Structural steel which is to be welded shall not be painted before welding is complete. If it is to be welded only in the fabricating shop and subsequently erected by bolting, it shall receive one coat of paint after shop welding is finished. Steel which is to be field welded shall be given one coat of boiled linseed oil or other approved protective coating after shop welding and shop fabrication is completed.

Surfaces of iron and steel castings, either milled or finished, shall be given one coat of paint.

With the exception of abutting joints and base plates, machine-finished surfaces shall be coated as soon as practicable after being accepted, with a hot mixture of white lead and tallow or other approved coating, before removal from the shop.

Erection marks for the field identification of members and weight marks shall be painted upon surface areas previously painted with the shop coat. Material shall not be loaded for shipment until it is thoroughly dry, and in any case not less than 24 hours after the paint has been applied.

2.14.12 — FIELD PAINTING

When the erection work is complete, including all riveting, welding, bolting and straightening of bent metal, all adhering rust, scale, dirt, grease or other foreign material shall be removed as specified under Article 2.14.10.

As soon as the inspector has examined and approved all field rivets and bolts, the heads of such rivets and bolts, all welds and any surfaces from which the shop or first coat of paint has become worn off or has otherwise become defective, shall be cleaned and thoroughly covered with one coat of shop-coat paint.

Contact surfaces to be riveted or bolted and surfaces which will be in contact with concrete shall not be painted. Surfaces which will be inaccessible after erection shall be painted with such field coats as are called for on plans or authorized. When the paint applied for retouching the shop coat has thoroughly dried and the field cleaning has been satisfactorily completed, such field coats as are called for on the plans or are authorized shall be applied. In no case shall a succeeding coat be applied until the previous coat has dried throughout the full thickness of the paint film. All small cracks and cavities which were not sealed in a watertight manner by the first field coat shall be filled with a pasty mixture of red lead and linseed oil before the second coat is applied.

The following provision shall apply to the application of both field coats. To secure a maximum coating on edges of plates or shapes, rivet heads and other parts subjected to special wear and attack, the edges shall first be striped with a longitudinal motion and the rivet heads with a rotary motion of the brush, followed immediately by the general painting of the whole surface, including the edges and rivet heads.

If, in the opinion of the Engineer, traffic produces an objectionable amount of dust, the Contractor shall, at his own expense, allay the dust for the necessary distance on each side of the bridge and take any other precautions necessary to prevent dust and dirt from coming in contact with freshly painted surfaces or with surfaces before the paint is applied.

The application of the second field coat shall be deferred until adjoining concrete work has been placed and finished. If concreting operations have damaged the paint, the surface shall be recleaned and repainted.

The Contractor shall protect pedestrian, vehicular and other traffic upon or underneath the bridge, and also all portions of the bridge superstructure and substructure, against damage or disfigurement by spatters, splashes and smirches of paint or paint materials.

Section 15—PROTECTION OF EMBANKMENTS AND SLOPES

2.15.1 — GENERAL

This work consists of furnishing and placing a protective covering of erosion resistant material for riprap as slope or pier foundation protection. The work shall be done in reasonably close conformity to the plans and specifications.

The areas to receive riprap or slope protection of any kind shall be dressed smooth to the slopes or shapes called for on the plans and shall be free from stumps, organic matter or waste material. Generally, a toe trench should be provided in which to key the bottom course of riprap. A filter blanket should be provided where it is anticipated that there may be migration of fines through the riprap. These items should be shown on the plans or called for in the project specifications.

All material, regardless of type or kind, shall be placed reasonably close to the lines called for on the plans.

MATERIAL

2.15.2 — MATERIALS

All stone, regardless of use, shall be clean (free from organic matter) durable, angular with fractured faces, nearly rectangular in shape with a breadth or thickness at least one-third its length and have a gradation dependent upon maximum size as determined by the Engineer.

Project plans or specifications will state the size and quality requirements of stone to be used for the various classes of work. Sizes shown in the following table are recognized by these specifications.

MINIMUM PERCENTAGE LARGER THAN
Classes

Rock Size	8 Ton	4 Ton	2 Ton	1 Ton	½ Ton	¼ Ton	Light	Facing	Filter No. 1	Filter No. 2
8 Ton	50	0								
4 Ton	95	50	0							
2 Ton	—	95	50	0						
1 Ton	—	—	95	50	0					
½ Ton	—	—	—	95	50	0				
¼ Ton	—	—	—	—	95	50	0			
200 Lb.	—	—	—	—	—	—	50	0		
75 Lb.	—	—	—	—	—	90	—	50		
5 Lb.	—	—	—	—	—	—	90	90	0	
No. 4	—	—	—	—	—	—	—	—	50	0
No. 200	—	—	—	—	—	—	—	—	95	90

The table above sets out minimum requirements for the large stone per class. It is understood that the Contractor will furnish material well graded with smaller stones to the extent that a homogeneous blanket of riprap will result with all interstices reasonably well filled with rock. Quality requirements for rock to be furnished under these specifications shall be checked prior to use by the stipulated tests and at appropriate times throughout the life of the project as determined by the Engineer.

CONSTRUCTION

2.15.3 — LOOSE RIPRAP FOR SLOPES

Stone for riprap shall be placed on the prepared or natural slope in a manner which will produce a reasonably well graded mass of stone with the minimum practicable percentage of voids, and shall be constructed to the lines, grades, and thicknesses shown on the drawing, or as directed. Riprap protection shall be placed to its full course thickness at one operation and in such a manner as to avoid displacing the underlying material. Placing of riprap protection in layers or by dumping into chutes or by similar methods likely to cause segregation will not be permitted. The larger stone shall be well distributed and the entire mass of stones shall be roughly graded to conform to approximate gradation specified in Article 2.15.2. All material going into riprap protection shall be so placed and distributed that there will be no large accumulations or areas composed largely of either the larger or smaller sizes of stone. It is the intent of the specifications to produce a fairly compact riprap protection in which all sizes of material are placed in their proper proportions. Hand placing or rearranging of individual stone by mechanical equipment may be required to the extent necessary to secure the results specified above. A tolerance of plus or minus 8 inches from the thicknesses shown on the drawing will be allowed in the finished surface of the riprap protection, except that either extreme of such tolerance shall not be continuous over an area greater than 200 square feet. The tolerance limit will be determined on the basis of the average surface elevation within two square feet. The desired distribution of the various sizes of stone throughout the mass may be obtained, at the option of the Contractor, either by selective loading at the quarry, or controlled dumping of successive loads during placing, or by a combination of these methods. Unless otherwise authorized by the Engineer, the riprap protection shall be placed in conjunction with the construction of the embankment with only sufficient lag in construction of the riprap protection as may be necessary to prevent mixture of embankment and riprap material.

2.15.4 — MORTAR RIPRAP FOR SLOPES

Stone for this purpose shall, as far as practicable, be selected as to size and shape in order to secure fairly large, flat-surfaced stone which will lay up with a true and even surface and a minimum of voids. These stones shall be placed first and roughly arranged in close contact, the largest stones being placed near the base of the slope. The spaces between the larger stones shall be filled with stones of suitable size, leaving the surface smooth, reasonably tight and conforming to the contour required. In general, the stone shall be laid with a degree of care that will insure for plane surfaces a maximum variation from a true plane of not more than 1½ inches in 4 feet. Warped and curved surfaces shall have the same general degree of accuracy as specified above for plane surfaces.

As each of the larger stones is placed, it shall be surrounded by fresh mortar and adjacent stones shall be shoved into contact. After the larger stones are in place all of the spaces or openings between them shall be filled with mortar and the smaller stones then placed by shoving them into position, forcing excess mortar to the surface and insuring that each stone is carefully and firmly bedded laterally.

After the work has been completed as above described, all excess mortar forced up shall be spread uniformly to completely fill all surface voids. All surface joints shall then be roughly pointed up either with flush joints or with shallow, smooth raked joints.

Weep holes shall be provided through the riprap cover as shown on the plans or as directed by the Engineer.

Mortar shall consist of one part cement complying with Article 2.4.2(A)(1) and three parts sand complying with Article 2.4.2(C)(2); thoroughly mixed with water to have a thick creamy consistency. Mortar shall not be placed in freezing weather. During hot, dry weather the work shall be protected from the sun and kept moist for a minimum of three days after placement.

Rock shall be kept wet during placing of the mortar.

2.15.5 — STONE RIPRAP FOR FOUNDATION PROTECTION

Stone shall be placed as nearly as is practical to the locations and areas called for on the plans.

When placed under water free dumping will not be permitted without written permission of the Engineer. Placement shall be by controlled methods using bottom dump buckets, or wire rope baskets lowered through the water to the point of placement.

2.15.6 — CONCRETE RIPRAP IN BAGS

In general this type material shall be used only where it can be placed in the dry.

Concrete for riprap in bags shall be Class "C" concrete complying with Article 2.4.5.

Heavy burlap bags 10 oz. about 19½" x 36" inside seams are preferred. If sacks of larger sizes are used the ends shall be folded to secure close contact with adjacent bags and to contain approximately ⅔ cu. ft. of concrete.

The bags shall be securely tied or folded over and immediately placed in the work. If bags are folded the fold shall be placed underneath the bag for headers and against the previously placed bag for stretchers. Each course shall be thoroughly tamped into place so that close contact with underlying and adjacent bags is obtained. No more than four horizontal courses shall be laid above concrete not yet having its initial set. At the start of each day's work, previously placed sacks shall be moistened and dusted heavily with cement.

Weep holes shall be provided through the riprap cover as shown on the plans or as directed by the Engineer.

2.15.7 — CONCRETE SLAB RIPRAP

(a) General—Concrete slab riprap may be either cast-in-place slabs or precast slabs manufactured on the job or at a regular masonry unit manufacturing plant. If reinforcement is required, it shall be furnished and placed as shown on the plans. All blocks shall be of the limiting dimensions shown on the plans.

(b) Manufacture—Cast-in-place slabs or precast slabs manufactured on the job shall be made of Class "A" concrete complying with the requirements of Section 2.4 except that the minimum compressive strength of the concrete shall be 1,800 pounds per square inch. All edges shall be tooled and the exposed surfaces shall be a wood float and fiber brush finish.

Plant manufactured slabs shall be uniform in texture with true sharp edges. Tooling of edges and brush finish will not be required. Plant manufactured blocks shall comply with ASTM Specification C 145, for grade U-II units.

Either conventional aggregate complying with Article 2.4.2(C)(1) and 2.4.2(D)(1) or light weight aggregate complying with ASTM C330 may be used in manufacture of the blocks.

(c) *Placing*—A trench of the dimensions shown on the plans or as given by the Engineer shall be dug at the toe of the slope and the slope shall be dressed to the lines and grades given by the Engineer. Excess material removed in dressing slopes shall be used in filling low areas and any excess not required for this purpose shall be spread on adjacent highway slopes as directed by the Engineer. Filled areas shall be compacted. If it is necessary to import material for filling depressions and low spots to bring slopes to lines and grades, importation of such material only will be paid for as extra work. Placing imported material will be considered a part of the work of dressing the slopes. The riprap shall be placed in blocks of the dimensions shown on the plans. Unless otherwise specified, blocks shall be laid in horizontal courses and successive courses shall break joints with preceding courses. Joint details shall be as shown on the plans, but if not shown, horizontal joints and joints up the slope in either type construction shall be ¾", filled with 1 to 3 grout and raked ¾".

Weep holes shall be provided through the riprap cover as shown on the plans or as directed by the Engineer.

The Contractor shall maintain the riprap protection as may be necessary to prevent mixture of embankment and riprap material. The Contractor shall maintain the riprap protection until accepted and any material displaced by any cause, shall be replaced at no additional cost to the State, to the lines and grades shown on the drawing.

FILTER MATERIAL

2.15.8 — FILTER OR BEDDING MATERIAL

Pit or quarry run material of the size specified and complying with Article 2.15.2 will be acceptable.

Stone shall be placed as nearly as is practical to the locations and areas called for on the plans.

When placed under water, free dumping will not be permitted without written permission of the Engineer. Placement shall be by controlled methods using bottom dump buckets, or wire rope baskets lowered through the water to the point of placement.

MEASUREMENT

2.15.9 — MEASUREMENT

(a) *Loose Riprap for Slopes.* Loose riprap for slopes may be measured for payment on either a square yard, cubic yard or weight basis as specified. If measured on a square yard basis, the quantity measured will be that actually placed to the limiting dimensions shown on the plans or as the plan dimensions may be revised by the Engineer.

(b) Mortar Riprap for Slopes. Mortar riprap for slopes will be measured for payment on a square yard basis. The quantity measured will be that actually placed to the limiting dimensions shown on the plans or as the plan dimensions may be revised by the Engineer.

(c) Stone Riprap for Foundation Protection. Unless otherwise specified, stone riprap will be measured for payment on a weight (ton) basis. Weighing shall be done in the presence of an inspector in the hauling vehicles either at the quarry or at or near the point of placement as may be directed by the Engineer. The quantity measured for payment will be that actually placed in accordance with instructions given by the Engineer.

(d) Concrete Riprap in Bags. Unless otherwise specified concrete riprap in bags will be measured for payment on a volume (cubic yard) basis. The quantity measured will be that actually placed to the limiting dimensions shown on the plans or as the plan dimensions may be revised by the Engineer.

(e) Concrete Slab Riprap. "Preparation of Slopes" will be measured for payment on a square yard unit. The quantity measured will be that actually prepared for coverage by riprap as required by plan dimensions or as plan dimensions may be revised by the Engineer. Excavated area for toe trenches or curbs, if required, will be included in these limiting dimensions.

"Concrete Slab Riprap" will be measured for payment on a square yard unit. The quantity measured will be that actually placed within the limiting dimensions called for on the plans or as plan dimensions may be revised by the Engineer. Curbing, if required, will be included in these limiting dimensions.

(f) Filter or Bedding Material. Filter or bedding material will be measured for payment on a cubic yard, square yard, or weight basis as called for in the project specifications. The quantity measured will be that actually placed to the limiting dimensions shown on the plans or as the plan dimensions may be revised by the Engineer. Measurements on a square yard basis shall be measured parallel to the finished surface.

PAYMENT

2.15.10 — PAYMENT

(a) Loose Riprap for Slopes—Loose riprap for slopes measured in accordance with Article 2.15.9(a) will be paid for at the price bid per square yard, per cubic yard or per ton as set forth in the project specifications.

(b) Mortar Riprap for Slopes—Mortar riprap for slopes measured in accordance with Article 2.15.9(b) will be paid for at the price bid per square yard.

(c) Stone Riprap for Foundation Protection—Stone riprap for foundation protection measured in accordance with Article 2.15.9(c) will be paid for at the price bid per ton unless otherwise specified.

(d) **Concrete Riprap in Bags**—Concrete riprap in bags measured in accordance with Article 2.15.9(d) will be paid for at the price bid per cubic yard unless otherwise specified.

(e) **Concrete Slab Riprap**—Payment for preparation of slopes will be made at the unit price bid per square yard therefor, which price and payment shall be full compensation for furnishing and placing all materials including labor, tools, equipment, and incidentals necessary to complete preparation of the area to be covered, including excavation and backfill of toe trenches and curbs if required. Borrow material, if required, will be paid for as extra work as set forth in Article 2.15.7(c).

Payment for concrete slab riprap will be at the unit price bid per square yard therefor, which price and payment shall be full compensation for furnishing all materials, including labor, equipment, tools and incidentals necessary to complete the work.

(f) **Filter or Bedding Material**—Filter or bedding material measured in accordance with Article 2.15.9(f) will be paid for on the cubic yard, square yard or weight basis as called for in the project specifications.

(g) **General**—Payment for riprap of the various classes at the unit prices bid will include full and complete compensation for all labor, materials, equipment, or other incidental expense in connection with preparation of subgrade, excavating and backfilling toe trenches where required, furnishing and placing the stone, slabs, grout, mortar, reinforcing steel, if required, and all other work and incidental material required to complete the work in accordance with the plans and specifications.

Section 16—CONCRETE CRIBBING

2.16.1 — GENERAL

The construction of concrete cribbing shall consist of the furnishing and installation of reinforced concrete crib members and the placing of the interior filling materials. The crib members shall be cast in the proportions and in conformance with the general requirements set forth for precast concrete bearing piles. Dowels, where used shall be of wrought-iron or galvanized steel not less than 1 inch in diameter and of the required length.

Casings for dowels shall be of galvanized steel or iron pipe not less than 1¼ inches in diameter.

The details of the crib members and their arrangement shall be as shown on the plans. If specific details for reinforcement are not shown on the plans, or if the Contractor is permitted to purchase the crib members from manufacturers, he shall submit detailed specifications and plans for the approval of the Engineer, and such plans must be approved before delivery of the material is begun.

All members shall be free from depressions and spalled, patched, or plastered surfaces or edges, or any other defects which may impair their strength or durability. Cracked or otherwise defective members will be rejected.

2.16.2 — CONSTRUCTION

The foundation or bed for the cribbing shall be firm and shall be approved by the Engineer before any of the crib work is placed. In general, transverse concrete sill members shall be used to support the lower cribbing course. Crib members shall be carefully handled and erected in such manner as to avoid any injury due to shock or impact. Each member shall be secured by approved interlocking details or by means of dowels passing through galvanized casings. Any members which become cracked or otherwise injured during erection shall be completely renewed and replaced.

The filling for the interior of the crib shall progress simultaneously with the erection of the cribbing, and shall be of approved material placed in layers not to exceed 12 inches in thickness and tamped or consolidated to the satisfaction of the Engineer.

2.16.3 — MEASUREMENT AND PAYMENT

Concrete cribbing will be paid for at the contract price per cubic foot for concrete cribbing complete in place. This price shall include all materials, equipment, tools, and labor incidental to the satisfactory erection of the cribbing, including necessary excavation. The volume to be paid for will be the actual net volume of the concrete in the crib members as shown on the plans. The filling for the interior of the crib will be paid for at the contract price per cubic yard for crib filling in place.

Section 17 — WATERPROOFING

2.17.1 — GENERAL

When specified on the plans or in the special provisions, surfaces shall be waterproofed as specified herein.

2.17.2 — MATERIALS

(A) Mortar

Mortar for the protective course shall conform to the provisions of Article 2.6.2(B).

(B) Asphalt

Waterproofing asphalt shall conform to the Specification for Asphalt for Dampproofing and Waterproofing, AASHO M 115 (ASTM D 449). NOTE: Type A is for use below ground and Type B for use above ground. Unless otherwise specified, Type B shall be used.

Primer for use with asphalt in waterproofing shall conform to the Specification for Primer for Use With Asphalt in Dampproofing and Waterproofing, AASHO M 116.

(C) Pitch

Waterproofing pitch shall conform to the Specification for Coal-Tar Pitch for Roofing, Dampproofing and Waterproofing, AASHO M 118 (ASTM D450). Type B pitch shall be furnished unless otherwise specified.

Primer for use with coal-tar pitch in dampproofing and waterproofing shall conform to the Specification for Creosote for Priming Coat with Coal-Tar Pitch in Dampproofing and Waterproofing, AASHO M 121 (ASTM D 43).

(D) Fabric

The fabric shall conform to the Specification for Woven Cotton Fabrics Saturated with Bituminous Substances for Use in Waterproofing, AASHO M 117 (ASTM D 173).

(E) Tar for Absorptive Treatment

Tar for absorptive treatment shall be a liquid water-gas tar which conforms to the following requirements:

Specific gravity, 25/25 C (77/77 F)	1.030 to 1.100
Specific viscosity at 40 C (104 F.) (Engler), not more than	3.0
Total distillate, per cent by weight, to 300 C (572 F), not more than	50.0
Bitumen (soluble in carbon disulphide), not less than, per cent	98.0
Water, not more than, per cent	3.0

(F) Tar Seal Coat

Tar seal coat shall conform to the Specification for Tar for Use in Road Construction, AASHO M 52, Grade RTCB-5.

(G) Joint Fillers

Filler for use in horizontal joints in waterproofing work shall be a straight refined oil asphalt conforming to the following requirements:

Flash point: Not less than 232 C (450 F).
Softening point: 48.9 C (120 F) to 54.4 C (130 F).
Penetration: at 0 C (32 F), 200 grams, 1 minute, not less than 15.
At 25 C (77 F), 100 grams, 5 seconds, 50 to 60.
At 46 C (115 F), 50 grams, 5 seconds, not more than 300.
Loss on heating: At 163 C (325 F), 50 grams, 5 hours, not more than 0.5 per cent.
Ductility: at 25 C (77 F), 5 centimeters per minute, not less than 85.
Total bitumen (soluble in carbon disulphide): Not less than 99.5 per cent.

Filler for use in vertical joints in waterproofing work shall be an asphalt conforming to the requirements specified above for horizontal joint filler, to which has been added 20 per cent, by weight, of asbestos fiber. The incorporation of the asbestos fiber with the asphalt shall be done at the factory of the manufacturer to insure a uniform distribution of the fiber throughout the mix.

(H) Inspection and Delivery

All waterproofing materials shall be tested before shipment. Unless otherwise ordered by the Engineer, they shall be tested at the place of manufacture, and, when so tested, a copy of the test results shall be sent to the Engineer by the chemist or inspection bureau which has been designated to make the tests, and each package shall have affixed to it a label, seal, or other mark of identification, showing that it has been tested and found acceptable, and identifying the package with the laboratory tests.

Factory inspection is preferred, but, in lieu thereof, the Engineer may order that representative samples, properly identified, be sent to him for test prior to shipment of the materials. After delivery of the materials, representative check samples shall be taken which shall determine the acceptability of the materials.

All materials shall be delivered on the work in original containers, plainly marked with the manufacturer's brand or label.

2.17.3 — STORAGE OF FABRIC

The fabric shall be stored in a dry, protected place. The rolls shall not be stored on end.

2.17.4 — PREPARATION OF SURFACE

All concrete surfaces which are to be waterproofed shall be reasonably smooth, and free from projections or holes which might cause puncture of the membrane. The surface shall be dry, so as to prevent the formation of steam when the hot asphalt or tar is applied, and, immediately before the application of the waterproofing, the surface shall be thoroughly cleaned of dust and loose materials.

No waterproofing shall be done in wet weather, nor when the temperature is below 35 F, without special authorization from the Engineer. Should the surface of the concrete become temporarily damp, it shall be covered with a 2-inch layer of hot sand, which shall be allowed to remain in place from one to two hours, or long enough to produce a warm and surface-dried condition, after which the sand shall be swept back, uncovering sufficient surface for beginning work, and the operation repeated as the work progresses.

2.17.5 — APPLICATION — GENERAL

Asphalt shall be heated to a temperature between 300 and 350 F, and tar for hot application shall be heated to a temperature between 200 and 250 F, with frequent stirring to avoid local overheating. The heating kettles shall be equipped with thermometers.

In all cases, the waterproofing shall begin at the low point of the surface to be waterproofed, so that water will run over and not against or along the laps.

The first strip of fabric shall be of half width; the second shall be full width, lapped the full width of the first sheet; and the third and each succeeding strip shall be full width and lapped so that there will be two layers of fabric at all points with laps not less than 2 inches wide. All end laps shall be at least 12 inches.

Beginning at the low point of the surface to be waterproofed, a coating of primer shall be applied and allowed to dry before the first coat of asphalt is applied. The waterproofing shall then be applied as follows:

Beginning at the low point of the surface to be waterproofed, a section about 20 inches wide and the full length of the surface shall be mopped with the hot asphalt or tar, and there shall be rolled into it, immediately following the mopping, the first strip of fabric, or half width, which shall be carefully pressed into place so as to eliminate all air bubbles and obtain close conformity with the surface. This strip and an adjacent section of the surface of a width equal to slightly more than half of the width of the fabric being used shall then be mopped with hot asphalt or tar, and a full width of the fabric shall be rolled into this, completely covering the first strip, and pressed into place as before. This second strip and an adjacent section of the concrete surface shall then be mopped with hot asphalt or tar and the third strip of fabric "shingled" on so as to lap the first strip not less than 2 inches. This process shall be continued until the entire surface is covered, each strip of fabric lapping at least 2 inches over the last strip but one. The entire surface shall then be given a final mopping of hot asphalt or tar.

The completed waterproofing shall be a firmly bonded membrane composed of two layers of fabric and three moppings of asphalt or tar, together with a coating of primer. Under no circumstances shall one layer of fabric touch another layer at any point or touch the surface, as there must be at least three complete moppings of asphalt or tar.

In all cases the mopping on concrete shall cover the surface so that no gray spots appear, and on cloth it shall be sufficiently heavy to completely conceal the weave. On horizontal surfaces not less than 12 gallons of asphalt or tar shall be used for each 100 square feet of finished work, and on vertical surfaces not less than 15 gallons shall be used. The work shall be so regulated that, at the close of a day's work, all cloth that is laid shall have received the final mopping of asphalt or tar. Special care shall be taken at all laps to see that they are thoroughly sealed down.

2.17.6 — APPLICATION — DETAILS

At the edges of the membrane and at any points where it is punctured by such appurtenances as drains or pipes, suitable provisions shall be made to prevent water from getting between the waterproofing and the waterproofed surface.

All flashing at curbs and against girders, spandrel walls, etc., shall be done with separate sheets lapping the main membrane not less than 12 inches. Flashing shall be closely sealed either with a metal counterflashing or by embedding the upper edges of the flashing in a groove poured full of joint filler.

Joints which are essentially open joints but which are not designed to provide for expansion shall first be caulked with oakum and lead wool and then filled with hot joint filler.

Expansion joints, both horizontal and vertical, shall be provided with sheet copper or lead in "U" or "V" form in accordance with the details. After the membrane has been placed, the joint shall be filled with hot joint filler. The membrane shall be carried continuously across all expansion joints.

At the ends of the structure the membrane shall be carried well down on the abutments and suitable provision made for all movement.

2.17.7 — DAMAGE PATCHING

Care shall be taken to prevent injury to the finished membrane by the passage over it of men or wheelbarrows, or by throwing any material on it. Any damage which may occur shall be repaired by patching. Patches shall extend at least 12 inches beyond the outermost damaged portion and the second ply shall extend at least 3 inches beyond the first.

2.17.8 — PROTECTION COURSE

Over the waterproofing membrane, constructed as specified above, there shall be constructed a protection course which, unless otherwise specified or shown on the plans, shall be a 2-inch course of mortar mixed in the proportion of one part portland cement and two parts sand. This mortar course shall be reinforced midway between its top and bottom surfaces with wire netting of 6-inch mesh and No. 12 gauge, or its equivalent. The top surface shall be troweled to a smooth, hard finish and, where required, true to grade.

The construction of the protection course shall follow the waterproofing so closely that the latter will not be exposed without protection for more than 24 hours.

2.17.9 — MEASUREMENT AND PAYMENT

Payment for waterproofing shall include the cost of furnishing all equipment, materials and labor necessary for the satisfactory completion of the waterproofing membrane and the protection course.

Payment will be made on the basis of the number of square yards of waterproofing complete in place.

Section 18 — DAMPPROOFING

2.18.1 — GENERAL

When specified on the plans or in the special provisions, surfaces shall be dampproofed as specified herein.

2.18.2 — MATERIALS

The material used for dampproofing shall be tar or asphalt as required by the special provisions.

Tar for absorptive treatment (or primer), tar seal coat, and asphalt for primer and seal coat shall conform to the requirements of Article 2.17.2.

2.18.3 — PREPARATION OF SURFACE

The surface to which the dampproofing coating is to be applied shall be cleaned of all loose and foreign material and dirt and shall be dry. When necessary the Engineer may require the surface to be scrubbed with water and a stiff brush, after which the surface shall be allowed to dry before application of the primer.

2.18.4 — APPLICATION

Concrete, brick or other surfaces which are to be protected by dampproofing shall be thoroughly clean before the primer is applied. They shall then be brush or spray painted with two or more coats (as indicated on the plans or in the special provisions) of tar or asphalt for absorptive treatment. Below ground not less than two coats shall be applied, using $\frac{1}{8}$ gallon for each square yard of surface. On the well-primed surface one application of tar or asphalt seal coat shall be applied by brush, using $\frac{1}{10}$ gallon per square yard.

Care shall be taken to confine all paints to the areas to be waterproofed and to prevent disfigurement of any other parts of the structure by dripping or spreading of the tar or asphalt.

2.18.5 — MEASUREMENT AND PAYMENT

Payment for dampproofing shall include the cost of furnishing all equipment, materials and labor necessary for the satisfactory completion of the work.

Payment will be made on the basis of the number of square yards of dampproofing complete in place.

Section 19 — NAME PLATES

2.19.1 — GENERAL REQUIREMENTS

When specified, the Contractor for the superstructure shall furnish and install name plates of such form, dimensions, material and design as may be shown on the plans. Unless otherwise provided, the contract price for the superstructure shall include the cost of such name plates.

No permanent plates or markers other than those shown on the plans or approved by the Engineer will be permitted on any structure.

Section 20—TIMBER STRUCTURES

2.20.1 — MATERIALS

(A) Lumber and Timber (Solid sawn or glued laminated)

Sawn lumber and timber shall conform to the Specifications for Structural Timber, Lumber and Piling, AASHO M 168.

Structural glued laminated timber shall conform to U.S. Commercial Standard CS 253-63 for Structural Glued Laminated Timber. The term structural glued laminated timber, as employed in CS 253-63, is an engineered, stress-rated product of a timber laminating plant, comprising assemblies of suitably selected and prepared wood laminations securely bonded together with adhesives. The grain of all laminations is approximately parallel longitudinally. The separate laminations may not exceed 2 inches in net thickness. They may be comprised of pieces end joined to form any length, of pieces placed or glued edge to edge to make wider ones, or of pieces bent to curved form during gluing.

For the various structural purposes, appropriate grades, or their equivalent, shall be selected in accordance with the design requirements for stress-grades given in Section 1.10.

Structural lumber and timber, solid sawn or glued laminated, shall not be used in exposed permanent structures without pressure preservative treatment. Temporary structures or lumber and timber with adequate heartwood requirements (see AASHO M 168) need not require preservatively treated lumber and timber.

(B) Structural Shapes

Rods, plates and shapes shall be of structural steel or wrought-iron, as specified, conforming to the requirements of Article 2.10.3. Eyebars shall conform to the requirements of Article 2.10.3(A)(2)(b).

(C) Castings

Castings shall be cast steel or gray-iron, as specified, conforming to the requirements of Article 2.10.3(D).

(D) Hardware

Machine bolts, drift-bolts and dowels may be either wrought-iron or medium steel. Washers may be cast iron ogee or malleable iron castings, or they may be cut from medium steel or wrought-iron plate, as specified.

Machine bolts shall have square heads and nuts, unless otherwise specified. Nails shall be cut or round wire of standard form. Spikes shall be cut or wire spikes, or boat spikes, as specified.

Nails, spikes, bolts, dowels, washers and lag screws shall be black or galvanized, as specified.

Unless otherwise specified all hardware for treated timber bridges, except malleable iron connectors, shall be galvanized or cadmium plated.

(E) Paint for Timber Structures

Paint for timber structures, except as otherwise provided herein, shall conform to the Specification for White and Tinted Ready-Mixed Paint (Lead and Zinc Base), AASHO M 70. The paint as specified is intended for use in covering previously painted surfaces. When it is applied to unpainted timber, turpentine and linseed oil shall be added as required by the character of the surface in an amount not to exceed one pint each per gallon of the paint as specified. The paint shall be either white or tinted as directed by the Engineer.

If aluminum or black paint is specified the first or prime coat shall be as specified above. The paint for additional coats shall be as follows:

(1) Aluminum Paint

Aluminum paint shall conform to the Specification for Aluminum Paint, AASHO M 69.

(2) Black Paint

(a) Composition

	Maximum per cent	Minimum per cent
Pigment	32	28
Liquid (containing at least 80 per cent linseed oil)	72	68
Water	0.5	—
Coarse particles and "skins" (total residue retained on No. 325 sieve based on pigment)	1.5	—
Weight per gallon, not less than 9.0 lbs.		

(b) Pigment

The pigment in both semipaste and ready-mixed paints shall consist of carbon, lead oxide, insoluble mineral material, and, at the option of the manufacturer, oxide of iron. The pigment shall show, on analysis, not less than 20 per cent of carbon and not less than 5 per cent of lead oxide calculated as Pb_3O_4. (Since oxide of lead may be dissolved by the oil in paint, in all cases when the amount of lead in the pigment calculated as Pb_3O_4 is found to be less than 5 per cent of the pigment, lead should be determined in the vehicle and the total lead in the paint computed to percentage of pigment.) The total of the lead oxide, iron oxide, insoluble mineral material, and loss on ignition shall be not less than 90 per cent.

(c) Vehicle

The liquid in semipaste paint shall be entirely linseed oil; in ready-mixed paint it shall contain not less than 80 per cent of linseed oil, the balance to be combined drier and thinner.

The thinner shall be turpentine, volatile mineral spirits, or a mixture thereof.

(d) Ready-Mixed Paint

Unless otherwise authorized by the **Engineer** the paint shall be "ready-mixed" (factory-mixed).

Ready-mixed paint shall be well-ground, shall not settle badly or cake in the container, shall be readily broken up with a paddle to a smooth uniform paint of good brushing consistency, and shall dry within 18 hours to a full oil gloss, without streaking, running, or sagging. The color and hiding power when specified shall be equal to those of a sample mutually agreed upon by buyer and seller.

(e) Methods of Analysis

Paint shall be analyzed in accordance with methods given in Federal Specifications Nos. TT-P-141a and TT-P-27.

(F) Timber Connectors

(1) General

Connectors for treated timber structures, except those of malleable iron, shall be galvanized in accordance with AASHO M 111 (ASTM A 123) and shall be of the type specified in Article 2.20.2. (See Table 2.20.1 for Typical Dimensions of Timber Connectors).

(2) Split Ring Connectors

Split rings of 2½-inch inside diameter, 4-inch inside diameter and 6-inch inside diameter shall be manufactured from hot rolled, low-carbon steel conforming to the Standard Specifications for **Carbon Steel Blooms, Billets, and Slab for Forgings, AASHO M 162** (ASTM A273), Grade 1015. Each ring shall form a closed true circle with the principal axis of the cross section of the ring metal parallel to the geometric axis of the ring. Except for the 6" ring (which is rectangular in cross section) the metal section shall be beveled from the central portion toward the edges to a thickness less than the mid section. It shall be cut through in one place in its circumference to form a tongue and slot.

Connector grooves in timber shall be cut concentric with the bolt hole, shall conform to the cross-sectional shape of the rings and shall provide a snug fit. Inside groove diameter shall be larger than nominal ring diameter in order that the ring will expand slightly during installation.

(3) Tooth-Ring Connectors

Toothed-ring timber connectors shall be stamped cold from U.S. Standard 16-gage hot rolled sheet steel conforming to the **Standard Specifications for Carbon Steel Blooms, Billets, and Slabs for Forgings, AASHO M 162** (ASTM A 273), Grade 1015, and shall be bent cold to form a circular, corrugated, sharp-toothed band and circle and shall be parallel to the axis of the ring. The central band shall be welded to fully develop the strength of the band. All sizes, 2-inch, 2⅝-inch, 3⅜-inch and 4-inch diameters, shall have an overall depth of .94 inch and a depth of fillet of .25 inch.

(4) Shear-Plate Connectors

Pressed Steel Type. Pressed steel shear-plates of 2⅝-inch diameter shall be manufactured from mild steel conforming to the Standard Specifications for Carbon Steel Blooms, Billets, and Slabs for Forgings, AASHO M 162 (ASTM A 273), Grade 1015. Each plate shall be a true circle with a flange around the edge, extending at right angles to the face of the plate and extending from one face only, the plate portion having a central bolt hole and two small perforations on opposite sides of the hole and midway from the center and circumference.

Malleable Iron Type. Malleable iron shear-plates of 4-inch diameter shall be manufactured according to AASHO M 106 (ASTM A 47), Grade No. 35018, for malleable iron castings. Each casting shall consist of a perforated round plate with a flange around the edge extending at right angles to the face of the plate and projecting from one face only, the plate portion having a central bolt hole reamed to size with an integral hub concentric to the bolt hole and extending from the same face as the flange.

(5) Spike-Grid Connectors

Spike-grid timber connectors shall be manufactured according to AASHO M 106 (ASTM A 47), Grade No. 35018, for malleable iron castings. They shall consist of four rows of opposing spikes forming a 4⅛-inch square grid with 16 teeth which are held in place by fillets. Fillets for the flat grid in cross section shall be diamond shaped. Fillets for the single curve grids shall be increased in depth to allow for curvature and shall maintain a thickness between the sloping faces of the fillets equal to the width of the fillet.

Circular grids of 3¼" diameter shall consist of 8 opposing spikes equally spaced around the outer circumference and held in place by connecting fillets around the outer diameter and radial fillets projecting to a central circular fillet which forms a bolt hole opening of 1¼". Fillets in cross section shall be diamond shaped except that the inner circular fillet may be flattened on one side to provide for manufacturer identification.

2.20.2 — TIMBER CONNECTORS

Timber connectors shall be one of the following types, as specified on the plans; the split ring, the toothed ring, the shear plate or the spike grid. The split ring and the shear plate shall be installed in precut grooves of dimensions as given herein or as recommended by the manufacturer. The toothed ring and the spike grid shall be forced into the contact surfaces of the timbers joined by means of pressure

TABLE 2.20.1
TYPICAL DIMENSIONS OF TIMBER CONNECTORS

SPLIT RINGS
Dimensions in Inches

	2½"	4"		2½"	4"
Split ring:			Washers, standard:		
Inside diameter at center when closed	2.500	4.000	Round, cast or malleable iron, diameter	2⅝	3
Thickness of metal at center	.163	.193	Round, wrought iron (minimum):		
Depth of metal (width of ring)	.750	1.000	Diameter	1¾	2
			Thickness	³⁄₃₂	⁵⁄₃₂
Groove:			Square plate:		
Inside diameter	2.56	4.08	Length of side	2	3
Width	.18	.21	Thickness	⅛	³⁄₁₆
Depth	.375	.50			
Bolt hole:			Projected area:		
Diameter	⁹⁄₁₆	¹³⁄₁₆	Portion of one ring within member, sq. in.	1.10	2.25

TOOTHED RINGS
Dimensions in Inches

	2"	2⅝"	3⅜"	4"		2"	2⅝"	3⅜"	4"
Toothed ring:					Washers, minimum:				
Diameter	2.000	2.625	3.375	4.000	Round, cast or malleable iron (diameter)	2	2⅝	3	3½
Thickness of metal	.061	.061	.061	.061	Square plate:				
Depth	.940	.940	.940	.940	Length of side	2	2½	3	3½
Depth of fillet (minimum)	.250	.250	.250	.250	Thickness	³⁄₁₆	¼	¼	⅜
Bolt hole:					Projected area:				
Diameter	⁹⁄₁₆	¹¹⁄₁₆	¹³⁄₁₆	¹³⁄₁₆	Portion of one ring within member, sq. in.	.94	1.23	1.59	1.89

Table 2.20.1 (Continued)

SHEAR PLATES

Dimensions in Inches

	2⅝"	2⅝"	4"	4"	2⅝"	2⅝"	4"	4"
Shear plate:								
Material	Pressed steel	Light gage	Malleable iron	Malleable iron				
Diameter of plate	2.62	2.62	4.02	4.02				
Diameter of bolt hole	0.81	0.81	0.81	0.94				
Thickness of plate	0.172	0.12	0.20	0.20				
Depth of flange	0.42	0.35	0.62	0.62				
Steel strap or shapes for use with shear plates:								
Steel straps or shapes, for use with shear plates, shall be designed in accordance with accepted engineering practices.								
Hole diameter in straps or shapes for bolts					13/16	13/16	13/16	15/16
Bolt hole—diameter in timber					13/16	13/16	13/16	15/16
Washers, standard:								
Round, cast or malleable iron, diameter					3	3	3	3½
Round, wrought iron, minimum:								
Diameter					2	2	2	2¼
Thickness					5/32	5/32	5/32	11/64
Square plate:								
Length of side					3	3	3	3
Thickness					¼	¼	¼	¼
Projected area:								
Portion of one shear plate within member, sq. in.					1.18	1.00	2.58	2.58

Circular Dap—dimensions: A, B, C, D, E, F, G, H, I

	2⅝"	2⅝"	4"	4"
A	2.63	2.63	4.03	4.03
B	0.81	1.07	1.55	1.55
C	0.81	0.81	0.81	0.81
D		0.65	0.97	0.97
E	0.19	0.13	0.27	0.27
F	0.45	0.38	0.64	0.64
G	0.25	0.14	0.22	0.22
H		0.34	0.50	0.50
I	2.25	2.37	3.49	3.49

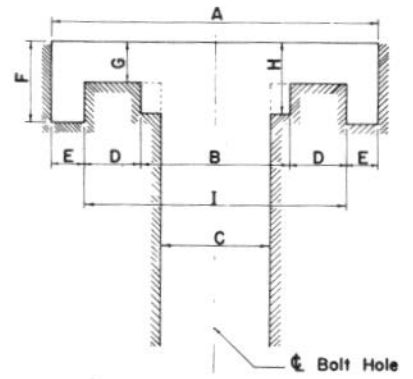

equipment. All connectors of this type at a joint shall be embedded simultaneously and uniformly.

Fabrication of all structures using connectors shall be done prior to treatment. When prefabricated from templates or shop details, bolt holes shall not be more than $\frac{1}{16}$ inch from required placement. Bolt holes shall be $\frac{1}{16}$ inch larger than bolt diameter. Bolt holes shall be bored perpendicular to the face of the timber.

Timber after fabrication shall be stored in a manner which will prevent changes in the dimensions of the members before assembly.

Dimensions of material and details not otherwise specified shall meet with the approval of the Engineer.

2.20.3 — STORAGE OF MATERIAL

Lumber and timber stored on the site shall be kept in orderly piles or stacks. Untreated material shall be open-stacked on supports at least 12 inches above the ground surface to avoid absorption of ground moisture and permit air circulation and it shall be so stacked and stripped as to permit free circulation of air between the tiers and courses. It will be advisable in particular cases for the Engineer to require protection from the weather by a suitable covering.

On glued laminated structural members that are not to be preservatively treated, an approved end sealer shall be applied after end trimming of each completed member.

2.20.4 — WORKMANSHIP

Workmanship shall be first class throughout. None but competent bridge carpenters shall be employed and all framing shall be true and exact. Unless otherwise specified, nails and spikes shall be driven with just sufficient force to set the heads flush with the surface of the wood. Deep hammer marks in wood surfaces shall be considered evidence of poor workmanship and sufficient cause for removal of the workman causing them. The workmanship on all metal parts shall conform to the requirement of Article 2.10.2.

2.20.5 — TREATED TIMBER

(A) Handling

Treated timber shall be carefully handled without sudden dropping, breaking of outer fibers, bruising or penetrating the surface with tools. It shall be handled with rope slings. Cant hooks, peaveys, pikes or hooks shall not be used.

(B) Framing and Boring

All cutting, framing, and boring of treated timbers shall be done before treatment in so far as is practicable. When treated timbers are to be placed in waters infested by marine borers, untreated cuts, borings or other joint framings below high water elevation shall be avoided.

(C) Cuts and Abrasions

All cuts in treated piles or timbers, and all abrasions, after having been carefully trimmed, shall be covered with 2 applications of a mixture of 60 per cent creosote oil and 40 per cent roofing pitch or brush coated with at least two applications of hot creosote oil and covered with hot roofing pitch.

For field treatment of other preservatives see AWPA Standard M 4-62 entitled "Standard for the Care of Pressure Treated Wood Products."

(D) Bolt Holes

All bolt holes bored after treatment shall be treated with creosote oil by means of an approved pressure bolt hole treater. Any unfilled holes, after being treated with creosote oil, shall be plugged with creosoted plugs.

(E) Temporary Attachment

Whenever, with the approval of the Engineer, forms or temporary braces are attached to treated timber with nails or spikes, the holes shall be filled by driving galvanized nails or spikes flush with the surface or plugging holes as required for bolt holes.

2.20.6 — UNTREATED TIMBER

In structures of untreated timber the following surfaces shall be thoroughly coated with two coats of hot creosote oil before assembling: ends, tops and all contact surfaces of sills, caps, floor beams and stringers; and all ends, joints, and contact surfaces of bracing and truss members. The back faces of bulkheads and all other timber which is to be in contact with earth, metal or other timber shall be similarly treated.

Bolts passing through non-resinous wood shall preferably be galvanized.

2.20.7 — TREATMENT OF PILE HEADS

(A) General

Pile heads, after cutting to receive the caps, and prior to placing the caps shall be treated to prevent decay.

Immediately after making final cut-off on treated timber foundation piles, the cut area shall be given two liberal applications of preservative followed by a heavy application of coal-tar pitch, or other approved sealer. Treated timber piles which will have the cut-off exposed in the structure shall be further protected by one of the following methods, as specified on the plans. If not otherwise specified, Method B shall be used.

(B) Method A — Zinc Covering

The sawed surface shall be covered with three applications of a mixture of 60 per cent creosote oil and 40 per cent roofing pitch or thoroughly brush coated with three applications of hot creosote oil and covered with hot roofing pitch. Before placing the cap, a sheet of 12 gauge (.028-inch) zinc shall be placed on each pile head. The sheet zinc shall be of sufficient size to project at least 4 inches outside of the pile, and it shall be bent down, neatly trimmed and securely fastened to the faces of the pile, with large headed galvanized roofing nails.

(C) Method B — Fabric Covering

The heads of all piles shall be covered with alternate layers of hot pitch and loosely woven fabric similar to membrane waterproofing, using four applications of pitch and three layers of fabric. The cover shall measure at least 6 inches more in dimension than the diameter of the pile and shall be neatly folded down over the pile and secured by large headed galvanized nails or by binding or serving with not less than seven complete turns of galvanized wire securely held in place by large-headed galvanized nails and staples. The edges of the fabric projecting below the wire wrapping shall be trimmed to present a workmanlike appearance.

The heads of untreated piles shall be given one of the following treatments, as may be specified or directed by the Engineer:

(1) The sawed surface shall be thoroughly brush coated with two applications of hot creosote oil.

(2) The sawed surface shall be heavily coated with red lead paint, after which it shall be covered with cotton duck, of at least 8-ounce weight, which shall be folded down over the sides of the pile and firmly secured thereto with large-headed roofing nails. The edges of the duck shall be trimmed to give a workmanlike appearance. The duck shall then be waterproofed by being thoroughly saturated and coated with one or more applications of red lead paint.

2.20.8 — HOLES FOR BOLTS, DOWELS, RODS AND LAG SCREWS

Holes for round drift-bolts and dowels shall be bored with a bit $\frac{1}{16}$ inch less in diameter than the bolt or dowel to be used. The diameter of holes for square drift-bolts or dowels shall be equal to the least dimension of the bolt or dowel.

Holes for machine bolts shall be bored with a bit the same diameter as the bolt, except as otherwise provided in Article 2.20.2.

Holes for rods shall be bored with a bit $\frac{1}{16}$ inch greater in diameter than the rod.

Holes for lag screws shall be bored with a bit not larger than the body of the screw at the base of the thread.

2.20.9 — BOLTS AND WASHERS

A washer, of the size and type specified, shall be used under all bolt heads and nuts which would otherwise come in contact with wood.

The nuts of all bolts shall be effectually locked after they have been finally tightened.

2.20.10 — COUNTERSINKING

All recesses in treated timber, formed for countersinking shall be painted with hot creosote oil. Recesses likely to collect injurious materials shall be filled with hot pitch.

2.20.11 — FRAMING

All lumber and timber shall be accurately cut and framed to a close fit in such manner that the joints will have even bearing over the entire contact surfaces. Mortises shall be true to size for their full depth and tenons shall fit snugly. No shimming will be permitted in making joints, nor will open joints be accepted.

2.20.12 — PILE BENTS

The piles shall be driven as indicated on the plans, with a variation of the portion above the ground of not more than ¼ inch per foot from the vertical or batter indicated, or so that the cap may be placed in its proper location without inducing excessive stresses in the piles. Excessive manipulation of the piles will not be permitted and the Contractor will be required to redrive or use other satisfactory methods to avoid such manipulations. No shimming on tops of piles will be permitted.

The piles for any one bent shall be carefully selected as to size, to avoid undue bending or distortion of the sway bracing. However, care shall be exercised in the distribution of piles of varying sizes to secure uniform strength and rigidity in the bents of any given structure.

Cut-offs shall be accurately made to insure perfect bearing between the cap and piles.

2.20.13 — FRAMED BENTS

(A) Mud Sills

Untreated timber used for mud sills shall be of heart cedar, heart cypress, redwood, or other durable timber. Mud sills shall be firmly and evenly bedded to solid bearing and tamped in place.

(B) Concrete Pedestals

Concrete pedestals for the support of framed bents shall be carefully finished so that the sills or posts will take even bearing on them. Dowels of not less than ¾-inch diameter and projecting at least 6 inches above the tops of the pedestals, shall be set in them when they are cast, for anchoring the sills or posts.

(C) Sills

Sills shall have true and even bearing on mud sills, piles or pedestals. They shall be drift-bolted to mud sills or piles with bolts of not less than ¾-inch diameter and extending into the mud sills or piles at least 6 inches. When possible, all earth shall be removed from contact with sills so that there will be free air circulation around them.

(D) Posts

Posts shall be fastened to pedestals with dowels of not less than ¾-inch diameter, extending at least 6 inches into the posts.

Posts shall be fastened to sills by one of the following methods, as indicated on the plans:

(1) By dowels of not less than ¾-inch diameter, extending at least 6 inches into posts and sills.

(2) By drift-bolts of not less than ¾-inch diameter driven diagonally through the base of the post and extending at least 9 inches into the sill.

(E) Design and Construction

Where framed structures will be subjected to earthquake, wind, tractive or centrifugal loads, the connections between members thereof shall be so designed and constructed as to resist the forces resulting therefrom. (See Article 1.2.1)

2.20.14 — CAPS

Timber caps shall be placed, with ends aligned, in a manner to secure an even and uniform bearing over the tops of the supporting posts or piles. All caps shall be secured by drift-bolts of not less than ¾-inch diameter, extending at least 9 inches into the posts or piles. The drift-bolts shall be approximately in the center of the post or pile. (See Article 2.20.13(E)).

2.20.15 — BRACING

The ends of bracing shall be bolted through the pile, post or cap with a bolt of not less than ⅝-inch diameter. Intermediate intersections shall be bolted, or spiked with wire or boat spikes, as indicated on the plans. In all cases spikes shall be used in addition to bolts.

2.20.16 — STRINGERS

Stringers shall be sized at bearings and shall be placed in position so that knots near edges will be in the top portions of the stringers.

Outside stringers may have butt joints with the ends cut on a taper, but interior stringers shall be lapped to take bearing over the full width of the floor beam or cap at each end. The lapped ends of untreated stringers shall be separated at least ½ inch for the circulation of air and shall be securely fastened by drift-bolting where specified. When stringers are two panels in length the joints shall be staggered.

Cross-bridging between stringers shall be neatly and accurately framed and securely toe-nailed with at least two nails in each end. All cross-bridging members shall have full bearing at each end against the sides of stringers. Unless otherwise specified in the contract, cross-bridging shall be placed at the center of each span.

2.20.17 — PLANK FLOORS

Plank shall be of the grade required as specified in Article 2.20.1(A). Unless otherwise specified, they shall be surfaced four sides (S 4 S).

Single plank floors shall consist of a single thickness of plank supported by stringers or joists. The planks shall be laid heart side down, with ¼-inch openings between them for seasoned material and with tight joints for unseasoned material. Each plank shall be securely spiked to each joist. The planks shall be carefully graded as to thickness and so laid that no two adjacent planks shall vary in thickness by more than $\frac{1}{16}$ inch.

Two-ply timber floors shall consist of two layers of flooring supported on stringers or joists. The top course may be laid either diagonal or parallel to the centerline of roadway, as specified and each floor piece shall be securely fastened to the lower course. Joints shall be staggered at least 3 feet. If the top flooring is placed parallel to the centerline of the roadway, special care shall be taken to securely fasten the ends of the flooring. At each end of the bridge these members shall be beveled.

2.20.18 — LAMINATED OR STRIP FLOORS

The strips shall be of the grade required as specified in Article 2.20.1(A). The strips shall be placed on edge, at right angles to the center line of roadway. Each strip shall be spiked to the preceding strip at each end and at approximately 18-inch intervals with the spikes driven alternately near the top and bottom edges. The spikes shall be of sufficient length to pass through two strips and at least half-way through the third strip.

If timber supports are used every other strip shall be toe-nailed to every other support. The size of the spikes shall be as shown on the plans. When specified on the plans, the strips shall be securely attached to steel supports by the use of approved galvanized metal clips. Care shall be taken to have each strip vertical and tight against the preceding one, and bearing evenly on all the supports.

2.20.19 — COMPOSITE WOOD-CONCRETE DECKS

(A) Slab Spans

Where the tensile strength of wood and the compressive strength of concrete are to be used compositely, the joining of the two materials shall be such as to resist all horizontal shear at that plane, and provision shall be made to prevent separation of the materials.

The horizontal shear may be resisted by metal devices set into and projecting above the top of the laminated strips, or by fabricating the upper edge of the strips in a serrated manner.

Separation of the materials may be resisted by nails driven at an angle in the upper edge of the strips, or by certain suitable devices, or by grooves or other working of upstanding strips.

(B) "T" Beams

Spans consisting of concrete slabs placed on wood stringers may be designed as "T" beams when the two materials are suitably joined so as to resist horizontal shear at their juncture and the materials are bonded permanently together.

A horizontal shear joint may be made using metal devices or by a serrated working of the tops of the stringers.

Separation of the concrete from the stringers may be prevented by driving nails in the top of the stringers at an angle, or by other suitable metal devices, or by grooving the sides of the stringers near the top, or other working of the wood, and then forming the concrete into the pattern worked in the wood.

2.20.20 — WHEEL GUARDS AND RAILING

Wheel guards and railing shall be accurately framed in accordance with the plans and erected true to line and grade.

Unless otherwise specified, wheel guards, rails and rail posts shall be surfaced four sides (S 4 S).

Wheel guards shall be laid in sections not less than 12 feet long.

2.20.21 — TRUSSES

Trusses, when completed, shall show no irregularities of line. Chords shall be straight and true from end to end in horizontal projection and, in vertical projection, shall show a smooth curve through panel points conforming to the correct camber. All bearing surfaces shall fit accurately. Uneven or rough cuts at the points of bearing shall be cause for rejection of the piece containing the defect.

2.20.22 — TRUSS HOUSINGS

The carpentry on truss housings shall be equal in all respects to the best house carpentry. The finished appearance of the housing is considered of primary importance and special care shall be taken to secure a high quality of workmanship and finish on this portion of the structure. Workmen wearing shoes with caulks will not be permitted on the roof.

2.20.23 — ERECTION OF HOUSING AND RAILINGS

Unless otherwise directed by the Engineer, housing and railings shall be built after the removal of the falsework and the adjustment of the trusses to correct alignment and camber.

2.20.24 — PAINTING

Rails and rail posts of untreated timber, shall be painted with three coats of paint.

Parts of the structure, other than rails and rail posts, which are to be painted, shall be designated on the plans or in the special provisions.

Metal parts, except hardware, shall be given one coat of shop paint and, after erection, two coats of field paint as specified in Section 2.14.

2.20.25 — MEASUREMENT AND PAYMENT

Payment for timber structures shall include the furnishing of materials, preservative treatment, equipment, tools and labor necessary for the erection and painting of the work in a satisfactory manner.

Lumber and timber, unless otherwise specified, shall be paid for at the contract price per 1,000 feet board-measure (M.B.M) for material remaining in the finished structure, including the cost of all hardware. Computations of the amount of lumber and timber in the structure shall be based on nominal sizes and the shortest commercial length which could be used. No other allowance for waste will be made.

Metal parts, other than hardware, shall be paid for at the contract price per pound, the weight being computed in the same manner as specified for steel structures, Article 2.10.63.

Section 21—PRESERVATIVE TREATMENTS FOR TIMBER

2.21.1 — GENERAL

The kind of preservative treatment required shall be as specified in the special provisions or as noted on the plans.

The preservatives specified herein are not intended to be used interchangeably, but the kind of preservative to be used shall be adopted for its suitability to the conditions of exposure to which it will be subjected. Some of the conditions to be considered are: effect of marine borers, effect of termites, action of exposure to water and leaching of the preservative, effect of contact with the ground, painting requirements and cleanliness requirements. Experience records for the particular exposure intended shall be given consideration in selecting the treatment to be used.

2.21.2 — MATERIALS

Piling shall conform to the requirements of AASHO M 168. Timber and lumber shall conform to the requirements of AASHO M168. Only wood species for which treatment requirements are listed in AASHO M 133 may be specified. All wood species so listed are not equally treatable and therefore all species are not equally acceptable under severe exposure conditions. Care should be exercised to select those species in AASHO M 133 which are acceptable for the intended application.

Timber preservatives and treatment method shall conform to AASHO M 133. The type of preservative furnished shall be in accordance with that specified in the special provisions or as noted on the plans. When selecting a preservative and a preservative retention, it should be noted that AASHO M 133 designates the preservatives and retentions recommended for Coastal Waters and in marine structures and further that timber for use in "ground or water contact" has requirements that differ from timbers for use "not in ground or water contact." In some instances there is a range of retentions offered which provides for different degrees of exposure based on climate or degree of insect infestation. Unless the higher retentions are specified, not less than the minimum retention is required.

Timbers expected to be painted should be treated with water-borne salts or pentachlorophenol carried in a volatile organic solvent.

2.21.3 — IDENTIFICATION AND INSPECTION

Each piece of treated timber shall bear a legible brand, mark or tag indicating the name of the treater and the specification symbol or specification requirements to which the treatment conforms. The Engineer shall be provided adequate facilities and free access to the necessary parts of the treating plant for inspection of material and workmanship to determine that the contract requirements are met. The Engineer reserves the right to retest all materials after delivery to the job site and to reject all materials which do not meet the requirements of the contract; provided that, at the job site reinspection, conformance within five percent of contract requirements shall be acceptable. Reinspection at the job site may include assay to determine retention of preservatives and extraction and analysis of preservative to determine its quality.

Section 22—TIMBER CRIBBING

2.22.1 — MATERIAL

(A) Timber

Timber used for cribbing shall conform to the requirements of Section 2.20, and unless otherwise specified shall be the same as for caps, posts, sills, etc. If treated timber is used, all hardware shall be galvanized or cadmium plated.

(B) Logs

Logs used for cribbing shall conform in quality to the requirements specified for timber piles in AASHO M 168.

2.22.2 — PREPARATION

When timber or logs are to be treated, all framing shall be completed before treatment and all surfaces cleaned of dirt and grease.

All timber and log framing shall be done in a workmanlike manner and true to line and angle.

2.22.3 — DIMENSIONS

(A) Timber

When cribs are constructed of sawed timber, no timber shall be less than 8 inches in least dimension. The face timber in the base tier shall be not less than 10 inches in least dimension.

(B) Logs

When cribs are constructed of logs, no face log shall have a diameter at the small end of less than 10 inches and tie logs shall be not less than 8 inches in diameter at the small end. The face log in the base tier shall be not less than 12 inches in diameter at the small end.

All logs for cribbing shall be selected from the logs available with as small an amount of taper as possible. The length of logs used shall be somewhat dependent upon the taper.

2.22.4 — CONSTRUCTION

(A) Foundation

The foundation or bed for the cribbing shall be excavated to exact grade and shall be approved as to bearing quality by the Engineer before any of the crib work is placed.

(B) Mud Sills

When mud sills are used, they shall be set at right angles to the face of the cribbing and firmly and evenly bedded in the foundation material.

Mud sills shall be not less than 12 by 12 inches in squared cross-sectional dimensions and not less than 3 feet in length. They shall be spaced not more than 4 feet apart.

Log or timber mud sills shall be leveled to fit the first tier resting upon them. In no case shall there be less than 100 square inches of flat contact surface between the face log and each mud sill.

Foundation material shall be thoroughly tamped around all mud sills.

(C) Face Logs or Timbers

The logs or timbers in the base tier and in alternate tiers above the base shall be as long as practicable and preferably, shall extend the full length of the face. In intermediate tiers they may have a length of not less than 8 feet, arranged to break joints. Crib faces shall be laid solid or with spacers as indicated on the plans.

All framed surfaces shall receive a heavy coat of approved preservative at the time of assembling.

Care shall be exercised in the erection of all cribs to produce a true face as shown on the plans and all timbers or logs in faces shall be horizontal.

(D) Ties

The length of ties shall be sufficient to develop the required anchorage against overturning, and in no case shall the length of tie extending into the fill be less than two-thirds of the height of fill above the tie in question.

Ties shall be anchored to the face walls by framing, either dovetailed or by sufficient projection beyond the face of the crib to form the proper anchorage. Ties shall be anchored at the fill end to cross pieces fastened to them at right angles by drift-bolts or other suitable means.

Ties shall be spaced not more than 8 feet center to center in any horizontal tier and shall be staggered with the next adjacent tier of ties. Tiers of ties shall be not more than 3 feet apart vertically.

(E) Fastening

Each successive tier of logs or timbers shall be drift-bolted to the one upon which it rests by drifts not less than ¾ inches in diameter and of sufficient length to extend through 2 tiers and not less than 4 inches into the third tier.

Drift-bolts shall be staggered and not more than 8 feet center to center in each tier.

All end joints and splices shall be half-lapped for 10 inches and drifted at the center.

Before assembling, all framed joints in contact shall be heavily coated with an approved preservative.

2.22.5 — FILLING

Filling inside and around cribs shall be of the material specified and shall be placed in a careful manner so as to avoid distortion of the crib. Filling shall be placed in even horizontal layers and compacted to reduce the voids to a minimum.

2.22.6 — MEASUREMENT AND PAYMENT

Payment for the construction of cribbing shall include the furnishing of all materials, equipment, tools and labor necessary for the excavation, crib erection, and filling, complete in place, in accordance with the plans and these specifications. Payment for timber and logs shall include the cost of drift-bolts and other miscellaneous hardware.

Excavation for cribbing shall be paid for at the contract price per cubic yard for material actually removed except that in no case shall this be computed to include material more than 1 foot outside of vertical planes through the extreme neat lines of the finished crib or its supports. The contract price for excavation shall include a yardage of back-fill equivalent to that excavated between the neat lines and the pay lines.

Timber shall be paid for at the contract price per 1,000 feet board measure (MBM) for material remaining in the finished structure.

Logs shall be paid for at the contract price per linear foot, for each size specified, for material remaining in the finished structure.

Filling material shall be paid for at the contract price per cubic yard for the actual volume placed.

Section 23—CONSTRUCTION AND INSTALLATION OF CORRUGATED METAL AND STRUCTURAL PLATE PIPES, PIPE-ARCHES AND ARCHES

2.23.1 — GENERAL

This item shall consist of furnishing corrugated metal or structural plate pipe, pipe-arches and arches conforming to these specifications and of the sizes and dimensions required on the plans, and installing such structures at the places designated on the plans or by the Engineer, and in conformity with the lines and grades established by the Engineer. Pipe shall be either circular or elongated as specified or shown on the plans.

The thickness of plates or sheets shall be as determined in Art. 1.8.2, Division I, and the radius of curvature shall be as shown on the plans. Each plate or sheet shall be curved to one or more circular arcs.

The plates at longitudinal and circumferential seams of structural plates shall be connected by bolts. Joints shall be staggered so that not more than three plates come together at any one point.

2.23.2 — FORMING AND PUNCHING OF CORRUGATED STRUCTURAL PLATES AND SHEETS FOR PIPE

(A) Structural Plate Pipe

Plates shall be formed to provide lap joints. The bolt holes shall be so punched that all plates having like dimensions, curvature, and the same number of bolts per foot of seam shall be interchangeable. Each plate shall be curved to the proper radius so that the cross-sectional dimensions of the finished structure will be as indicated on the drawings or as specified.

Unless otherwise specified, bolt holes along those edges of the plates that form longitudinal seams in the finished structure shall be in two rows. Bolt holes along those edges of the plates that form circumferential seams in the finished structure shall provide for a bolt spacing of not more than 12 inches. The minimum distance from center of hole to edge of the plate shall be not less than 1¾ times the diameter of the bolt. The diameter of the bolt holes in the longitudinal seams shall not exceed the diameter of the bolt by more than one-eighth (⅛) inch.

Plates for forming skewed or sloped ends shall be cut so as to give the angle of skew or slope specified. Burned edges shall be free from oxide and burrs and shall present a workmanlike finish. Legible identification numerals shall be placed on each plate to designate its proper position in the finished structure.

(B) Corrugated Metal Pipe

Punching and forming of sheets shall conform to AASHO M 36.

(C) Elongation

If elongated structural plate or corrugated metal pipe is specified or called for on the plans, the plates or pipes shall be formed so that the finished pipe is elliptical in shape with the vertical diameter approximately five (5) per cent greater than the nominal diameter of the pipe. Pipe-arches shall not be elongated. Elongated pipes shall be installed with the longer axis vertical.

2.23.3 — ASSEMBLY

Corrugated metal pipe, and structural plate pipe shall be assembled in accordance with the manufacturer's instructions. All pipe shall be unloaded and handled with reasonable care. Pipe or plates shall not be rolled or dragged over gravel or rock and shall be prevented from striking rock or other hard objects during placement in trench or on bedding.

Corrugated metal pipe shall be placed on the bed starting at downstream end with the inside circumferential laps pointing downstream and with the longitudinal laps at the side or quarter points.

Bituminous coated pipe and paved invert pipe shall be installed in a similar manner to corrugated metal pipe with special care in handling to avoid damage to coatings. Paved invert pipe shall be installed with the invert pavement placed and centered on the bottom.

The pipe sections shall be joined by coupling bands of like material. One-piece or two-piece bands may be used. Coupling bands for annular and helical corrugated metal pipe shall provide circumferential and longitudinal strength to preserve the culvert alignment, prevent separation of the pipe sections, and prevent infiltration of sidefill material.

Structural plate pipe, pipe-arches and arches shall be installed in accordance with the plans and detailed erection instructions.

2.23.4 — BEDDING

When, in the opinion of the Engineer, the natural soil does not provide a suitable bedding, a bedding blanket conforming to Figure 2.23A, shall be provided. Bedding shall be uniform for the full length of the pipe.

2.23.5 — PIPE FOUNDATION

The foundation material under the pipe shall be investigated for its ability to support the load. If rock strata or boulders are closer than 12 inches under the pipe, the rock or boulders shall be removed and replaced with suitable ganular material as shown in Figure 2.23B. Where, in the opinion of the Engineer, the natural foundation soil is such as to require stabilization, such material shall be replaced by a layer of suitable granular material as shown in Figure 2.23C. Where an unsuitable material (peat, muck, etc.) is encountered at or below invert elevation during excavation, the necessary subsurface exploration and analysis shall be made and corrective treatment shall be as directed by the Engineer.

2.23.6 — SIDEFILL

Sidefill material within one pipe diameter of the sides of pipe and not less than one foot over the pipe shall be fine readily compactible soil or granular fill material. Sidefill beyond these limits may be regular embankment fill. Job-excavated soil used as backfill shall not contain stones retained on a 3-inch ring, frozen lumps, chunks of highly plastic clay, or other objectionable material. Sidefill material shall be noncorrosive.

Sidefill material shall be placed as shown in Figure 2.23D, in layers not exceeding 6 inches in compacted thickness at near optimum moisture content by engineer-approved equipment to the density required for superimposed embankment fill. Other approved compacting equipment may be used for sidefill more than 3 feet from sides of pipe. The sidefill shall be placed and compacted with care under the haunches of the pipe and shall be brought up evenly and simultaneously on both sides of the pipe to not less than 1 foot above the top for the full length of the pipe. Fill above this elevation may be material for embankment fill. The width of trench shall be kept to the minimum width required for placing pipe, placing adequate bedding and sidefill, and safe working conditions. Ponding or jetting of sidefill will not be permitted except upon written permission by the Engineer.

2.23.7 — BRACING

Temporary bracing shall be installed and shall remain in place as required to protect workmen during construction.

2.23.8 — CAMBER

The invert grade of the pipe shall be cambered, when required, by an amount sufficient to prevent the development of a sag or back slope in the flow line as the foundation under the pipe settles under the weight of embankment. The amount of camber shall be based on consideration of the flow-line gradient, height of fill, compressive characteristics of the supporting soil, and depth of supporting soil stratum to rock.

2.23.9 — ARCH SUBSTRUCTURES AND HEADWALLS

Substructures and headwalls shall be designed in accordance with the requirements of Division I.

Each side of each arch shall rest in a groove formed into the masonry or shall rest on a galvanized angle or channel securely anchored to or embedded in the substructure. Where the span of the arch is greater than fifteen (15) feet or the skew angle is more than twenty (20) degrees, a metal bearing surface, having a width at least equal to the depth of the corrugation, shall be provided for all arches.

Metal bearings may be either rolled structural or cold formed galvanized angles or channels, not less than three-sixteenths ($3/16$) inch in thickness with the horizontal leg securely anchored to the substructure on a maximum of twenty-four (24) inch centers. When the metal bearing is not embedded in a groove in the substructure, one vertical leg should be punched to allow bolting to the bottom row of plates.

Where an invert slab is provided which is not integral with the arch footing, the invert slab shall be continuously reinforced.

When backfilling arches before headwalls are placed, the first material shall be placed midway between the ends of the arch, forming as narrow a ramp as possible until the top of the arch is reached. The ramp shall be built evenly from both sides and the backfilling material shall be thoroughly compacted as it is placed. After the two ramps have been built to depth specified to the top of the arch, the remainder of the backfill shall be deposited from the top of the arch both ways from the center to the ends, and as evenly as possible on both sides of the arch.

If the headwalls are built before the arch is backfilled, the filling material shall first be placed adjacent to one headwall, until the top of the arch is reached, after which the fill shall be dumped from the top of the arch toward the other headwall, with care being taken to deposit the material evenly on both sides of the arch.

In multiple installations the procedure above specified shall be followed, but extreme care shall be used to bring the backfill up evenly on each side of each arch so that unequal pressure will be avoided.

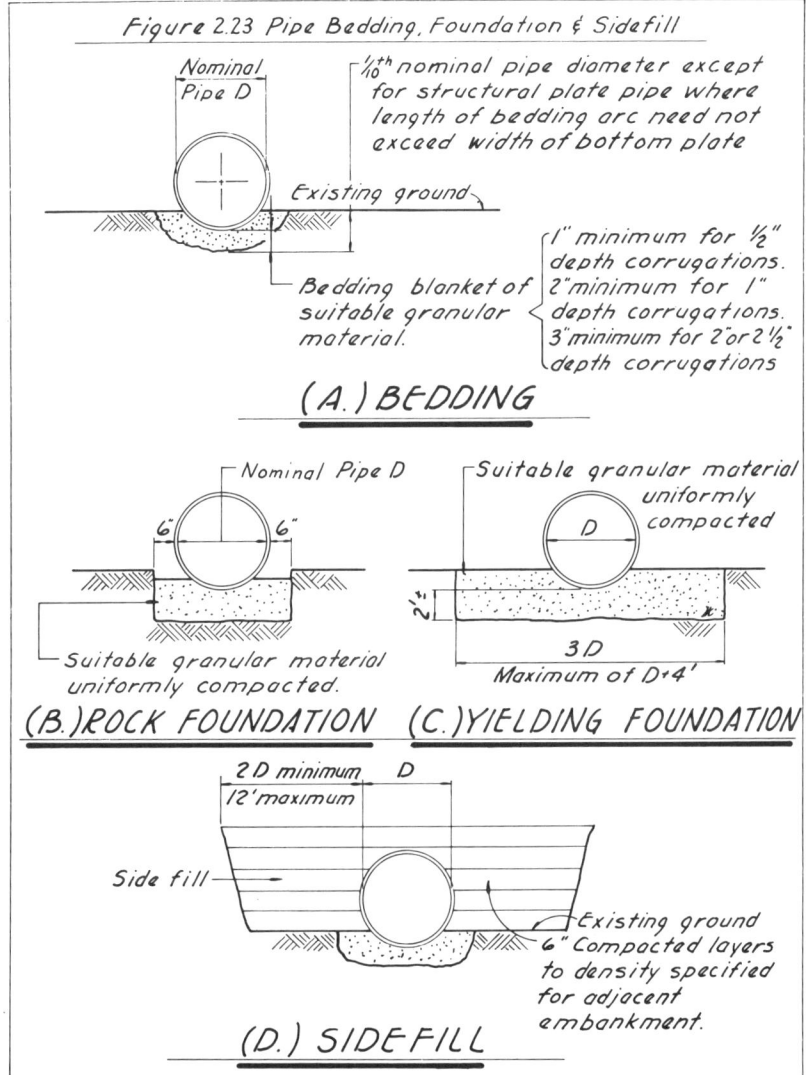

Figure 2.23 Pipe Bedding, Foundation & Sidefill
(A.) BEDDING
(B.) ROCK FOUNDATION
(C.) YIELDING FOUNDATION
(D.) SIDEFILL

In all cases the filling material shall be thoroughly but not excessively tamped. Puddling the backfill will not be permitted.

2.23.10 — COVER OVER PIPE DURING CONSTRUCTION

All pipe shall be protected by sufficient cover before permitting heavy construction equipment to pass over them during construction.

2.23.11 — WORKMANSHIP AND INSPECTION

In addition to compliance with the details of construction, the completed structure shall show careful finished workmanship in all particulars. Structures on which the spelter coating has been bruised or broken either in the shop or in shipping, or which shows defective workmanship, shall be rejected unless repaired to the satisfaction of the Engineer. The following defects are specified as constituting poor workmanship and the presence of any or all of them in any individual culvert plate or in general in any shipment shall constitute sufficient cause for rejection unless repaired:

1. Uneven laps.
2. Elliptical shaping (unless specified).
3. Variation from specified alignment.
4. Ragged edges.
5. Loose, unevenly lined or spaced bolts.
6. Illegible brand.
7. Bruised, scaled, or broken spelter coating.
8. Dents or bends in the metal itself.

2.23.12 — METHOD OF MEASUREMENT

Corrugated metal and structural plate pipe, pipe-arches or arches shall be measured in linear feet, installed in place, completed, and accepted. The number of linear feet shall be the average of the top and bottom centerline lengths for pipe, the bottom centerline length for pipe-arches, and the average of springing line lengths for arches.

2.23.13 — BASIS OF PAYMENT

The footages, determined as herein given shall be paid for at the contract unit prices per linear foot bid for corrugated metal and structural plate pipe, pipe-arch or arches of the several sizes, as the case may be, which prices and payments shall constitute full compensation for furnishing, handling, erecting, and installing the pipe, pipe-arches or arches and for all materials, labor, equipment, tools, and incidentals necessary to complete this item, but for arches shall not constitute payment for concrete or masonry headwalls and foundations, or for excavation.

Section 24—WEARING SURFACES

2.24.1 — DESCRIPTION

Separate wearing surfaces, when required, shall conform to details shown on the plans, and construction shall conform to that specified on the plans or in the special provisions.

2.24.2 — ORTHOTROPIC DECK BRIDGES

(A) Material

The Engineer shall specify or approve the wearing-surface materials. The material shall be of uniform quality and void of foreign matter.

(B) Placement

Careful and competent workmen shall be used in preparing and placing the wearing surface. The wearing surface preferably shall be placed in two courses to assure a smooth riding surface. The Engineer shall specify or approve the methods of cleaning and preparing the deck, applying the bond and tack coats, grading the aggregate, measuring the proportions, mixing the ingredients, regulating the temperatures, placing and compacting the material, and selecting suitable atmospheric conditions for the work.

Prior to placement, the surface of the deck shall be thoroughly cleaned to ensure the complete removal of all mill scale, dirt, debris, oil, grease, salt, and moisture. Air, water vapor, and other gases shall not be entrapped in or under the wearing surface.

(C) Inspection

The Engineer shall observe the manufacture and placement of the wearing surface to assure that all requirements of paragraphs A and B are met. The Engineer shall specify or approve suitable control tests to establish the acceptability of the wearing surface.

Section 25—ELASTOMERIC BEARINGS

2.25.1 — DESCRIPTION

Elastomeric bearings as herein specified shall include plain bearings (consisting of elastomer only) and laminated bearings (consisting of layers of elastomer restrained at their interfaces by bonded laminates).

2.25.2 — MATERIALS

The elastomer portion of the elastomeric compound shall be 100% virgin natural polyisoprene (natural rubber) meeting the requirements of Table A or 100% virgin chloroprene (neoprene) meeting the requirements of Table B, as specified by the Engineer. Compounds of nominal hardness between the values shown may be used and the test requirements interpolated. When test specimens are cut from the finished product a 10% variation in "Physical Properties" will be allowed.

TABLE A

ASTM Standard	Physical Properties	50 Duro	60 Duro	70 Duro
	Hardness ASTM D2240	50±5	60±5	70±5
	Tensile strength, min. psi ASTM D 412	2500	2500	2500
	Ultimate elongation, min. %	450	400	300
	Heat Resistance			
D573 70 hr. @ 158 F	Change in durometer hardness, max. points	+10	+10	+10
	Change in tensile strength, max. %	−25	−25	−25
	Change in ultimate elongation, max. %	−25	−25	−25
	Compression Set			
D395 Method B	22 hours @ 158 F, max. %	25	25	25
	Ozone			
D1149	25 pphm ozone in air by volume, 20% strain 100 F ±2 F, 48 hours mounting procedure D518, Procedure A	No Cracks	No Cracks	No Cracks
	Adhesion			
D429,B	Bond made during vulcanization, lbs. per inch	40	40	40
	Low Temperature Test			
ASTM D746 Procedure B	Brittleness at −40F	No Failure	No Failure	No Failure

TABLE B

ASTM Standard	Physical Properties	50 Duro 50±5	60 Duro 60±5	70 Duro 70±5
	Hardness ASTM D2240			
	Tensile strength, min. psi ASTM D 412	2500	2500	2500
	Ultimate elongation, min. %	400	350	300
	Heat Resistance			
D573 70 hr. @212 F	Change in durometer hardness, max. points	+15	+15	+15
	Change in tensile strength, max. %	−15	−15	−15
	Change in ultimate elongation, max. %	−40	−40	−40
	Compressive Set			
D395 Method B	22 hours @ 212 F, max. %	35	35	35
	Ozone			
D1149	100 pphm ozone in air by volume, 20% strain 100 F ±2 F, 100 hours mounting procedure D518, Procedure A	No Cracks	No Cracks	No Cracks
	Adhesion			
D429,B	Bond made during vulcanization lbs. per inch	40	40	40
	Low Temperature Test			
ASTM D746 Procedure B	Brittleness at −40F	Duro No Failure	No Failure	No Failure

Laminates shall be rolled mild steel sheets conforming to AASHO M183 (ASTM A36) or A245, Grade C or D unless otherwise specified by the Engineer.

2.25.3 — MANUFACTURING REQUIREMENTS

Plain bearings may be molded individually, cut from previously molded strips or slabs or extruded and cut to length. Cut edges shall be at least as smooth as ANSI 250 finish. Unless otherwise shown on the plans, all components of a laminated bearing shall be molded together into an integral unit, and all edges of the nonelastic laminations shall be covered by a minimum of $\frac{1}{8}''$ of elastomer except at laminate restraining devices and around holes that will be entirely closed on the finished structure.

2.25.4 — TOLERANCES

Tolerances, relative dimensions, finishes and appearance, flash, and rubber-to-metal bonding, shall meet the requirements below as defined in the Rubber Handbook, 2nd Edition, published by the Rubber Manufacturers Association, Inc., 444 Madison Avenue, New York 22, N. Y.

Symbol	Requirement
A3	Commercial dimensional tolerances, Table III, Page 15
F3	Commercial Finish, Table V, Page 20.
T.063	Tear trim tolerance no hand trimming required, Table VI, Page 23.
B2	Class 2, Method B, Minimum bond destructive value Table VII, Page 25.
Grade 2	Bond destructive value at 40 lbs. per inch width, Table VIII, Page 25.

2.25.5 — QUALITY ASSURANCE

Whenever practical, the mechanical properties of the finished bearing shall be verified by laboratory test.

The following values shall be met under laboratory testing conditions of full size bearings:

(a) Compressive strain of any layer of an elastomeric bearing shall not exceed 7% at 800 psi average unit pressure, or at the design dead load plus live load pressure if so indicated on the plans.

(b) The shear resistance of the bearing shall not exceed 30 psi for 50 durometer, 40 psi for 60 durometer or 50 psi for 70 durometer, TABLE A compounds; nor 50 psi for 50 durometer, 75 psi for 60 durometer or 110 psi for 70 durometer TABLE B compounds at 25% strain of the total effective rubber thickness after an extended four-day ambient temperature of -20 F.

Section 26 — CONSTRUCTION OF TUNNEL USING STEEL TUNNEL LINER PLATES

2.26.1 — SCOPE

These specifications are intended to cover the installation of tunnel liner plates in tunnels constructed by conventional tunnel methods. For the purposes of these specifications, tunnels excavated by full face, heading and bench, or multiple drift procedures are considered conventional methods. Liner plates used with any construction procedure utilizing a full or partial shield, a tunneling machine, or other piece of equipment which will exert a force upon the liner plates for the purpose of propelling, steering, or stabilizing the equipment are considered special cases and are *not* covered by these specifications.

2.26.2 — DESCRIPTION

This item shall consist of furnishing cold formed steel tunnel liner plates conforming to these specifications and of the sizes and dimensions required on the plans, and installing such plates at the locations designated on the plans by the Engineer, and in conformity with the lines and grades established by the Engineer. The completed liner shall consist of a series of steel liner plates assembled with staggered longitudinal joints. Liner plates shall be fabricated to fit the cross section of the tunnel. Liner plates herein described must meet the Sectional Properties —thickness, area, and moment of inertia—as listed in Article 1.13.9.

All plates shall be connected by bolts on both longitudinal and circumferential seams or joints and shall be so fabricated as to permit complete erection from the inside of the tunnel.

Grout holes 2 inches or larger in diameter shall be provided as shown on the plans to permit grouting as the erection of tunnel liner plates progresses.

2.26.3 — FORMING AND PUNCHING OF LINER PLATES

All plates shall be formed to provide circumferential flanged joints. Longitudinal joints may be flanged or of the offset lap seam type. All plates shall be punched for bolting on both longitudinal and circumferential seams or joints. Bolt spacing in circumferential flanges shall be in accordance with the manufacturer's standard spacing and shall be a multiple of the plate length so that plates having the same curvature shall be interchangeable and will permit staggering of the longitudinal seams. Bolt spacing at flanged longitudinal seams shall be in accordance with the manufacturer's standard spacing. For lapped longitudinal seams, bolt size and spacing shall be in accordance with the manufacturer's standard but not less than that required to meet the longitudinal seam strength requirements of Article 1.13.4.

2.26.4 — INSTALLATION

All liner plates for the full length of a specified tunnel shall be of one type only, either the flanged or the lapped seam type of construction.

Liner plates shall be assembled in accordance with the manufacturer's instructions.

Coated plates shall be handled in such a manner as to prevent bruising, scaling, or breaking of the coating. Any plates that are damaged during handling or placing, shall be replaced by the Contractor at his expense, except that small areas with minor damage may be repaired by the Contractor as directed by the Engineer.

When and as designated by the Engineer, voids occuring between the liner plate and the tunnel wall shall be force-grouted. The grout shall be forced through the grouting holes in the plates with such pressure that all voids will be completely filled.

Full compensation for back packing or grouting shall be considered as included in the contract price paid for tunnel and no separate payment will be made therefor.

2.26.5 — MEASUREMENT

The length of tunnel to be paid for will be the length measured on the tunnel liner plate invert.

2.26.6 — PAYMENT

Payment for the footage of each size of tunnel as determined under measurement shall be at the contract unit prices per lineal foot bid for the various sizes, which payment shall include full compensation for furnishing all labor, materials, tools, equipment and incidentals to complete this item, including removal and disposal of material resulting from the excavation of the bore and force-grouting voids.

APPENDIX A

LOADING — H 15-44

TABLE OF MAXIMUM MOMENTS, SHEARS AND REACTIONS.—SIMPLE SPANS, ONE LANE

Spans in feet; moments in thousands of foot-pounds; shears and reactions in thousands of pounds.
These values are subject to specification reduction for loading of multiple lanes.
Impact not included.

Span	Moment	End shear and end reaction (a)	Span	Moment	End shear and end reaction (a)
1	6.0(b)	24.0(b)	42	274.4(b)	29.6
2	12.0(b)	24.0(b)	44	289.3(b)	30.1
3	18.0(b)	24.0(b)	46	304.3(b)	30.5
4	24.0(b)	24.0(b)	48	319.2(b)	31.0
5	30.0(b)	24.0(b)	50	334.2(b)	31.5
6	36.0(b)	24.0(b)	52	349.1(b)	32.0
7	42.0(b)	24.0(b)	54	364.1(b)	32.5
8	48.0(b)	24.0(b)	56	379.1(b)	32.9
9	54.0(b)	24.0(b)	58	397.6	33.4
10	60.0(b)	24.0(b)	60	418.5	33.9
11	66.0(b)	24.0(b)	62	439.9	34.4
12	72.0(b)	24.0(b)	64	461.8	34.9
13	78.0(b)	24.0(b)	66	484.1	35.3
14	84.0(b)	24.0(b)	68	506.9	35.8
15	90.0(b)	24.0(b)	70	530.3	36.3
16	96.0(b)	24.8(b)	75	590.6	37.5
17	102.0(b)	25.1(b)	80	654.0	38.7
18	108.0(b)	25.3(b)	85	720.4	39.9
19	114.0(b)	25.6(b)	90	789.8	41.1
20	120.0(b)	25.8(b)	95	862.1	42.3
21	126.0(b)	26.0(b)	100	937.5	43.5
22	132.0(b)	26.2(b)	110	1,097.3	45.9
23	138.0(b)	26.3(b)	120	1,269.0	48.3
24	144.0(b)	26.5(b)	130	1,452.8	50.7
25	150.0(b)	26.6(b)	140	1,648.5	53.1
26	156.0(b)	26.8(b)	150	1,856.3	55.5
27	162.7(b)	26.9(b)	160	2,076.0	57.9
28	170.1(b)	27.0(b)	170	2,307.8	60.3
29	177.5(b)	27.1(b)	180	2,551.5	62.7
30	185.0(b)	27.2(b)	190	2,807.3	65.1
31	192.4(b)	27.3(b)	200	3,075.0	67.5
32	199.8(b)	27.4(b)	220	3,646.5	72.3
33	207.3(b)	27.5(b)	240	4,266.0	77.1
34	214.7(b)	27.7	260	4,933.5	81.9
35	222.2(b)	27.9	280	5,649.0	86.7
36	229.6(b)	28.1	300	6,412.5	91.5
37	237.1(b)	28.4			
38	244.5(b)	28.6			
39	252.0(b)	28.9			
40	259.5(b)	29.1			

(a) Concentrated load is considered placed at the support. Loads used are those stipulated for shear.
(b) Maximum value determined by Standard Truck Loading. Otherwise the Standard Lane Loading governs.

LOADING — H 20-44

Table of Maximum Moments, Shears and Reactions.—Simple Spans, One Lane

Spans in feet; moments in thousands of foot-pounds; shears and reactions in thousands of pounds.

These values are subject to specification reduction for loading of multiple lanes.

Impact not included.

Span	Moment	End shear and end reaction (a)	Span	Moment	End shear and end reaction (a)
1	8.0 (b)	32.0 (b)	42	365.9 (b)	39.4
2	16.0 (b)	32.0 (b)	44	385.8 (b)	40.1
3	24.0 (b)	32.0 (b)	46	405.7 (b)	40.7
4	32.0 (b)	32.0 (b)	48	425.6 (b)	41.4
5	40.0 (b)	32.0 (b)	50	445.6 (b)	42.0
6	48.0 (b)	32.0 (b)	52	465.5 (b)	42.6
7	56.0 (b)	32.0 (b)	54	485.5 (b)	43.3
8	64.0 (b)	32.0 (b)	56	505.4 (b)	43.9
9	72.0 (b)	32.0 (b)	58	530.1	44.6
10	80.0 (b)	32.0 (b)	60	558.0	45.2
11	88.0 (b)	32.0 (b)	62	586.5	45.8
12	96.0 (b)	32.0 (b)	64	615.7	46.5
13	104.0 (b)	32.0 (b)	66	645.5	47.1
14	112.0 (b)	32.0 (b)	68	675.9	47.8
15	120.0 (b)	32.5 (b)	70	707.0	48.4
16	128.0 (b)	33.0 (b)	75	787.5	50.0
17	136.0 (b)	33.4 (b)	80	872.0	51.6
18	144.0 (b)	33.8 (b)	85	960.5	53.2
19	152.0 (b)	34.1 (b)	90	1,053.0	54.8
20	160.0 (b)	34.4 (b)	95	1,149.5	56.4
21	168.0 (b)	34.7 (b)	100	1,250.0	58.0
22	176.0 (b)	34.9 (b)	110	1,463.0	61.2
23	184.0 (b)	35.1 (b)	120	1,692.0	64.4
24	192.0 (b)	35.3 (b)	130	1,937.0	67.6
25	200.0 (b)	35.5 (b)	140	2,198.0	70.8
26	208.0 (b)	35.7 (b)	150	2,475.0	74.0
27	216.9 (b)	35.9 (b)	160	2,768.0	77.2
28	226.8 (b)	36.0 (b)	170	3,077.0	80.4
29	236.7 (b)	36.1 (b)	180	3,402.0	83.6
30	246.6 (b)	36.3 (b)	190	3,743.0	86.8
31	256.5 (b)	36.4 (b)	200	4,100.0	90.0
32	266.5 (b)	36.5 (b)	220	4,862.0	96.4
33	276.4 (b)	36.6 (b)	240	5,688.0	102.8
34	286.3 (b)	36.9	260	6,578.0	109.2
35	296.2 (b)	37.2	280	7,532.0	115.6
36	306.2 (b)	37.5	300	8,550.0	122.0
37	316.1 (b)	37.8			
38	326.1 (b)	38.2			
39	336.0 (b)	38.5			
40	346.0 (b)	38.8			

(a) Concentrated load is considered placed at the support. Loads used are those stipulated for shear.

(b) Maximum value determined by Standard Truck Loading. Otherwise the Standard Lane Loading governs.

LOADING — HS 15-44

TABLE OF MAXIMUM MOMENTS, SHEARS AND REACTIONS.—SIMPLE SPANS, ONE LANE

Spans in feet; moments in thousands of foot-pounds; shears and reactions in thousands of pounds.

These values are subject to specification reduction for loading of multiple lanes.

Impact not included.

Span	Moment	End shear and end reaction (a)	Span	Moment	End shear and end reaction (a)
1	6.0(b)	24.0(b)	42	364.0(b)	42.0(b)
2	12.0(b)	24.0(b)	44	390.7(b)	42.5(b)
3	18.0(b)	24.0(b)	46	417.4(b)	43.0(b)
4	24.0(b)	24.0(b)	48	444.1(b)	43.5(b)
5	30.0(b)	24.0(b)	50	470.9(b)	43.9(b)
6	36.0(b)	24.0(b)	52	497.7(b)	44.3(b)
7	42.0(b)	24.0(b)	54	524.5(b)	44.7(b)
8	48.0(b)	24.0(b)	56	551.3(b)	45.0(b)
9	54.0(b)	24.0(b)	58	578.1(b)	45.3(b)
10	60.0(b)	24.0(b)	60	604.9(b)	45.6(b)
11	66.0(b)	24.0(b)	62	631.8(b)	45.9(b)
12	72.0(b)	24.0(b)	64	658.6(b)	46.1(b)
13	78.0(b)	24.0(b)	66	685.5(b)	46.4(b)
14	84.0(b)	24.0(b)	68	712.3(b)	46.6(b)
15	90.0(b)	25.6(b)	70	739.2(b)	46.8(b)
16	96.0(b)	27.0(b)	75	806.3(b)	47.3(b)
17	102.0(b)	28.2(b)	80	873.7(b)	47.7(b)
18	108.0(b)	29.3(b)	85	941.0(b)	48.1(b)
19	114.0(b)	30.3(b)	90	1,008.3(b)	48.4(b)
20	120.0(b)	31.2(b)	95	1,074.9(b)	48.7(b)
21	126.0(b)	32.0(b)	100	1,143.0(b)	49.0(b)
22	132.0(b)	32.7(b)	110	1,277.7(b)	49.4(b)
23	138.0(b)	33.4(b)	120	1,412.5(b)	49.8(b)
24	144.5(b)	34.0(b)	130	1,547.3(b)	50.7
25	155.5(b)	34.6(b)	140	1,682.1(b)	53.1
26	166.6(b)	35.1(b)	150	1,856.3	55.5
27	177.8(b)	35.6(b)	160	2,076.0	57.9
28	189.0(b)	36.0(b)	170	2,307.8	60.3
29	200.3(b)	36.6(b)	180	2,551.5	62.7
30	211.6(b)	37.2(b)	190	2,807.3	65.1
31	223.0(b)	37.7(b)	200	3,075.0	67.5
32	234.4(b)	38.3(b)	220	3,646.5	72.3
33	245.8(b)	38.7(b)	240	4,266.0	77.1
34	257.7(b)	39.2(b)	260	4,933.5	81.9
35	270.9(b)	39.6(b)	280	5,649.0	86.7
36	284.2(b)	40.0(b)	300	6,412.5	91.5
37	297.5(b)	40.4(b)			
38	310.7(b)	40.7(b)			
39	324.0(b)	41.1(b)			
40	337.4(b)	41.4(b)			

(a) Concentrated load is considered placed at the support. Loads used are those stipulated for shear.

(b) Maximum value determined by Standard Truck Loading. (One HS truck). Otherwise the Standard Lane Loading governs.

LOADING — HS 20-44

TABLE OF MAXIMUM MOMENTS, SHEARS AND REACTIONS.—SIMPLE SPANS, ONE LANE

Spans in feet; moments in thousands of foot-pounds; shears and reactions in thousands of pounds.

These values are subject to specification reduction for loading of multiple lanes.

Impact not included.

Span	Moment	End shear and end reaction (a)	Span	Moment	End shear and end reaction (a)
1	8.0(b)	32.0(b)	42	485.3(b)	56.0(b)
2	16.0(b)	32.0(b)	44	520.9(b)	56.7(b)
3	24.0(b)	32.0(b)	46	556.5(b)	57.3(b)
4	32.0(b)	32.0(b)	48	592.1(b)	58.0(b)
5	40.0(b)	32.0(b)	50	627.9(b)	58.5(b)
6	48.0(b)	32.0(b)	52	663.6(b)	59.1(b)
7	56.0(b)	32.0(b)	54	699.3(b)	59.6(b)
8	64.0(b)	32.0(b)	56	735.1(b)	60.0(b)
9	72.0(b)	32.0(b)	58	770.8(b)	60.4(b)
10	80.0(b)	32.0(b)	60	806.5(b)	60.8(b)
11	88.0(b)	32.0(b)	62	842.4(b)	61.2(b)
12	96.0(b)	32.0(b)	64	878.1(b)	61.5(b)
13	104.0(b)	32.0(b)	66	914.0(b)	61.9(b)
14	112.0(b)	32.0(b)	68	949.7(b)	62.1(b)
15	120.0(b)	34.1(b)	70	985.6(b)	62.4(b)
16	128.0(b)	36.0(b)	75	1,075.1(b)	63.1(b)
17	136.0(b)	37.7(b)	80	1,164.9(b)	63.6(b)
18	144.0(b)	39.1(b)	85	1,254.7(b)	64.1(b)
19	152.0(b)	40.4(b)	90	1,344.4(b)	64.5(b)
20	160.0(b)	41.6(b)	95	1,434.1(b)	64.9(b)
21	168.0(b)	42.7(b)	100	1,524.0(b)	65.3(b)
22	176.0(b)	43.6(b)	110	1,703.6(b)	65.9(b)
23	184.0(b)	44.5(b)	120	1,883.3(b)	66.4(b)
24	192.7(b)	45.3(b)	130	2,063.1(b)	67.6
25	207.4(b)	46.1(b)	140	2,242.8(b)	70.8
26	222.2(b)	46.8(b)	150	2,475.1	74.0
27	237.0(b)	47.4(b)	160	2,768.0	77.2
28	252.0(b)	48.0(b)	170	3,077.1	80.4
29	267.0(b)	48.8(b)	180	3,402.1	83.6
30	282.1(b)	49.6(b)	190	3,743.1	86.8
31	297.3(b)	50.3(b)	200	4,100.0	90.0
32	312.5(b)	51.0(b)	220	4,862.0	96.4
33	327.8(b)	51.6(b)	240	5,688.0	102.8
34	343.5(b)	52.2(b)	260	6,578.0	109.2
35	361.2(b)	52.8(b)	280	7,532.0	115.6
36	378.9(b)	53.3(b)	300	8,550.0	122.0
37	396.6(b)	53.8(b)			
38	414.3(b)	54.3(b)			
39	432.1(b)	54.8(b)			
40	449.8(b)	55.2(b)			

(a) Concentrated load is considered placed at the support. Loads used are those stipulated for shear.

(b) Maximum value determined by Standard Truck Loading. (One HS truck). Otherwise the Standard Lane Loading governs.

APPENDIX B

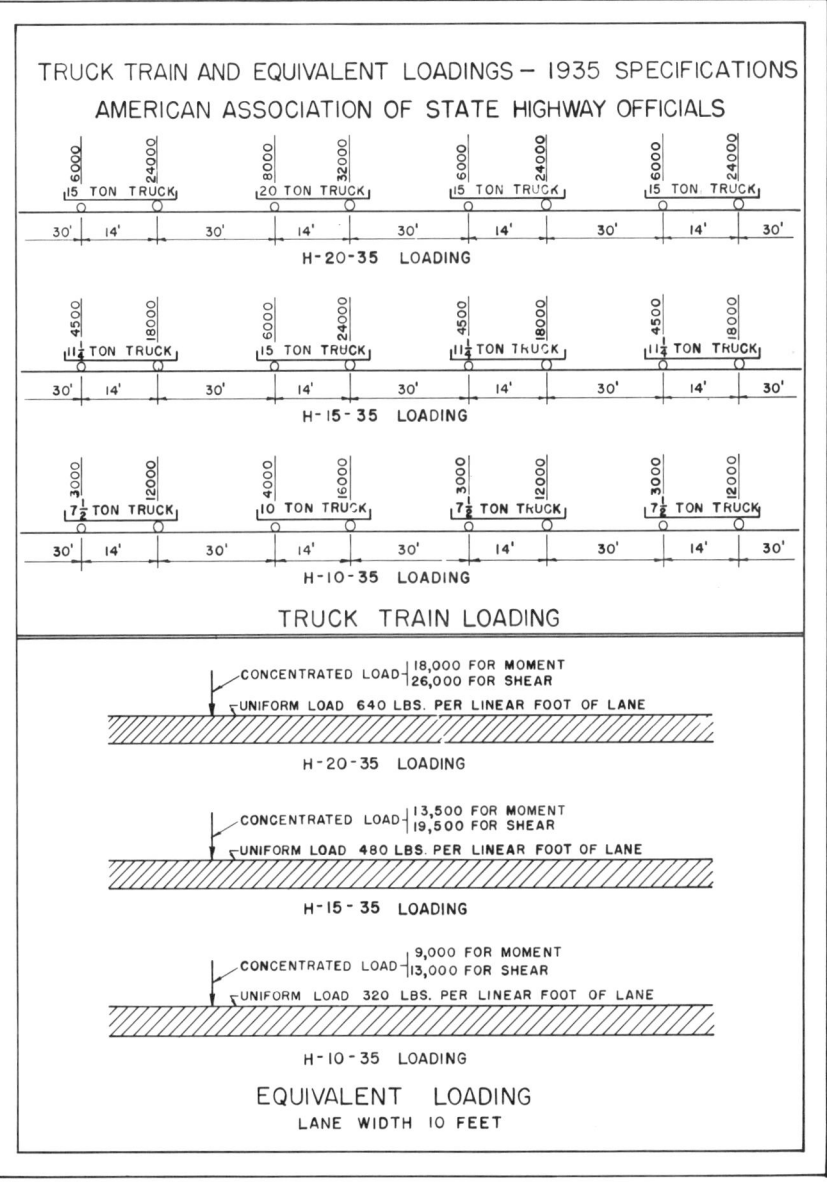

APPENDIX C

FORMULAS FOR STEEL COLUMNS *

The permissible average unit stress for steel columns shall be:

$$^{**} \quad f_s = \frac{\frac{f_y}{\eta}}{1 + \left(.25 + \frac{e_g c}{r^2}\right) B \operatorname{Cosec} \Phi} = \frac{P}{A} \quad (A)$$

$P =$ load parallel to the axis of the member in lbs.
$A =$ gross cross-sectional area of column in sq. in.
$f_y =$ yield point or yield strength (See Article 1.7.1)
$\eta =$ factor of safety based on yield point or yield strength
  $= 1.80$ for carbon steel (A36), low alloy steel (A572 with 45,000 and 65,000 yield points), and high yield strength quenched & tempered steel (A514/A517 with 90,000 yield strength).
  $= 1.82$ for low alloy steel (A242, A440, A441, A572 and A588 with 50,000 yield point) and high yield strength quenched & tempered steel (A514/A517 with 100,000 yield strength).
  $= 1.84$ for low alloy steel (A242, A440, A441 and A588 with 46,000 yield point and A572 with 55,000 and 60,000 yield points).
  $= 1.87$ for low alloy steel (A242, A440, A441, A572, and A588 with 42,000 yield points).
$c =$ distance from neutral axis to the extreme fiber in compression.
$r =$ radius of gyration in the plane of bending.

$$\Phi = \frac{L}{r} \sqrt{\frac{\eta f_s}{E}} \quad \text{radians}$$

$L =$ effective length of the column
  $= 75\%$ of the total length of a column having riveted end connections.
  $= 87.5\%$ of the total length of a column having pinned end connections.
$E =$ modulus of elasticity of steel
  $= 29{,}000{,}000$ lbs. per sq. in.
$B = \sqrt{a^2 - 2a \cos \Phi + 1}$

$a = \dfrac{\dfrac{e_s c}{r^2} + 0.25}{\dfrac{e_g c}{r^2} + 0.25}$ When e_g and e_s lie on the same side of the column axis, a is positive; when on opposite sides, a is negative.

$e_g =$ eccentricity of applied load at the end of column having the greater computed moment, in inches.
$e_s =$ eccentricity at opposite end.

For values of $\dfrac{L}{r}$ equal to or less than $\arccos \alpha \left[\dfrac{E \left(1 + .25 + \dfrac{e_g c}{r^2}\right)}{f_y} \right]^{1/2}$ (B)

---

* Refer also to the column formulas given in Article 1.7.1.
** When the radius of gyration perpendicular to the plane of bending is less than "r", the column shall be investigated for the case of a long column concentrically loaded, having a greater value of $\dfrac{L}{r}$

the permissible f_s shall be determined from the formula:

$$f_s = \frac{\frac{f_y}{\eta}}{1 + .25 + \frac{e_g c}{r^2}} \quad \text{(C)}$$

For $a = -1$ with values of $\frac{L}{r}$ greater than determined by formula B, the permissible f_s shall be determined by the formula:

$$f_s = \frac{\pi^2 E}{\eta \left(\frac{L}{r}\right)^2} \quad \text{(D)}$$

When the values of end moments are not computed but considered negligible in amount, a shall be assumed equal to $+1$.

a shall be assumed equal to $+1$ for a member subject to bending stresses induced by the components of externally applied loads acting perpendicular to its axis. For this case the general formula becomes:

$$f_s = \frac{\frac{f_y}{\eta} - \frac{Mc}{I}}{1 + \left[.25 + (e_g + d)\frac{c}{r^2}\right]\text{Sec } \frac{1}{2}\Phi} \quad \text{(E)}$$

d = deflection due to the transverse components of externally applied loads, in inches.
I = moment of inertia of section about an axis perpendicular to the plane of bending, in (inches)[4].
M = moment due to the transverse components of externally applied load, in inch pounds.

Note: The value of 0.25 in the above formulas provides for inherent crookedness and unknown eccentricity.

GRAPHS

Permissible unit stresses, in accordance with the formulas A and B given above, may be obtained from the following graphs. Graphs are provided for carbon, low alloy, and high yield strength, quenched and tempered steels, and for values of $a = +1, 0$, and -1.

Straight line interpolation between $a = +1$ and 0, and between $a = 0$ and -1, may be used for intermediate values of a.

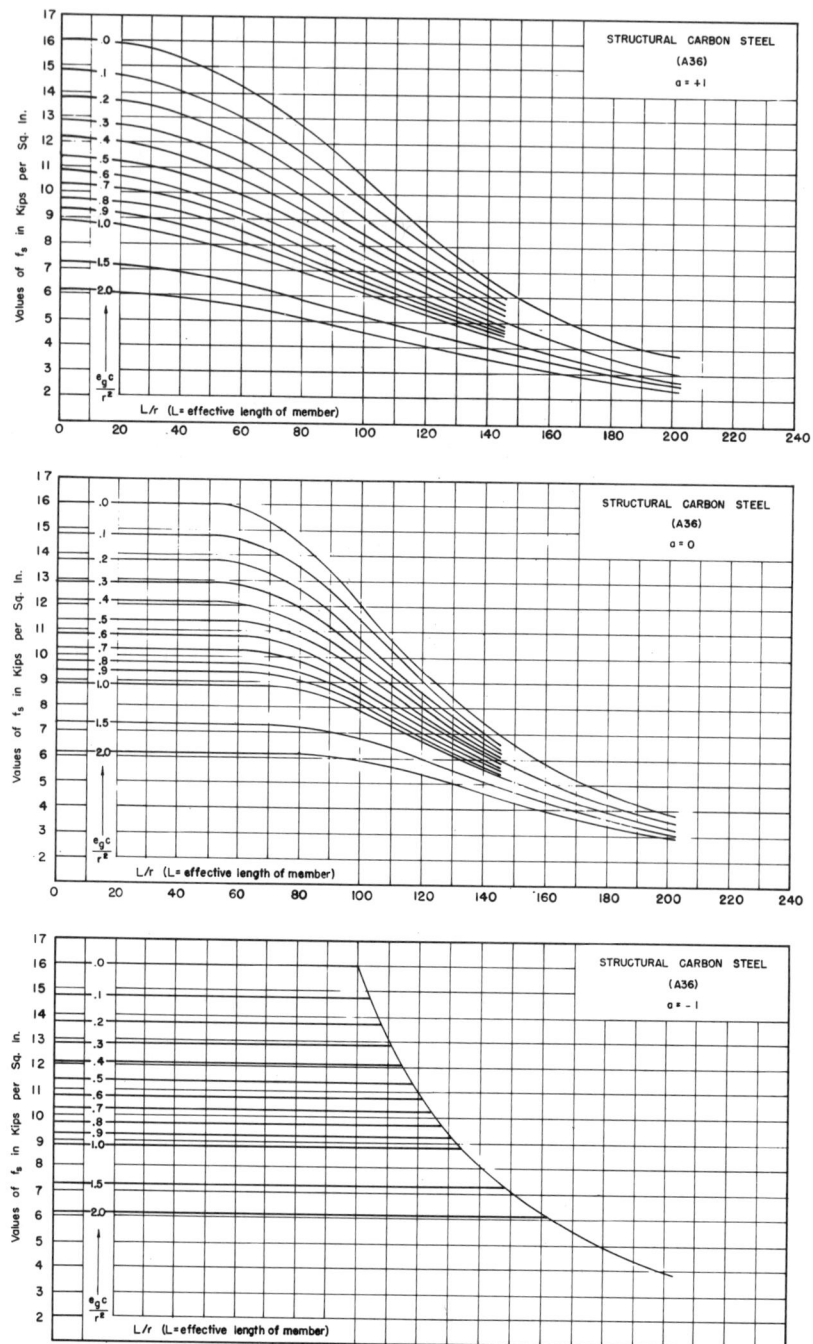

App. C

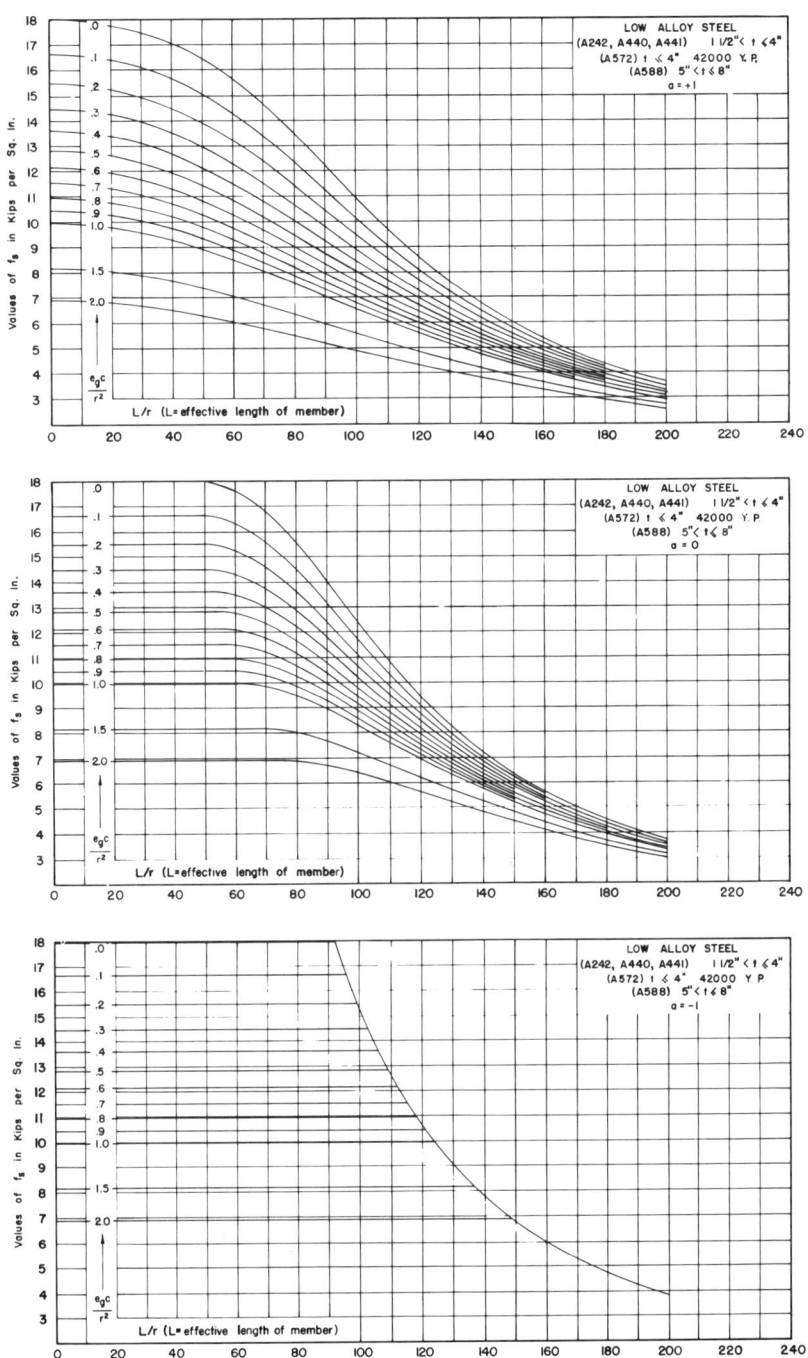

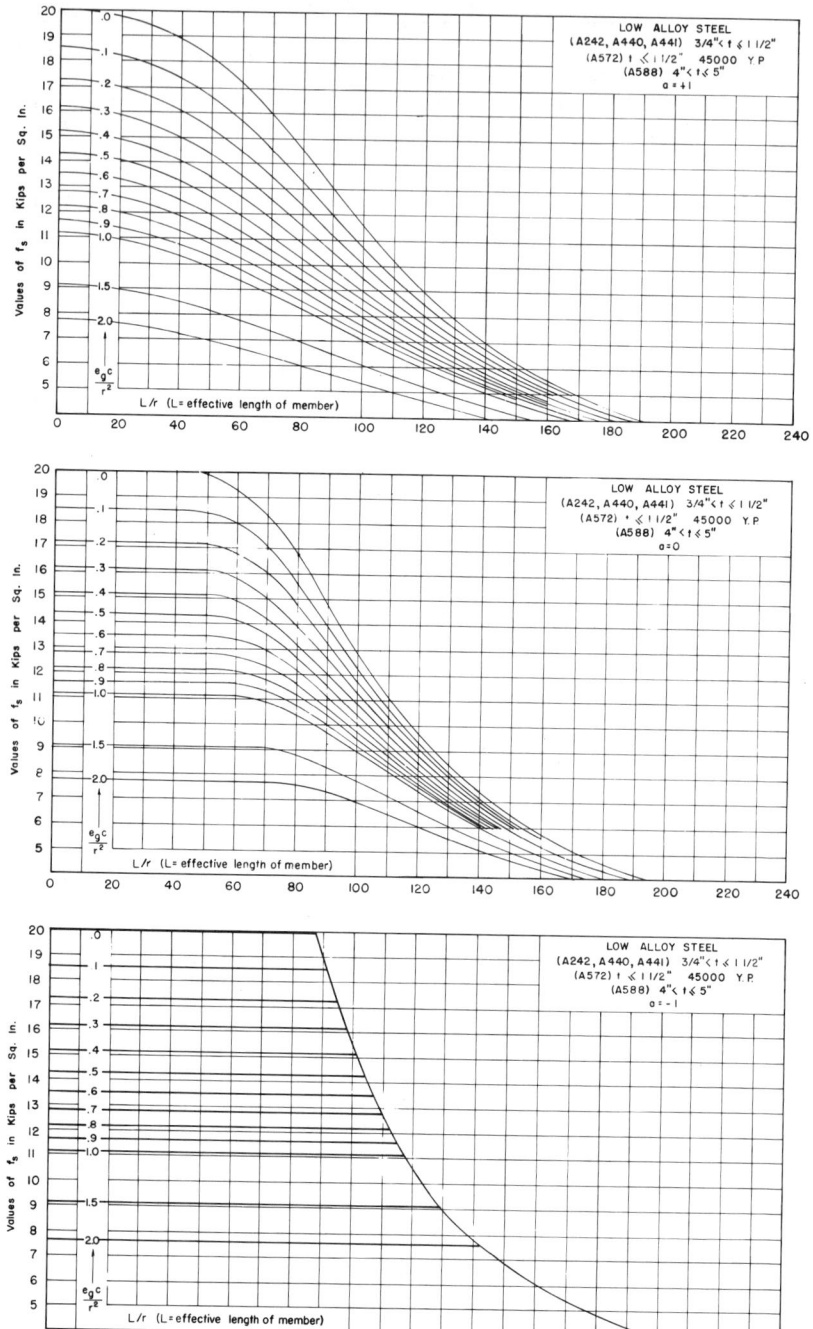

App. C

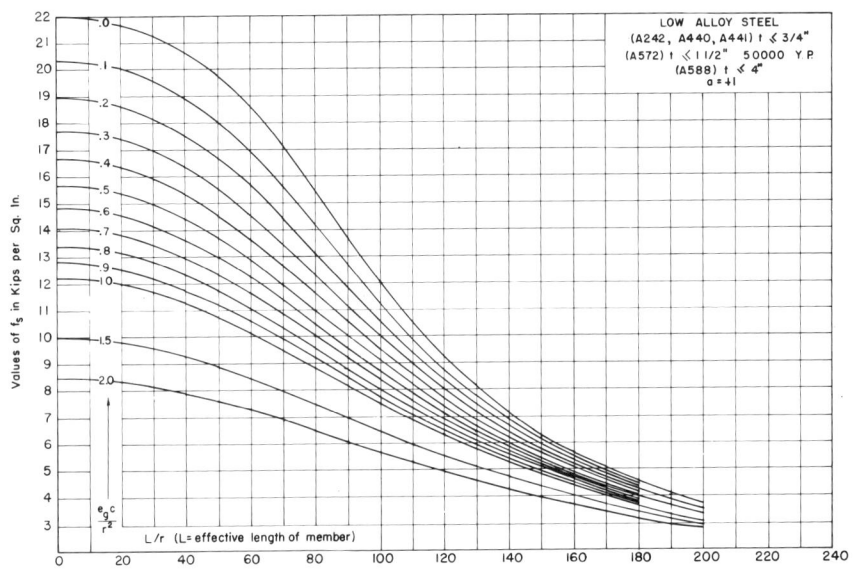

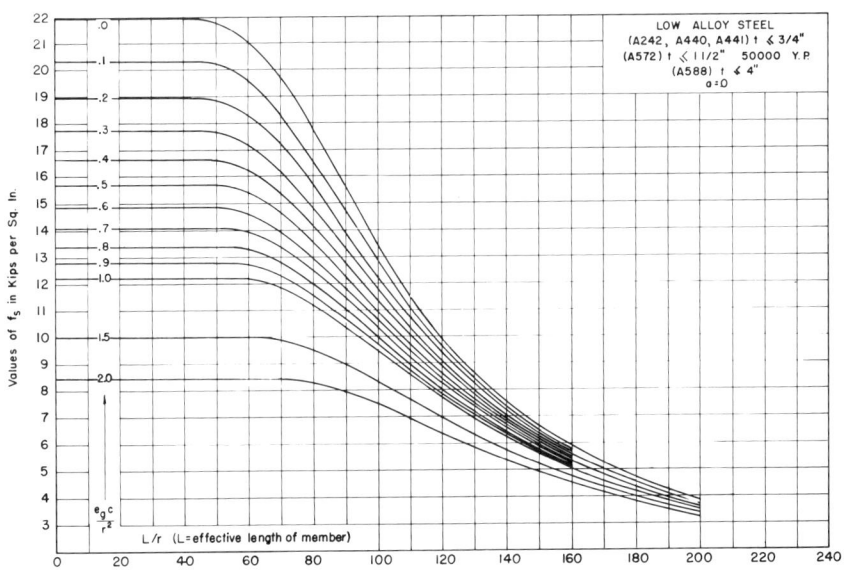

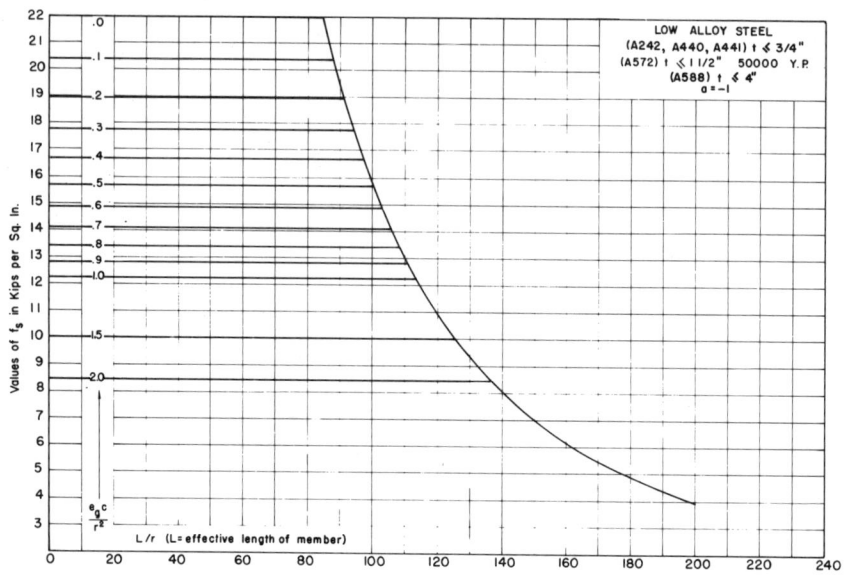
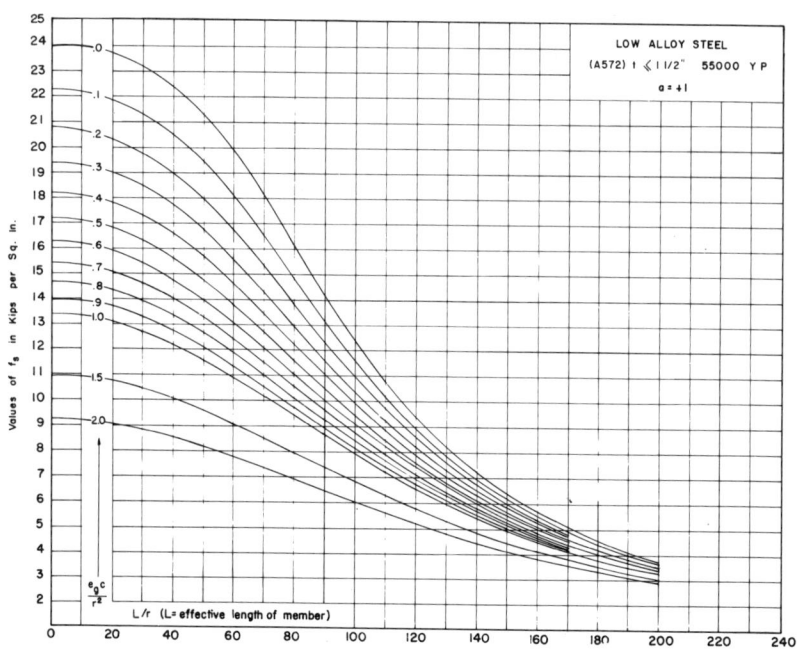

App. C

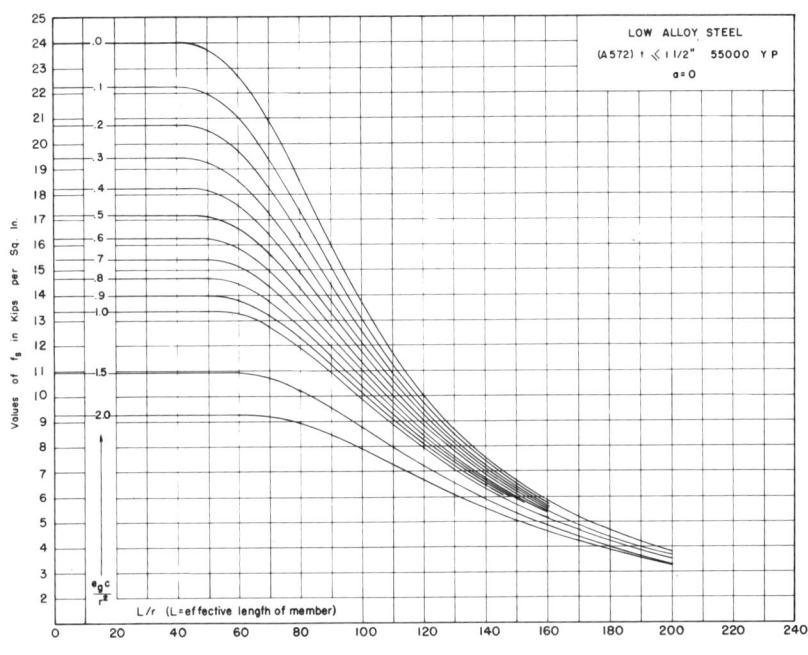

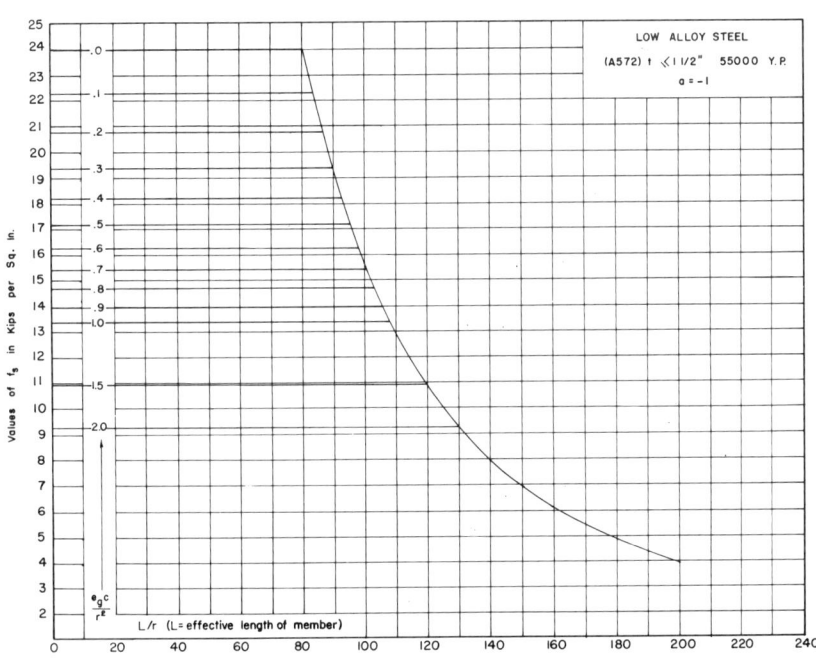

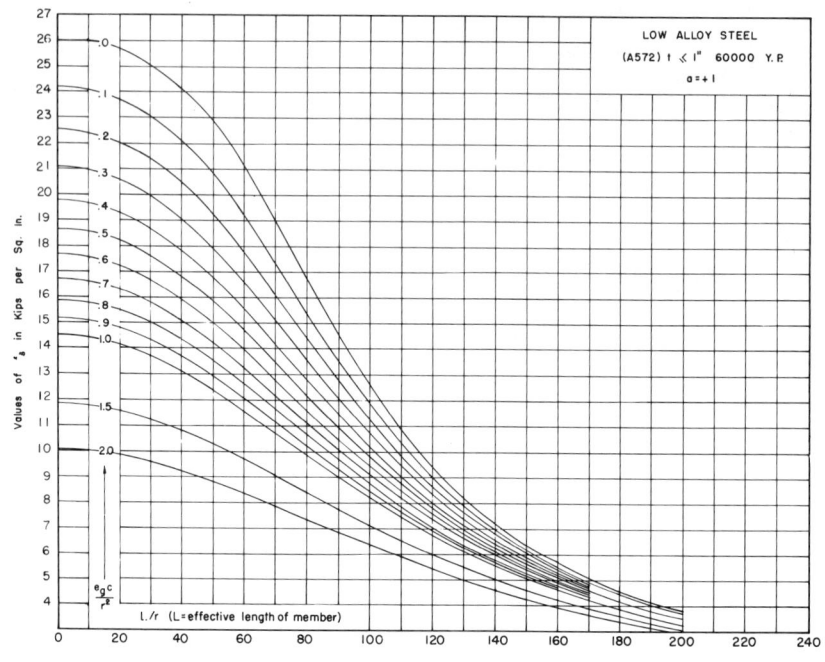

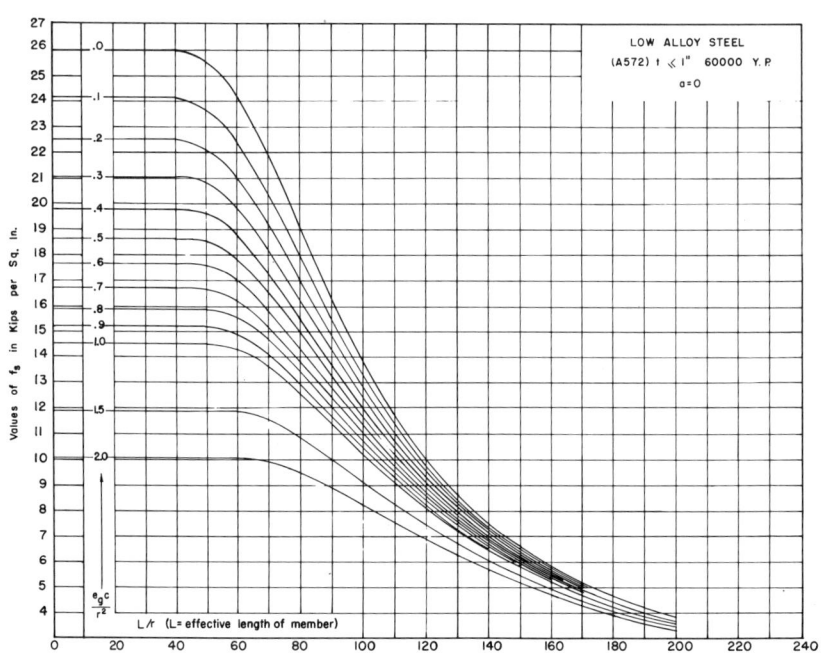

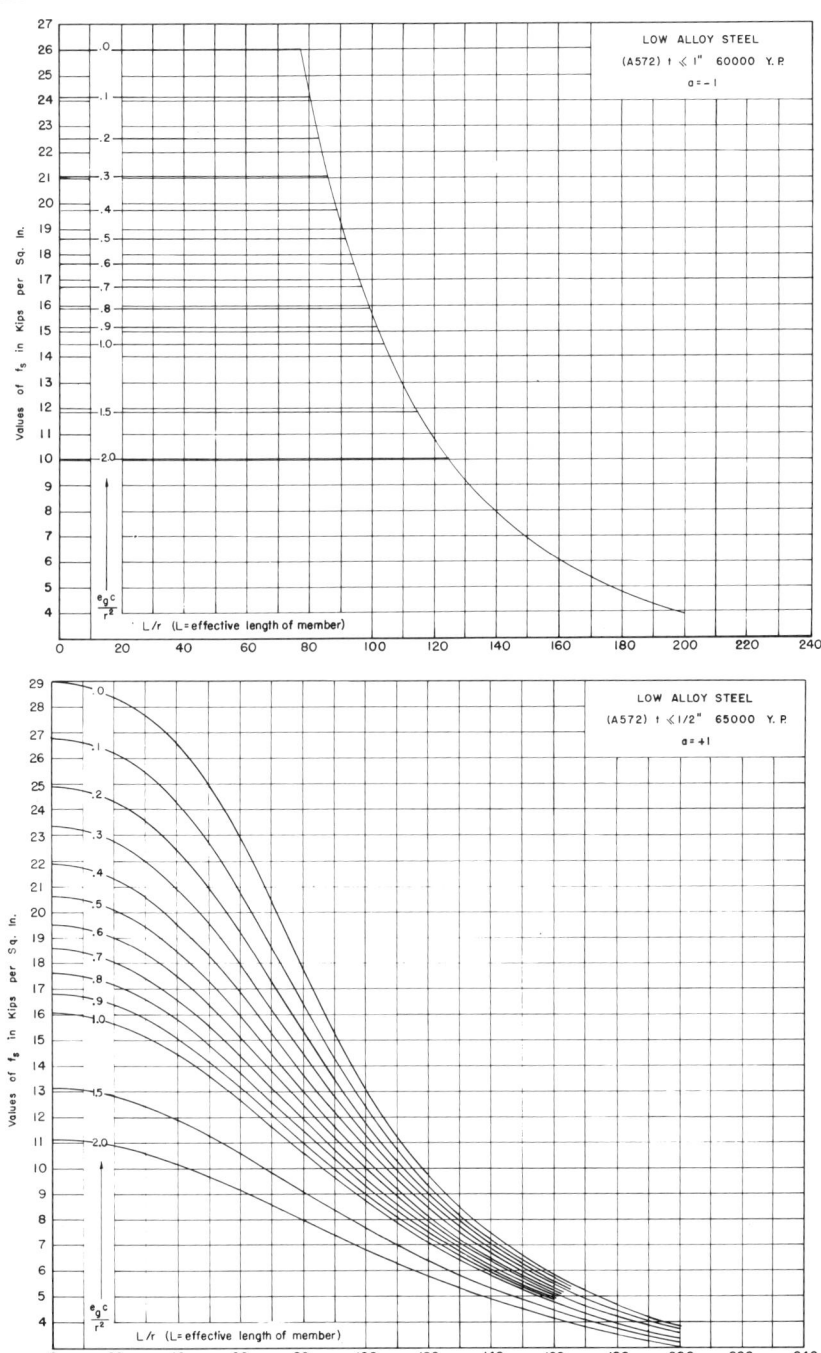

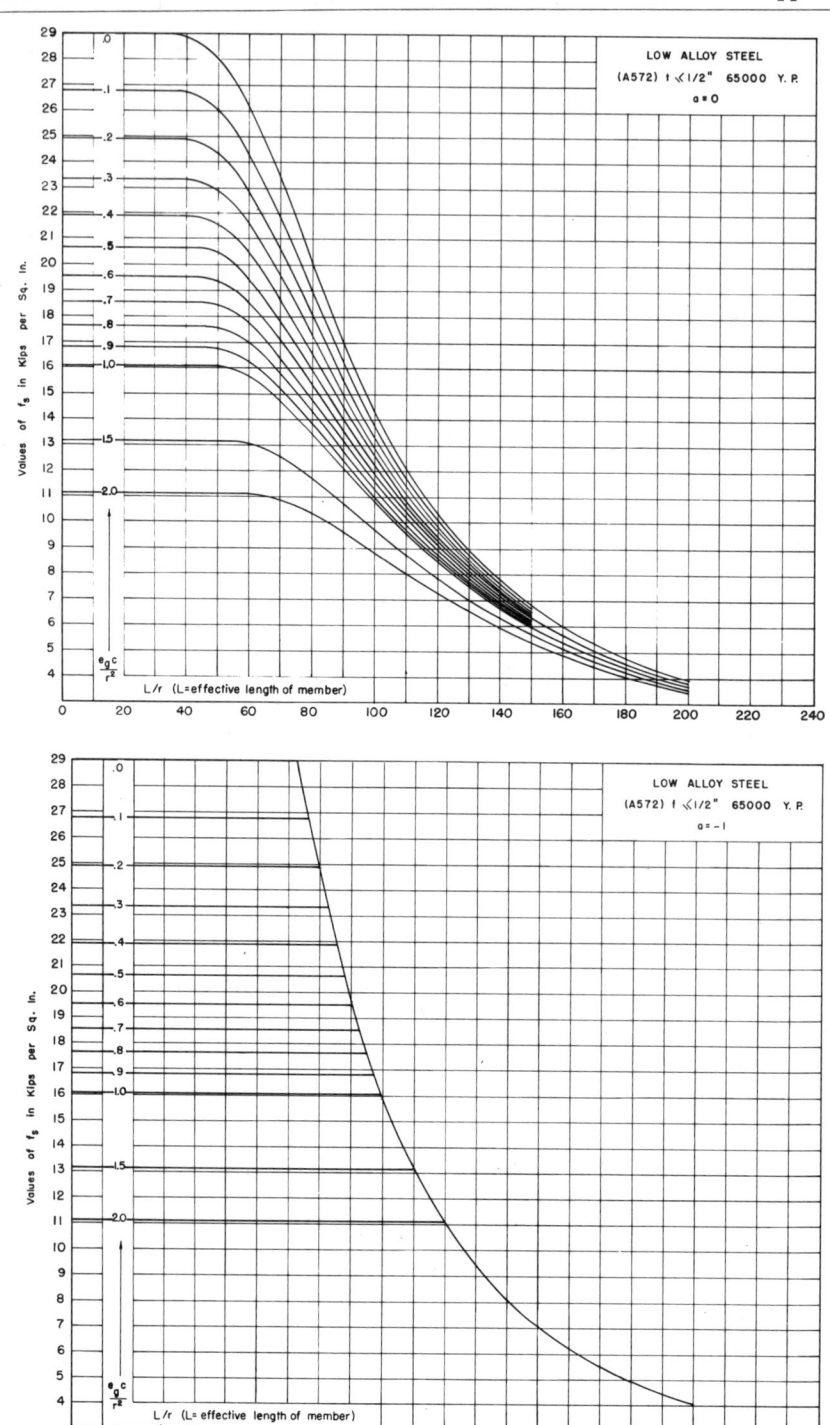

App. C

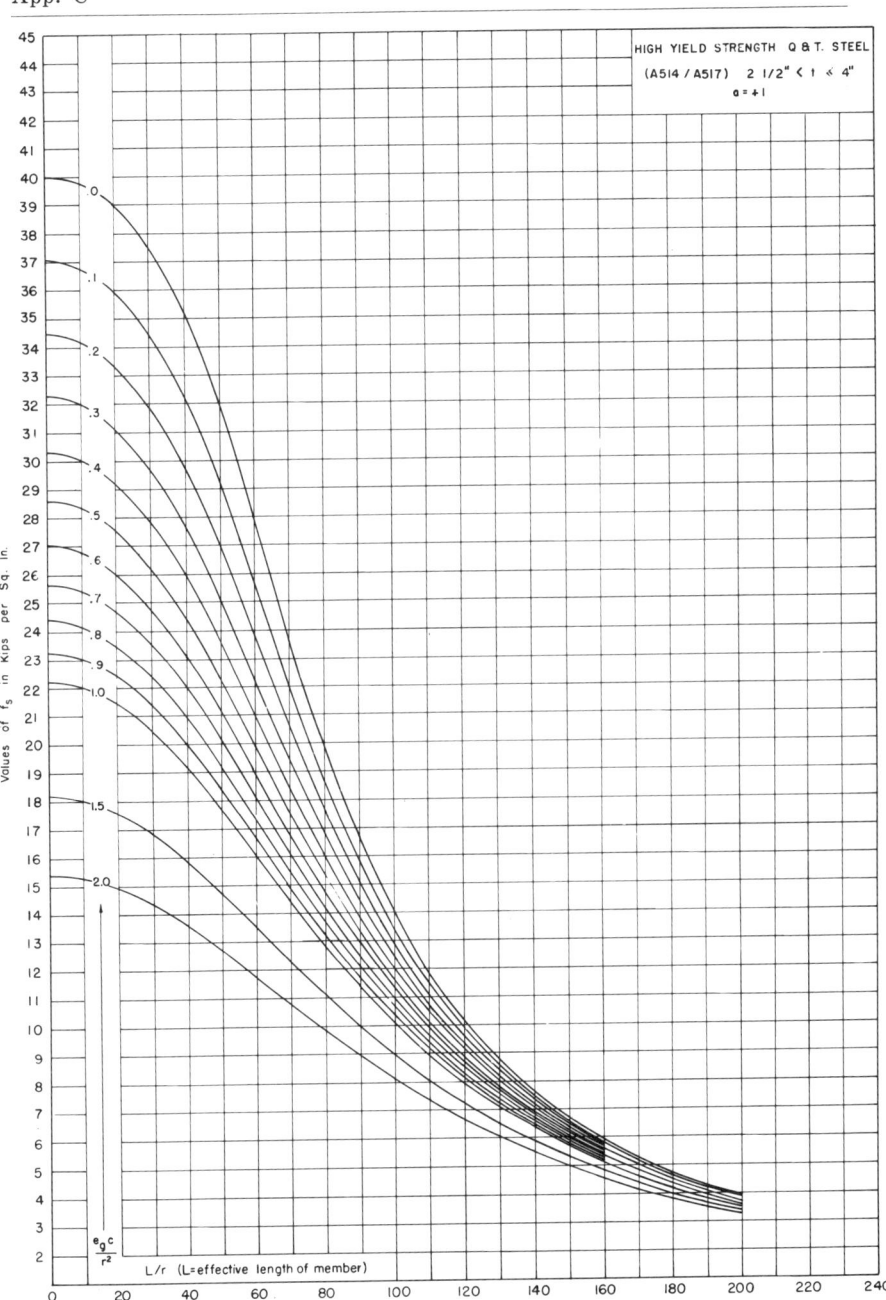

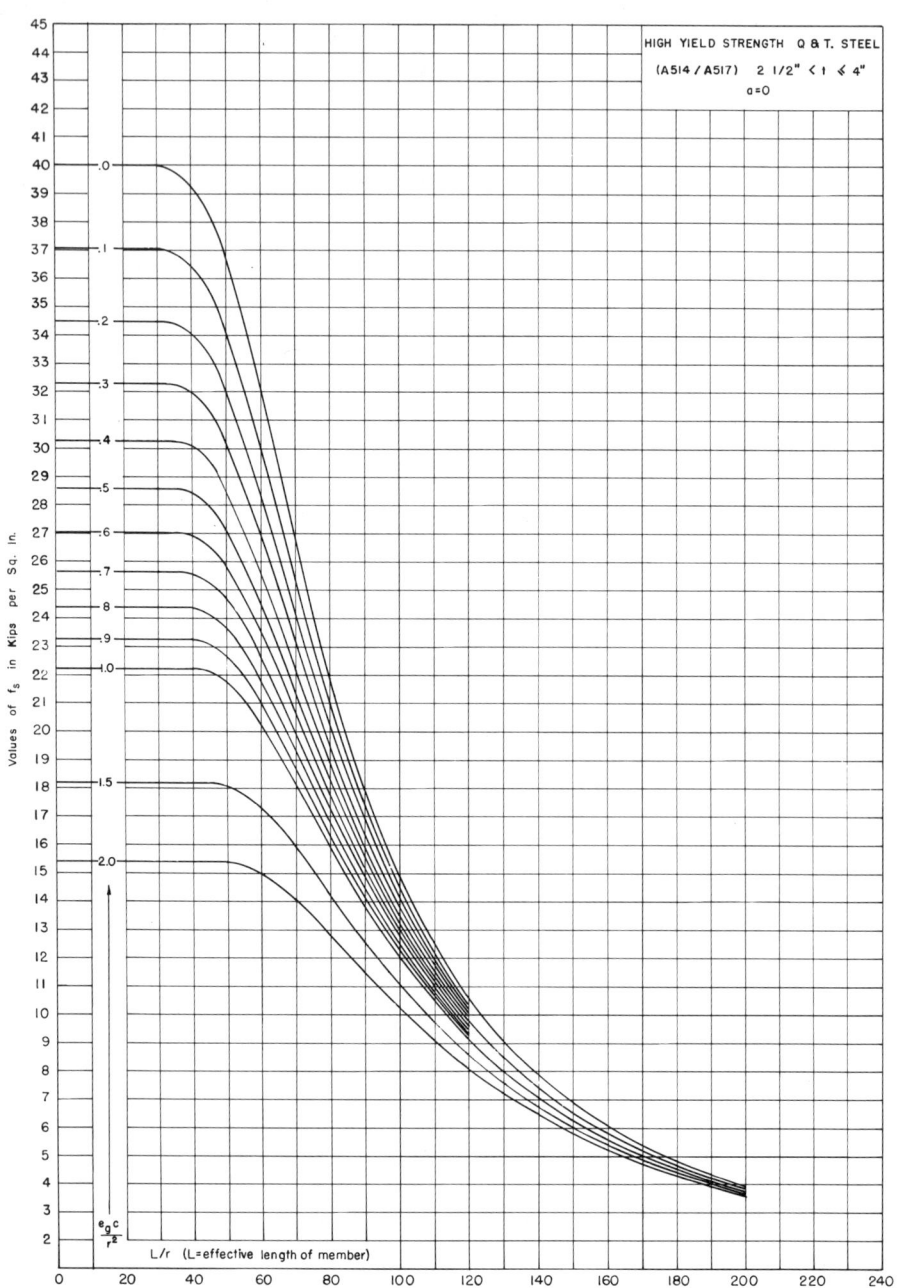

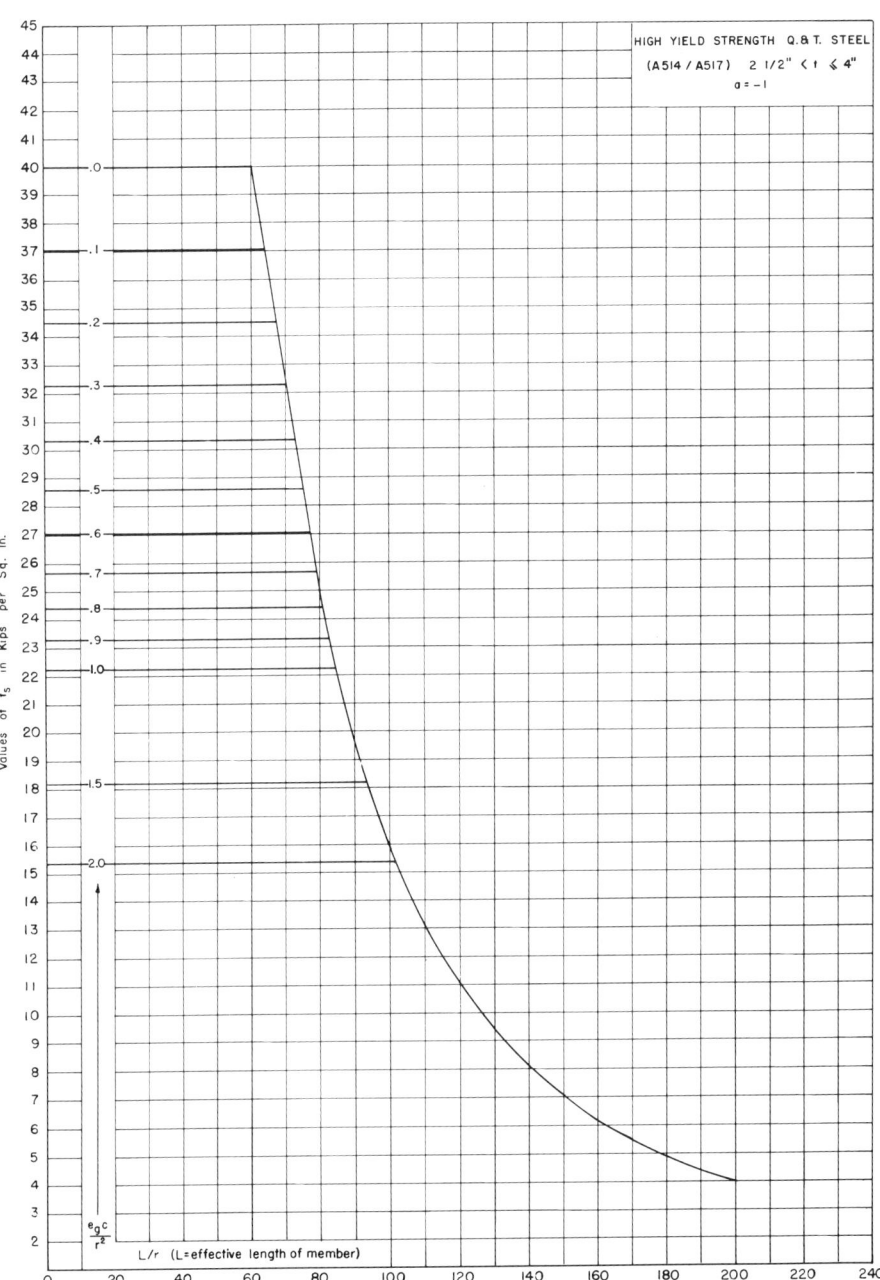

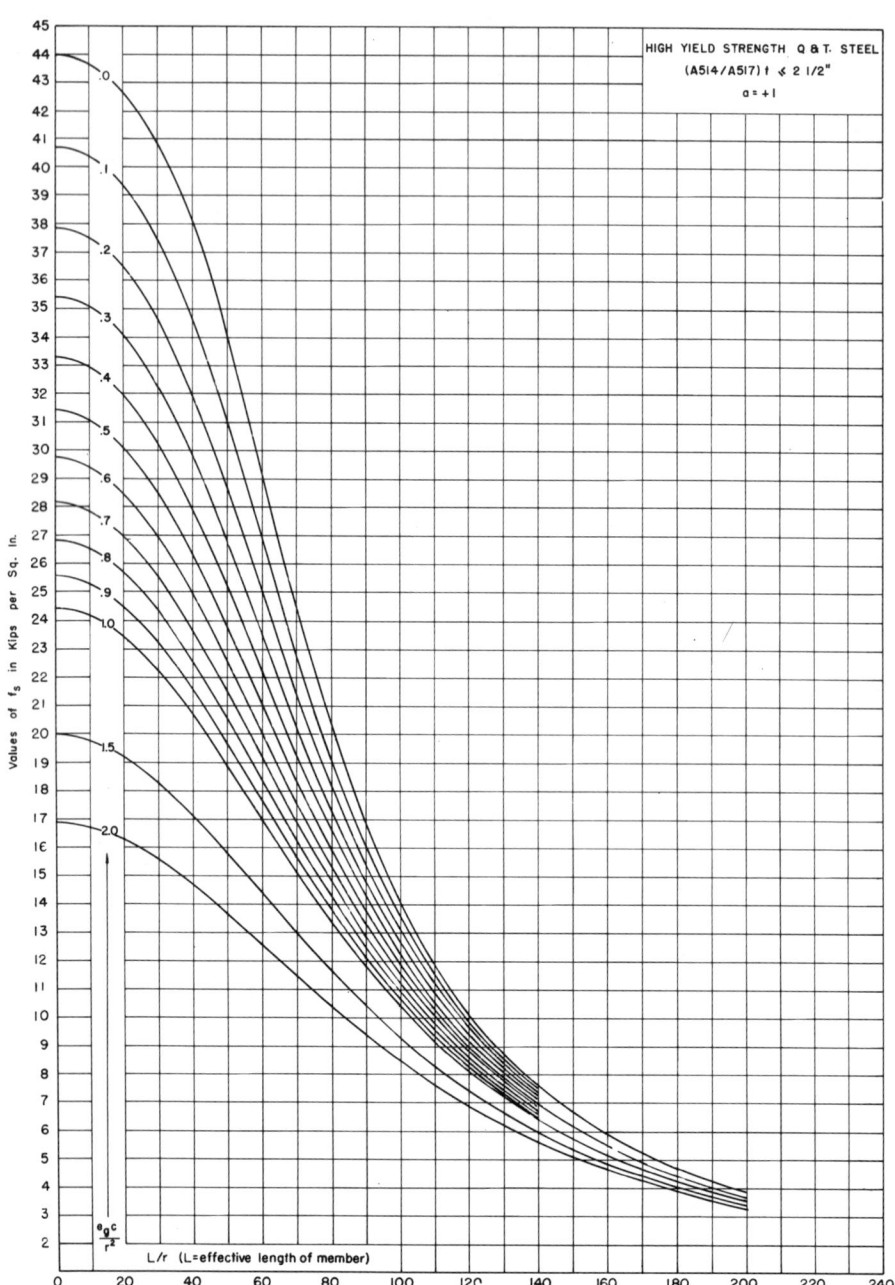

App. C

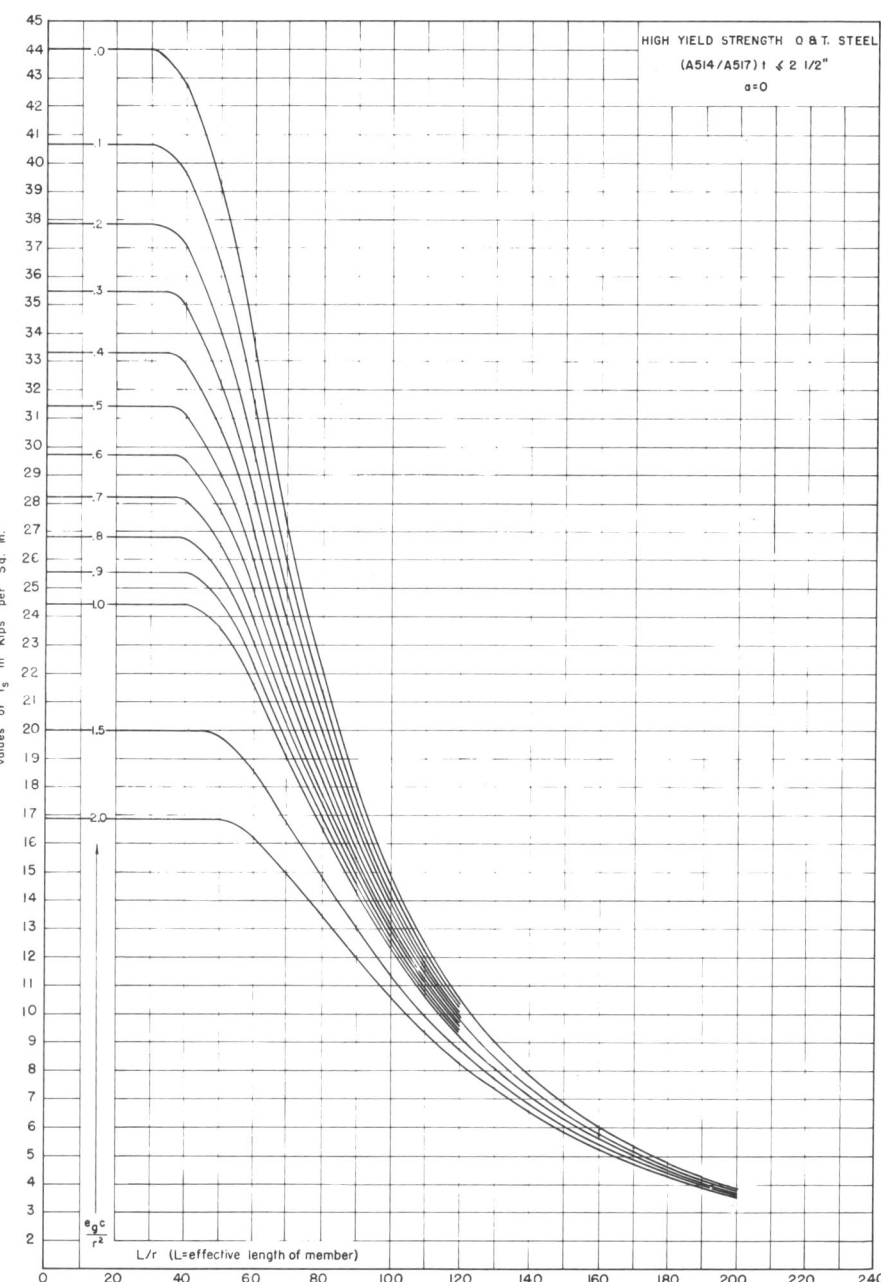

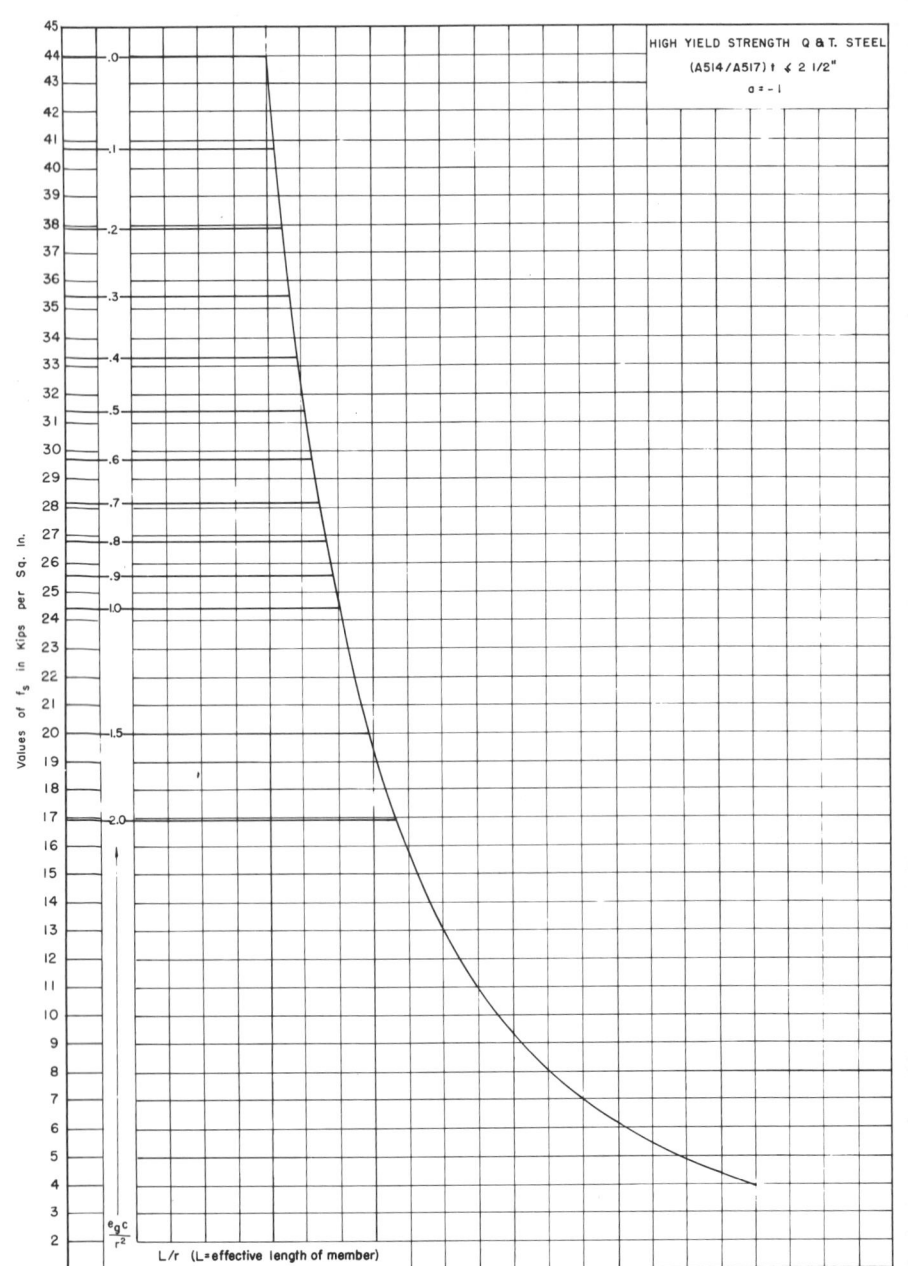

APPENDIX D

POSITION AND DIRECTION OF NEUTRAL AXIS *

When the plane of bending does not lie on a principal axis of the column section or when the point of application of the resultant load does not lie within the kern area of the gross transformed section, the position and direction of the neutral axis may be determined by the following formula:

$$\frac{P}{A} + \frac{M'_y}{I'_y} X_o + \frac{M'_x}{I'_x} Y_o = 0$$

P = load parallel to the axis of the column in pounds
A = transformed area of cracked section in square inches

$$M'_y = M_y - \frac{I_{xy}}{I_x} M_x \qquad M'_x = M_x - \frac{I_{xy}}{I_y} M_y$$

$$I_y' = I_y - \frac{(I_{xy})^2}{I_x} \qquad I'_x = I_x - \frac{(I_{xy})^2}{I_y}$$

M_y = moment of external forces about centroidal Y axis
M_x = moment of external forces about centroidal X axis
X_o, Y_o = coordinates referred to axes passing through the centroid of the section
I_y = moment of inertia of cracked transformed section about the centroidal Y axis
I_x = moment of inertia of cracked transformed section about the centroidal X axis
I_{xy} = product of inertia of cracked transformed section about centroidal axes, X and Y

In solving the above formula it is necessary to assume a value for either X_o or Y_o.

FORMULAS FOR STRESSES

With the position and direction of the neutral axis determined, the maximum unit stress in the concrete shall be computed with the formula:

$$f = \frac{M'_x}{I'_x} Y_n \quad \text{or} \quad f = \frac{M'_y}{I'_y} X_n, \quad \text{in which}$$

Y_n = distance from the neutral axis to the extreme fiber in compression measured parallel to the centroidal Y axis
X_n = distance from the neutral axis to the extreme fiber in compression measured parallel to the centroidal X axis

---
* Article by David M. Goodall, Public Roads, Vol. 25, No. 4, June 1948.

INDEX

A

	Article	Page
Abutments:		
Backfill	2.1.8	249
Design	1.4.7	52
Drainage	1.4.7D	53
Overturning	1.4.7A	52
Pile bent	1.10.8	230
Plain concrete	1.4.7A	52
Reinforcement cover	1.5.6B	61
Scour protection	1.1.2C	2
Stone masonry	1.4.7A	52
Temperature reinforcement	1.4.7B	53
Wing walls	1.4.7C	53
Admixtures for concrete	2.4.2B	264
Aggregates:		
Care and storage	2.4.3	265
Coarse	2.4.2D	265
Fine	2.4.2C	264
Heating in cold weather	2.4.21	280
Rubble or cyclopean	2.4.2D	265
Specific gravity	2.4.6	267
Weighing	2.4.8	268
Air-entrained concrete	2.4.5, 2.4.6	266, 267
Alkali exposed concrete	2.4.17	278
Allowable loads:		
Columns, concrete	1.5.9	67
Piles	1.4.4	41
Shear connectors	1.7.100	162
Unreinforced piers and pedestals	1.5.9B	68
Allowable stresses:		
Bearing on masonry	1.7.8	125
Bearing type connections	1.7.5	121
Bond, concrete	1.5.1C	57
Bridge seats	1.7.8	125
Bronze	1.7.7	125
Cast iron	1.7.6	124
Cast steel	1.7.6	124
Columns, concrete	1.5.9	67
Columns, steel	1.7.1, App. C	111, 416
Compression reinforcement	1.5.7	64
Concrete	1.5.1	56
Copper-alloy	1.7.7	125
Ductile iron castings	1.7.6	124
Elastomeric bearing pads	1.12.2	238
Existing bridges	1.11.5	235
Expansion rockers	1.7.4	117
Fasteners	1.7.5	121
Fatigue	1.7.3	115
Friction type connections	1.7.5	121
Malleable castings	1.7.6	124
Pins	1.7.4	117
Prestressed concrete	1.6.6	97
Reinforcement	1.5.1D	57

# INDEX

	Article	Page
Allowable stresses, cont'd.		
Reinforcement, compression	1.5.7	64
Retaining walls	1.4.1	40
Rollers	1.7.4	117
Round timber piles	1.4.4B	41
Steel piles	1.4.4C	43
Steel	1.7.1	111
Substructures	1.4.1	40
Timber	1.10.1	207
Weld metal	1.7.2	115
Anchorage:		
Expansion bearings	1.7.50	138
Fixed bearings	1.7.49	138
Steel structures	1.7.55	139
Anchorage Zones	1.6.15	106
Anchor bolts	1.7.55	139
Anchor bolts, erection	2.10.56	352
Angles:		
Effective area in tension	1.7.14	127
End connections of floorbeams and stringers	1.7.61, 2.10.27	141, 342
Flange	1.7.69B	143
Minimum size for bracing	1.7.21	131
Outstanding leg, width	1.7.15, 1.7.69B, 1.7.88	128, 143, 156
Outstanding leg of bearing stiffeners	1.7.73B	150
Outstanding legs of tranverse stiffeners	1.7.71	146
Placement of vertical leg	1.7.24, 1.7.76	132, 151
Shelf	1.7.61	141
Size of fastener	1.7.35	134
Thickness of metal	1.7.15, 1.7.69, 1.7.88	128, 143, 156
Angles of repose	1.4.3	41
Annealing and stress relieving	2.10.34	344
Application of loadings	1.2.8	19
Approach embankment	2.1.10	250
Apron plates	1.7.60	140
Arches:		
Ashlar masonry	2.6.15	299
Mortar rubble masonry	2.7.8	302
Placement of concrete	2.4.10D	275
Voussoirs	2.6.15	299
Arches, concrete	1.5.10	71
Arches—pipe		
(*See* Pipe-structural Plate)		
Arches—ribbed steel	1.7.90	158
Arches, structural plate		
Assembly	2.23.3	400
Backfill	2.23.9	402
Bearings	2.23.9	402
Bracing	2.23.7	402
Cover	1.9.3	206
Headwalls	2.23.9	402
Inspection	2.23.11	404
Measurement	2.23.12	404
Multiple	1.9.5, 2.23.9	207, 402
Payment	2.23.13	404
Rise to span ratio	1.9.2	206
Substructures	2.23.9	402

	Article	Page
Ashlar masonry	Section 2.6	295
Arches	2.6.15	299
Copings	2.6.14	299
Cores and backing	2.6.8	297
Dowels and cramps	2.6.13	298
Dressing stone	2.6.5	296
Headers	2.6.7	296
Laying stone	2.6.10	297
Leveling courses	2.6.11	298
Materials	2.6.2	295
Measurement and payment	2.6.17	300
Mixing mortar	2.6.9	297
Pointing	2.6.16	299
Resetting	2.6.12	298
Size of stone	2.6.3	295
Stretchers	2.6.6	296
Surface finishes	2.6.4	296

B

	Article	Page
Back-fill	2.1.8	249
Bar reinforcement	2.5.1A	291
Batter piles	1.4.5E	47
Bearings:		
Anchorage	1.7.49, 1.7.51, 1.7.55	138, 139
Expansion	1.7.50, 1.7.51, 1.7.52, 2.4.23G	138, 139, 282
Fixed	1.7.49, 2.4.23G	138, 282
Height	1.7.54	139
Inclined	1.7.53	139
Masonry	1.7.54	139
Masonry plates	1.7.53	139
Masonry, allowable stresses	1.7.8	125
Pedestals and shoes	1.7.56	140
Sole plates	1.7.53	139
Bearing area:		
Fasteners	1.7.5	121
Pins	1.7.4	117
Bearing devices, concrete structures	1.5.13, 1.6.20, 2.4.23G	75, 108, 282
Bearing piles	1.4.4	41
(*See also* Piles)		
Bearing values	2.3.6C	256
Capping, timber	2.3.12	257
Caps, driving	2.3.3B	252
Cast-in-place concrete piles	2.3.15	258
Collars	2.3.3C	253
Cutoff	2.3.10, 2.3.11, 2.3.18	257, 259
Defective piles	2.3.5	254
Extensions	2.3.16	259
Hammers	2.3.4B, C	253
Loading tests	2.3.6A	255
Measurement and payment	2.3.18, 2.3.19, 2.3.20	259, 262
Methods of driving	2.3.4	253
Order list	2.3.8	256
Painting	2.3.17	259
Pointing, timber	2.3.3D	253
Precast concrete piles	2.3.13	257
Preparation for driving	2.3.3	252

INDEX

	Article	Page
Bearing piles, cont'd.		
Splicing	2.3.3E	253
Steel	2.3.1	252
Storage and handling	2.3.9, 2.3.14	256, 258
Test piles	2.3.7	256
Timber pile formulas	2.3.6B	255
Bearing plates:		
(See also Bearings)		
Allowable stresses	1.7.6, 1.7.7	124, 125
Bronze or copper-alloy	Section 2.11	358
Erection	2.10.56	352
Measurement and payment	2.11.6	358
Placement	2.11.5	358
Surface finish	2.10.25	342
Bearing stiffeners:		
Plate girders, design	1.7.73	149
Rolled beams	1.7.66	142
Bearing type connections	1.7.5	121
(See also Connections)		
Bearing value:		
Foundation soils	1.4.2	40
Piles	1.4.4	41
Bedding Material	2.15.8	374
Bending moments, standard loading	Appendix A	411
Bents:		
Caps	1.4.5D, 2.20.14	46, 393
Concrete	1.5.11	72
Framed	2.20.13	392
Timber	1.10.8	230
Bents and towers, steel	1.7.91	159
Bent plates	2.10.31	343
Blast cleaning, metal surfaces	2.14.10C	367
Blast plates and protection	1.1.13	8
Bolts:		
(See also Anchor bolts)		
(See also Fasteners)		
Connections	2.10.19, 2.10.20	320
High strength	2.10.3A	306
Holes	2.10.7	316
Pipe-structural plate	1.8.5	205
Timber structures	1.10.4, 2.20.9	230, 391
Weight	2.10.63	355
Bond:		
Allowable stresses	1.5.1	56
Slabs	1.3.2F	35
Spread footings	1.4.6E	50
Box girders:		
Load distribution	1.3.1, 1.7.103	29, 166
(See also Concrete box girders and composite box girders, steel)		
Bracing:		
Bents and towers	1.7.94	159
Columns, concrete	1.5.9A	67
Double system	1.7.21	131
Eccentric connections	1.7.18	128
Fasteners, minimum number	1.7.21	131
Limiting length	1.7.11	126
Minimum size of angle	1.7.21	131

INDEX

	Article	Page
Bracing, cont'd.		
Minimum size of fillet weld	1.7.26	132
Outstanding legs of angles	1.7.15	128
Pile and framed bents	1.10.8	230
Portal, sway	1.7.81	152
Steel box girders	1.7.108	170
Timber structures	1.10.9E, 2.20.15	232, 393
Tubular steel piers	1.4.10F	55
Brick masonry	Section 2.9	304
Construction	2.9.3	304
Materials	2.9.2	304
Measurement and payment	2.9.5	305
Bridge locations	1.1.1	1
Bridge seats, allowable stresses	1.7.8	125
Bronze or copper-alloy bearing and expansion plates:		
Allowable stress	1.7.7	125
Materials	2.10.3G, 2.11.2	312, 358
Measurement and payment	2.11.6	358
Bronze plates:		
Allowable stress	1.7.7	125
Sliding expansion bearings	1.7.51	138
Bundled bars	2.5.5	293
Prestressing steel	1.6.16C	107
Reinforcing steel	1.5.6I	64
Buoyancy	1.2.18	27
Piles	1.4.4F	45
Buttresses	1.4.8D	54
Butt welds, allowable stresses	1.7.2, 1.7.3	115
(*See also* Welding)		
Butt-welded splices		
(*See* Splices)		

C

	Article	Page
Camber:		
Plate girders	1.7.74	150
Steel grid flooring	2.12.4	360
Trusses, steel	1.7.79	152
Trusses, timber	1.10.9F	233
Cantilever slabs:		
Design	1.3.2H, 1.7.102	35, 165
Railing loadings	1.3.2H	35
Canvas, for bedding masonry plates	2.10.3K	313
Caps, bent	1.4.5D	46
Caps, timber bents	2.20.14	393
Castings, steel	1.7.6, 2.10.3C, D, E, F	124, 310, 311, 312
Cast-in-place concrete piles	1.4.5H, 2.3.15	48, 258
Cast iron, allowable stresses	1.7.6	124
Cast steel, allowable stresses	1.7.6	124
Cement:		
Types	2.4.2A(1)	262
Sampling and testing	2.4.2A(5)	263
Storage	2.4.4	265
Water for	2.4.2B	264
Centrifugal forces	1.2.21	28
Channel changes	1.1.1	1
Channel, preservation	2.1.2	247

INDEX

	Article	Page
Cleaning metal surfaces	2.14.10	367
Clearances:		
Bridges	1.1.7	3
Depressed roadway	1.1.16	10
Navigational	1.1.7A	3
Tunnels	1.1.15	9
Underpasses	1.1.17	10
Vehicular	1.1.7B	4
Vertical	1.1.7, 1.1.15, 1.1.17	3, 9, 10
Clearing and grubbing, compensation for	2.1.1	247
Coefficients:		
Concrete	1.5.1	56
Structural steel	1.7.1	111
Coefficient of friction, bents and towers	1.7.95	160
Cofferdams and cribs	2.1.5	248
Columns, concrete	1.5.9	67
Allowable load	1.5.9B, C, D, E, F	68, 69, 70
Bending about two axes	1.5.9F	70
Bending moment	1.5.9E	70
Bracing	1.5.9A	67
Bundled reinforcement	1.5.6I	64
Combined stresses	1.5.9F	70
Degree of restraint	1.5.2	58
Lateral ties	1.5.9D	69
Least lateral dimension	1.5.9A	67
Long	1.5.9C, D	68, 69
Minimum reinforcement	1.5.9A	67
Minimum size	1.5.9F	70
Modular ratio "n"	1.5.2, 1.5.9F	58, 70
Moment of inertia	1.5.2	58
Placement of concrete	2.4.10C	274
Reinforcement	1.5.6	61
Short	1.5.9C, D	68, 69
Spiral	1.5.9A, C	67, 68
Tied	1.5.9A, D	67, 69
Tie spacing	1.5.9D	69
Unsupported length	1.5.9A	67
Columns, steel:		
(*See also* Compression members)		
Bents and towers	1.7.91	159
Limiting length	1.7.11	126
Splices	1.7.19	128
Columns, Timber	1.10.2, 1.10.7	209, 230
Combined stresses:		
Columns, concrete	1.5.9F	70
Fasteners	1.7.5	121
Steel members	1.7.17	128
Composite box girders, steel	1.7.102 to 1.7.109	165-170
Access	1.7.109	170
Bottom flange plates	1.7.105	167
Diaphragms	1.7.107	170
Distribution of loads	1.7.103	166
Drainage	1.7.109	170
Flange to web welds	1.7.106	170
Lateral bracing	1.7.108	170
Web plates	1.7.104	166

INDEX

	Article	Page
Composite I-girders, steel	1.7.96 to 1.7.101	160-165
Continuous	1.7.99	161
Creep effect	1.7.96	160
Deflection	1.7.101	165
Depth ratio	1.7.10	125
Effective flange width	1.7.98	161
Fatigue	1.7.100	162
Horizontal shear	1.7.100	162
Modular ratio, "n"	1.7.96	160
Shear connectors	1.7.97, 1.7.100	161, 162
Stresses	1.7.99	161
Composite structures, prestressed concrete	1.6.14	105
Composite wood-concrete members	1.3.5	38
Compression chords:		
Lateral bracing	1.7.21	131
Trusses	1.7.76	151
Compression members:		
Fasteners	1.7.37	135
Limiting length	1.7.11	126
Pitch of fasteners	1.7.86	155
Splices	1.7.19	128
Thickness of metal	1.7.88	156
Concrete:		
(*See also* Concrete structures)		
Abrasion	1.2.2	12
Admixtures for	2.4.2B	264
Aggregates	2.4.2C, D	264, 265
Air content	2.4.6	267
Air-entrained	2.4.5	266
Alkali exposed	2.4.17	278
Allowable stresses	1.5.1	56
Cement content	2.4.6	267
Class	1.5.1, 2.4.5	56, 266
Cold weather placing	2.4.21	280
Composition	2.4.6	267
Compressive strength	1.5.1	56
Construction joints	2.4.14	276
Cribbing	Sec. 2.16	376
Cyclopean	2.4.15	277
Curing	2.4.22	280
Decks	1.3.2	32
Delivery	2.4.9H	271
Exposed to sea water	2.4.16	277
Falsework	2.4.18	278
Finishing	2.4.24	282
Forms	2.4.19	278
Handling and placing	2.4.10	271
Lightweight	1.6.6B	98
Materials	2.4.2	262
Measurement and payment	2.4.34	290
Mixing	2.4.9	269
Mixing water	2.4.8	268
Modular ratio, "n"	1.5.2	58
Placement by pumping	2.4.12	275
Pneumatic placing	2.4.11	275
Precast piles	1.4.5G	47

INDEX 441

	Article	Page
Concrete, cont'd.		
Prestressed	Sec. 1.6, 2.4.33	95, 287
(*See also* Prestressed concrete)		
Reinforcement	Sec. 2.5	291
(*See also* Reinforcement)		
Removal of falsework	2.4.20	279
Retempering	2.4.9I	271
Rubble	2.4.15	277
Sampling and testing	2.4.7	268
Slump	2.4.6	267
Storage of cement	2.4.4	265
Strength	1.5.1	56
Thermal stresses	1.5.4	60
Time of hauling and placing mixed concrete	2.4.9F	270
Underwater placement	2.4.13	275
Wearing surface	1.2.2	12
Concrete Arches	1.5.10	71
Concrete Bents	1.5.11	72
Concrete box girders:		
Dead load distribution	1.3.1	29
Diaphragms	1.5.12G	74
Effective flange width	1.5.12A	73
Flange reinforcement	1.5.12F	74
Flange thickness	1.5.12B	73
Flexure stresses	1.5.12C	73
Live load distribution	1.3.1B	30
Minimum reinforcement	1.5.12F	74
Placement of Concrete	2.4.10C	274
Shear	1.5.12D	74
Shrinkage reinforcement	1.5.6H	63
Stem taper	1.5.12D	74
Temperature reinforcement	1.5.6H	63
Web reinforcement	1.5.8, 1.5.12D	65, 74
Wheel load distribution	1.3.1B	30
Concrete cribbing	Sec. 2.16	376
Concrete design	Sec. 1.5	56
Concrete flooring, wheel load distribution	1.3.1B	30
Concrete masonry	Sec. 2.4	262
Concrete piles:		
(*See* Piles)		
Concrete slabs:		
(*See* Slabs)		
Concrete structures:		
Arches	1.5.10	71
Bearings	1.5.13	75
Bents	1.5.11	72
Box girders:		
(*See* Concrete box girders)		
Coefficients	1.5.1	56
Columns	1.5.9	67
(*See also*, Columns, concrete)		
Deflection	1.5.2	58
Expansion	1.5.4	60
Footings	1.4.6	50
Moment of inertia	1.5.2	58
Piers	1.4.9, 1.5.2	54, 58
Placing sequence of concrete	1.2.1	12

	Article	Page
Concrete Structures, cont'd.		
Placement of Concrete	2.4.10	271
Prestressed	Sec. 1.6	95
(*See also* Prestressed concrete)		
Slabs:		
(*See* Slabs)		
Span length	1.5.3	59
T-beams	1.5.5	60
(*See also* Concrete T-beams)		
Temperature range	1.2.15	26
Temperature provision	1.5.4	60
Concrete T-beams:		
Construction joints	1.5.5E	61
Dead load distribution	1.3.1	29
Depth at support	1.5.2	58
Diaphragms	1.5.5D	61
Effective flange width	1.5.5A	60
Isolated beams	1.5.5C	60
Live load distribution	1.3.1	29
Negative moment reinforcement	1.5.6G	63
Placement of concrete	2.4.10C	274
Shear	1.5.5B	60
Shear keys	1.5.5E	61
Web reinforcement	1.5.6G, 1.5.8A	63, 65
Wheel load distribution	1.3.1	29
Connections:		
Bearing type	1.7.5	121
Bolted	2.10.19, 2.10.20	320
Eccentric	1.7.18	128
Edge distance of fasteners	1.7.38	135
Eyebars	1.7.46, 1.7.47	137, 138
Fillers	1.7.82	152
Forked ends	1.7.48	138
Friction type	1.7.5	121
Hangers	1.7.40	136
Location of pins	1.7.41	137
Pin connected tension members	1.7.40	136
Minimum number of fasteners	1.7.20	130
Pins	1.7.41 to 1.7.44	137
Pin plates	1.7.43	137
Steel members	1.7.19	128
Strength of	1.7.20	130
Subject to stress reversal	1.7.5	121
Subpunching and reaming field connections	2.10.10	317
To tensile members by welding	1.7.3	115
Upset ends	1.7.45	137
Welded:		
(*See* Welding)		
Connectors:		
Stud shear connectors	2.10.3A	306
Timber connectors	2.20.1F	385
Timber structures	2.20.1F, 2.20.2	385, 386
Construction joints	2.4.14	276
Continuous steel structures:		
Allowable increase in stress	1.7.1	111
Composite girders	1.7.99	161

INDEX 443

	Article	Page
Continuous steel structures, cont'd.		
Depth ratio	1.7.10	125
Span length	1.7.10	125
Contraction, steel structures	1.7.16	128
Contraction joints, retaining walls	1.4.8F	54
Copings:		
Ashlar masonry	2.6.14	299
Brick masonry	2.9.4	305
Dry rubble masonry	2.8.7	303
Mortar rubble masonry	2.7.7	301
Copper-alloy plates:		
Allowable stress	1.7.7	125
Sliding expansion bearings	1.7.51	138
Copper bearing steels	2.10.3	306
Copper sheet	2.4.23F	282
Corrugated metal and structural plate pipes and pipe arches	Sec. 1.8 and 2.23	200, 399
Assembly	2.23.3	400
Bedding	2.23.4	401
Bracing	2.23.7	402
Camber	2.23.8	402
Chemical and mechanical requirements	1.8.3	204
Cover	1.8.8, 2.23.10	206, 403
Design	1.8.2	200
Forming and punching	2.23.2	400
Foundations	2.23.5	401
Headwalls	2.23.9	402
Inspection	2.23.11	404
Materials	1.8.1	200
Measurement	2.23.12	404
Minimum gage	1.8.1, 1.8.2, 1.8.4	200, 205
Multiple structures	1.8.6	205
Payment	2.23.13	404
Rivets and bolts	1.8.5	205
Sidefill	2.23.6	401
Sloped ends-skewed	1.8.7	206
Substructure	2.23.9	402
Workmanship	2.23.11	404
Counterforts:		
Concrete arches	1.5.10B	72
Retaining walls	1.4.8D	54
Counters	1.7.76	151
Countersinking, timber structures	2.20.10	391
Countersinking, fasteners	1.7.5	121
Cover plates:		
Built-up compression members	1.7.88	156
Plate girders	1.7.69	143
Rolled beams	1.7.67	142
Cover plates, perforated:		
Effective area	1.7.11	126
Fasteners	1.7.83	153
Radius of gyration	1.7.11	126
Thickness of metal	1.7.88	156
Creep effect, composite girders	1.7.96	160
Cribbing, concrete	Sec. 2.16	376
Cribbing, timber	Sec. 2.22	397

INDEX

	Article	Page
Cross frames:		
Design	1.7.21	131
Spacing	1.7.21, 1.7.59	131, 140
Culverts:		
Concrete handling and placing	2.4.10B	273
Dead loads	1.2.2A	13
Distribution reinforcement	1.3.2E	35
Distribution of wheel loads through earth fills	1.3.3	37
Footings	1.4.6	50
General features of design	Sec. 1.1	1
Location and length	1.1.5	3
Shear in slabs	1.2.2B	14
Waterway opening	1.1.4	2
Curbs	1.1.8	5
Depressed roadways	1.1.16C	10
Height	1.1.8, 1.1.15, 1.1.16, 1.1.17	5, 9, 10
Loading	1.2.11B	21
Tunnels	1.1.15C	10
Underpasses	1.1.17C	12
Width	1.1.8, 1.1.15, 1.1.16, 1.1.17	5, 9, 10
Cycles, fatigue design	1.7.3	115
Cyclopean concrete	2.4.15	277

D

	Article	Page
Dampproofing	Sec. 2.18	381
Dead loads	1.2.2	12
Deck trusses:		
(*See* Trusses)		
Deflection:		
Composite girders	1.7.101	165
Concrete structures, modular ratio "n"	1.5.2	58
Elastomeric bearing pads	1.7.50, Sec. 1.12	138, 238
Fixed bearings	1.7.49	138
Steel structures	1.7.12	126
Depressed roadways	1.1.16	10
Depth ratio, steel structures	1.7.10	125
Details of design, structural steel	1.7.9 to 1.7.56	125-140
Diaphragms:		
Concrete box girders	1.5.12G	74
Concrete T-beams	1.5.5D	61
Design, steel structures	1.7.21	131
Minimum size of fillet welds	1.7.26	132
Spacing, steel structures	1.7.21, 1.7.59, 1.7.107	131, 140, 170
Trusses	1.7.78	152
Distribution reinforcement, slabs	1.3.2E	35
Distribution of loads:		
Cantilever slabs	1.3.2H	35
Composite wood-concrete members	1.3.5	38
Composite box girders, steel	1.7.103	166
Concrete slabs	1.3.2	32
Exterior stringers and beams	1.3.1	29
Floor beams	1.3.1	29
Interior stringers and beams	1.3.1B	30
Longitudinal beams	1.3.1	30
Multi-beam precast concrete bridges	1.3.2	32

	Article	Page
Distribution of loads, cont'd.		
Slabs supported on 4 sides	1.3.2I	36
Steel grid flooring	1.3.6	39
Through earth fills	1.3.3	37
Timber flooring	1.3.4	37
Dowels, embedment	1.4.6G, 1.5.6E	52, 63
Dowels and cramps, ashlar masonry	2.6.13	298
Drainage:		
Abutments	1.4.7D	53
Retaining walls	1.4.8G	54
Rigid frames	1.2.19	27
Roadway	1.1.10	8
Spandrel fills, concrete arches	1.5.10F	72
Steel box girders	1.7.109	170
Driving nuts	2.10.39	346
Dry rubble masonry	Sec. 2.8	303

E

Earth fills:		
Dead load on culverts	1.2.2	12
Distribution of wheel loads through	1.3.3	37
Earth pressure	1.2.19	27
Abutments	1.4.7A	52
Culverts	1.2.2A	13
Retaining walls	1.4.8A	53
Earthquake stresses	1.2.20	27
Edge beams, longitudinal	1.3.2D	35
Effective width:		
Flanges, box girders	1.5.12A	73
Flanges, composite girders	1.7.98	161
Flanges, T-beams	1.5.5A	60
Flanges, wood-concrete T-beams	1.3.5A	38
Slabs	1.3.2	32
Elastomeric bearing pads		
Design	1.12.2	238
Manufacturing requirements	2.25.3	408
Materials	2.25.2	405
Quality assurance	2.25.5	408
Tolerances	2.25.4	408
Electric railway loading	1.2.10	20
End blocks	1.6.15	106
Erection:		
Assembling steel	2.10.58	353
Basis of payment	2.10.64	357
Bearings and anchorages	2.10.56	352
Delivery of materials	2.10.52	351
Falsework	2.10.54	351
Handling and storing materials	2.10.53	351
Methods and equipment	2.10.55	352
Method of measurement	2.10.63	355
Misfits	2.10.61	354
Pin connections	2.10.60	354
Plans	2.10.50	351
Removal of old structure and falsework	2.10.62	354
Riveting	2.10.59	353

446   INDEX

	Article	Page
Erection, cont'd.		
Straightening bent material	2.10.57	352
Structural plate pipes, pipe-arches, and arches	2.23.3	400
Structural steel	2.10.49 to 2.10.64	351-357
Excavation:		
Classification	2.1.11	250
Foundation and substructure	2.1.1	247
Inspection	2.1.7	249
Measurement and payment	2.1.12	250
Preparation for driving piles	2.3.3A	252
Existing bridges, rating	Sec. 1.11	234
Expansion:		
Concrete structures	1.5.4	60
Steel, coefficient	1.7.1	111
Steel structures	1.7.16	128
Expansion bearings	1.7.50, 1.7.51, 1.7.52	138, 139
Deflection provision	1.7.50	138
Rollers	1.7.4, 1.7.52, 2.10.35	117, 139, 345
Sliding	1.7.50, 1.7.51	138
Expansion joints:		
Concrete railings	2.13.11	363
Retaining walls	1.4.8F	54
Spandrel walls	1.5.10C	72
Steel structures	1.7.60, 2.4.23D	140, 281
Expansion plates	Sec. 2.11	358
Allowable stresses	1.7.7	125
Bronze or copper-alloy	2.11.2	358
Measurement and payment	2.11.6	358
Placing	2.11.5	358
Expansion rockers, allowable stresses	1.7.4	117
Expansion rollers	1.7.4, 1.7.52, 2.10.35	117, 139, 345
Exterior stringers, load distribution	1.3.1	29
Eyebars:		
Design	1.7.46	137
Manufacture	2.10.33	344
Materials	2.10.3A	306
Packing	1.7.47	138
Timber trusses	1.10.9D	232

F

	Article	Page
Fabrication, structural steel	2.10.1 to 2.10.47	305-349
Fabric pads, preformed	2.10.3L	313
Falsework:		
Concrete	2.4.18	278
Removal	2.4.20	279
Steel structures	2.10.54	351
Fasteners:		
Allowable stresses	1.7.5	121
Blastplates	1.1.13	8
Built-up members	1.7.83	153
Countersunk rivets	1.7.5	121
Edge distance	1.7.38	135
End connections of floorbeams and stringers	1.7.61	141
Effective bearing area	1.7.5	121
End distance	1.7.38	135
Fatigue design	1.7.3	115
Gage	1.7.36, 1.7.37	134, 135

INDEX

Fasteners, cont'd.	Article	Page
Lacing bars	1.7.83	153
Long rivets	1.7.39	136
Number in connections	1.7.19, 1.7.20, 1.7.21	128-131
Pitch	1.7.36, 1.7.37	134, 135
Pitch in compression members	1.7.86	155
Riveted flange angles	1.7.69	143
Size	1.7.35	134
Sealing	1.7.37	135
Spacing	1.7.36	134
Stitch	1.7.37	135
Through filler plates	1.7.19	128
Weight, pay	2.10.63	355
Fatigue, composite girders	1.7.100	162
Fatigue design	1.7.3	115
Field coat, metal structures	2.14.2, 2.14.12	364, 369
Field connections, subpunching and reaming	2.10.10	317
Fillers	1.7.19	128
Fillets, concrete	1.5.3	59
Fillet welds, allowable stress	1.7.2	115
(*See also* Welding)		
Filter Material	2.15.8	374
Finishing concrete surface	2.4.24 to 2.4.31	282-285
Fire stops, timber structures	1.10.11	234
Flame cleaning, metal surfaces	2.14.10D	367
Flanges, plate girders	1.7.69	143
Flange angles	1.7.69	143
Flange plates, built-up compression members	1.7.88	156
Flange splices	1.7.19	128
Flange width:		
Composite girders	1.7.98	161
Box girders, concrete	1.5.12A	73
T-beams, concrete	1.5.5A	60
T-beams, wood-concrete	1.3.5A	38
Floor system:		
Floorbeams		
(*See* Floorbeams)		
Sidewalk brackets	1.7.64	141
Steel structures	1.7.57 to 1.7.64	140-141
Stringers	1.7.57	140
Floorbeams:		
Bending moment	1.3.1C	32
End	1.7.62	141
End connection	1.7.61	141
End shear and reactions	1.3.1A	29
Floor system	1.7.58	140
Load distribution	1.3.1	29
Timber trusses	1.10.9B	232
Floors:		
Composite wood-concrete desks	2.20.19	394
Concrete	1.3.2	32
Laminated or strip	2.20.18	394
Plank	2.20.17	393
Steel grid	1.3.6	39
Footings:		
Abutments	1.4.7A	52
Anchorage	1.4.6B	50

	Article	Page
Footings, cont'd.		
Bond stress	1.5.1	56
Critical sections	1.4.6E	50
Culverts	1.4.6A	50
Depth	1.4.6A	50
Design	1.4.6E	50
Distribution of pressure	1.4.6C	50
Elevation of bottoms	2.1.3	248
Preparation of foundations	2.1.4	248
Reinforcement	1.4.6F	52
Retaining wall	1.4.8B	53
Slabs	1.4.6F	52
Spread	1.4.6D, E	50
Stepped	1.4.6D	50
Stress transfer, columns	1.4.6G	52
Forgings and shafting, steel	2.10.3B	310
Forked end truss member	1.7.48	138
Forms:		
Concrete	2.4.19	278
Removal	2.4.20	279
Foundation, bearing value	1.4.2	40
Foundation seals	2.1.5	248
Framed bents	2.20.13	392
Friction losses, prestressing steel	1.6.7A	99
Friction piles	1.4.4C, D, F, G	43, 44, 45
(*See also* Piles)		
Friction type connections	1.7.5	121
(*See also* Connections)		

G

	Article	Page
Galvanizing:		
Painting galvanized surfaces	2.14.9	367
Repairing damaged galvanized coatings	2.12.8	360
Structural steel	2.10.3J	312
Girders:		
Concrete:		
Box	1.5.12	73
Composite, prestressed	1.6.14	105
T-beam	1.5.5	60
Steel:		
Composite, box	1.7.102 to 1.7.109	165-170
Composite, "I"	1.7.96 to 1.7.101	160-165
Hybrid	1.7.110 to 1.7.113	170-173
"I"-beam	1.7.65 to 1.7.67	142
Plate	1.7.68 to 1.7.74	143-150
Glued laminated timber	1.10.1B	207
(*See also* Timber structures)		
Group loading, piles	1.4.4G	45
Group loadings	1.2.22	28
Gusset plates	1.7.84	154
Stiffening required	1.7.21, 1.7.84	131, 154
Thickness of metal	1.7.13, 1.7.84	127, 154

H

Half-through truss spans
  (*See* Trusses)

	Article	Page
Hammers:		
Gravity	2.3.4B	253
Steam	2.3.4C	253
Hand cleaning, metal surfaces	2.14.10B	367
Headers:		
Ashlar masonry	2.6.7	296
Mortar rubble masonry	2.7.4	300
Heat Curing	1.7.114	173
High strength bolts:		
(*See also* Fasteners)		
Bolt and nut dimensions	Table 2.10.3A	309
Bolted parts	2.10.20C	321
Bolt tension	2.10.20D	321
Calibrated wrench tightening	2.10.20D	321
Inspection	2.10.20E	323
Installation	2.10.20D	321
Materials	2.10.3A	306
Turn-of-nut tightening	2.10.20D	321
Washer dimensions	Table 2.10.3B	309
High strength low alloy structural steel	2.10.3A	306
Highway loading	1.2.5	14
H loadings	1.2.5B	15
HS loadings	1.2.5C	16
Load Factors		
Prestressed Concrete	1.6.5	97
Reinforced Concrete	1.5.17	79
Steel	1.7.123	176
Load Factor Design		
Concrete	1.5.14	75
Steel	1.7.117	174
Load lane width	Fig. 1.2.5A	15
Minimum loading	1.2.5F	18
Position of live load	1.2.6	18
Traffic lanes	1.2.6	18
High-tensile-strength prestressing reinforcement	2.4.33J	289
H loads	1.2.5B	15
Holes:		
Accuracy	2.10.11, 2.10.12	317
Bolt holes	2.10.7, 8 & 9	316
Bolts, timber members	2.20.8	391
Drifting	2.10.15	319
Punched	2.10.8	316
Reamed or drilled	2.10.9	316
Rivet holes	2.10.7, 8 & 9	316
Subpunching and reaming	2.10.10	317
Hooks, capacity	1.5.8E	66
HS loads	1.2.5C	16
Hydraulic studies	1.1.2C	2
Hydrologic studies	1.1.2B	2

I

	Article	Page
Ice pressure, piers	1.2.17	27
Impact	1.2.12	22
Impact, composite wood-concrete members	1.3.5B	38
Inspection of structural steel, mill and shop	2.10.41	346

INDEX

	Article	Page
Inspector's authority	2.10.42	346
Interior stringers, load distribution	1.3.1	29
Intermediate stiffeners, transverse	1.7.71	146

J

Jetting bearing piles ....2.3.4G............ 254
Joints:
  Bearing type
    (*See* Connections)
  Brick masonry ....2.9.3............. 304
  Contraction, retaining walls ....1.4.8F............ 54
  Expansion and fixed in concrete structures..........2.4.23............ 281
  Expansion type
    (*See* Expansion joints)
  Friction type
    (*See* Connections)
  Mortar rubble masonry ....2.7.6, 2.7.8........ 301, 302
  Structural steel ....2.10.26........... 342
  Timber members ....1.10.9A............ 231
Joint fillers:
  Premolded expansion ....2.4.23C............ 281
  Refined oil-asphalt ....2.17.2G............ 378

K

Keys, shear ....1.5.5E, 1.6.14A......... 61, 105

L

Lacing bars ....1.7.83, 2.10.28........ 153, 343
Laminated timber, allowable stresses................1.10.1B............ 207
Lane loading:
  Continuous span ....1.2.7, 1.2.8C.......... 18, 19
  Maximum stress ....1.2.8D............. 19
  Moments, shears ....1.2.7, App. A......... 18, 411
  Position of ....1.2.6............. 18
Lane loads ....1.2.5............. 14
Lanes, number ....1.2.6............. 18
Lap joints ....1.7.31............ 134
Lateral bracing:
  Steel structures ....1.7.21............ 131
  Minimum size of fillet welds ....1.7.26............ 132
  Timber trusses ....1.10.9E............ 232
Lead, sheet ....2.10.3H............ 312
Leads, pile driving ....2.3.4E............ 254
Live load ....1.2.3............. 14
  (*See also* Highway loadings and Loadings)
  Application of ....1.2.8............. 19
  Classes ....1.2.5D............. 16
  Designation ....1.2.5E............. 18
  H loading ....1.2.5B............. 15
  HS loading ....1.2.5C............. 16
  Impact ....1.2.12............. 22
  Interior stringers ....1.3.1B............. 30
  Lane loads ....1.2.5, 6, 7......... 14, 18

	Article	Page
Live load, cont'd.		
Minimum	1.2.5F	18
Outside roadway stringers	1.3.1B	30
Position	1.2.6	18
Sidewalk	1.2.11A	20
Slab design	1.3.2C	34
Load capacity rating	Sec. 1.11	234
Load length, fatigue design	1.7.3	115
Loading combinations	1.2.22	28
Loading tests, pile	1.4.2	40
Loadings:		
Application	1.2.8	19
Buoyancy	1.2.18	27
Centrifugal forces	1.2.21	28
Combination railing	1.2.11C	21
Combinations	1.2.22	28
Continuous spans	1.2.8D	19
Curb	1.2.11B	21
Dead loads	1.2.2	12
Drift	1.2.17	27
Earth pressure	1.2.19	27
Earthquake stresses	1.2.20	27
Electric railway	1.2.10	20
Ice	1.2.17	27
Impact	1.2.12	22
Live load	1.2.3	14
(See also Highway loading and Live load)		
Longitudinal forces	1.2.13	23
Overload provision	1.2.4	14
Pedestrian railing	1.2.11C	21
Railings	1.2.11C	21
Reduction in intensity	1.2.9	20
Sidewalk	1.2.11A	20
Stream current	1.2.17	27
Thermal forces	1.2.15	26
Timber structures	1.10.1	207
Traffic railing	1.2.11C	21
Uplift	1.2.16	26
Wind	1.2.14	24
Loads	Sec. 1.2	12
Cantilever slab	1.3.2H	35
Distribution of	Sec. 1.3	29
(See also Distribution of loads)		
General	1.2.1	12
Piers	1.4.9A	54
Spread Footings	1.4.6E	50
Longitudinal beams, load distribution	1.3.1	29
Longitudinal forces	1.2.13	23
Longitudinal stiffeners, plate girder	1.7.72	148
Loop bars	1.7.76	151
Lumber		
(See Timber)		

M

	Article	Page
Magnetic particle inspection	2.10.23F	333
Main members, limiting length	1.7.11	126
Malleable castings, allowable stresses	1.7.6B	124

	Article	Page
Masonry, bearing-allowable stresses	1.7.8	125
Masonry bearings	1.7.54	139
Masonry cement	2.4.2A	262
Masonry plates, thickness	1.7.53	139
Materials:		
Angles of repose	1.4.3	41
Pin nuts	1.7.44	137
Railings	1.1.9	5
Splices	1.7.19	128
Structural steel	2.10.3	306
Weights	1.2.2	12
Median slabs	1.3.2J	37
Modular ratio "n":		
Columns, concrete; bending & axial loads	1.5.9F	70
Composite I-girders	1.7.96	160
Concrete—steel	1.5.2	58
Concrete—wood	1.3.5C	38
Steel—Wood	1.3.5C	38
Moment of inertia; concrete structures	1.5.2	58
Mortar:		
Ashlar masonry	2.6.2B, 2.6.9	295, 297
Measurement and payment, pneumatically applied	2.4.34	290
Pneumatically applied	2.4.32	285
Mortar rubble masonry	Sec. 2.7	300
Arches	2.7.8	302
Headers	2.7.4	300
Laying stone	2.7.6	301
Materials	2.7.2	300
Measurement and payment	2.7.10	302
Pointing	2.7.9	302
Mud sills	1.10.8C, 2.20.13A, 2.22.4B	231, 392, 398
Multi-beam precast concrete bridges	1.3.2	32

N

Name plates	2.19.1	382
Natural cement	2.4.2A	262

O

Orthotropic Deck Bridges	1.7.139	196
Overload permit	1.11.1	234
Overload provision	1.2.4	14
Overturning, abutments	1.4.7A	52
Overturning forces	1.2.14C	26
Oxygen Cutting	2.10.24	341

P

Paint for Timber structures	2.20.1E, 2.20.24	384, 395
Painting metal structures	Sec. 2.14	364
Application	2.14.6	366
Cleaning of surfaces	2.14.10	367
Field painting	2.14.12	369
Galvanized surfaces	2.14.9	367
Materials	2.14.2	364

	Article	Page
Painting metal structures, cont'd.		
Mixing of paint	2.14.4	365
Number of coats and color	2.14.3	365
Removal of paint	2.14.7	366
Shop painting	2.14.11	369
Thinning paint	2.14.8	366
Weather conditions	2.14.5	365
Parapets, transverse force distribution	1.2.11C	21
Pedestals, allowable load	1.5.9B	68
Perforated cover plates	1.7.11	126
(*See also* Cover plates, perforated)		
Pier nose	1.4.9B	55
Piers:		
Abrasion	1.4.9A	54
Allowable load	1.5.9	67
Backfill	2.1.8	249
Concrete, moment of inertia	1.5.2	58
Footings		
(*See* Footings)		
Loading	1.2.17	27
Orientation	1.1.3	2
Reinforcement cover	1.5.6B	61
Scour protection	1.1.2C	2
Skewed	1.5.2	58
Spacing	1.1.3	2
Tubular steel	1.4.10	55
Type	1.1.3	2
Unreinforced	1.5.9B	68
Pile bents	1.10.8, 2.20.12	230, 392
Pile driving methods	2.3.4	253
Piles:		
(*See also* Bearing piles and Sheet piles)		
Allowable loads	1.4.4C	43
Batter	1.4.5E	47
Bearing value	1.4.4	41
Bond (in seals)	1.5.1C	57
Buoyancy	1.4.5F	47
Capacity of ground	1.4.4D	44
Capacity of pile	1.4.4B, C	41, 43
Caps	1.4.5I	48
Cast-in-place concrete	1.4.5H	48
Clearance	1.4.5D	46
Combination	1.4.4C	43
Concrete	1.4.4E, 1.4.5G	45, 47
Concrete filled pipe	1.4.4C	43
Corestoppers	1.4.5I	48
Corrosion	1.4.5K	49
Cover in cap	1.4.5D	46
Designed as column	1.4.4B	41
Design load	1.4.4A, E, 1.4.5C	41, 45, 46
Embedment in cap	1.4.5D	46
Friction	1.4.4C	43
Group loading	1.4.4G	45
Load carrying capacity	1.4.4C, D	43, 44
Loading tests	1.4.2	40
Lugs	1.4.5I	48
Overloading due to backfill settlement	1.4.2	40

	Article	Page
Piles, cont'd.		
Point bearing	1.4.4C	43
Precast concrete	1.4.5G	47
Preservative treatment	2.21.2	396
Reinforcement cover	1.5.6B	61
Scabs	1.4.5I	48
Scour	1.4.5I	48
Spacing	1.4.5D, 1.4.6C	46, 50
Splices	1.4.5I	48
Spread footings	1.4.6E	50
Steel	1.4.4E, 1.4.5I	45, 48
(*See also* Steel piles)		
Tapered	1.4.5G	47
Thickness of metal	1.4.5I	48
Timber	1.4.4E	45
(*See also* Timber piles)		
Tip reinforcement	1.4.4C, E	43, 45
Trestle construction	1.4.5G	47
Tubular steel piers	1.4.10C	55
Untreated timber	1.4.5B	46
Uplift	1.4.4F	45
Pilot nuts	2.10.39	346
Pin nuts	1.7.44	137
Pins:		
Allowable stress	1.7.4	117
Boring holes	2.10.36	345
Clearances	2.10.37	345
Effective bearing area	1.7.4	117
Erection	2.10.58	353
Length	1.7.44	137
Location of	1.7.41	137
Manufacture	2.10.35	345
Pedestals and shoes	1.7.56	140
Pilot and driving nuts	2.10.39	346
Pin plates	1.7.43	137
Size	1.7.42	137
Size, eyebars	1.7.46	137
Pipe, Corrugated metal		
(*See also* Corrugated metal and Structural plate pipes and Pipe-arches)		
Pipe, structural plate		
(*See also* Corrugated metal and Structural plate pipes and Pipe-arches)		
Bolts	1.8.5	205
Reinforcing skewed ends	1.8.7	207
Thickness of metal	1.8.1	200
Pipe-arches		
(*See also* Corrugated metal and Structural plate pipes and Pipe-arches)		
Plank floors	2.20.17	393
Plate arches		
(*See* Arches, structural plate)		
Plate cut edges	2.10.22	325
Plate girders		
(*See also* Steel structures)		
Bearing stiffeners	1.7.73	149
Camber	1.7.74	150
Cover plates	1.7.69	143
Design	1.7.68	143

INDEX

	Article	Page
Plate girders, cont'd.		
Flanges	1.7.69	143
Longitudinal stiffeners	1.7.72	148
Transverse stiffeners	1.7.71, 2.10.32	146, 344
Web plates	1.7.10, 2.10.30	144, 343
Plug welds, allowable stresses	1.7.2	115
Pneumatic placing of concrete	2.4.11	275
Pointing stone joints	2.6.16, 2.7.9	299, 302
Portal bracing:		
Trusses, steel	1.7.81	152
Trusses, timber	1.10.9E	232
Portland blast-furnace slag cement	2.4.2A	262
Portland cement	2.4.2A	262
Post tensioning ducts, diameter	1.6.16B	107
Post-tensioned members		
(See Prestressed concrete)		
Posts, framed bents	2.20.13D	392
Precast concrete bridges	1.3.2	32
Precast concrete piles	1.4.5G	47
Manufacture	2.3.13	257
Storage and handling	2.3.14	258
Precast concrete railing	2.13.9	363
Preservative treatments:		
Identification and inspection	2.21.3	396
Materials	2.21.2	396
Prestressed concrete	Sec. 1.6	95
Allowable stresses	1.6.6	97
Anchorage zones	1.6.15	106
Assumptions	1.6.4	97
Bonded members—steel stress	1.6.9C	102
Composite structures	1.6.14	105
Concrete strength, maximum	1.6.6	97
Cracking load	1.6.10B	102
Critical section, shear	1.6.13	104
Design theory	1.6.3	96
Flexure	1.6.8	101
Grouting of bonded steel	2.4.33I	289
High tensile strength steel	2.4.33J	289
Jacking equipment	2.4.33C	287
Load factors	1.6.5	97
Measurement and payment	2.4.34	290
Notation	1.6.2	95
Placement and type of concrete	2.4.33D	287
Post-tensioning method	2.4.33H	289
Prestressing steel	1.6.6A	98
(See also Prestressing steel)		
Pretensioning method	2.4.33G	288
Reinforcement, non-prestressed	1.6.11	102
Reinforcement, prestressing	2.4.33J	289
Shear	1.6.13	104
Shrinkage, composite structures	1.6.14E	105
Steam curing	2.4.33E	288
Stirrups, cover	1.6.16A	106
Strength at stress transfer	1.6.19	108
Testing prestressing reinforcement and anchorages	2.4.33K	290
Transportation and storage	2.4.33F	288
Ultimate strength	1.6.9	101

INDEX

	Article	Page
Prestressed concrete, cont'd.		
Unbonded members—steel stress	1.6.9C	102
Web reinforcement, spacing	1.6.13	104
Wheel load distribution	1.3.1B	30
Prestressing steel:		
Bundling	1.6.16C	107
Cover	1.6.16	106
Embedment	1.6.18	107
Friction losses	1.6.7A	99
Loss of prestress	1.6.7	99
Overstressing	1.6.6A	98
Percentage, maximum and minimum	1.6.10	102
Post-tensioning ducts, diameter	1.6.16B	107
Spacing	1.6.16	106
Stress transfer	1.6.19	108
Temporary stress	1.6.6	97
Ultimate stress	1.6.9C	102
Procedure qualification	2.10.23E	332

Q

	Article	Page
Qualification of welding operators	2.10.23D	331

R

	Article	Page
Radiographic inspection, butt-welded splices	1.7.3, 2.10.23E, F	115, 332, 333
Railings:		
Combination	1.1.9, 1.2.11C	5, 21
Concrete	2.13.6 to 2.13.11	362-363
Expansion joints	2.13.11	363
Materials	2.13.2	362
Metal	2.13.4	362
Pedestrian	1.1.9, 1.2.11C	5, 21
Precast	2.13.9	363
Stone and brick	2.13.12	364
Timber structures	2.20.20	395
Traffic	1.1.9, 1.2.11C	5, 21
Wood	1.10.10H, 2.13.13	234, 364
Reactions, standard loading	App. A	411
Reactions, end, position of loads	1.3.1A	29
Red lead	2.10.3K, 2.14.2	313, 364
Reinforcement:		
(*See also* Reinforcing steel)		
Arches, concrete	1.5.10D	72
Bars	2.5.1A	291
Bar mats	2.5.1C	292
Columns	1.5.6I, 1.5.9C, D	64, 68, 69
Concrete	1.5.6	61
Concrete piles	1.4.5G	47
Construction	1.5.6G	63
Counterforts	1.4.8D	54
Compression	1.5.7	64
Distribution, slabs	1.3.2E	35
Fabrication	2.5.4	292
Flanges—box girders	1.5.12F	74
Footings	1.4.6F, G	52
Lapping, wire mesh or bar mat	2.5.7	293

INDEX

	Article	Page
Reinforcement, cont'd.		
Measurement	2.5.9	294
Minimum, columns	1.5.9A	67
Negative	1.5.6E	63
Negative, T-beams	1.5.6G	63
Order lists	2.5.2	292
Payment	2.5.10	294
Pile tip	1.4.4C	43
Placing and fastening	2.5.5	293
Prestressed members	1.6.6, 1.6.10, 1.6.11	97, 102
Protection of material	2.5.3	292
Retaining walls	1.4.8	53
Shrinkage	1.5.6H	63
Slabs supported on 4 sides	1.3.2I	36
Splicing	2.5.6	293
Spiral columns	1.5.9C	68
Stems, T-beams	1.5.6G	63
Structural shapes	2.5.1D	292
Substitutions	2.5.8	294
Temperature	1.4.7B, 1.4.8E, 1.5.6H	53, 54, 63
Tied columns	1.5.9D	69
Web	1.5.8	65
Wire and wire mesh	2.5.1B	291
Reinforcing steel:		
(*See also* Reinforcement)		
Allowable stress	1.5.1	56
Anchorage	1.5.8E	66
Bending	1.5.6D	62
Bent up web reinforcement	1.5.8C	65
Bond stress	1.5.1	56
Bundled	1.5.6I, 2.5.5	64, 293
Cold drawn wire	1.5.9C, 2.5.1B	68, 291
Covering	1.5.6B	61
Cover, prestressing steel	1.6.16	106
Embedment	1.5.6E	63
End anchorage	1.5.6D	62
Extension	1.5.6E	63
Extension, box girder flange	1.5.12F	74
Hook capacity	1.5.8E	66
Hooks	1.5.6D	62
Hoops	1.5.9D	69
Lapped splices	1.5.6C	61
Maximum size	1.5.6F	63
Post-tensioning		
(*See* Prestressing steel)		
Prestressing		
(*See* Prestressing steel)		
Radii of Bend	1.5.6D, 2.5.4	62, 292
Shear stress	1.5.8B	65
Spacing	1.5.6A	61
Spacing, bent up bars	1.5.8C	65
Spacing, box girder flange	1.5.12F	74
Spacing, prestressing steel	1.6.16	106
Spiral	1.5.9C	68
Spirals, pitch	1.5.9C	68
Splices	1.5.6C	61
Splices, columns	1.5.9F	70

	Article	Page
Reinforcing steel, cont'd.		
Stirrup embedment	1.5.8E	66
Stirrups	1.5.6D, 1.5.8D	62, 66
Ties	1.5.6D	62
Web, Spacing, Prestressed	1.6.13	104
Welded splices	1.5.6C	61
Yield strength	1.5.1	56
Relief bridges, waterway opening	1.1.2	1
Retaining walls	Sec. 1.4	40
Base or footing slab	1.4.8B	53
Buttresses	1.4.8D	54
Contraction joints	1.4.8F	54
Counterforts	1.4.8D	54
Drainage	1.4.8G	54
Earth pressure	1.4.8A	53
Expansion joints	1.4.8F	54
Reinforcement cover	1.5.6B	61
Temperature reinforcing	1.4.8E	54
Vertical wall design	1.4.8C	54
Ribbed arches, thickness of web plates	1.7.90	158
Rigid frame:		
Sidesway	1.5.2	58
Span length	1.5.3	59
Riprap	2.15.3	371
Concrete, in bags	2.15.6	373
Concrete slab	2.15.7	373
Materials	2.15.2	370
Measurement	2.15.9	374
Mortared, for slopes	2.15.4	372
Payment	2.15.10	375
Stone	2.15.5	373
Rivet steel	2.10.3A	306
Riveted girders		
(*See* Plate girders)		
Rivets:		
(*See also* Fasteners)		
Field riveting	2.10.59	353
Furnishing field rivets	2.10.18	319
Heating and driving	2.10.21	324
Holes	2.10.7	316
Size	2.10.17	319
Weight of rivets and rivet heads	2.10.63	355
Roadway:		
Bridges	1.1.6, 1.1.7	3
Clearances	Figs. 1.1.7, 1.1.15, 1.1.17	4, 9, 11
Depressed roadway	1.1.16	10
Drainage	1.1.10	8
Superelevation	1.1.11	8
Tunnels	1.1.15	9
Underpasses	1.1.17	10
Rolled beams	1.7.65 to 1.7.67	142
(*See also* Steel structures)		
Bearing stiffeners	1.7.66	142
Cover plates	1.7.67	142
Lateral support by timber floors	1.7.65	142
Minimum thickness	1.7.13	127
Rollers, expansion	1.7.4, 1.7.52, 2.10.35	117, 139, 345

	Article	Page

S

Sandblasting	2.14.10C	367
Scour	1.1.1, 1.1.2, 1.1.3, 1.1.4	1, 2
Scour, piles	1.4.5I	48
Seal, foundation	2.1.5	248

Seal weld
 (*See* Welding)
Secondary members:
Limiting length	1.7.12	126
Thickness of metal	1.7.88	156

Secondary stresses:
Curved laminated timber members	1.10.2B	217
Trusses	1.7.77	151

Shear:
Calculation, concrete	1.5.8B	65
Composite girders	1.7.100	162
Composite, prestressed	1.6.14B, C	105
Composite wood-concrete members	1.3.5	38
Concrete box girders	1.5.12D	74
Concrete T-beams	1.5.5B	60
Culvert slabs	1.2.2B	14
Prestressed concrete	1.6.13	104
Slabs	1.3.2F	35
Spread footings	1.4.6E	50
Standard loadings	App. A	411
Steel, allowable stresses	1.7.1	111
Through earth fills	1.3.3	37

Shear connectors:
Allowable loads	1.7.100	162
Composite girders	1.7.97	161
Composite wood-concrete members	1.3.5	38
Concrete cover, penetration	1.7.97	161
Design	1.7.100	162
Edge distance	1.7.97	161
Pitch	1.7.100	162
Spacing	1.7.100	162

Shear, end, position of loads	1.3.1	29
Shear keys	1.5.5E, 2.4.10C	61, 274
Shear-plate connectors	2.20.1F	385
Shear stress, fasteners	1.7.5	121

Sheet piles:
Concrete	2.2.3	251
Measurement and payment	2.2.5	252
Steel	2.2.4	251
Timber	2.2.2	251

Shop assembly:
Match marking	2.10.16	319
Structural steel	2.10.14	318

Shop coat, metal structures	2.14.2A, 2.14.11	364, 369
Shrinkage, composite structures, prestressed	1.6.14E	105
Shrinkage, coefficient	1.5.1	56
Shrinkage, reinforcement	1.5.6H	63
Sidewalk brackets	1.7.64	141
Sidewalk loading	1.2.11A	20

INDEX

	Article	Page
Sidewalks	1.1.6, 1.1.8	3, 5
Cantilever	1.2.11A	20
Finish	2.4.31	285
Live load	1.2.11A	20
Sills and mud sills	1.10.8C, 2.20.13C	231, 392
Skid resistance, floors	1.1.12, 2.12.2D	8, 359
Slab design	Fig. 1.2.5A, C, 1.3.2	15, 17, 32
Slabs:		
Bending moment	1.3.2C	34
Bond	1.3.2F	35
Bond stress	1.5.1	56
Cantilever		
(See Cantilever slabs)		
Depth	1.5.2	58
Distribution of loads	1.3.2	32
Distribution reinforcement	1.3.2E	35
Edge beam, longitudinal	1.3.2D	35
Edge distance of wheel loads	1.3.2B	33
Effective width, cantilever	1.3.2H	35
Footings	1.4.6F	52
Median	1.3.2J	37
Placement of concrete	2.4.10C	274
Railing loads	1.3.2H	35
Reinforcing cover	1.5.6B, 1.6.16A	61, 106
Shear	1.3.2F	35
Span lengths	1.3.2A	32
Supported on 4 sides	1.3.2I	36
Truck loads, cantilever	1.3.2H	35
Unsupported edges	1.3.2G	35
Sleeve nuts	1.7.76	151
Slenderness ratio	1.7.11	126
Soils:		
Angles of repose	1.4.3	41
Bearing power	1.4.2	40
Sole plates	1.7.53	139
Spaced columns, timber, design	1.10.2E	224
Span length:		
Concrete structures	1.5.3	59
Steel structures	1.7.9	125
Spandrel arches, filled	1.5.10, 2.1.9	71, 250
Spike-grid connectors	2.20.1F	385
Splices:		
Butt-welded	1.7.3, 1.7.19	115, 128
Fasteners, number of	1.7.19	128
Reinforcement	1.5.6C, 2.5.6	61, 293
(See also Reinforcing steel)		
Steel members	1.7.19	128
Timber members	1.10.9A	231
Spiral columns		
(See Columns, concrete)		
Spiral reinforcing steel	1.5.9C	68
Split ring connectors	2.20.1F	385
Spread footings	1.4.6D	50
Standard trucks		
(See Highway loadings)		
Stay plates, thickness of metal	1.7.89	158

INDEX

	Article	Page
Steam curing, prestressed concrete	2.4.33E	288
Steel beams, wheel load distribution	1.3.1B	30

Steel box girders
  (*See* Composite box girders, steel)
Steel design
  (*See* Structural steel)

	Article	Page
Steel grid flooring	1.3.6	39
Arrangement of sections	2.12.3	359
Concrete filler	2.12.9	361
Connection to supports	2.12.6	360
Field assembly	2.12.5	360
Materials	2.12.2	359
Measurement and payment	2.12.11	361
Painting	2.12.10	361
Protective treatment	2.12.2B	359
Provision for camber	2.12.4	360
Repairing galvanized coatings	2.12.8	360
Skid resistance	2.12.2D	359
Welding	2.12.7	360
Wheel load distribution	1.3.6	39
Steel piers	1.4.10	55
Steel piles	1.4.4B	41

  (*See also* Piles)

	Article	Page
Allowable stress	1.4.4C	43
Bond stress	1.5.1C	57
Caps	1.4.5I	48
Corrosion	1.4.5K	49
Design load	1.4.4E	45
Splices	1.4.5I	48
Thickness of metal	1.4.5I	48
Tip reinforcement	1.4.4E	45
Steel structures	Sec. 1.7, 2.10	111, 305

  (*See also* Structural steel)

	Article	Page
Accessibility of parts	1.7.23	132
Anchor bolts	1.7.55	139
Bents and towers	1.7.91 to 1.7.95	159-160

    (*See also* Bents and towers)

	Article	Page
Cleaning surfaces for painting	2.14.10	367
Closed sections and pockets	1.7.24	132
Combined stresses	1.7.17	128
Composite I-girders	1.7.96 to 1.7.101	160-165
Composite box girders, steel	1.7.102 to 1.7.109	165-170
Connections		

    (*See* Connections)
  Continuous
    (*See* Continuous steel structures)

	Article	Page
Contraction	1.7.16	128
Cross frames	1.7.21	131
Cross-frame spacing (timber floors)	1.7.59	140
Deflection	1.7.12	126
Depth ratio	1.7.10	125
Design	Sec. 1.7	111
Diaphragms	1.7.21	131
Diaphragm spacing (timber floors)	1.7.59	140
Eccentric connections	1.7.18	128
Expansion bearings	1.7.50, 1.7.51, 1.7.52	138, 139

INDEX

	Article	Page
Steel structures, cont'd.		
Expansion joints	1.7.60	140
Expansion	1.7.16	128
Fatigue due to wind load	1.7.3	115
Fillers	1.7.82	152
Fixed bearings	1.7.49	138
Floor system	1.7.57 to 1.7.64	140-141
(*See also* Floor system)		
Forked ends	1.7.48	138
Gusset plates	1.7.21	131
Hanger	1.7.40	136
Identification of steel	2.10.43	346
Lateral bracing	1.7.21	131
(*See also* Bracing)		
Limiting length of members	1.7.11	126
Links	1.7.40	136
Location of pins	1.7.41	137
Masonry bearings and plates	1.7.53 to 1.7.54	139
Number of main members	1.7.22	131
Outstanding legs of angles	1.7.15	128
Painting		
(*See* Painting metal structures)		
Pedestals	1.7.56	140
Perforated cover plates	1.7.11, 1.7.83	126, 153
Plate girders	1.7.68 to 1.7.74	143-150
(*See also* Plate girders)		
Ribbed arches	1.7.90	158
Rolled beams	1.7.65 to 1.7.67	142
(*See also* Rolled beams)		
Rollers	1.7.52	139
Shoes	1.7.56	140
Sidewalk brackets	1.7.64	141
Size of pins	1.7.42	137
Skewed bridges	1.7.63	141
Sole plates	1.7.53	139
Span length	1.7.9	125
Strength of connections	1.7.20	130
Temperature range	1.2.15	26
Tension members	1.7.14, 1.7.40	127, 136
Thickness of metal	1.7.13	127
(*See also* Thickness of metal)		
Through spans	1.7.22	131
Trusses	1.7.75 to 1.7.89	150-158
(*See also* Trusses)		
Upset ends	1.7.45	137
Welding	1.7.25 to 1.7.34, 2.10.23	132-134, 328
(*See also* Welding)		
Welding to tension members	1.7.3	115
Working drawings	2.10.43	346
Stiffeners:		
(*See also* Bearing stiffeners)		
Fit	2.10.32	344
Longitudinal	1.7.72	148
Minimum size of fillet weld	1.7.26	132
Transverse	1.7.71	146
Stirrups:		
Anchorage	1.5.8E	66

INDEX

	Article	Page
Stirrups, cont'd.		
Cover	1.5.6B	61
Cover, prestressed members	1.6.16A	106
Embedment	1.5.8E	66
Radii of bends	1.5.6D, 2.5.4	62, 292
Spacing	1.5.8D	66
Stone:		
Ashlar masonry	2.6.2A	295
Dry rubble masonry	2.7.2A	300
Laying	2.6.10, 2.7.6. 2.8.6	297, 301, 303
Mortar rubble masonry	2.7.2A	300
Resetting	2.6.12	298
Shaping	2.7.5	301
Straightening material	2.10.5, 2.10.57	313, 352
Stress-grade lumber, allowable stresses	1.10.1A	207
Stress relieving	2.10.34	344
Stress sheets	1.2.1	12
Stretchers	2.6.6	296
Stringers:		
Bending moment	1.3.1B	30
End connection—floor system	1.7.61	141
Floor system	1.7.57	140
Timber	2.20.16	393
Total capacity	1.3.1B	30
Structural steel design	Sec. 1.7, 2.10.3A	111, 306
Structural plate arches		
(*See* Arches, structural plate)		
Structural plate pipes		
(*See* Corrugated metal and structural plate pipes and pipe arches)		
Structural steel:		
(*See also* Steel structures)		
Abutting joints	2.10.26	342
Allowable stresses	1.7.1, 1.7.3	111, 115
Annealling and stress relieving	2.10.34	344
Assembly	2.10.14, 2.10.58	318, 353
Basis of payment	2.10.64	357
Bent plates	2.10.31	343
Butt welded splices	1.7.3	115
Camber diagram	2.10.14	318
Castings	1.7.6, 2.10.3C, D, E, F	124, 310, 311, 312
Coefficient of expansion	1.7.1	111
Delivery of materials	2.10.52	351
Details of design	1.7.9 to 1.7.56	125-140
Edge planing	2.10.22	325
End connection angles	2.10.27	342
Erection	2.10.48 to 2.10.64	349-357
Eyebars	1.7.46, 1.7.47, 2.10.3A, 2.10.33	137, 138, 306, 344
Fabrication	2.10.1 to 2.10.46	305-348
Facing of bearing surfaces	2.10.25	342
Fasteners	1.7.5	121
(*See also* Fasteners)		
Fatigue stresses	1.7.3	115
Finish	2.10.6	315
Finished members	2.10.29	343
Fit of stiffeners	2.10.32	344
Fitting for riveting and bolting	2.10.13	317
Forgings and shaftings	2.10.3B	310

	Article	Page
Structural steel, cont'd.		
Galvanizing	2.10.3J	312
Handling and storing materials	2.10.53	351
Inspection, mill and shop	2.10.40 to 2.10.44	346-348
Lacing bars	2.10.28	343
Marking and shipping	2.10.46	348
Match marking	2.10.16	319
Materials	2.10.3	306
Method of measurement	2.10.63	355
Misfits	2.10.61	354
Modulus of elasticity	1.7.1	111
Oxygen cutting	2.10.24	341
Painting		
(See Painting metal structures)		
Pins and rollers	1.7.4, 1.7.43, 1.7.44, 2.10.35	117, 137, 345
Protection against corrosion	1.7.13	127
Removal of old structure and falsework	2.10.62	354
Riveting	2.10.21	324
Splices	1.7.3, 1.7.19	115, 128
Storage	2.10.4	313
Straightening material	2.10.5	313
Tests, full size	2.10.45	348
Threads for bolts and pins	2.10.38	345
Web plates	2.10.30	343
Weldable grades	1.7.25	132
Weighing of members	2.10.44	348
Welds	2.10.23	328
Structural steel design	Sec. 1.7	111
Strutting; structural plate pipe	2.23.7	402
Stud shear connectors:		
Material	2.10.3A	306
Tensile properties	2.10.3A	306
Welding	2.10.23G	335
Substructures	Sec. 1.4	40
Superelevation	1.1.11	8
Surface finish of bearing surfaces	2.10.25	342
Sway bracing:		
Trusses, steeel	1.7.81	152
Trusses, timber	1.10.9E	232

T

T-beams		
(See also Concrete T-beams)		
Composite wood-concrete members	1.3.5	38
Load distribution	1.3.1	29
Retaining wall counterforts	1.4.8D	54
Tee sections, effective area in tension	1.7.14	127
Temperature range	1.2.15	26
Temperature reinforcement:		
Abutments	1.4.7B	53
Beams	1.5.6H	63
Retaining walls	1.4.8E	54
Tension members:		
Effective area, angles and tees	1.7.14	127
Eyebars	1.7.46, 1.7.47	137, 138
Fasteners	1.7.37	135

INDEX 465

	Article	Page
Tension members, cont'd.		
Limiting length	1.7.11	126
Net section	1.7.40, 1.7.87	136, 155
Pin connected	1.7.40, 1.7.46	136, 137
Splices, timber members	1.10.9A	231
Upset ends	1.7.45	137
Tests, full size on structural steel	2.10.45	348
Thermal coefficient	1.5.1	56
Thermal forces	1.2.15	26
Thermal stresses, concrete	1.5.4	60
Thermal movement, steel structures	1.7.16	128
Thickness of metal:		
Angles in compression	1.7.88	156
Bearing stiffeners	1.7.73	149
Compression members	1.7.88	156
End connection angles	1.7.61	141
Flanges, plate girders	1.7.69	143
Gusset plates	1.7.84	154
Lacing bars	1.7.83	153
Longitudinal stiffeners	1.7.72	148
Masonry plates	1.7.53	139
Perforated cover plates	1.7.88	156
Piles	1.4.5I, J	48, 49
Plates in compression	1.7.88	156
Sole plates	1.7.53	139
Stay plates	1.7.89	158
Steel pedestals	1.7.56	140
Structural plate pipe and pipe arch	1.8.1	200
Structural steel	1.7.13	127
Transverse stiffeners	1.7.71	146
Tubular steel piers	1.4.10D	55
Webs, ribbed arches	1.7.90	158
Web plates, plate girders	1.7.70, 1.7.71	144, 146
Threads, screw	2.10.38	345
Through span, number of main members	1.7.22	131
Through trusses		
(*See* Trusses)		
Tied columns		
(*See* Columns, concrete)		
Ties:		
Radii of bends	1.5.6D, 2.5.4	62, 292
Spacing in columns	1.5.6I, 1.5.9D	64, 69
Timber:		
Allowable stresses	1.10.1, 1.10.2	207, 209
Preservative treatments for	Sec. 2.21	396
Timber beams, horizontal shear	1.3.1A, 1.10.2A	29, 209
Timber connectors	1.10.2I	226
(*See also* Connectors)		
Timber cribbing	Sec. 2.22	397
Timber floors	1.10.10	233
Bending moment	1.3.4	37
Cross frame spacing	1.7.59	140
Used with plate girders	1.7.68	143
Used with rolled beams	1.7.65	142
Wheel loads	1.3.4	37
Timber flooring, wheel load distribution	1.3.1B, 1.3.4	30, 37

INDEX

	Article	Page
Timber piles:		
(*See also* Piles)		
Bond stress	1.5.1	56
Designed as column	1.4.4B	41
Design load	1.4.4E	45
Limitations	1.4.5B	46
Round	1.4.4B	41
Treatment of pile heads	2.20.7	390
Untreated	1.4.5B	46
Timber stringers:		
Framing into steel floorbeams	1.7.61	141
Wheel load distribution	1.3.1B	30
Timber structures	Sec. 1.10, 2.20	207, 383
Bolts and washers	1.10.4, 1.10.5, 2.20.9	230, 391
Bearing stress	1.10.2H	226
Bracing	1.10.9E, 2.20.15	232, 393
Camber	1.10.9F	233
Caps	2.20.14	393
Composite wood-concrete decks	1.3.5, 2.20.19	38, 394
Computation of stresses	1.10.2	209
Connectors	2.20.1F, 2.20.2	385, 386
Countersinking	2.20.10	391
Fire stops	1.10.11	234
Floors and railings	1.10.10	233
Framing	2.20.11	392
Framed bents	1.10.8, 2.20.13	230, 392
Glued-laminated	1.10.1B	207
Hardware	1.10.6, 2.20.1D	230, 383
Holes for Bolts, Dowels	2.20.8	391
Laminated or strip floors	2.20.18	394
Materials	2.20.1	383
Measurement and Payment	2.20.25	395
Paint	22.0.1E, 2.20.24	384, 395
Pile bents	1.10.8, 2.20.12	230, 392
Pile bent abutments	1.10.8	230
Plank floors	2.20.17	393
Posts	2.20.13D	392
Railing	1.10.10, 2.20.20	233, 395
Stringers	1.10.10A, 2.20.16	233, 393
Storage of materials	2.20.3	389
Treatment of pile heads	2.20.7	390
Treated timber	2.20.5	389
Trusses	1.10.9, 2.20.21	231, 395
(*See also* Timber trusses)		
Untreated timber	2.20.6	390
Wheel guards	2.20.20	395
Workmanship	2.20.4	389
Timber trusses	1.10.9	231
Tooth-ring connectors	2.20.1F	385
Towers, steel	1.7.91 to 1.7.95	159-160
Traffic lane units	1.2.8A	19
Traffic lanes:		
Number	1.2.6, 1.2.8B	18, 19
Position	1.2.8B	19
Simultaneous loading	1.2.9	20
Transverse stiffeners, plate girders	1.7.71	146

	Article	Page
Treated timber	2.20.5	389
Tremie concrete	2.4.13	275
Truck loading, maximum stress	1.2.8D	19

Trusses:
	Article	Page
Built-up compression members	1.7.88	156
Camber	1.7.79	152
Compression chords	1.7.76, 1.7.80	151, 152
Compression chords, bracing	1.7.21	131
Deflection	1.7.12	126
Depth ratio	1.7.10	125
Diaphragms	1.7.78	152
Fastener pitch in ends of compression members	1.7.86	155
Fillers	1.7.19	128
Floorbeams	1.7.58	140
Floor system	1.7.57 to 1.7.64	140-141
Gusset plates	1.7.84	154
Half through spans	1.7.85	155
Jacking provision	1.7.62	141
Lacing bars	1.7.83	153
Members	1.7.76	151
Net section of tension members	1.7.87	155
Perforated cover plates	1.7.11, 1.7.83	126, 153
Portal bracing	1.7.81	152
Secondary stresses	1.7.77	151
Spacing	1.7.75	150
Sway bracing	1.7.81	152
Thickness of metal-compression members	1.7.88	156
Timber structures	1.10.9, 2.20.21	231, 395
Top chord unsupported length	1.7.11	126
Working lines and gravity axes	1.7.80	152

	Article	Page
Tubular steel piers	1.4.10	55
Tunnels	1.1.15	9

U

	Article	Page
Underpasses	1.1.17	10

Unit stresses
(*See* Allowable stresses)

	Article	Page
Unit weights of metal	2.10.63	355
Uplift	1.2.16	26
Anchorage of structures	1.7.55	139
Piles	1.4.4F	45
Upset ends	1.7.45	137
Utilities	1.1.14	9

V

	Article	Page
Vibration of concrete	2.4.10	271
Vibrators	2.4.10	271
Visual inspection	2.10.22B	325

Voussoirs
	Article	Page
Ashlar masonry arches	2.6.15	299
Mortar rubble masonry	2.7.8	302

	Article	Page

W

Washers, steel	2.10.3A	306
Waterproofing	Sec. 2.17	377
Application	2.17.5, 2.17.6	379, 380
Concrete arches	1.5.10E	72
Joint fillers	2.17.2G	378
Materials	2.17.2	377
Measurement and payment	2.17.9	381
Preparation of surface	2.17.4	379
Protection course	2.17.8	381
Storage of fabric	2.17.3	379
Water pressure, piers	1.2.17	27
Water stops	2.4.23E	281
Waterway openings:		
Bridges	1.1.2	1
Culverts	1.1.4	2
Relief bridges	1.1.1	1
Wearing surfaces	1.2.2, 2.24.1	12, 404
Web members, trusses	1.7.76	151
Web plates:		
Box girders, steel	1.7.104	166
Built-up compression members	1.7.88	156
Plate girders	1.7.70, 1.7.71, 1.7.72, 2.10.30	144, 146, 148, 343
Ribbed arches	1.7.90	158
Splices	1.7.19	128
Web reinforcement	1.5.8	65
Weights		
Computed	2.10.63	355
Materials	1.2.2	13
Welded girders		
(*See* Plate girders and Steel structures)		
Welding:		
Butt-welded splices	1.7.3, 1.7.19	115, 128
Cover plates to rolled beams	1.7.67	142
Effective length of fillet welds	1.7.29	133
Effective weld areas	1.7.28	133
End returns of fillet welds	1.7.30	133
Fatigue stresses	1.7.3	115
Flanges	1.7.69, 1.7.106	143, 170
General	1.7.25	132
Holes and slots	1.7.34	134
Lap joints	1.7.31	134
Maximum effective size of fillet welds	1.7.27	133
Minimum size of fillet welds	1.7.26	132
Minimum size of seal weld	1.7.26	132
Qualification of welding operators	2.10.23D	331
Seal welds	1.7.32	134
Skewed tee joints	1.7.33	134
Splices	1.7.3, 1.7.19	115, 128
Structural steel	1.7.25, 2.10.3A, 2.10.23	132, 306, 328
Stud shear connectors	2.10.23G	335
To tension members	1.7.3	115
Weld metal, allowable stresses	1.7.2, 1.7.3	115
Wheel guards, timber structures	1.10.10F, 2.20.20	234, 395

INDEX 469

	Article	Page
Wheel load distribution	1.3.1	29
Composite wood-concrete members	1.3.5	38
Steel grid flooring	1.3.6	39
Through earth fills	1.3.3	37
Timber flooring	1.3.4	37
Wheel load, position:		
Edge distance	1.3.2B	33
End reaction and shear	1.3.1A	29
Slab design	1.3.2C	34
Wind loads	1.2.14	24
Fatigue design	1.7.3	115
Wire and wire mesh	2.5.1B	291
Wood-concrete decks, composite	1.3.5, 2.20.19	38, 394
Woven cotton fabric	2.17.2D, 2.17.3	378, 379

Y

Yield strength:		
Reinforcing steel	1.5.1D	57
Spiral reinforcement	1.5.9C	68
Structural steel	1.7.1	111

Z

| Zinc, sheet | 2.10.3I | 312 |